Ecological Studies

Analysis and Synthesis

Volume 234

Ecological Studies is Springer's premier book series treating all aspects of ecology. These volumes, either authored or edited collections, appear several times each year. They are intended to analyze and synthesize our understanding of natural and managed ecosystems and their constituent organisms and resources at different scales from the biosphere to communities, populations, individual organisms and molecular interactions. Many volumes constitute case studies illustrating and synthesizing ecological principles for an intended audience of scientists, students, environmental managers and policy experts. Recent volumes address biodiversity, global change, landscape ecology, air pollution, ecosystem analysis, microbial ecology, ecophysiology and molecular ecology.

More information about this series at http://www.springer.com/series/86

Takayuki Ohgushi • Susanne Wurst •
Scott N. Johnson
Editors

Aboveground–Belowground Community Ecology

Springer

Editors
Takayuki Ohgushi
Center for Ecological Research
Kyoto University
Shiga, Japan

Susanne Wurst
Functional Biodiversity, Institute of Biology
Freie Universität
Berlin, Germany

Scott N. Johnson
Hawkesbury Institute for the Environment
Western Sydney University
Penrith, New South Wales, Australia

ISSN 0070-8356 ISSN 2196-971X (electronic)
Ecological Studies
ISBN 978-3-030-06266-8 ISBN 978-3-319-91614-9 (eBook)
https://doi.org/10.1007/978-3-319-91614-9

This Springer imprint is published by the registered company Springer Nature Switzerland AG
The registered company address is: Gewerbestrasse 11, 6330 Cham, Switzerland

Foreword

For long, community ecologists have considered the world from either an aboveground or a belowground perspective. During most of the twentieth century, aboveground and belowground studies had their own specific research questions while apparently ignoring conceptual developments and approaches in the neighboring subsystem. Certain aspects have been almost exclusively studied in one of the two subsystems. For example, evolutionary studies and multi-trophic interactions have been studied mostly aboveground, whereas belowground studies have been strongly focused on decomposition, mutualistic symbionts, and determining flows and fluxes of carbon and nutrients through feeding guilds in soil food webs. Already in 1960, the Green World Hypothesis by Hairston, Smith, and Slobodkin included both aboveground and belowground components, and in the end of the 1980s, Valerie Brown, Alan Gange, and coworkers showed how belowground and aboveground insecticides had differential effects on secondary succession in restored grasslands. At the same time, research on plant–soil feedback interactions generated interest in unraveling the contribution of belowground biota to plant community dynamics and ecosystem development.

Then came the turn of the millennium and work on aboveground–belowground interactions took off, first mainly driven by soil ecologists and soon joined in by plant ecologists and entomologists. First studies involved relatively simple experiments with plants, insects aboveground, and insects belowground, soon expanding complexity with numbers of insect species, types of functional groups, and also other taxa, such as nematodes, mycorrhizal fungi, and, later, plant pathogens. Most of these studies were undertaken first by ecologists, followed by molecular biologists using their model species to unravel how signal transduction pathways and other molecular mechanisms make aboveground and belowground biota interact. Agronomists stepped in relatively late, so that many of the aboveground–belowground interactions still remain to be tested under farmers' field conditions. Most likely, interest will grow and the concept of plant phytobiomes, which has been successfully coined by phytopathologists, may further boost the application of aboveground–belowground interactions in production ecosystems.

The current multi-authored book, edited by Takayuki Ohgushi, Susanne Wurst, and Scott Johnson, combines a great variety of highly interesting chapters on aspects of aboveground–belowground interactions, from plant–soil feedbacks to ecological, evolutionary, and theoretical aspects of interactions between plants, aboveground and belowground herbivores, pathogens, mutualistic symbionts, and decomposers. This is a very timely overview of current approaches of aboveground–belowground ecology that will stimulate both ecologists to consider these interactions in their own studies, as well as that it will promote application in crop protection in both agri- and horticulture, as well as in grassland management. The book also provides interesting study material to students, thereby stimulating integral thinking about biodiversity conservation, eco-evolutionary dynamics, and sustainable food and feed production.

Netherlands Institute of Ecology
Wageningen, The Netherlands

Wim H. van der Putten

Contents

1 **Linking Aboveground–Belowground Ecology: A Short Historical Perspective** 1
Richard D. Bardgett

2 **Belowground Experimental Approaches for Exploring Aboveground–Belowground Patterns** 19
Scott N. Johnson, Felicity V. Crotty, James M. W. Ryalls, and Philip J. Murray

3 **Modeling Aboveground–Belowground Interactions** 47
Katrin M. Meyer

4 **Intraspecific Plant–Soil Feedbacks Link Ecosystem Ecology and Evolutionary Biology** 69
Jennifer A. Schweitzer, Michael Van Nuland, and Joseph K. Bailey

5 **Fungal-Mediated Above–Belowground Interactions: The Community Approach, Stability, Evolution, Mechanisms, and Applications** 85
Alison E. Bennett, Peter Orrell, Antonino Malacrino, and Maria José Pozo

6 **Interactions Involving Rhizobacteria and Foliar-Feeding Insects** 117
Kiran R. Gadhave and Alan C. Gange

7 **Belowground–Aboveground Interactions Between Pathogens and Herbivores** 135
Bastien Castagneyrol, Pilar Fernandez-Conradi, Pil U. Rasmussen, Cécile Robin, and Ayco J. M. Tack

8 **Soil Macro-Invertebrates: Their Impact on Plants and Associated Aboveground Communities in Temperate Regions** 175
Susanne Wurst, Ilja Sonnemann, and Johann G. Zaller

9 The Feedback Loop Between Aboveground Herbivores and Soil Microbes via Deposition Processes 201
Cari A. Ritzenthaler, Caitlin E. Maloney, Audrey M. Maran, Eric A. Moore, Amanda Winters, and Shannon L. Pelini

10 Eco-evolutionary Factors Driving Plant-Mediated Above–Belowground Invertebrate Interactions Along Elevation Gradients 223
Alan Kergunteuil, Moe Bakhtiari, and Sergio Rasmann

11 Cross-Compartment Herbivory Effects on Antagonists and Mutualists and Their Consequences for Plant Fitness 247
Nicholas A. Barber

12 Eco-evolutionary Dynamics of Above- and Belowground Herbivores and Invasive Plants 271
Wei Huang, Evan Siemann, and Jianqing Ding

13 Soil Biota as Drivers of Plant Community Assembly 293
Paul Kardol, Jonathan R. De Long, and Pierre Mariotte

14 Application and Theory of Plant–Soil Feedbacks on Aboveground Herbivores 319
Ian Kaplan, Ana Pineda, and Martijn Bezemer

15 Current Knowledge and Future Challenges of Aboveground and Belowground Community Ecology 345
Takayuki Ohgushi, Susanne Wurst, and Scott N. Johnson

Index 363

Chapter 1
Linking Aboveground–Belowground Ecology: A Short Historical Perspective

Richard D. Bardgett

1.1 Introduction

The last few decades have seen an explosion of interest in the topic of aboveground–belowground interactions, to the point that it is now a major theme in terrestrial ecology. It is now well established that linkages between aboveground and belowground communities play a fundamental role in regulating the structure and function of terrestrial ecosystems, as well as their response to human-induced global change. Also, the role of aboveground–belowground interactions as drivers of ecosystem services is gaining much attention, as is the potential to harness this new understanding to address major global challenges, such as sustainable food production, land restoration, and mitigation of climate change. Put simply, aboveground–belowground ecology is now a central theme in community and ecosystem ecology and has an important role to play in addressing many of the major environmental challenges that face our planet.

So, how did the topic of aboveground–belowground ecology emerge to be a major theme in ecology? This is what this chapter is about, in that I explore the historical developments of this topic, spanning some 30 years, and how the topic rose in prominence within community and ecosystem ecology. It is in no way an exhaustive overview; I have missed many key papers, as I simply cannot include them all. But it is more a personal perspective on how the topic has developed over the last three decades and of some of the challenges that lie ahead. First, I consider how the two historically distinct fields of aboveground and belowground ecology initially came together during the late 1980s and 1990s. Second, I consider some key developments that served to take the field forward during first decade of the twenty first century. And third I explore more recent developments, especially in the context

R. D. Bardgett (✉)
School of Earth and Environmental Sciences, The University of Manchester, Manchester, UK
e-mail: richard.bardgett@manchester.ac.uk

T. Ohgushi et al. (eds.), *Aboveground–Belowground Community Ecology*, Ecological Studies 234, https://doi.org/10.1007/978-3-319-91614-9_1

of the role of aboveground–belowground interactions as drivers of ecosystem responses to global change. Finally, I wrap up with a few thoughts on what the future holds and the challenges that lie ahead for aboveground and belowground community ecology.

1.2 The Merging of Aboveground–Belowground Ecology

Historically, the disciplines of aboveground and belowground (soil) ecology have largely been considered independently of each other. There are of course notable exceptions to this. For example, the classic work of the Danish forester P.E. Müller in the late 1800s demonstrated the powerful influence that plants can have on soils and their associated animal communities, forming contrasting mull and mor soils (Müller 1884). And around the same time, the Russian soil scientist Vasily Dokuchaev recognized vegetation to be one of the main soil-forming factors, along with geology, climate, topography, and time (Dokuchaev 1883). Nevertheless, it is fair to say that these two realms of ecology have traditionally been studied apart, and it wasn't until the 1980s and early 1990s when ecologists began to link the two together in experimental studies. From a belowground perspective, for example, it was in the late 1980s and early 1990s when soil ecologists first started to carry out manipulative experiments to explore the role of belowground trophic interactions in regulating aboveground properties, including plant nutrient supply (Ingham et al. 1985; Setälä and Huhta 1991), plant community dynamics (Brown and Gange 1989; Gange and Brown 1989), and non-nutritional effects on plant growth via animal grazing affecting bacterial production of plant growth-promoting hormones (Jentschke et al. 1995; Alphei et al. 1996). And while it has long been known that plants form symbiotic relationships with soil-borne microorganisms, a wave of new experimental studies began to demonstrate the importance of mycorrhizal fungi (Grime et al. 1987; Allen and Allen 1990; Gange et al. 1993; Newsham et al. 1995) and nitrogen fixing symbionts (Vitousek and Walker 1989; Olff et al. 1993; Chapin et al. 1994) for plant community dynamics and ecosystem processes. Further, around this time, the first attempts to incorporate information on soil food webs into ecosystem models of carbon and nutrient cycling were made (Hunt et al. 1987; Moore and Hunt 1988).

Spurred by growing interest in the role of species as drivers of ecosystem processes (Lawton 1994; Jones and Lawton 1995), ecologists also began to focus their attention on the issue of biodiversity–function relationships (Naeem et al. 1994; Tilman et al. 1996). Most of these studies focussed on aboveground processes, especially primary productivity, but a small group of ecologists also began to look belowground, testing how plant diversity and composition (including litter) influences soil organisms and processes, such as decomposition and nutrient cycling (e.g., Hooper and Vitousek 1998; Bardgett and Shine 1999; Wardle et al. 1997, 1999). Studies also explored how diversity of soil biota influences plant diversity (Van der Heijden et al. 1998) and how multi-trophic interactions in soil influence plant nutrient acquisition and growth (Bardgett and Chan 1999; Laakso and Setälä 1999). There was also growing interest in the role of plant litter quality as driver of ecosystem processes (Cadisch and Giller 1997). It was well known that litter decomposition rates depend on physicochemical properties of leaves (Swift et al. 1979), but a key development was placing this understanding in the context of plant functional classification. In particular, comparative studies revealed that variation in litter decomposition rates across species could be explained on the basis of plant functional traits related to resource acquisition strategy (Cornelissen 1996; Cornelissen and Thompson 1997; Wardle et al. 1998), which also served to put a focus on the importance of plant traits as drivers of ecosystem processes.

Ecologists also began to explore aboveground–belowground interactions from a multi-trophic perspective. Early studies showed that aboveground insect herbivory could have indirect consequences for belowground organisms and vice versa (Brown and Gange 1989; Masters and Brown 1992), and that foliar herbivory could stimulate soil microbes due to enhanced root carbon flux to soil (Holland et al. 1996; Mawdsley and Bardgett 1997), but, in general, reduce root colonization by arbuscular mycorrhizal fungi (Gehring and Whitham 1994). Studies also showed that aboveground herbivores indirectly affect belowground processes through selective foraging, which increases the dominance of less nutritious species that produce low quality litter, thereby reducing rates of nutrient cycling (Pastor et al. 1993; Ritchie et al. 1998), or induces the production of secondary metabolites in foliage which reduce litter quality and decomposability (Rhoades 1985; Findlay et al. 1996). These studies not only served to increase recognition of the importance of herbivores as mediators of aboveground–belowground interactions (Masters et al. 1993; Bardgett et al. 1998), but also they stimulated a new generation of research on multi-trophic linkages between aboveground and belowground food webs.

The last major development of this time that I want to highlight concerns the concept of plant–soil feedback, which is now commonplace in the ecological literature as a determinant of vegetation dynamics. The general idea of plant–soil feedback is that a given plant species or genotype may increase (positive feedback) or decrease (negative feedback) its growth rate relative to other plants via its impact on the soil community (Bever 1994). Agriculturalists had long known about the importance of negative feedback, in that repeated cropping of a particular crop leads to the accumulation of specific soil-borne pathogens, which cause declines in crop yield; such knowledge formed the basis of crop rotation. It was also known that there is a high degree of specificity between plant species and soil organisms, including mycorrhizal fungi (Smith and Read 1997; van der Heijden et al. 1998) and root pathogens and herbivores (Yeates 1979), and that this can affect plant–plant interactions (Turkington et al. 1988; Chanway et al. 1989). But the importance of plant–soil feedback for vegetation dynamics in natural systems hadn't been explicitly considered. Two notable studies redressed this. The first, by van der Putten et al. (1993), showed in dune systems that the accumulation of soil-borne pathogens under the pioneer plant species *Ammophila arenaria* caused a decline in its growth and enhancement of the coexisting grass species, *Festuca rubra*, thereby accelerating vegetation succession. The second, by Bever (1994), showed that "culturing" of the soil community by some old-field plants had negative effects specific to that species, thereby potentially affecting local spatial patterns of vegetation in grassland.

1.3 The Rise of Aboveground–Belowground Ecology

With the new millennium came a period of rapid development for aboveground–belowground ecology, fueled by an abundance of conceptual reviews and books on the topic (Wolters et al. 2000; Mikola et al. 2002; Scheu and Setälä. 2002; Wardle 2002; Bardgett and Wardle 2003; Wardle et al. 2004a; De Deyn and van der Putten 2005; Bardgett 2005; Bardgett et al. 2005; Bezemer and van Dam 2005). A number of research themes emerged, which were placed into three broad, but complementary, categories by Bardgett and Wardle (2010). The first involved aboveground–belowground interactions at the local scales and their consequences for plant–plant interactions and ecosystem processes; the second involved the characterization of aboveground–belowground interactions over space and time, including contrasting ecosystems; and the third involved aboveground–belowground interactions in relation to human-induced global change phenomena, especially climate change.

Regarding the local scale, a key development of this decade was embedding the concept of plant–soil feedback as a driver of plant community dynamics, including plant species replacement (Packer and Clay 2000; De Deyn et al. 2003; Kardol et al. 2006), plant species invasion (Klironomos 2002; Reinhart et al. 2005; Callaway et al. 2004), and plant range expansion under global change (Engelkes et al. 2008). While most studies from this time point to the occurrence of negative feedbacks due to species-specific accumulation of soil-borne antagonists (Kulmatiski et al. 2008), some also showed that positive feedback occurs, for example when a given plant species benefits from interactions with mycorrhizal fungi (Klironomos 2002). Such positive feedbacks were also shown to result from "*home-field advantage*," when a plant species selects for decomposer communities that preferentially breakdown and mineralize nutrients from their own litter versus that of another plant species (Vivanco and Austin 2008; Ayres et al. 2009), or when litter chemistry of a given plant species influences nitrogen cycling to maximize nitrogen acquisition by its mycorrhizal roots, while hindering nitrogen uptake by roots of co-occurring plants (Wurzburger and Hendrick 2009). Studies also demonstrated that invasive plant species can modify belowground properties in ways that promote their growth, for instance by producing high quality litter which decomposes rapidly and accelerates rates of nutrient cycling (Kourtev et al. 2002) as well as through escape from specialist soil-borne pathogens (Blumenthal et al. 2009). Research also revealed new mechanisms by which changes in the diversity and composition of soil animal and microbial communities influence ecosystem processes and plant growth (Bradford et al. 2002; Heemsbergen et al. 2004; Setälä and McLean 2004), and how plant species effects on belowground communities influence species coexistence. For instance, studies demonstrated that certain plants produce litter rich in soluble polyphenols, which alters microbial nitrogen mineralization dynamics and reduces nitrogen availability to coexisting plants (e.g., Bowman et al. 2004; Meier et al. 2008).

Another development during this period was the recognition that aboveground–belowground interactions operate over a hierarchy of temporal scales, with differing consequences for ecosystem structure and function (Bardgett et al. 2005). For instance, studies using tracer techniques (e.g., ^{13}C labeling) showed that large quantities of photosynthetic carbon are transferred to root symbionts and other soil organisms over short timescale of hours to days, with consequences for belowground communities and the processes they drive (Högberg et al. 2001; Högberg and Read 2006; Pollierer et al. 2007). Further, studies showed that such root carbon transfer to soil microbes was enhanced by aboveground biotic interactions, such as foliar herbivory, with feedback consequences for plant growth (Hamilton and Frank 2001; Mikola et al. 2001; Ayres et al. 2004). It also became clear that seasonal cycling of labile N pools relies on intimate, temporal coupling between plants and microbes and their resource demands (Bardgett et al. 2005), and that resource pulses (e.g., from animal carcasses) influence aboveground and belowground interactions over interannual timescales (Yang 2004). Mechanistic advances were also made concerning the role of aboveground–belowground interactions in vegetation succession (De Deyn et al. 2003; Kardol et al. 2006) and ecosystem development over millennial timescales (Wardle et al. 2004b; Peltzer et al. 2010). It also became clear that aboveground–belowground interactions operate over a hierarchy of spatial scales (Ettema and Wardle 2002) and are strongly context dependent (Jonsson and Wardle 2008; Wall et al. 2008), and also that plant traits act as important drivers of spatial patterns in belowground communities and ecosystem processes at the community and landscape scale (Porazinska et al. 2003; Wardle and Zackrisson 2005; Cornwell et al. 2008; Fortunel et al. 2009).

Another key development was the recognition that human-induced global change, including climate change, nitrogen deposition, and species invasions, have both direct and indirect effects on the soil biological communities with feedback consequences for ecosystem processes. Regarding direct effects, literature accumulated showing that warming (Cole et al. 2002; Gange et al. 2007; Bradford et al. 2008) and nitrogen deposition (Frey et al. 2004; Allison et al. 2008; Treseder 2008; Ramirez et al. 2010) have strong direct impacts on different components of soil biological communities, as do extreme climate events such as drought and freezing (Fierer and Schimel 2002; Freeman et al. 2004; Sharma et al. 2006). Regarding indirect effects, studies also showed impacts of global change (e.g., elevated atmospheric carbon dioxide concentrations, warming, and nitrogen deposition) on belowground communities and processes can be indirect, driven by changes in plant production and vegetation composition, which alter the amount and quality of organic matter entering soil from roots, symbionts, and aboveground litter (e.g., Hu et al. 2001; Heath et al. 2005; Pollierer et al. 2007; Hawkes et al. 2008). Further, researchers began to tease apart direct and indirect effects of global change drivers on belowground processes (Manning et al. 2006; Suding et al. 2008) and demonstrate the potential for multiple global change drivers to have additive or antagonistic effects on belowground communities and processes (Mikkelsen et al. 2008;

Tylianakis et al. 2008). It also became evident that there is interplay between global change and aboveground consumers with impacts on belowground communities and processes (Van der Wal et al. 2003; Rinnan et al. 2009). These studies, along with several others, helped to establish the notion that to understand how terrestrial ecosystems respond to global change requires a combined aboveground–belowground approach.

1.4 Consolidation of Aboveground and Belowground Ecology

The current decade has been a period of consolidation for aboveground and belowground ecology. A new generation of studies have focused on resolving the mechanisms that underpin interactions between aboveground and belowground communities. But also studies have sought to explore the importance of these biotic interactions for community dynamics across spatial and temporal scales and in the context of global change. With regard to belowground communities and their aboveground effects, an area that has attracted much recent attention is testing how soil biodiversity influences ecosystem processes and plant growth (Bardgett and van der Putten 2014). Past work had shown that diversity effects on soil processes of nutrient and carbon cycling are highly variable, and that effects of species loss are greatest at the low end of the diversity spectrum (Nielsen et al. 2011). But recent studies have gone a step further, showing that the selective loss of certain groups of organisms from complex soil communities impairs multiple soil functions, including plant growth (Wagg et al. 2014), and also that rare soil microbes can have important roles, for instance in plant defense (Hol et al. 2010). Recent studies have also shown that impacts of changes in soil faunal complexity differ for different ecosystem processes (Bradford et al. 2014) and that the structure of competitive networks within belowground communities ultimately determines whether diversity effects on function are positive or negative (Maynard et al. 2017).

At larger spatial scales, cross-biome studies have also shown that the loss of key components of decomposer communities have strong, consistent negative effects on ecosystem processes, especially litter decomposition (Handa et al. 2014). Cross-site studies have also shown that the structure of soil food webs determines soil processes of carbon and nutrient cycling across European countries (De Vries et al. 2013), and that decomposer (microbial) biomass strongly regulates decomposition at regional scales (Bradford et al. 2017). Ecologists are also increasingly using molecular tools to interrogate soil communities, with studies showing in situ that key groups of soil organisms, such as root-associated fungi, play a key role in regulating forest soil carbon dynamics (Clemmensen et al. 2013), and that soil microbial diversity positively relates to multifunctionality across broad spatial scales, even when accounting for effects of climate and soil abiotic factors on ecosystem processes (Delgado-Baquerizo et al. 2016). The use of molecular tools has also revealed an enormous diversity of microorganisms associated with plant roots, which provide a "frontline defense" for plant roots against attack by soil-borne pathogens through mechanisms such as the production of metabolites that inhibit competing microbes (e.g., antimicrobial compounds and volatile organic compounds) and through modulating the plant immune system (Mendes et al. 2011; Berendsen et al. 2012). Together, these studies have not only advanced understanding of the role of complex belowground communities in regulating ecosystem processes and plant growth, but also they have shown that they are of importance across spatial scales.

Another area that has seen significant development during the current decade concerns plant diversity effects on belowground communities and understanding the mechanisms involved. Studies have shown, for instance, that plant diversity can predict beta diversity of grassland soil microbial communities worldwide (Prober et al. 2015), although the diversity of fungi has been shown to be unrelated to plant diversity at global scales, being related more to latitude and mean annual precipitation (Tedersoo et al. 2014). At local scales, however, plant diversity effects on belowground communities have been shown to be relatively strong, and in some cases stronger than effects of global change drivers, such as elevated CO_2 and nitrogen deposition (Eisenhauer et al. 2013). Plant diversity effects on belowground communities have also be shown to operate across trophic levels, cascading from decomposers and herbivores to carnivores and omnivores (Scherber et al. 2010), and it is now known that certain microbial groups, especially AM fungi, are more sensitive to plant species loss than others, such as decomposer fungi and bacteria (De Deyn et al. 2011). However, it is also evident that effects of species loss on belowground communities depend strongly on species identity (Leff et al. 2018), and that subordinate species can have disproportionate effects on belowground properties and processes (Mariotte et al. 2013). On a related theme, recent studies have also demonstrated that variation in soil microbial communities and biogeochemical cycles can be explained, in part, by plant functional traits in ways that are broadly consistent with the leaf economics spectrum (Laughlin 2011; De Vries et al. 2012a; Freschet et al. 2012; Makkonen et al. 2012; Grigulis et al. 2013), although this is not always the case (Barberán et al. 2015; Leff et al. 2018). And evidence is growing that root traits, in particular, have important effects on belowground communities and processes (Moreau et al. 2015; Prieto et al. 2016; Thion et al. 2016; Bardgett 2017), and that mycorrhizal status can predict the nutrient economy of terrestrial ecosystems (Phillips et al. 2013) and amounts of carbon stored in soil (Averill et al. 2014) at large spatial scales.

The use of plant trait-based approaches has also extended to the study of plant–soil feedback, especially root traits that play an important role in determining whether feedbacks are positive or negative. For instance, Baxendale et al. (2014) found that plant traits were able to predict plant–soil feedback across a wide range of grassland species, in that species performed better in soil conditioned by species with similar traits when grown in a competitive environment. Also, in a study of grasslands, plant–soil feedback was shown to correlate positively with arbuscular mycorrhizal fungal colonization (Cortois et al. 2016), as is also the case in Australian shrublands, where plants with ectomycorrhizal fungi displayed positive feedback, and nitrogen-fixers and non-mycorrhizal plants displayed negative feedback (Teste et al. 2017). Also, in a study of a wide range of North American tree populations and species, Bennett et al. (2017) discovered that plant–soil feedbacks consistently depend on mycorrhizal type: AM trees experienced negative feedback, whereas ECM trees displayed positive feedback, which contributed to forest vegetation dynamics. While recent research shows that plant–soil feedback and cascading effects on herbivorous insects are explained by plant community composition rather than plant traits (Heinen et al. 2017), and also that intraspecific genetic diversity modulates plant–soil feedback and ecosystem function (Semchenko et al. 2017), these studies shed new light on the mechanisms by which interactions between aboveground and belowground communities contribute to the dynamics of terrestrial ecosystems.

Much recent research has also focused on aboveground–belowground interactions and climate change (Blankinship et al. 2011; Bardgett et al. 2013; Classen et al. 2015). Particular features of recent work have been the use of landscape-level climatic gradients (e.g., elevation gradients) to explore the role of plant–soil biotic interactions as drivers of future plant species distributions and ecosystem dynamics (Sundqvist et al. 2013; Mayor et al. 2017; Van Nuland et al. 2017), and field experiments to show that effects of climate warming on belowground communities can be moderated by vegetation diversity and composition (Ward et al. 2015; Thakur et al. 2017). Another area of focus has been extreme climate events, such as droughts, floods, and heat waves, which are expected to increase in frequency and severity in coming years (Reichstein et al. 2013). Not only have recent studies shown that extreme climate events can strongly impact the structure of soil microbial communities (Hawkes et al. 2011; De Vries et al. 2012b; Barnard et al. 2013) but also that microbial responses to changing moisture depend on legacies of historic climate (Hawkes et al. 2017), which is consistent with research showing that climate history shapes contemporary belowground functioning (Strickland et al. 2015).

New research has also shown that belowground microbial responses to climate extremes can feedback to plant growth and community dynamics. For example, studies have shown that adaptation of soil microbial communities to recurrent droughts improves plant fitness and their ability to withstand subsequent droughts (Lau and Lennon 2012). Also, drought has been shown to favor mycorrhizal associations and mutualistic soil bacteria that enhance the tolerance of plants to drought via improved access to water and nutrients and hormonal signaling (Mariotte et al. 2017; Rubin et al. 2017). Further, recent work has shown that drought can have long lasting legacy effects on soil microbial communities that can change the direction of plant–soil feedback and affect plant competitive interactions (Kaisermann et al. 2017), and that belowground responses to drought impact plant invasiveness, promoting exotics over natives (Meisner et al. 2013; Schrama and Bardgett 2016). These are just a few examples, but they demonstrate, along with new studies on flooding (Sarneel and Veen 2017; Barnes et al. 2018), that extreme climate events can shape microbial-mediated plant–soil feedbacks with consequences for plant community dynamics and ecosystem function.

The final topic that I want to mention concerns the application of aboveground–belowground ecology to major global challenges, such as sustainable food production, climate mitigation, and the restoration of degraded land. Indeed, awareness is growing that there is high potential to integrate knowledge of aboveground–belowground interactions in complex natural systems to increase resource use efficiency in agriculture (Mariotte et al. 2018), while also bringing wider benefits for soil functioning, such as soil carbon storage, soil physical properties, and the retention of nutrients in soil (Isbell et al. 2017; Bardgett and Gibson 2017). Studies also suggest that manipulation of aboveground–belowground interactions could provide a tool for managing plant communities to control plant invasions (de Voorde et al. 2014; Kulmatiski 2018), and to restore botanical diversity (Wubs et al. 2016; Fry et al. 2017) and the functioning of degraded soils (Gould et al. 2016), which is especially important given the extent that soils are degraded worldwide.

1.5 Concluding Remarks and Future Directions

The study of aboveground–belowground interactions has come a long way over the last three decades, and the ecological literature is now replete with studies showing how they regulate the structure and functioning of terrestrial ecosystems, as well as their response to global change across a spectrum of temporal and spatial scales. But many challenges remain. One of the biggest concerns the need to integrate this new understanding into theoretical frameworks and models of aboveground–belowground interactions and their importance for ecosystem dynamics, especially under global change. Arguably the biggest hurdle here concerns the context dependency of aboveground–belowground interactions, and the implications of this for understanding their role, relative to other factors, as drivers of plant community dynamics, ecosystem function, and biogeochemical feedbacks, which remains poorly understood. For instance, the sensitivity of aboveground–belowground feedbacks to climate varies markedly across ecosystems, and, as such, understanding the underpinning mechanisms behind this variability represents a continued challenge. Another challenge concerns emerging evidence that plant–soil feedbacks have evolutionary implications via changes in plant fitness (Schweitzer et al. 2014), and also that global change can trigger altered eco-evolutionary dynamics between aboveground and belowground communities with potential consequences for ecosystem function (Fischer et al. 2014; terHorst and Zee 2016). As such, studies are needed to advance understanding of the eco-evolutionary consequences of aboveground–belowground interactions under global change, which requires incorporation of such responses into existing long-term experiments and studies using climatic gradients. Finally, and as noted above, ecology is becoming increasingly focused on optimizing land management for the delivery of ecosystem services, such as food production, climate mitigation, and the storage of nutrients and water. Given this, a major challenge for the future is to integrate new understanding of aboveground–belowground interactions into sustainable land management strategies for food production and the delivery of multiple ecosystem services.

References

Allen EB, Allen MF (1990) The mediation of competition by mycorrhizae in successional and patchy environments. In: Grace JB, Tilman D (eds) Perspectives in plant competition. Academic, San Diego, pp 367–389

Allison SD, Czimczik CI, Treseder KK (2008) Microbial activity and soil respiration under nitrogen addition in Alaskan boreal forest. Glob Change Biol 14:1156–1168

Alphei J, Bonkowski M, Scheu S (1996) Protozoa, Nematoda and Lumbricidae in the rhizosphere of *Hordelymus europaeus* (Poaceae): faunal interaction, response of microorganisms and effects on plant growth. Oecologia 106:111–126

Averill C, Turner BL, Finzi AC (2014) Mycorrhiza-mediated competition between plants and decomposers drives soil carbon storage. Nature 505:543–545

Ayres E, Heath J, Possell M et al (2004) Tree physiological responses to above-ground herbivory directly modify below-ground processes of soil carbon and nitrogen cycling. Ecol Lett 7:469–479

Ayres E, Steltzer H, Berg S et al (2009) Soil biota accelerate decomposition in high-elevation forests by specializing in the breakdown of litter produced by the plant species above them. J Ecol 97:901–912

Barberán A, McGuire KL, Wolf JA et al (2015) Relating belowground microbial composition to the taxonomic, phylogenetic, and functional trait distributions of trees in a tropical forest. Ecol Lett 18: 1397–1405

Bardgett RD (2005) The biology of soil: a community and ecosystem approach. Oxford University Press, Oxford

Bardgett RD (2017) Plant trait-based approaches for interrogating belowground function. Biol Environ Proc R Irish Acad 117B(1):13

Bardgett RD, Chan KF (1999) Experimental evidence that soil fauna enhance nutrient mineralization and plant nutrient uptake in montane grassland ecosystems. Soil Biol Biochem 31: 1007–1014

Bardgett RD, Gibson DJ (2017) Plant ecological solutions to global food security. J Ecol 105: 859–864

Bardgett RD, Shine A (1999) Linkages between plant litter diversity, soil microbial biomass and ecosystem function in temperate grasslands. Soil Biol Biochem 31:317–321

Bardgett RD, van der Putten WH (2014) Belowground biodiversity and ecosystem functioning. Nature 515:505–511

Bardgett RD, Wardle DA (2003) Herbivore mediated linkages between aboveground and belowground communities. Ecology 84:2258–2268

Bardgett RD, Wardle DA (2010) Aboveground-belowground linkages: biotic interactions, ecosystem processes, and global change. Oxford University Press, Oxford

Bardgett RD, Wardle DA, Yeates GW (1998) Linking above-ground and below-ground food webs: how plant responses to foliar herbivory influence soil organisms. Soil Biol Biochem 30: 1867–1878

Bardgett RD, Bowman WD, Kaufmann R et al (2005) A temporal approach to linking aboveground and belowground ecology. Trends Ecol Evol 20:634–641

Bardgett RD, Manning P, Morrien E et al (2013) Hierarchical responses of plant–soil interactions to climate change: consequences for the global carbon cycle. J Ecol 101:334–343

Barnard RL, Osborne CA, Firestone MK (2013) Responses of soil bacterial and fungal communities to extreme desiccation and rewetting. ISME J 7:2229–2241
Barnes CJ, van der Gast CJ, McNamara NP et al (2018) Extreme rainfall affects assembly of the root-associated fungal community. New Phytol. https://doi.org/10.1111/nph.14990
Baxendale C, Orwin KH, Poly F et al (2014) Are plant–soil feedback responses explained by plant traits? New Phytol 204:408–423
Bennett JA, Maherali H, Reinhart KO et al (2017) Plant-soil feedbacks and mycorrhizal type influence temperate forest population dynamics. Science 355: 181–184
Berendsen RL, Pieterse CM, Bakker PA (2012) The rhizosphere microbiome and plant health. Trends Plant Sci 17:478–486
Bever JD (1994) Feedback between plants and their soil communities in an old field community. Ecology 75:1965–1977
Bezemer TM, van Dam NM (2005) Linking aboveground and belowground interactions via induced plant defenses. Trends Ecol Evol 20:617–624
Blankinship J, Niklaus P, Hungate B (2011) A meta-analysis of responses of soil biota to global change. Oecologia 165:553–565
Blumenthal D, Mitchell CE, Pyšek P et al (2009) Synergy between pathogen release and resource availability in plant invasion. Proc Natl Acad Sci USA 106:7899–7904
Bowman WD, Steltzer H, Rosenstiel TN et al (2004) Litter effects of two co-occurring alpine species on plant growth, microbial activity and immobilization of nitrogen. Oikos 104:336–344
Bradford MA, Jones TH, Bardgett RD et al (2002) Impacts of soil faunal community composition on model grassland ecosystems. Science 298:615–618
Bradford MA, Davies CA, Frey SD et al (2008) Thermal adaptation of soil microbial respiration to elevated temperature. Ecol Lett 11:1316–1327
Bradford MA, Wood SA, Bardgett RD et al (2014) Discontinuity in the responses of ecosystem processes and multifunctionality to altered soil community composition. Proc Natl Acad Sci USA 111: 14478–14483
Bradford MA, Veen GC, Bonis A et al (2017) A test of the hierarchical model of litter decomposition. Nat Ecol Evol 1:1836–1845
Brown VK, Gange AC (1989) Differential effects of above-and below-ground insect herbivory during early plant succession. Oikos 54:67–76
Cadisch G, Giller KE (eds) (1997) Driven by nature – plant litter quality and decomposition. CAB International, Wallingford
Callaway RM, Thelen GC, Rodriguez A et al (2004) Soil biota and exotic plant invasion. Nature 427:731–737
Chanway CP, Holl FB, Turkington R (1989) Effect of *Rhizobium leguminosarum* biovar *trifolii* genotype on specificity between *Trifolium repens* and *Lolium perenne*. J Ecol 77:1150–1160
Chapin FS, Walker LR, Fastie CL et al (1994) Mechanisms of primary succession following deglaciation at Glacier Bay, Alaska. Ecol Monogr 64:149–175
Classen AT, Sundqvist MK, Henning JA et al (2015) Direct and indirect effects of climate change on soil microbial and soil microbial-plant interactions: what lies ahead? Ecosphere 6:1–21
Clemmensen KE, Bahr A, Ovaskainen O et al (2013) Roots and associated fungi drive long-term carbon sequestration in boreal forest. Science 339:1615–1618
Cole L, Bardgett RD, Ineson P et al (2002) Relationships between enchytraeid worms (Oligochaeta), temperature, and the release of dissolved organic carbon from blanket peat in northern England. Soil Biol Biochem 34:599–607
Cornelissen JHC (1996) An experimental comparison of leaf decomposition rates in a wide range of temperate plant species and types. J Ecol 84:573–582
Cornelissen JHC, Thompson K (1997) Functional leaf attributes predict litter decomposition rate in herbaceous plants. New Phytol 135:109–114
Cornwell WK, Cornelissen JHC, Amatangelo K et al (2008) Plant species traits are the predominant control on litter decomposition rates within biomes worldwide. Ecol Lett 11:1065–1071
Cortois R, Schröder-Georgi T, Weigelt A et al (2016) Plant–soil feedbacks: role of plant functional group and plant traits. J Ecol 104:1608–1617

De Deyn GB, van der Putten WH (2005) Linking aboveground and belowground ecology. Trends Ecol Evol 20:625–633
De Deyn GB, Raaijmakers CE, Zoomer HR et al (2003) Soil invertebrate fauna enhances grassland succession and diversity. Nature 422:711–713
De Deyn GB, Quirk H, Bardgett RD (2011) Plant species richness, identity and productivity differentially influence key groups of microbes in grassland soils of contrasting fertility. Biol Lett 7:75–78
de Voorde TF, Bezemer TM, Van Groenigen JW et al (2014) Soil biochar amendment in a nature restoration area: effects on plant productivity and community composition. Ecol Appl 24:1167–1177
De Vries FT, Manning P, Tallowin JRB et al (2012a) Abiotic drivers and plant traits explain landscape-scale patterns in soil microbial communities. Ecol Lett 15:1230–1239
De Vries FT, Liiri ME, Bjørnlund L et al (2012b) Land use alters the resistance and resilience of soil food webs to drought. Nat Clim Change 2:276–280
De Vries FT, Thébault E, Liiri M et al (2013) Soil food web properties explain ecosystem services across European land use systems. Proc Natl Acad Sci USA 110:14296–14301
Delgado-Baquerizo M, Maestre FT, Reich PB et al (2016) Microbial diversity drives multifunctionality in terrestrial ecosystems. Nat Commun 7:10541
Dokuchaev VV (1883) The Russian Chernozem. Report to the free economic society. Imperial University of St. Petersburg, St. Petersburg
Eisenhauer N, Dobies T, Cesarz S et al (2013) Plant diversity effects on soil food webs are stronger than those of elevated CO_2 and N deposition in a long-term grassland experiment. Proc Natl Acad Sci USA 110:6889–6894
Engelkes T, Morriën E, Verhoeven KJ et al (2008) Successful range-expanding plants experience less above-ground and below-ground enemy impact. Nature 456:946–948
Ettema C, Wardle DA (2002) Spatial soil ecology. Trends Ecol Evol 17:177–183
Fierer N, Schimel JP (2002) Effects of drying-rewetting frequency on soil carbon and nitrogen transformations. Soil Biol Biochem 34:777–787
Findlay S, Carreiro M, Krischik V et al (1996) Effects of damage to living plants on leaf litter quality. Ecol Appl 6:269–275
Fischer DG, Chapman SK, Classen AT et al (2014) Plant genetic effects on soils under climate change. Plant Soil 379:1–19
Fortunel C, Garnier E, Joffre R et al (2009) Leaf traits capture the effects of land use changes and climate on litter decomposability of grasslands across Europe. Ecology 90: 598–611
Freeman C, Fenner N, Ostle NJ et al (2004) Export of dissolved organic carbon from peatlands under elevated carbon dioxide levels. Nature 430:195–198
Freschet GT, Aerts R, Cornelissen JHC (2012) A plant economics spectrum of litter decomposability. Funct Ecol 26:56–65
Frey SD, Knorr M, Parrent JL et al (2004) Chronic nitrogen enrichment affects the structure and function of the soil microbial community in temperate hardwood and pine forests. Forest Ecol Manage 196:159–171
Fry EL, Pilgrim ES, Tallowin JRB et al (2017) Plant, soil and microbial controls on grassland diversity restoration: a long-term, multi-site mesocosm experiment. J Appl Ecol 54:1320–1330
Gange AC, Brown VK (1989) Effects of root herb-ivory by an insect on a foliar-feeding species, mediated through changes in the host plant. Oecologia 81:38–42
Gange AC, Brown VK, Sinclair GS (1993) Vesicular-arbuscular mycorrhizal fungi: a determinant of plant community structure in early succession. Funct Ecol 7:616–622
Gange AC, Gange EG, Sparks TH et al (2007) Rapid and recent changes in fungal fruiting patterns. Science 316:71–71
Gehring CA, Whitham TG (1994) Interactions between aboveground herbivores and the mycorrhizal mutualists of plants. Trends Ecol Evol 9:251–255
Gould IJ, Quinton JN, Weigelt A et al (2016) Plant diversity and root traits benefit physical properties key to soil function in grasslands. Ecol Lett 19:1140–1149

Grigulis K, Lavorel S, Krainer U et al (2013) Relative contributions of plant traits and soil microbial properties to mountain grassland ecosystem services. J Ecol 101:47–57
Grime JP, Mackey JML, Hillier SH et al (1987) Floristic diversity in a model system using experimental microcosms. Nature 328:420–422
Hamilton EW, Frank DA (2001) Can plants stimulate soil microbes and their own nutrient supply? Evidence from a grazing tolerant grass. Ecology 82:239–244
Handa IT, Aerts R, Berendse F et al (2014) Consequences of biodiversity loss for litter decomposition across biomes. Nature 509:218–221
Hawkes CV, Hartley IP, Ineson P et al (2008) Soil temperature affects carbon allocation within arbuscular mycorrhizal networks and carbon transport from plant to fungus. Glob Change Biol 14:1181–1190
Hawkes CV, Kivlin SN, Rocca JD et al (2011) Fungal community responses to precipitation. Glob Change Biol 17:1637–1645
Hawkes CV, Waring BG, Rocca JD et al (2017) Historical climate controls soil respiration responses to current soil moisture. Proc Natl Acad Sci USA 114:6322–6327
Heath J, Ayres E, Possell M et al (2005) Rising atmospheric CO_2 reduces sequestration of root-derived soil carbon. Science 309: 1711–1713
Heemsbergen DA, Berg MP, Loreau M et al (2004) Biodiversity effects on soil processes explained by interspecific functional dissimilarity. Science 306: 1019–1020
Heinen R, van der Sluijs M, Biere A et al (2017) Plant community composition but not plant traits determine the outcome of soil legacy effects on plants and insects. J Ecol. https://doi.org/10.1111/1365-2745.12907
Högberg P, Read DJ (2006) Towards a more plant physiological perspective on soil ecology. Trends Ecol Evol 21:548–554
Högberg P, Nordgren A, Buchmann N et al (2001) Large-scale forest girdling shows that current photosynthesis drives soil respiration. Nature 411:789–792
Hol WH, De Boer W, Termorshuizen AJ et al (2010) Reduction of rare soil microbes modifies plant-herbivore interactions. Ecol Lett 13:292–301
Holland JN, Cheng W, Crossley DA (1996) Herbivore-induced changes in plant carbon allocation: assessment of below-ground C fluxes using carbon-14. Oecologia 107:87–94
Hooper DU, Vitousek PM (1998) Effects of plant composition and diversity on nutrient cycling. Ecol Monogr 68:121–149
Hu S, Chapin FS III, Firestone MK et al (2001) Nitrogen limitation of microbial decomposition in a grassland under elevated CO_2. Nature 409:188–191
Hunt HW, Coleman DC, Ingham ER et al (1987) The detrital food web in a shortgrass prairie. Biol Fertil Soil 3:57–68
Ingham RE, Trofymow JA, Ingham ER et al (1985) Interactions of bacteria, fungi, and their nematode grazers: effects on nutrient cycling and plant growth. Ecol Monogr 55:119–140
Isbell F, Adler PR, Eisenhauer N et al (2017) Benefits of increasing plant diversity in sustainable agroecosystems. J Ecol 105:871–879
Jentschke G, Bonkowski M, Godbold DL et al (1995) Soil protozoa and forest growth: non-nutritional effects and interaction with mycorrhizae. Biol Fertil Soil 20:263–269
Jones CG, Lawton JH (eds) (1995) Linking species and ecosystems. Chapman and Hall, New York
Jonsson M, Wardle DA (2008) Context dependency of litter-mixing effects on decomposition and nutrient release across a long-term chronosequence. Oikos 117:1674–1682
Kaisermann A, Vries FT, Griffiths RI et al (2017) Legacy effects of drought on plant–soil feedbacks and plant–plant interactions. New Phytol 215:1413–1424
Kardol P, Bezemer TM, van der Putten WH (2006) Temporal variation in plant-soil feedback controls succession. Ecol Lett 9:1080–1088
Klironomos JN (2002) Feedback with soil biota contributes to plant rarity and invasiveness in communities. Nature 417:67–70
Kourtev PS, Ehrenfeld JG, Haggblom M (2002) Exotic plant species alter the microbial community structure and function in the soil. Ecology 83:3152–3166
Kulmatiski A (2018) Community-level plant–soil feedbacks explain landscape distribution of native and non-native plants. Ecol Evol 8:2041–2049

Kulmatiski A, Beard KH, Stevens JR et al (2008) Plant–soil feedbacks: a meta-analytical review. Ecol Lett 11:980–992

Laakso J, Setälä H (1999) Sensitivity of primary production to changes in the architecture of belowground food webs. Oikos 87:57–64

Lau JA, Lennon JT (2012) Rapid responses of soil microorganisms improve plant fitness in novel environments. Proc Natl Acad Sci USA 109:14058–14062

Laughlin DC (2011) Nitrification is linked to dominant leaf traits rather than functional diversity. J Ecol 99:1091–1099

Lawton JH (1994) What do species do in ecosystems? Oikos 71:367–374

Leff JW, Bardgett RD, Wilkinson A et al (2018) Predicting the structure of soil communities from plant community taxonomy, phylogeny, and traits. ISME J 12:1794–1805.

Makkonen M, Berg MP, Handa IT et al (2012) Highly consistent effects of plant litter identity and functional traits on decomposition across a latitudinal gradient. Ecol Lett 15:1033–1041

Manning P, Newington JE, Robson HR et al (2006) Decoupling the direct and indirect effects of nitrogen deposition on ecosystem function. Ecol Lett 9:1015–1024

Mariotte P, Vandenberghe C, Meugnier C et al (2013) Subordinate plant species impact on soil microbial communities and ecosystem functioning in grassland: findings from a removal experiment. Perspect Plant Ecol Evol Syst 15:77–85

Mariotte P, Canarini A, Dijkstra FA (2017) Stoichiometric N:P flexibility and mycorrhizal symbiosis favour plant resistance against drought. J Ecol 105:958–967

Mariotte P, Mehrabi Z, Bezemer TM et al (2018) Plant–soil feedback: bridging natural and agricultural sciences. Trends Ecol Evol 33:129–142

Masters GJ, Brown VK (1992) Plant-mediated interactions between two spatially separated insects. Funct Ecol 6:175–179

Masters GJ, Brown VK, Gange AC (1993) Plant mediated interactions between above- and belowground insect herbivores. Oikos 66:148–151

Mawdsley JL, Bardgett RD (1997) Continuous defoliation of perennial ryegrass (*Lolium perenne*) and white clover (*Trifolium repens*) and associated changes in the composition and activity of the microbial population of an upland grassland soil. Biol Fertil Soil 24:52–58

Maynard DS, Crowther TW, Bradford MA (2017) Competitive network and diversity–function. Proc Natl Acad Sci USA 114:11464–11469

Mayor JR, Sanders NJ, Classen AT et al (2017) Elevation alters ecosystem properties across temperate treelines globally. Nature 542: 91–95

Meier CL, Suding KN, Bowman WD (2008) Carbon flux from plants to soil: roots are a belowground source of phenolic secondary compounds in an alpine ecosystem. J Ecol 96:421–430

Meisner A, De Deyn GB, de Boer W et al (2013) Soil biotic legacy effects of extreme weather events influence plant invasiveness. Proc Natl Acad Sci USA 110:9835–9838

Mendes R, Kruijt M, de Bruijn I et al (2011) Deciphering the rhizosphere microbiome for disease-suppressive bacteria. Science 332:1097–1100

Mikkelsen TN, Beier C, Jonasson S et al (2008) Experimental design of multifactor climate change experiments with elevated CO_2, warming and drought: the CLIMAITE project. Funct Ecol 22:185–195

Mikola J, Yeates GW, Barker GM et al (2001) Effects of defoliation intensity on soil food-web properties in an experimental grassland community. Oikos 92:333–343

Mikola J, Bardgett RD, Hedlund K (2002) Biodiversity, ecosystem functioning and soil decomposer food webs. In: Loreau M, Naeem S, Inchausti P (eds) Biodiversity and ecosystem functioning: synthesis and perspectives. Oxford University Press, Oxford, pp 169–180

Moore JC, Hunt WH (1988) Resource compartmentation and the stability of real ecosystems. Nature 333:261–263

Moreau D, Pivato B, Bru D et al (2015) Plant traits related to nitrogen uptake influence plant-microbe competition. Ecology 96:2300–2310

Müller PE (1884) Forest soil studies: on mull and mor in oak-forests and on heaths with some physical and chemical investigations of forest and heath soils by CFA Tuxen. Tidsskr Skovbrug Copenhagen 7(1)

Naeem S, Thompson LJ, Lawler SP et al (1994) Declining biodiversity can alter the performance of ecosystems. Nature 368:734–737

Newsham KK, Watkinson AR, Fitter AH (1995) Symbiotic fungi determine plant community structure—changes in a lichen-rich community induced by fungicide application. Funct Ecol 9:442–447

Nielsen U, Ayres E, Wall D et al (2011) Soil biodiversity and carbon cycling: a review and synthesis of studies examining diversity-function relationships. Eur J Soil Sci 62:105–116

Olff H, Huisman J, van Tooren BF (1993) Species dynamics and nutrient accumulation during early succession in coastal sand dunes. J Ecol 81:693–706

Packer A, Clay K (2000) Soil pathogens and spatial patterns of seedling mortality in a temperate tree. Nature 404:278–281

Pastor J, Dewey B, Naima RJ et al (1993) Moose browsing and soil fertility in the boreal forests of Isle Royale National Park. Ecology 74:467–480

Peltzer DA, Wardle DA, Allison VJ et al (2010) Understanding ecosystem retrogression. Ecol Monogr 80:509–529

Phillips RP, Brzostek E, Midgley MG (2013) The mycorrhizal-associated nutrient economy: a new framework for predicting carbon–nutrient couplings in temperate forests. New Phytol 199:41–51

Pollierer MM, Langel R, Körner C et al (2007) The underestimated importance of belowground carbon input for forest soil animal food webs. Ecol Lett 10:729–736

Porazinska DL, Bardgett RD, Blaauw MB et al (2003) Relationships at the aboveground–belowground interface: plants, soil biota, and soil processes. Ecol Monogr 73:377–395

Prieto I, Stokes A, Roumet C (2016) Root functional parameters predict fine root decomposability at the community level. J Ecol 104:725–733

Prober SM, Leff JW, Bates ST et al (2015) Plant diversity predicts beta but not alpha diversity of soil microbes across grasslands worldwide. Ecol Lett 18:85–95

Ramirez KS, Lauber CL, Knight R et al (2010) Consistent effects of nitrogen fertilization on soil bacterial communities in contrasting systems. Ecology 91: 3463–3470

Reichstein M, Bahn M, Ciais P et al (2013) Climate extremes and the carbon cycle. Nature 500: 287–295

Reinhart KO, Royo AA, van der Putten WH et al (2005) Soil feedback and pathogen activity in *Prunus serotina* throughout its native range. J Ecol 93:890–898

Rhoades DF (1985) Offensive-defensive interactions between herbivores and plants: their relevance in herbivore population dynamics and ecological theory. Am Nat 125:205–238

Rinnan R, Stark S, Tolanen A (2009) Response of vegetation and soil microbial communities to warming and simulated herbivory in a subarctic heath. J Ecol 97:788–800

Ritchie ME, Tilman D, Knops JMH (1998) Herbivore effects on plant and nitrogen dynamics in oak savanna. Ecology 79:165–177

Rubin RL, van Groenigen KJ, Hungate BA (2017) Plant growth promoting rhizobacteria are more effective under drought: a meta-analysis. Plant Soil 416:309–323

Sarneel JMJ, Veen GFC (2017) Legacy effects of altered flooding regimes on decomposition in a boreal floodplain. Plant Soil 421:57–66

Scherber C, Eisenhauer N, Weisser WW, Schmid B, Voigt W, Fischer M, Schulze ED, Roscher C, Weigelt A, Allan E, Beßler H et al (2010) Bottom-up effects of plant diversity on multitrophic interactions in a biodiversity experiment. Nature 468:553–556

Scheu S, Setälä H (2002) Multitrophic interactions in decomposer food webs. In: Tscharntke T, Hawkins B (eds) Multitrophic level interactions. Cambridge University Press, Cambridge, pp 223–264

Schrama M, Bardgett RD (2016) Grassland invasibility varies with drought effects on soil functioning. J Ecol 104:1250–1258

Schweitzer JA, Juric I, Voorde TF et al (2014) Are there evolutionary consequences of plant–soil feedbacks along soil gradients? Funct Ecol 28:55–64

Semchenko M, Saar S, Lepik A (2017) Intraspecific genetic diversity modulates plant–soil feedback and nutrient cycling. New Phytol 216:90–98

Setälä H, Huhta V (1991) Soil fauna increase Betula pendula growth: laboratory experiments with coniferous forest floor. Ecology 72:665–671
Setälä H, McLean MA (2004) Decomposition rate of organic substrates in relation to the species diversity of soil saprophytic fungi. Oecologia 139:98–107
Sharma S, Szele Z, Schilling R et al (2006) Influence of freeze-thaw stress on the structure and function of microbial communities and denitrifying populations in soil. Appl Environ Microbiol 72:2148–2154
Smith SE, Read DJ (1997) Mycorrhizal symbiosis, 2nd edn. Academic, London
Strickland MS, Keiser AD, Bradford MA (2015) Climate history shapes contemporary leaf litter decomposition. Biogeochemistry 122:165–174
Suding KN, Ashton IW, Bechtold H et al (2008) Plant and microbe contribution to community resilience in a directionally changing environment. Ecol Monogr 78:313–329
Sundqvist MK, Sanders NJ, Wardle DA (2013) Community and ecosystem responses to elevational gradients: processes, mechanisms, and insights for global change. Annu Rev Ecol Evol Syst 44: 261–280
Swift MJ, Heal OW, Anderson JM (1979) Decomposition in terrestrial ecosystems. Blackwell, Oxford
terHorst CP, Zee PC (2016) Eco-evolutionary dynamics in plant–soil feedbacks. Funct Ecol 30: 1062–1072
Tedersoo L, Bahram M, Põlme S et al (2014) Global diversity and geography of soil fungi. Science 346:1256688
Teste FP, Kardol P, Turner BJ et al (2017) Plant-soil feedback and the maintenance of diversity in Mediterranean-climate shrublands. Science 355: 173–176
Thakur MP, Tilman D, Purschke O et al (2017) Climate warming promotes species diversity, but with greater taxonomic redundancy, in complex environments. Sci Adv 3:e1700866
Thion CE, Poirel JD, Cornulier T et al (2016) Plant nitrogen-use strategy as a driver of rhizosphere archaeal and bacterial ammonia oxidiser abundance. FEMS Microbiol Ecol 92:fiw091
Tilman D, Wedin D, Knops J (1996) Productivity and sustainability influenced by biodiversity in grassland ecosystems. Nature 379:718–720
Treseder KK (2008) Nitrogen additions and microbial biomass: a meta-analysis of ecosystem studies. Ecol Lett 11:1111–1120
Turkington R, Holl FB, Chanway CP et al (1988) The influence of microorganisms, particularly Rhizobium, on plant competition in grass-legume communities. In: Davey AJ, Hutchinson MJ, Watkinson AR (eds) Plant population ecology. Blackwell Scientific, Boston, pp 343–366
Tylianakis JM, Didham RK, Bascompte J et al (2008) Global change and species interactions in terrestrial ecosystems. Ecol Lett 11:1351–1363
van der Heijden MG, Klironomos JN, Ursic M et al (1998) Mycorrhizal fungal diversity determines plant biodiversity, ecosystem variability and productivity. Nature 396:72–75.
van der Putten WH, Van Dijk C, Peters BAM (1993) Plant-specific soil-borne diseases contribute to succession in foredune vegetation. Nature 362:53–56
van der Wal R, Pearce I, Brooker R et al (2003) Interplay between nitrogen deposition and grazing causes habitat degradation. Ecol Lett 6:141–146
van Nuland ME, Bailey JK, Schweitzer JA (2017) Divergent plant–soil feedbacks could alter future elevation ranges and ecosystem dynamics. Nat Ecol Evol 1:0150. https://doi.org/10.1038/s41559-017-0150
Vitousek PM, Walker LR (1989) Biological invasion by Myrica faya in Hawaiíi: plant demography, nitrogen fixation, ecosystem effects. Ecol Monogr 59:247–265
Vivanco L, Austin AT (2008) Tree species identity alters forest litter decomposition through long term plant and soil interactions in Patagonia, Argentina. J Ecol 96:727–736
Wagg C, Bender SF, Widmer F et al (2014) Soil biodiversity and soil community composition determine ecosystem multifunctionality. Proc Natl Acad Sci USA 111:5266–5270
Wall DH, Bradford MA, St John MG et al (2008) Global decomposition experiment shows soil animal impacts on decomposition are climate-dependent. Glob Change Biol 14:2661–2677
Ward SE, Orwin K, Ostle NJ et al (2015) Vegetation exerts a greater control on litter decomposition than climate warming in peatlands. Ecology 96:113–123

Wardle DA (2002) Communities and ecosystems: linking the aboveground and belowground components. Princeton University Press, Princeton
Wardle DA, Zackrisson O (2005) Effects of species and functional group loss on island ecosystem properties. Nature 435:806–810
Wardle DA, Zackrisson O, Hörnberg G et al (1997) The influence of island area on ecosystem properties. Science 277:1296–1299
Wardle DA, Barker GM, Bonner KI et al (1998) Can comparative approaches based on plant ecophysiological traits predict the nature of biotic interactions and individual plant species effects in ecosystems? J Ecol 86:405–420
Wardle DA, Bonner KI, Barker GM et al (1999) Plant removals in perennial grassland: vegetation dynamics, decomposers, soil biodiversity, and ecosystem properties. Ecol Monogr 69:535–568
Wardle DA, Bardgett RD, Klironomos JN et al (2004a) Ecological linkages between aboveground and belowground biota. Science 304:1629–1633
Wardle DA, Walker LR, Bardgett RD (2004b) Ecosystem properties and forest decline in contrasting long-term chronosequences. Science 305:509–513
Wolters V, Silver WL, Bignell DE et al (2000) Effects of global changes on above-and belowground biodiversity in terrestrial ecosystems: implications for ecosystem functioning. BioScience 50:1089–1098
Wubs ERJ, van der Putten WH, Bosch M et al (2016) Soil inoculation steers restoration of terrestrial ecosystems. Nat Plants 2:16107
Wurzburger N, Hendrick RL (2009) Plant litter chemistry and mycorrhizal roots promote a nitrogen feedback in a temperate forest. J Ecol 97:528–536
Yang LH (2004) Periodical cicadas as resource pulses in North American forests. Science 306: 1565–1567
Yeates GW (1979) Soil nematodes in terrestrial ecosystems. J Nematol 11:213–229

Chapter 2
Belowground Experimental Approaches for Exploring Aboveground–Belowground Patterns

Scott N. Johnson, Felicity V. Crotty, James M. W. Ryalls, and Philip J. Murray

2.1 Introduction

Experiments involving aboveground and belowground components are now commonplace in community ecology but they remain challenging compared to those that focus solely on either component. Challenges are manifold but mainly stem from the difficulty in observing and manipulating belowground components. Overcoming these challenges relies, in part, on techniques and approaches used by soil scientists whose interest in the aboveground compartment was traditionally confined to plant growth and health. This is even illustrated in the title of the excellent *Soil Conditions and Plant Growth* edited by Gregory and Nortcliff (2013) which is the twelfth incarnation of *Soil Conditions and Plant Growth*, first published in 1912 (Russell 1912). Contributors to these volumes rarely considered plant-mediated interactions between organisms beyond those interactions between microbial communities. Similarly, community ecologists were either largely unaware of techniques used in

S. N. Johnson (✉)
Hawkesbury Institute for the Environment, Western Sydney University, Penrith, NSW, Australia
e-mail: Scott.Johnson@westernsydney.edu.au

F. V. Crotty
Game and Wildlife Conservation Trust, Leicester, UK

J. M. W. Ryalls
Hawkesbury Institute for the Environment, Western Sydney University, Penrith, NSW, Australia

Centre for Agri-Environmental Research, School of Agriculture, Policy and Development, University of Reading, Reading, UK
e-mail: j.ryalls@reading.ac.uk

P. J. Murray
Rothamsted Research, North Wyke, Okehampton, UK

T. Ohgushi et al. (eds.), *Aboveground–Belowground Community Ecology*, Ecological Studies 234, https://doi.org/10.1007/978-3-319-91614-9_2

soil science or unclear of how they could be incorporated experimentally. With growing emphasis on multidisciplinary research, these barriers have broken down and researchers have incorporated techniques from different disciplines when designing and executing experiments.

In this chapter, we consider a range of experimental approaches which have helped characterise the belowground component of aboveground–belowground interactions. These experimental approaches have helped advance the field of aboveground–belowground community ecology. The enormous diversity of interactions between microbes, plants and animals described in the 15 chapters in this volume prevents us from covering all experimental approaches. Instead, we focus on particular experimental approaches that may allow researchers to better characterise and manipulate belowground components of their experiments. Facilitating this aspect, in particular, should enable researchers to answer key questions in aboveground–belowground community ecology. We provide examples of each experimental approach which we readily acknowledge focus on study systems familiar to the authors (e.g. root herbivores) rather than comprehensive coverage of all groups that comprise soil communities. Nonetheless, many of the approaches we discuss are applicable to organisms we don't specifically mention and we encourage readers to keep this in mind.

2.2 The Ubiquitous Pot Study

The vast majority of studies in aboveground–belowground community ecology still rely on pot experiments, whether conducted in controlled or field environments. The issue of pot size in experiments has been considered extensively, with the problems and remedies being equally germane to aboveground–belowground experimentation. These are discussed in general terms elsewhere and we encourage readers to consult such papers (e.g. Passioura 2006; Poorter et al. 2012) when designing experiments. The ubiquitous nature of pot studies defies any meaningful classification in the context of aboveground–belowground community ecology, but some experimental issues should be mentioned here. Pot studies are simple, inexpensive and lend themselves to the widest range of research questions in community ecology. Nonetheless, pot studies possibly have the most limitations for investigating aboveground–belowground interactions because they have been well documented to impose experimental artefacts on both components. A big problem is that the soil or growing medium becomes saturated with water towards the bottom of the pot creating hypoxic conditions, particularly in short pots containing field soil which have fewer, if any, large air pores than in the field (Passioura 2006). In addition to affecting the plant, these conditions will clearly affect organisms in the soil and, therefore, how they interact with aboveground organisms. Moreover, pots in glasshouse experiments are likely to have warmer soil temperatures than organisms would experience under field conditions (Passioura 2006).

While not concerned specifically with aboveground–belowground community ecology, a meta-analysis of pot studies suggested that around 65% of experiments used pots that were too small to permit plants to reach their full growth potential (Poorter et al. 2012). In their meta-analysis, Poorter et al. (2012) recommended that researchers use an appropriate pot size in which the plant biomass does not exceed 1–2 g L^{-1}. This value was arrived at by considering plant biomass per given volume of rooting space from experiments. Researchers wishing to include soil macroinvertebrates, such as root herbivores or earthworms, should be additionally aware of avoiding small pots because of negative interactions, such as intra- and interspecific competition. Confining several root herbivores in small pots, for instance, can lead to carnivorous behaviour between individuals which would be much less likely under field conditions (Barnett and Johnson 2013).

2.3 Experiments in Controlled Environments

2.3.1 Manipulating Plant Factors Belowground

Controlled-environment experiments allow researchers to isolate specific environmental factors, which, in natural soil profiles, interact to influence root growth, chemistry and other belowground characteristics. The ease of handling, manipulating and replicating the conditions of growth in controlled environments offer a major advantage over field-based approaches. Three common methods are used to manipulate belowground factors, although their use and application depends on the hypotheses to be tested.

2.3.1.1 Simulated Root Damage Experiments

Artificial wounding experiments are commonly used to simulate root herbivory by chewing herbivores. Root boring, galling and sucking insects are considerably more difficult to simulate (Lehtilä and Boalt 2008). In pot experiments, artificial root damage has been imposed by pruning parts of the root system or piercing roots to trigger a defensive plant response (Erb and Lu 2013). Other studies (e.g. Hol et al. 2004; Hatch and Murray 1994; Ryalls et al. 2015) have used a sharp steel blade inserted into a narrow opening cut in plastic pots to sever the root system at a specific point below the base of the stem. The degree of damage imposed can depend on its ecological relevance (i.e. the type of herbivory that the study intends to simulate). Root nodule pruning, for example, has been used to simulate herbivory by *Sitona* weevil larvae that specifically target legume root nodules (Quinn and Hall 1992). The effects of artificial root damage on interactions between aboveground–belowground plant responses have demonstrated increases in basal levels of shoot defences, including terpenoid aldehydes (Bezemer et al. 2004), pyrrolizidine alkaloids (Hol et al. 2004) and extrafloral nectar (Wäckers and Bezemer 2003). Whether such studies accurately imitate natural herbivory, however, remains unclear and while many studies have compared plant responses to artificial and natural herbivory aboveground (see Lehtilä and Boalt 2008), relatively few have compared the two belowground.

Experiments involving simulated root damage have been criticised for not being able to precisely simulate the distribution and timing of natural damage by herbivores (Johnson et al. 2016b; Steinger and Müller-Schärer 1992). Simulations may fail to induce biochemical responses or capture complex biotic interactions (Hjältén 2004). Severing whole sections of the root, in particular, could limit the roots' ability to supply water and nutrients to the shoots more than natural herbivores would (Blossey and Hunt-Joshi 2003). In general, simulated root herbivory may lead to greater aboveground growth reductions than similar root loss imposed by insect feeding, as demonstrated in a meta-analysis by Zvereva and Kozlov (2012). In particular, the deleterious effects of simulated herbivory on aboveground traits were found to be twice as strong as the effects of insect herbivory, associated with changes in spatial and temporal patterns of root damage as opposed to changes in damage severity between simulated and natural damage. Griffiths et al. (1994) compared the responses of forage rape plants under natural and simulated herbivory by either inoculating plants with turnip root fly (*Delia floralis*) larvae or clipping one-third of their root volume. Root clipping and natural herbivory reduced and increased root glucosinolate content, respectively, suggesting that artificial root damage was not a reliable proxy for root fly larval herbivory. They also suggested that the continuous damage caused by larval feeding was significantly less traumatic than the sudden clipping of roots, demonstrating the importance of mimicking the intensity of root herbivory. When maize roots were pierced with a needle to simulate herbivory by western corn rootworm (*Diabrotica virgifera virgifera*), (E)-β-caryophyllene (EBC) increased in both mechanically wounded and naturally damaged roots (Erb et al. 2012). This suggested that herbivore-derived cues were not required to trigger an EBC response, although the difficulty in effectively simulating the slow accumulative increase in EBC in herbivore-fed roots over time was clear from the sudden surge in EBC concentrations in artificially damaged roots.

Natural herbivory and artificial damage have produced similar effects on plant growth and chemical responses in many other studies (Lehtilä and Boalt 2008; Erb et al. 2008). van Dam et al. (2012) demonstrated increases in a number of volatile organic damage-elicited compounds in multiple *Brassica* species, and any differences between the two types of damage enabled them to identify compounds that acted as markers for root damage. This demonstrates the benefit of incorporating both natural and artificial root damage in identifying herbivore- and damage-elicited compounds. Given the inaccessibility of the belowground environment, root clipping experiments are considerably more tractable and often less time-intensive. By damaging plants artificially, confounding effects can be minimised, as damage is standardised. The variability in damage output associated with root herbivore feeding can often make it difficult to assess, for example, the effect of the plant genotype on the impact of root infestation on leaf-feeders. This can be circumvented by damaging the roots. While artificial herbivory may not mimic natural damage exactly (Blossey and Hunt-Joshi 2003), it enables control of the type, timing and intensity of damage, which can be especially important in complex systems with multiple interacting environmental variables (Ryalls 2016). Simulating herbivory can be especially valuable for selecting effective biological control agents quickly by narrowing the prospective list of biological control agents (Ehler 1998) and enabling different types and levels of damage to be integrated into statistical models for comparison against the performance of biological control agents (Raghu and Dhileepan 2005). Given the difficulty and time required to assess the damage caused by belowground herbivores that are hidden in the soil profile, implementing quick comparisons between biotic agents and artificial damage may serve to limit potential outbreaks.

Using herbivore elicitors in combination with root clipping may be an even more reliable way to mimic herbivory. Aboveground studies have applied saliva to damaged plants with careful imitation of the timing and spatial pattern of damage to more accurately simulate natural damage (Lehtilä and Boalt 2008). Oral secretion by herbivores is a major cue that regulates the induction of specific tolerance (i.e. growth and reproduction after damage) or resistance responses of the plant (Kafle et al. 2017; Walling 2000). However, plant responses to oral secretions by belowground herbivores have received scant attention despite the role of oral secretions in manipulating the plant's gene regulation and metabolism (Vanholme et al. 2004; Johnson et al. 2016b), and suppressing plant defence responses (Chung et al. 2013). Erb et al. (2012) suggested the application of insect oral secretions to the roots in combination with artificial root wounding to determine whether plant responses are due to the specific perception of herbivores (i.e. herbivore elicitors) or due to different patterns of damage (i.e. damage elicitors).

In many cases, artificial damage can provide a reliable proxy for herbivory, although studies should use real herbivores whenever possible (Lehtilä and Boalt 2008; Hjältén 2004). It may be difficult to infer complex biotic interactions from root clipping experiments (Rasmann et al. 2005; Ali et al. 2010). Where both artificial and natural root herbivory show similar plant responses (e.g. Quinn and Hall 1992), however, it may be possible to infer more complex biotic and environmental interactions (Borowicz 2010; Ryalls et al. 2015).

2.3.1.2 Split-Root Experiments

Split-root experiments divide the root system of an intact plant into multiple containers and are commonly used to study the influence of different fertiliser treatments or stress conditions (e.g. abiotic factors and root herbivory) on plant growth (Böhm 1979). Incorporating root herbivory and other stress factors simultaneously in a split-root design eliminates the direct effects of the stress factor on the herbivore but only systemic effects can be investigated (Erb and Lu 2013). Most studies (e.g. Thibaud et al. 2010; Pedrotti et al. 2013; Cosme et al. 2016), therefore, combine split-root designs with other experimental techniques (e.g. sequential inoculation or transcriptomic analysis) to differentiate local plant responses from systemic responses. de la Peña et al. (2006), for example, identified a reduction in nematode (*Pratylenchus penetrans*) abundance in response to arbuscular mycorrhizal fungal (AMF) inoculation of *Ammophila arenaria* grass roots. This was followed by a split-root experiment to determine the importance of the presence of AMF and *P. penetrans* in the same root compartment of *A. arenaria* for the outcome of this interaction. They showed that nematode suppression by AMF did not occur through a systemic plant response but via local mechanisms.

By inoculating one-half of the roots of rice plants with the wound response hormone jasmonic acid (in the form of methyl jasmonate, or MeJA) and infesting both sides of the roots with larvae of the rice water weevil (*Lissorhoptrus oryzophilus*), Lu et al. (2015) demonstrated a reduction in larval performance in response to MeJA irrespective of whether the larvae were in direct contact with the chemical or not. In other words, MeJA-induced resistance was mediated through the plant. Some specialist herbivores may manipulate plant resistance and take advantage of induced metabolic changes, leading to 'induced susceptibility'. Maize roots, for example, became more susceptible to feeding by western corn rootworm (*Diabrotica virgifera virgifera*) when conspecifics were already feeding on the roots (Robert et al. 2012a), and split-root experiments revealed that this induced susceptibility was attributable to plant-mediated (i.e. systemic) effects rather than the physical interaction with conspecifics (Robert et al. 2012b). Systemic changes in plant defence can alter plant susceptibility to aboveground herbivores depending on the temporal nature of herbivory. Schöning and Wurst (2016), for example, demonstrated that continuous (i.e. long-term) herbivory by root-knot nematodes (*Meloidogyne incognita*) indirectly increased nitrogen availability and shoot biomass of coyote tobacco (*Nicotiana attenuata*) and reduced the reproductive output of a specialist aboveground herbivore (*Manduca sexta*), whereas transient (i.e. short-term) belowground herbivory had no effect. However, they argued that the split-root design prevented them from detecting interaction effects of the nematode and *M. sexta*. Root herbivory on the entire root system may elicit different aboveground responses compared with root herbivory on only half of the root system (Kaplan et al. 2008a).

The expression of herbivore-induced plant responses may also depend on the degree of vascular connectivity that links damaged to undamaged tissues. In a split-root experiment using aboveground (*Spodoptera exigua*) and belowground herbivores (*M. incognita*) feeding on cultivated tobacco (*Nicotiana tabacum*), Kaplan et al. (2008b) demonstrated that foliar herbivory-induced changes in root chemicals were more pronounced in roots that were vertically aligned with insect-defoliated leaves. Manipulating belowground factors may concentrate resources into specific sectors of the shoot with, for example, split-roots of tomato grown in nutrient-rich soil preferentially supplying orthostichous leaves and branches (Orians et al. 2002). This connectivity is mediated by groups of xylem vessels that branch and fuse together (Price et al. 1996). Differential sectoriality in trees, mediated by vessel-to-vessel connectivity, has also been demonstrated by split-root experiments (Ellmore et al. 2006). If damage is concentrated within one vascularly connected sector, damage-induced increases in chemical and morphological heterogeneity would be enhanced. Heterogeneity is also likely to vary with leaf age and herbivore guild, with herbivores feeding on older leaves generating more belowground heterogeneity than those that feed on younger leaves and leaf-chewers generating greater heterogeneity than phloem-feeders (Orians and Jones 2001).

Split-root designs may prove difficult for some plant species (e.g. milkweed or lucerne) with a single main tap root, which cannot be subdivided (Erwin et al. 2013). While this technique may also be time-intensive and costly (Ellmore et al. 2006), its ability to facilitate the identification of systemic-versus-localised effects and its application for comparative sectoriality is particularly valuable. Combining isotope tracers or transcriptomic analysis with split-root experiments while simultaneously manipulating both aboveground and belowground factors would be particularly useful for examining the consequences of spatial and temporal variation in plant resource availability and biotic interactions (Kaplan et al. 2008b; Thorn and Orians 2011; Schöning and Wurst 2016).

2.3.1.3 Aboveground–Belowground Olfactometer Experiments

Olfactometer experiments, in which invertebrates are given a choice between two or more odour sources, allow identification of behavioural preferences for specific chemical substances (Ballhorn and Kautz 2013). Six-arm olfactometers have been designed to identify and quantify the effects of up to six odour sources simultaneously and have proven to be highly effective in identifying key attractants involved in multitrophic interactions (Turlings et al. 2005). Hiltpold et al. (2011), for example, combined a split-root design with a six-arm olfactometer (Fig. 2.1) to assess the spatial and temporal activity of (E)-β-caryophyllene within maize roots

Fig. 2.1 A good example of combining multiple experimental design procedures (split-root and olfactometer) reproduced from Hiltpold et al. (2011) with permission

and its systemic effects on the multitrophic interactions between root-feeding larvae (*D. virgifera*) and entomopathogenic nematodes (*Heterorhabditis megidis*). Herbivory triggered the systemic production of (E)-β-caryophyllene within the root system, and nematodes were found to be more attracted towards local feeding sites than systemically induced roots.

Both aboveground and belowground olfactometers (used to determine shoot- and root-attractants, respectively) can be connected or used individually depending on the hypotheses being tested. Aboveground plant defence by herbivore-induced plant volatiles (HIPVs) is now broadly understood, and the role of HIPVs belowground has become increasingly evident (War et al. 2011; Ali and Davidson-Lowe 2015). Neveu et al. (2002) studied the plant volatiles involved in host selection in a multitrophic system belowground. Using a four-arm olfactometer, they showed that larval endoparasitoids (*Trybliographa rapae*) were attracted to undamaged roots of turnip plants (*Brassica rapa* var. *rapa*) that were infested with cabbage root fly (*Delia radicum*) larvae, potentially due to herbivore-induced isothiocyanates. Interestingly, artificially damaged plants were only attractive to *T. rapae* when they were treated with crushed *D. radicum* larvae. Moreover, *T. rapae* were not attracted to plants that were treated with crushed salivary glands from *D. radicum* larvae alone, suggesting that the volatile released originated from another part of the larvae. Other organisms, including herbivores, hyperparasitoids, and neighbouring plants can perceive and use HIPVs. Some are even able to exploit these plant signals to identify competitors and determine the defensive state of the plant (Dicke and van Loon 2000). Robert et al. (2012a), for example, used dual-choice olfactometers to determine whether the root feeder *D. virgifera* could detect whether a potential host plant is already under herbivore attack (described briefly in Sect. 2.3.1.2). They showed that the root feeder *D. virgifera* was more attracted to plants that were already infested with conspecifics, associated with the release of (E)-β-caryophyllene. Moreover, *D. virgifera* avoided plants that were attacked by larvae of the foliar feeding moth (*Spodoptera littoralis*) associated with the suppression of ethylene by *S. littoralis*. Following these results, Robert et al. (2012b) demonstrated that *D. virgifera* increased in small groups compared with larger groups and were able to use (E)-β-caryophyllene in a dose-dependent manner to select host plants with a suitable density of conspecifics. Identifying specific volatiles as attractants combined with their site of release is key to understanding their wide-ranging effects on plants, herbivores and natural enemies. They can serve as multitrophic, multifunctional signals that confer protection to plants (Rostas et al. 2015) and act as herbivore traps for biological control (Degenhardt et al. 2009). However, they may have contrasting effects on organisms by, for example, simultaneously increasing and decreasing resistance to plant pathogens and herbivores, respectively (D'Alessandro et al. 2014).

In recent years, olfactometers have been used to determine how belowground defences mediate higher trophic interactions (i.e. natural enemy responses) aboveground, and vice versa. Filgueiras et al. (2016a, b), for example, demonstrated how maize plants fed upon by an adult folivore (cucurbit beetle, *Diabrotica speciosa*) attracted the belowground natural enemy (entomopathogenic nematode,

Heterorhabditis amazonensis) of the root-feeding larval stage of *D. speciosa*. This suggests that belowground predators can be highly attuned to plant-mediated changes in aboveground herbivory. They also showed that *H. amazonensis* was attracted to foliar applications of methyl salicylate (MeSA) and MeJA, suggesting similarities between aboveground herbivory and induction of plant defence pathways. By acting as a proxy for aboveground herbivory, herbivore elicitors and the associated release of plant volatiles could, therefore, be used to augment biological control of belowground herbivores via the recruitment of natural enemies. The response of belowground predators or parasitoids to volatile cues, however, may be genotype and species-specific, with, for example, nematodes showing no attraction to citrus plants (*Citrus paradisi* × *Poncirus trifoliata*) in response to aboveground herbivory by weevils (*Diaprepes abbreviatus*). Biological control attempts should also be considered with caution since the release of nematode-attracting cues may invoke ecological costs associated with the attraction of other plant-parasitic species (Ali et al. 2011). Simultaneous feeding by both aboveground and belowground herbivores may also attenuate the attraction of natural enemies and thwart biological control strategies (Rasmann and Turlings 2007), even if both aboveground and belowground herbivores belong to the same species (Moujahed et al. 2014).

Olfactometer experiments, like most controlled-environment studies, can be used to determine the effects of multiple factors, including climate change (Tariq et al. 2013), on aboveground–belowground herbivore dynamics, although logistical constraints associated with the experimental design may not accurately reflect spatial, temporal or seasonal changes in the life cycle of the plant, herbivore or natural enemy and their interactions with other organisms (Ballhorn and Kautz 2013). Insect age, for example, may affect decision making, which could be an important oversight in olfactometer experiments in general and should be considered in future behavioural studies. While olfactometer studies provide useful information on long-distance orientation by insects, their exclusive use for analysing insect behaviour is not recommended for drawing ecologically meaningful conclusions (Ballhorn and Kautz 2013). As always, an integrated approach combining controlled-environment and field experiments provides greater insights into the complex interactions between aboveground and belowground systems.

2.3.2 Observing Belowground Factors

Belowground interactions between soil fauna and the environment have often been referred to as a "black box" (Wilkinson 2008) or an "enigma" (Anderson 1975), mainly due to our inability to see these interactions occurring. Direct observational studies of soil fauna in the field are rare. Gunn and Cherrett (1993) created in situ rhizotrons (see Sect. 3.2.3) to attempt to see soil fauna feeding interactions, but the size range of organisms and patchy distribution in time and space has meant this is not a preferred method. Soil has been referred to as "the poor man's tropical rainforest" (Giller 1996) due to the abundance and diversity of organisms living within it. These organisms exist at a range of scales from microfauna and microflora (bacteria, fungi and protozoa) to mesofauna (springtails and mites) and macrofauna (earthworms, Coleoptera, etc.). Relationships between soil organisms are complex, and the cryptic nature of the soil environment makes disentangling belowground interactions between soil fauna problematic. There is a huge diversity of functional groups coexisting together, but there is also a large amount of functional redundancy. This baffling complexity has been a focus for ecologists over the last century and still requires greater study.

One of main drawbacks of manipulating belowground components discussed in Sect. 3.1 is that it is often difficult to validate the nature and success of the manipulation because it is not readily visualised. For example, experimentalists do not usually know whether belowground inoculation has been successful until the end of the experiment when they recover the herbivore destructively (e.g. Torode et al. 2016; Power et al. 2016). Moreover, it's difficult to establish when root herbivory is initiated, how long it lasts for, the extent of its effects (e.g. amount of tissue removed) or which parts of the roots are being attacked. These issues can be easily validated for shoot herbivores with simple observations. Observing or at least characterising belowground factors is, therefore, of some importance.

2.3.2.1 Slant Boards, Rhizotrons and Rhizotubes

Because of the opacity of the soil, it is always difficult to fully investigate the behaviour of both the plant roots and the soil biota. Researchers have developed a number of techniques to address this in controlled conditions.

Nutrient slant boards (Kendall and Leath 1974) and their derivatives (e.g. Baker and Byers 1977; Dawson et al. 2002; Murray and Clements 1994) have been used extensively to study root growth and architecture, to observe the behaviour of soil dwelling biota, including the effects of insect root herbivory. The concept of this technique is to force plants to grow in what is essentially a two-dimensional form in a soilless environment, which allows easy visualisation of the roots and associated insects. For example, Murray and Clements (1992) used 300 mm × 150 mm trays lined with capillary matting and polyester cloth. The seedling plant was placed on the cloth and covered with a second layer of cloth and capillary matting and finished by adding a stiff plastic cover, clamping the whole assembly together and irrigating with a nutrient solution at a rate of 2 L h^{-1}. This apparatus allowed regular measurements to be made of root growth rate and nodulation and the impact root feeders had on them.

A development of the slant board is the thin-plate rhizotron, which is assembled in a similar way and allows a thin layer of soil or other growth medium to be held between two glass plates. This allows the plants to be grown in a more 'natural' root environment whilst still allowing regular observation and measurement. For example, Alcántara et al. (2016) used rhizotrons to determine the interactions between a soil fungus and plant growth. The rhizotrons were constructed from two plates, one glass and one plastic (each 120 mm × 240 mm) separated by 5 mm with 10 mm wide plastic strips and filled with 100 g of air-dried, sieved (to 2 mm) soil, re-wetted with the top surface of each left open to allow plant growth. The plant species used in the study were perennial ryegrass (cv. AberMagic) and white clover (cv. AberDai). These were grown alone and in mixtures. Several seeds were sown in each rhizotron and covered with a thin layer of soil. After germination, two seedlings among the best developed per rhizotron were selected and allowed to grow, while the remainder were removed. The treatments were GR (ryegrass + ryegrass), CL (clover + clover) and MX (ryegrass + clover) in soil inoculated with *Trichoderma hamatum* (+T) or un-inoculated (−T). (Clover plants in the MX treatment are referred to as MX CL and ryegrass plants as MX GR). The rhizotrons were wrapped in aluminium foil to keep the roots in the dark and stood in water-filled trays with approximately 20 mm depth of water that was periodically replaced. The technique allowed the numbers of *Rhizobium* nodules on the roots of the clover plants to be counted periodically and regular measurements of root growth made. The results from these experiments demonstrated that *Trichoderma* confers a competitive advantage on white clover when grown in clover–ryegrass mixtures.

Tubes (often called ‘rhizotubes’) with a detachable side have been used to gain direct access to roots during an experiment, either to inoculate plants with a root herbivore or verify the activity of the herbivore. Typically, rhizotubes are constructed from plastic cable conduits which are used in construction (Fig. 2.2). For example, rhizotubes were used to establish the plant-mediated effects of a root herbivore (vine weevil; *Otiorhynchus sulcatus*) on the European raspberry aphid feeding aboveground (*Amphorophora idaei*) (McKenzie et al. 2013). This approach provided a novel demonstration of reciprocal feeding facilitation between the two herbivores, whereby both benefitted from each other’s presence on the plant (McKenzie et al. 2013).

2.3.2.2 X-ray Tomography

Researchers have deployed a number of non-destructive observational techniques, including X-ray tomography for studying soil organisms (Taina et al. 2008). X-ray computed tomography utilises X-rays to create cross-sections of a physical object which then recreates a virtual 3D model. Variations include high-resolution X-ray tomography and micro-computed tomography (micro-CT or μCT), which basically reflects the size of pixels in scanned cross-sections (Elliott and Dover 1982). X-ray tomography has its origins in medical imaging and in industrial computed tomography, where it is widely used. Broadly speaking, there are two types of scanners. In stationary configurations, the specimen is rotated around the X-ray source and detector whereas the second usually involves a gantry system that involves the X-ray tube and detector moving around a stationary specimen.

Using X-ray tomography for studying soil organisms, such as root herbivores, is fairly uncommon but has been successful in the past (Taina et al. 2008). In an early example, Harrison et al. (1993) was able to observe movement patterns of the fourth

Fig. 2.2 Example of rhizotron used by McKenzie et al. (2013) which allowed non-destructive access to the roots to monitor feeding impacts of vine weevils on aphids aboveground. Photograph provided by Scott Johnson

instar pecan weevil (*Curculio caryae*). The digging behaviour of the Argentine ant (*Linepithema humile*) was similarly quantified using X-ray computed tomography (Halley et al. 2005). Using higher resolution X-ray micro-tomography, first instar clover root weevil (*Sitona lepidus*, now *Sitona obsoletus*) could be visualised locating specific parts of the root system (Johnson et al. 2004b) and to distinguish between host and non-host root systems (Johnson et al. 2004a). In particular, larvae could be seen to target root nodules of their host plant (white clover, *Trifolium repens*) which house N-fixing bacteria (Johnson et al. 2004b) (Fig. 2.3). Early instars of this species were known to benefit from feeding on this nitrogen-rich tissue (Gerard 2001), and it has been suggested that they were able to locate these nodules using chemical cues (Johnson et al. 2005). X-ray tomography was also used to establish the duration of larval herbivory on root nodules and to determine how this affected oviposition behaviour of maternal insects living aboveground (Johnson and Gregory 2012). Intriguingly, maternal insects initially laid eggs equally on plants with and without larvae, but started to show a preference for plants without larvae 12 h after larvae had started to feed. This simple study suggested that maternal insects were able to distinguish between occupied and unoccupied plants and potentially maximise the fitness of their offspring by avoiding future intra-specific competition between siblings (Johnson et al. 2006).

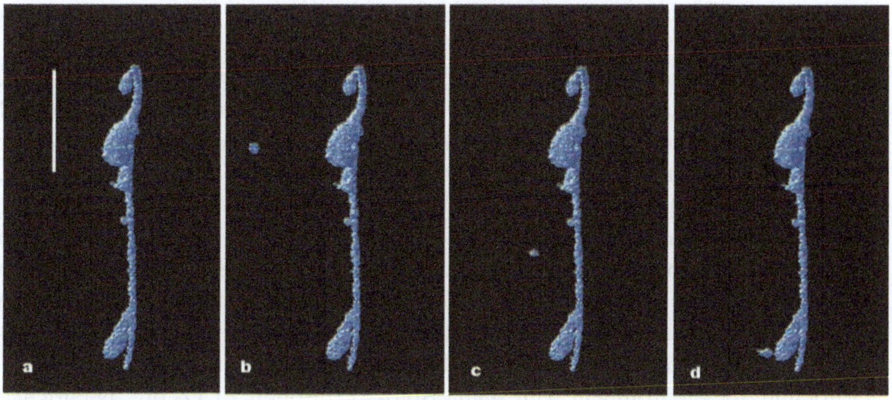

Fig. 2.3 X-ray tomographic images of the sequential movement of a neonatal *Sitona lepidus* (now *Sitona obsoletus*) larva towards the lower clover root nodule. Location at (**a**) 0 h, (**b**) 3 h, (**c**) 6 h and (**d**) 9 h. The larva was subsequently recovered from the nodule when the column was dismantled. The white bar represents 1 cm. Reproduced from Johnson et al. (2004b) with permission

While not used explicitly in the context of aboveground–belowground interactions, X-ray tomography has also been used to study microbial communities and earthworm behaviour (Zappala et al. 2013; Taina et al. 2008; Ritz 2011). These organisms have important impacts on organisms aboveground (see chapters in this volume), so X-ray tomography has the potential to identify linkages between subtle aspects of microbial communities, earthworm behaviour and the ecosystem aboveground. The biggest constraints to using X-ray tomography to study aboveground–belowground interactions are that it is expensive, laboratory-based and can handle only relatively small experimental units (e.g. pots). Crucially, the size of the experimental unit and the resolution of images that can be obtained is a compromise between scanning time and X-ray dosage (Mankin et al. 2008; Johnson et al. 2007; Zappala et al. 2013).

2.3.2.3 Isotope Labelling

Over 50 years ago, radioisotopes were used to measure recolonisation and consumption of fungi by soil fauna (Coleman and MacFadyen 1966; Coleman and McGinnis 1970) and to follow the movement of ^{14}C photosynthate into roots and exudates (McDougall 1970) and the impact of aboveground herbivory on rhizosphere microbial growth (Coleman et al. 2002). However, due to concern over health and safety and improvements in measurement capabilities, stable isotopes are now the preferred labelling technique. Stable isotopes are non-radioactive, do not decay (Hood-Nowotny and Knols 2007) and occur naturally in the environment at different levels, increasing the simplicity of studies. Stable isotopes are used to study the diet of animals as an organism's tissues retain a fixed isotopic enrichment (or depletion) in relation to diet and are a measure of assimilated (not just ingested) food (Crotty et al.

2012b). Different ecosystems can also be distinguished by their isotopic composition (Tiunov 2007).

Carbon isotopes are thought to be ecosystem specific (France and Peters 1997), whilst nitrogen isotopes have been shown to increase regularly per trophic level (Minagawa and Wada 1984). Utilising the step-wise enrichment of ^{15}N to infer trophic position for individual species has occurred for a number of taxa, e.g. Collembola (Chahartaghi et al. 2005), Oribatid mites (Erdmann et al. 2007), Elaterid larvae (Traugott et al. 2007) and earthworms (Schmidt et al. 1997). Stable isotopes at natural abundance have also been used to define the whole soil food web (Ponsard and Arditi 2000; Scheu and Falca 2000). Stable isotopes have also been used to differentiate between C_3 and C_4 plants—defining whether translocated leaf litter is being consumed compared to root exudates (Kramer et al. 2012).

Enriching the environment with stable isotopes (through $^{13}CO_2$ or ^{15}N-enriched nutrients) is another method to define the soil food web (Setälä and Aarnio 2002; Pollierer et al. 2007). These methods do not trace the passage of energy through the bacterial or fungal energy channels within the soil food web. However, Crotty et al. (2011, 2012a) used isotopically enriched organisms (bacteria, then protozoa) to trace consumption through the different trophic levels within the soil food web, with results suggesting that omnivory is much more common than expected. While the majority of work with isotope labelling has largely focussed on belowground food webs, there is scope for incorporating aboveground components and potentially shed light on aboveground–belowground interactions.

2.4 Experiments in the Field

2.4.1 Manipulating Belowground Factors in the Field

While reductionist experiments conducted in controlled conditions are essential for gaining a mechanistic understanding of aboveground–belowground community ecology, they lack the realism and stochastic variation in environmental conditions that organisms experience. Field experimentation is, therefore, an important component of aboveground–belowground community ecology, particularly considering that observations from controlled environment studies do not necessarily translate to field situations (e.g. Vandegehuchte et al. 2010).

2.4.1.1 Supplementation Experiments

Conducting research into aboveground–belowground community interactions in the field is challenging, especially in terms of manipulating belowground factors. Broadly speaking, such field experiments can either involve supplementing natural populations of organisms in the soil (providing ‘ambient’ and ‘increased’ treatments) or else attempt to remove organisms completely. In the latter case, discussed in

Sect. 4.1.2, most examples seem to be a hybrid of field and pot approaches whereby soil is excavated, subjected to some treatment (e.g. sieving and inoculation) and then reintroduced to contained field plots isolated from surrounding field soil (e.g. submerged pots or root barriers).

There are examples of supplementation studies using root herbivores in particular. One study conducted in North East Scotland established 4 field sites of 250 blackcurrant bushes (Johnson et al. 2013). Six plants were selected at random for monitoring, 3 were inoculated with 50 vine weevil (*O. sulcatus*) eggs in late Spring. Bushes were subsequently monitored for occurrence of aboveground invertebrates, including herbivores and natural enemies, in a systematic fashion. Soil cores were taken at the end of the season to establish that inoculated plants hosted more vine weevils than plants that were naturally colonised by vine weevils. Weevils increased field populations of aphids by ca. 700%, which was followed by an increase in the abundance of aphid natural enemies. In contrast, sawfly populations were 77% smaller during mid-June and adult emergence delayed by >14 days on plants with weevils.

A more recent supplementation study, also involving root herbivores, simultaneously manipulated rainfall regimes. The experimental platform named DRI-Grass (Drought and Root herbivore Impacts on Grasslands) comprised 60 grassland plots in South East Australia, 48 of which were under rain exclusion shelters that simulated 5 rainfall regimes (Power et al. 2016). Herbivore supplementation was achieved by applying 27 g of locally collected (with light traps) adult scarab beetles (Coleoptera: Scarabaeidae) to the herbivore addition plots in December 2013 and an additional 9 g of adult beetles in February–March 2014. Beetles were placed in mesh enclosures in the plots and allowed to oviposit for 3 days after which they were removed. Mesh enclosures were also applied to the insect-free plots. The efficacy of herbivore treatments was verified 18 months after beetle additions by excavating two holes (25 cm × 10 cm) per plot to a depth of 20 cm (see Power et al. 2016 for full details). The study demonstrated that root herbivory induced foliar silicon-based defences in two grass species *Cynodon dactylon* and *Eragrostis curvula* (Power et al. 2016). Induction of silicon defences aboveground by belowground herbivores had not previously been reported (Hartley and DeGabriel 2016). A number of correlations between the abundances of above- and below-ground invertebrate groups under ambient rainfall were also observed, suggesting that these communities were linked, although the mechanisms were unclear (Torode et al. 2016). However, these correlations dissipated under altered rainfall regimes, which was termed 'climatic decoupling' (Torode et al. 2016).

Field studies have also included additions of earthworms, usually by first extracting worms from a different site and supplementing experimental plots (Rhea-Fournier and González 2017). This may appear straightforward, but inoculations are frequently unsuccessful because earthworms have been harmed during extraction and, if applied to the soil surface, they are vulnerable to desiccation and predation. Five months after inoculation, Subler et al. (1997) found no difference in the abundances of earthworms from control and supplemented field plots, highlighting that manipulating this aspect of the belowground compartment is more challenging than it first appears. Butt (1999) considered a range of factors that may affect supplementation success, suggesting that researchers establish a "starter culture" in containers of soil for several months so that inocula include adults, cocoons and hatchlings.

2.4.1.2 Exclusion Experiments

Field studies that have sought to exclude belowground organisms often use approaches that reduce the abundance of target organisms, while reducing impacts on non-target organisms. For belowground herbivores, this usually means applying soil insecticides (Brown and Gange 1989; Masters 1995). This can be problematic because insecticides often have compounding effects on plants and non-target organisms including those living aboveground. Masters (2004) advocated the use of a granular formation of chlorpyrifos to remove root herbivores, citing evidence for minimal effects on bacteria, fungi, nematodes, earthworms and small mammals and birds. Many other studies have excluded above- and belowground herbivores using insecticides in different field plots, having first assessed their confounding effects on plants in pilot experiments (e.g. Schädler et al. 2004). In that experiment, the effects of the insecticides on plant growth were deemed minimal (Schädler et al. 2004). Nonetheless, chemical interventions may also affect the chemical signalling pathways that drive or influence interactions between above- and belowground organisms.

Studies involving aboveground and belowground components often involve exclusion of earthworms, which usually relies on electroshock or application of faunacides (e.g. naphthalene or carbofuran). There are, however, dozens of different techniques for removing or expelling earthworms from manipulated field plots (see Rhea-Fournier and González 2017 for a recent and comprehensive review). Chemical interventions are problematic in terms of affecting non-target organisms, as discussed above, and electroshock methods are generally favoured in field experiments. Designs vary, but basically involve metal electrodes or probes being dug 50–60 cm into the ground and a current >0.5 A being applied for 30–60 min. Octet configurations for the probes are very popular (Rushton and Luff 1984; Bohlen et al. 1995; Eisenhauer et al. 2008), though these studies were not necessarily concerned with manipulative field experiments (i.e. electroshock was used for sampling rather than earthworm exclusion). Rhea-Fournier and González (2017) emphasise that these are unlikely to exclude all earthworms and are best considered reductions rather than eliminations or exclusions in the strictest sense. In many cases, field plots are subsequently enclosed with physical barriers buried to some depth with part of the barrier emerging aboveground (Blair et al. 1995).

Some experiments take this one stage further by first excavating field soil to remove micro- and macro-arthropods, then returning the soil to holes in the field site which have been lined with submerged pots. The soil is then inoculated with the experimental organism under investigation in specific plots, which are then caged to confine or exclude aboveground organisms. These field mesocosm experiments seem especially popular in aboveground–belowground ecological research, with researchers being able to manipulate both communities. Field mesocosm experiments established that the earthworm *Aporrectodea caliginosa* promoted populations of the aphid *Rhopalosiphum padi* and its parasitoid *Aphidius ervi* (Johnson et al. 2011). Root herbivores (*Agriotes* spp. wireworms), in contrast, reduced populations of *Sitobion avenae*, *Metopolophium dirhodum* and *R. padi* (Wade et al. 2017). While soil-dwelling dung beetles (*Bubas bison*) had significant impacts on plant (*Brassica oleracea*) growth and chemistry in field mesocosm experiments, they did not have any impacts on diamondback moth (*Plutella xylostella*) (Johnson et al. 2016c).

2.4.2 Observing Patterns in Field Populations

It is important to understand the interactions that are occurring within natural populations and not just assess biodiversity (Crotty et al. 2014). Soil organisms play a key role in ecosystem function—cycling nutrients, modifying soil structure, regulating water levels and often provide resilience to environmental perturbations within the belowground system (Brussaard et al. 1997). The large diversity of species that occurs within the soil has created a paradox to some extent, with the belief that there are not enough individual niches for all the different species found within the soil (Coleman 2008). It is only by observing patterns in natural populations that we can start to assess and understand how these organisms coexist utilising the same resources. Without this understanding, we will not be able to identify soil fauna at risk of extinction or environments that have deteriorated and lost the functional redundancy that is one of the main features of soil biodiversity.

2.4.2.1 Chemical and Destructive Sampling

There is a limited number of approaches for sampling soil-dwelling invertebrates, reviewed in more detail by Coleman et al. (1999) and Gange (2005). In particular, Coleman et al. (1999) provide very detailed protocols for extracting invertebrates ranging from microfauna (e.g. protozoa, rotifers), mesofauna (e.g. nematodes), microarthropods (e.g. collembola) and macrofauna (e.g. earthworms). While chemical application to soils can either force invertebrates to the soil surface (e.g. potassium permanganate, formalin, mustard) or possibly attract them in the case of baits (Gange 2005), these are not usually used for addressing research questions in aboveground–belowground community ecology. Instead, virtually all approaches start with collection of appropriately sized soil cores, which are returned to the laboratory for extraction. Invertebrates are then removed from the soil using different extraction procedures, the exact method being mainly dictated by the size and behavioural characteristics of the organism. This can range from hand-sorting (Gange 2005), flotation (Walter et al. 1987) or, more commonly, high-gradient extraction which induces invertebrates to leave the soil by subjecting soil to artificially high temperature and moisture gradients. Various designs of apparatus are discussed by Coleman et al. (1999), but the preferred 'Tullgren-type' extraction method is that based on Crossley and Blair (1991) which is frequently used in studies exploring aboveground–belowground community linkages (e.g. Torode et al. 2016; Nielsen et al. 2010a, b).

2.4.2.2 Non-destructive Methods in the Field

There are several non-destructive methods for assessing the behaviour and abundance of belowground organisms especially those that damage roots (Johnson et al. 2007; Mankin et al. 2008; Dawson and Byers 2008). Rhizotrons, similar to those described in Sect. 3.2.1, have occasionally been used in field situations. These are typically much bigger than those used in controlled conditions and installed into an excavated field plot, often with a transparent interface pressed vertically against the soil profile. Studying soil food webs, for example, was achieved by establishing a large rhizotron at the soil interface (Gunn and Cherrett 1993). Including the aboveground component as well, at least in terms of plant productivity, Wilson et al. (1995) investigated soil pesticide effects on belowground communities of a grassland.

Acoustic detection of belowground herbivores has been successfully applied in both pot and field studies (Mankin et al. 2008). The approach has been widely used for detecting invertebrate infestations in wood and food, but has now been tested successfully for soil fauna. Instruments typically involve sensors (e.g. microphones, accelerometer, piezoelectric probes and geophones) being inserted into the soil. The soil strongly attenuates vibrations above 200 Hz, so the sensors are exposed to lower levels of mid- to high-frequency background 'noise' than aboveground; experimentalists are, therefore, able to detect low amplitude (500–1800 Hz) sounds produced by many invertebrates above a certain size (Mankin et al. 2008). The attenuation, however, means that invertebrates need to be 10–30 cm from the sensor to be reliably detected. Nonetheless, acoustic detection can even allow different types of behaviour (e.g. feeding or burrowing) to be detected (e.g. Zhang et al. 2003).

Another approach used in the field to track the movement and location of *Melolontha melolontha* was to develop a tag-and-trace system (Bont et al. 2017). This involved fitting larvae with a copper ring tag and monitoring their movements in relation to a preferred host plant *Taraxacum officinale*. The authors reported that the ring weighed only 105 mg, which represented only a 5% increase in mass, which while larvae moved more slowly when tagged, foraging behaviour was largely unaffected (Bont et al. 2017).

Moving more towards a field scale, experimentalists have been able to use remote sensing to detect patterns of root herbivore populations, with reasonable levels of success. For example, Johansen et al. (2014) used object-based image analysis and high spatial resolution satellite imagery to map canegrub damage in sugarcane plantations. Essentially, they used the amount of green leaves and image texture within image blocks to identify areas infested with canegrubs. This relationship was then validated with field observations of canegrub damage, which were made at the time of satellite image capture. The method had accuracy rates of 53–80% (Johansen et al. 2014).

Workers have also been able to create 'risk maps' which identify high incidences of root herbivores such as redheaded cockchafer (*Adoryphorus couloni*; RHC) (Cosby et al. 2016); these could potentially be useful for identifying linkages with aboveground fauna. In this particular study, electromagnetic surveys of soil properties, optical sensor derived estimates of pasture biomass and topography via GPS elevation survey were combined to predict densities of *A. couloni*. A combination of these variables was used to produce 'risk maps' with an accuracy of 88% at predicting likely RHC density-categories (Cosby et al. 2016).

2.4.2.3 Natural Abundance of Isotopes in Aboveground and Belowground Organisms

Stable isotopes are used to investigate the food source, functional group and trophic level of organisms, particularly when it is difficult to observe this directly. Stable isotopes at natural abundance can assess whether faunal communities with similar taxa are utilising the same or different basal resources. For example, Crotty et al. (2014) investigated the different feeding preferences within the soil food web of organisms residing in an agricultural grassland compared to a woodland and found that the woodland soil fauna utilised litter resources to a greater extent than the grassland fauna which utilised root derived resources. The isotopic composition can highlight whether they are consuming algae and lichens (Schneider et al. 2004; Tiunov 2007) or whether they are predatory (Ponsard and Arditi 2000). Linking aboveground animals to belowground food sources, Hobson (1999) found ^{15}N natural abundance signatures splitting songbirds into those consuming invertebrates within agricultural wetlands from those utilising invertebrates in boreal forests. A greater understanding of the environment being investigated is necessary to group soil fauna into different feeding groups. For example, Vanderklift and Ponsard (2003) postulated that the activity of microorganisms may lead to a progressive shift in the isotopic composition of detritus—potentially affecting the decomposers consuming them. Isotope studies have found that primary and secondary decomposers within the soil appear to form a continuum (Scheu and Falca 2000), from those feeding on litter to those feeding on the microorganisms utilising the litter (e.g. Chahartaghi et al. 2005; Erdmann et al. 2007). Unfortunately, the small size of soil fauna leads to a bulking of many individuals depending on mass spectrometry methods used (Crotty et al. 2013), which can affect assessments of feeding preferences.

2.5 Closing Remarks

This survey of belowground approaches that may assist community ecologists interested in aboveground–belowground interactions is not exhaustive. We aimed to consider those techniques that either had been used in aboveground–belowground community ecology studies or at least had significant potential to contribute to this topic. A recent review highlighted how molecular approaches could be useful for studying root herbivores (Benefer and Blackshaw 2013); such techniques are already used by soil ecologists. Molecular approaches offer significant potential and we finish by identifying two promising areas of research.

2.5.1 Belowground Invertebrate Gut Contents

One of the key problems in studying soil invertebrates is identifying their feeding preferences. Crotty et al. (2014) used stable isotope signatures to determine trophic interactions, but there is relatively little work on identifying the actual species that are preyed upon. Despite the importance of these interactions, our knowledge of belowground herbivory is limited to relatively few species of invertebrates that are significant crop pests (Johnson and Murray 2008; Johnson et al. 2016a). Many other species inhabit the soil and directly impact plants. Clements et al. (1990) showed average forage yield increases of 11% when insecticides were applied to permanent grassland, despite the absence of the major pest species. Many other species inhabit the soil and directly impact plants; the lack of knowledge of the feeding habits of the majority of the soil fauna is mainly due to the methodological constraints for studying soil biota in situ, as described earlier. Many of the experimental systems available utilise sieved soils and are artificial in their construction, and very few experiments (e.g. Murray et al. 2009; Crotty et al. 2012a) use intact soils with their full complement of biota in feeding studies. One sure way to determine the provenance of food material would be to identify the material in the gut of the insect. However, this has been extremely challenging. Advances in molecular techniques have used identification of DNA in gut contents to reveal the feeding activity in natural conditions. However, the literature is dominated by studies of animal prey (e.g. King et al. 2008; Wallinger et al. 2013). It is only now that there is a body of work that is focussed on the identification of ingested plants. One of the more robust methodologies is the PCR assay developed by Wallinger et al. (2012) which may be used to screen large numbers of samples. Wallinger et al. (2013) describe an optimised PCR assay to determine how detectable ingested fresh plant or decaying plant DNA was in the click beetle *Agriotes* spp. Both types of litter were detectable, demonstrating that their consumption could be verified by this technique.

2.5.2 *Meta-Barcoding*

The study of Collembola and mites is severely constrained by the lack of taxonomic expertise needed for identification. High resource demands and lack of experts result in the grouping of taxa at higher taxonomic levels, e.g. super-families and sub-orders, rather than at the species level. The development of molecular techniques based on DNA extraction, over the last 10 years, has the potential to revolutionise the identification of soil fauna (Hebert and Gregory 2005). DNA meta-barcoding refers to the automated identification of multiple species from a bulk sample containing entire organisms or from environmental samples containing parts of organisms (Taberlet et al. 2012), although taxonomic reference libraries need to be created. Metagenomics has been the focus of microbiology, bypassing the need for isolation or cultivation of microorganisms (Simon and Daniel 2011), but only classifies the DNA/rRNA sequences as 'OTU's—operational taxonomic units. Advances in high throughput sequencing for meta-barcoding and metagenomics now allow soil mesofauna to be identified along with other soil organisms. For example, nematode molecular identification has been established since 2006 (Griffiths et al. 2006) and this technique has been developed further since (Sapkota and Nicolaisen 2015). Arribas et al. (2016) have recently shown that integrating standard mesofauna extraction techniques with high-throughput sequencing allowed the characterisation of soil biodiversity in a phylogenetic and community ecology context. However, a common database of soil biodiversity is still needed, and it will take a combined effort of researchers to fully characterise soil biodiversity (Orgiazzi et al. 2015).

Acknowledgements The authors are very grateful to Mattias Erb and Peter Gregory for reviewing this book chapter and providing valuable insights for its improvement. SNJ and JMWR acknowledge financial support from the Australian Research Council (Discovery Grants DP14100363 and DP17102278) and a Future Fellowship (FT170100342) awarded to SNJ.

References

Alcántara C, Thornton CR, Pérez-de-Luque A et al (2016) The free-living rhizosphere fungus *Trichoderma hamatum* GD12 enhances clover productivity in clover–ryegrass mixtures. Plant Soil 398:165–180

Ali JG, Davidson-Lowe E (2015) Plant cues and factors influencing the behaviour of beneficial nematodes as a belowground indirect defense. In: Bais H, Sherrier J (eds) Plant microbe interactions (Advances in Botanical Research vol 75). Academic, New York, pp 191–214

Ali JG, Alborn HT, Stelinski LL (2010) Subterranean herbivore-induced volatiles released by citrus roots upon feeding by *Diaprepes abbreviatus* recruit entomopathogenic nematodes. J Chem Ecol 36:361–368

Ali JG, Alborn HT, Stelinski LL (2011) Constitutive and induced subterranean plant volatiles attract both entomopathogenic and plant parasitic nematodes. J Ecol 99:26–35

Anderson JM (1975) The enigma of soil animal species. In: Vaněk L (ed) Progress in soil zoology: proceedings of the fifth international colloquium on progress in soil zoology. W Junk and Prague Academia, Dordrecht, pp 51–58

Arribas P, Andújar C, Hopkins K et al (2016) Metabarcoding and mitochondrial metagenomics of endogean arthropods to unveil the mesofauna of the soil. Methods Ecol Evol 7:1071–1081
Baker PB, Byers RA (1977) A laboratory technique for rearing the clover root curculio. Melsheimer Entomol Ser 23:8–10
Ballhorn DJ, Kautz S (2013) How useful are olfactometer experiments in chemical ecology research? Commun Integr Biol 6:e24787
Barnett K, Johnson SN (2013) Living in the soil matrix: abiotic factors affecting root herbivores. Adv Insect Physiol 45:1–52
Benefer CM, Blackshaw RP (2013) Molecular approaches for studying root herbivores. Adv Insect Physiol 45:220–255
Bezemer TM, Wagenaar R, van Dam NM et al (2004) Above- and below-ground terpenoid aldehyde induction in cotton, *Gossypium herbaceum*, following root and leaf injury. J Chem Ecol 30:53–67
Blair JM, Bohlen PJ, Edwards CA et al (1995) Manipulations of earthworm populations in field experiments in agroecosystems. Acta Zool Fenn 169:48–51
Blossey B, Hunt-Joshi TR (2003) Belowground herbivory by insects: influence on plants and aboveground herbivores. Annu Rev Entomol 48:521–547
Bohlen PJ, Parmelee RW, Blair JM et al (1995) Efficacy of methods for manipulating earthworm populations in large-scale field experiments in agroecosystems. Soil Biol Biochem 27:993–999
Böhm W (1979) Methods of studying root systems. Ecological studies, vol 33. Springer, Berlin
Bont Z, Arce C, Huber M et al (2017) A herbivore tag-and-trace system reveals contact- and density-dependent repellence of a root toxin. J Chem Ecol 43:295–306
Borowicz VA (2010) The impact of arbuscular mycorrhizal fungi on strawberry tolerance to root damage and drought stress. Pedobiologia 53:265–270
Brown VK, Gange AC (1989) Differential effects of above-ground and below-ground insect herbivory during early plant succession. Oikos 54:67–76
Brussaard L, Behan-Pelletier VM, Bignell DE et al (1997) Biodiversity and ecosystem functioning in soil. Ambio 26:563–570
Butt KR (1999) Inoculation of earthworms into reclaimed soils: the UK experience. Land Degrad Dev 10:565–575
Chahartaghi M, Langel R, Scheu S et al (2005) Feeding guilds in Collembola based on nitrogen stable isotope ratios. Soil Biol Biochem 37:1718–1725
Chung SH, Rosa C, Scully ED et al (2013) Herbivore exploits orally secreted bacteria to suppress plant defenses. Proc Natl Acad Sci USA 110:15728–15733
Clements RO, Murray PJ, Bentley BR et al (1990) The impact of pests and diseases on the herbage yield of permanent grassland at eight sites in England and Wales. Ann Appl Biol 117:349–357
Coleman DC (2008) From peds to paradoxes: linkages between soil biota and their influences on ecological processes. Soil Biol Biochem 40:271–289
Coleman DC, MacFadyen A (1966) Recolonization of gamma-irradiated soil by small arthropods – a preliminary study. Oikos 17:62–70
Coleman DC, McGinnis JT (1970) Quantification of fungus – small arthropod food chains in soil. Oikos 21:134–137
Coleman DC, Blair JM, Elliott ET et al (1999) Soil invertebrates. In: Robertson GP, Coleman DC, Bledsoe CS et al (eds) Standards soil methods for long-term ecological research. Oxford University Press, Cary, pp 349–377
Coleman D, Fu SL, Hendrix P et al (2002) Soil foodwebs in agroecosystems: impacts of herbivory and tillage management. Eur J Soil Biol 38:21–28
Cosby AM, Falzon GA, Trotter MG et al (2016) Risk mapping of redheaded cockchafer (*Adoryphorus couloni*) (Burmeister) infestations using a combination of novel k-means clustering and on-the-go plant and soil sensing technologies. Precis Agric 17:1–17
Cosme M, Lu J, Erb M et al (2016) A fungal endophyte helps plants to tolerate root herbivory through changes in gibberellin and jasmonate signaling. New Phytol 211:1065–1076

Crossley DA, Blair JM (1991) A high-efficiency, low-technology Tullgren-type extractor for soil microarthropods. Agric Ecosyst Environ 34:187–192

Crotty FV, Blackshaw RP, Murray PJ (2011) Tracking the flow of bacterially derived ^{13}C and ^{15}N through soil faunal feeding channels. Rapid Commun Mass Spectrom 25:1503–1513

Crotty FV, Adl SM, Blackshaw RP et al (2012a) Protozoan pulses unveil their pivotal position within the soil food web. Microb Ecol 63:905–918

Crotty FV, Adl SM, Blackshaw RP et al (2012b) Using stable isotopes to differentiate trophic feeding channels within soil food webs. J Eukaryot Microbiol 59:520–526

Crotty FV, Stocki M, Knight JD et al (2013) Improving accuracy and sensitivity of isotope ratio mass spectrometry for $\delta\ ^{13}C$ and $\delta\ ^{15}N$ values in very low mass samples for ecological studies. Soil Biol Biochem 65:75–77

Crotty FV, Blackshaw RP, Adl SM et al (2014) Divergence of feeding channels within the soil food web determined by ecosystem type. Ecol Evol 4:1–13

D'Alessandro M, Erb M, Ton J et al (2014) Volatiles produced by soil-borne endophytic bacteria increase plant pathogen resistance and affect tritrophic interactions. Plant Cell Environ 37:813–826

Dawson LA, Byers RA (2008) Methods for studying root herbivory. In: Johnson SN, Murray PJ (eds) Root Feeders – an ecosystem perspective. CABI, Wallingford, pp 3–19

Dawson LA, Grayston SJ, Murray PJ et al (2002) Root feeding behaviour of *Tipula paludosa* (Meig.) (Diptera : Tipulidae) on *Lolium perenne* L. and *Trifolium repens* L. Soil Biol Biochem 34:609–615

de la Peña E, Echeverria SR, van der Putten WH et al (2006) Mechanism of control of root-feeding nematodes by mycorrhizal fungi in the dune grass *Ammophila arenaria*. New Phytol 169:829–840

Degenhardt J, Hiltpold I, Köllner TG et al (2009) Restoring a maize root signal that attracts insect-killing nematodes to control a major pest. Proc Natl Acad Sci USA 106:13213–13218

Dicke M, van Loon JJA (2000) Multitrophic effects of herbivore-induced plant volatiles in an evolutionary context. Entomol Exp Appl 97:237–249

Ehler LE (1998) Invasion biology and biological control. Biol Control 13:127–133

Eisenhauer N, Straube D, Scheu S (2008) Efficiency of two widespread non-destructive extraction methods under dry soil conditions for different ecological earthworm groups. Eur J Soil Biol 44:141–145

Elliott JC, Dover SD (1982) X-ray microtomography. J Microsc 126:211–213

Ellmore GS, Zanne AE, Orians CM (2006) Comparative sectoriality in temperate hardwoods: hydraulics and xylem anatomy. Bot J Linn Soc 150:61–71

Erb M, Lu J (2013) Soil abiotic factors influence interactions between belowground herbivores and plant roots. J Exp Bot 64:1295–1303

Erb M, Ton J, Degenhardt J et al (2008) Interactions between arthropod-induced aboveground and belowground defenses in plants. Plant Physiol 146:867–874

Erb M, Glauser G, Robert CAM (2012) Induced immunity against belowground insect herbivores-activation of defenses in the absence of a jasmonate burst. J Chem Ecol 38:629–640

Erdmann G, Otte V, Langel R et al (2007) The trophic structure of bark-living oribatid mite communities analysed with stable isotopes (^{15}N, ^{13}C) indicates strong niche differentiation. Exp Appl Acarol 41:1–10

Erwin AC, Geber MA, Agrawal AA (2013) Specific impacts of two root herbivores and soil nutrients on plant performance and insect–insect interactions. Oikos 122:1746–1756

Filgueiras CC, Willett DS, Junior AM et al (2016a) Stimulation of the salicylic acid pathway aboveground recruits entomopathogenic nematodes belowground. PLoS One 11:e0154712

Filgueiras CC, Willett DS, Pereira RV et al (2016b) Eliciting maize defense pathways aboveground attracts belowground biocontrol agents. Sci Rep 6:36484

France RL, Peters RH (1997) Ecosystem differences in the trophic enrichment of ^{13}C in aquatic food webs. Can J Fish Aquat Sci 54:1255–1258

Gange AC (2005) Sampling insects from roots. In: Leather SL (ed) Insect sampling in forest ecosystems. Blackwell Scientific Publishing, Oxford, pp 16–36

Gerard PJ (2001) Dependence of *Sitona lepidus* (Coleoptera: Curculionidae) larvae on abundance of white clover *Rhizobium* nodules. Bull Entomol Res 91:149–152

Giller PS (1996) The diversity of soil communities, the 'poor man's tropical rainforest'. Biodivers Conserv 5:135–168

Gregory PJ, Nortcliff S (2013) Soil conditions and plant growth. Wiley, Chichester

Griffiths DW, Birch ANE, Macfarlane-Smith WH (1994) Induced changes in the indole glucosinolate content of oilseed and forage rape (*Brassica napus*) plants in response to either turnip root fly (*Delia floralis*) larval feeding or artificial root damage. J Sci Food Agric 65:171–178

Griffiths BS, Donn S, Neilson R et al (2006) Molecular sequencing and morphological analysis of a nematode community. Appl Soil Ecol 32:325–337

Gunn A, Cherrett JM (1993) The exploitation of food resources by soil meso-invertebrates and macro-invertebrates. Pedobiologia 37:303–320

Halley JD, Burd M, Wells P (2005) Excavation and architecture of Argentine ant nests. Insect Soc 52:350–356

Harrison RD, Gardner WA, Tollner WE et al (1993) X-ray computed-tomography studies of the burrowing behavior of 4th-instar Pecan weevil (Coleoptera, Curculionidae). J Econ Entomol 86:1714–1719

Hartley SE, DeGabriel JL (2016) The ecology of herbivore-induced silicon defences in grasses. Funct Ecol 30:1311–1322

Hatch DJ, Murray PJ (1994) Transfer of nitrogen from damaged roots of white clover (*Trifolium repens* L.) to closely associated roots of intact perennial ryegrass (*Lolium perenne* L). Plant Soil 166:181–185

Hebert PDN, Gregory TR (2005) The promise of DNA barcoding for taxonomy. Syst Biol 54:852–859

Hiltpold I, Erb M, Robert CAM et al (2011) Systemic root signalling in a belowground, volatile-mediated tritrophic interaction. Plant Cell Environ 34:1267–1275

Hjältén J (2004) Simulating herbivory: problems and possibilities. In: Weisser WW, Siemann E (eds) Insects and ecosystem function, Ecological studies, vol 173. Springer, Berlin, pp 243–255

Hobson KA (1999) Stable-carbon and nitrogen isotope ratios of songbird feathers grown in two terrestrial biomes: implications for evaluating trophic relationships and breeding origins. Condor 101:799–805

Hol WHG, Macel M, van Veen JA et al (2004) Root damage and aboveground herbivory change concentration and composition of pyrrolizidine alkaloids of *Senecio jacobaea*. Basic Appl Ecol 5:253–260

Hood-Nowotny R, Knols BGJ (2007) Stable isotope methods in biological and ecological studies of arthropods. Entomol Exp Appl 124:3–16

Johansen K, Robson A, Samson P et al (2014) Mapping canegrub damage from high spatial resolution satellite imagery. Proc Aust Soc Sugar Cane Tech 36:62–70

Johnson SN, Gregory PJ (2012) Breaking open the black box – can non-invasive imaging help answer questions in aboveground-belowground ecology? Paper presented at the aboveground-belowground interactions: technologies and new approaches. Joint Symposium of British Ecological Society, the Biochemical Society and the Society for Experimental Biology, Charles Darwin House, London, 8–10 October 2012

Johnson SN, Murray PJ (eds) (2008) Root Feeders – an ecosystem perspective, 1st edn. CABI, Wallingford

Johnson SN, Gregory PJ, Murray PJ et al (2004a) Host plant recognition by the root feeding clover weevil, *Sitona lepidus* (Coleoptera: Curculionidae). Bull Entomol Res 94:433–439

Johnson SN, Read DB, Gregory PJ (2004b) Tracking larval insect movement within soil using high resolution X-ray microtomography. Ecol Entomol 29:117–122

Johnson SN, Gregory PJ, Greenham JR et al (2005) Attractive properties of an isoflavonoid found in white clover root nodules on the clover root weevil. J Chem Ecol 31:2223–2229
Johnson SN, Birch ANE, Gregory PJ et al (2006) The 'mother knows best' principle: should soil insects be included in the preference–performance debate? Ecol Entomol 31:395–401
Johnson SN, Crawford JW, Gregory PJ et al (2007) Non-invasive techniques for investigating and modelling root-feeding insects in managed and natural systems. Agric For Entomol 9:39–46
Johnson SN, Staley JT, McLeod FAL et al (2011) Plant-mediated effects of soil invertebrates and summer drought on above-ground multitrophic interactions. J Ecol 99:57–65
Johnson SN, Mitchell C, McNicol JW et al (2013) Downstairs drivers – root herbivores shape communities of above-ground herbivores and natural enemies via changes in plant nutrients. J Anim Ecol 82:1021–1030
Johnson SN, Benefer CM, Frew A et al (2016a) New frontiers in belowground ecology for plant protection from root-feeding insects. Appl Soil Ecol 108:96–107
Johnson SN, Erb M, Hartley SE (2016b) Roots under attack: contrasting plant responses to below- and aboveground insect herbivory. New Phytol 210:413–418
Johnson SN, Lopaticki G, Barnett K et al (2016c) An insect ecosystem engineer alleviates drought stress in plants without increasing plant susceptibility to an above-ground herbivore. Funct Ecol 30:894–902
Kafle D, Hänel A, Lortzing T et al (2017) Sequential above- and belowground herbivory modifies plant responses depending on herbivore identity. BMC Ecol 17:5
Kaplan I, Halitschke R, Kessler A et al (2008a) Constitutive and induced defenses to herbivory in above- and belowground plant tissues. Ecology 89:392–406
Kaplan I, Halitschke R, Kessler A et al (2008b) Effects of plant vascular architecture on aboveground–belowground-induced responses to foliar and root herbivores on *Nicotiana tabacum*. J Chem Ecol 34:1349–1359
Kendall WA, Leath KT (1974) Slant-board culture methods for root observations of red clover. Crop Sci 14:317–320
King RA, Read DS, Traugott M et al (2008) Molecular analysis of predation: a review of best practice for DNA-based approaches. Mol Ecol 17:947–963
Kramer S, Marhan S, Ruess L et al (2012) Carbon flow into microbial and fungal biomass as a basis for the belowground food web of agroecosystems. Pedobiologia 55:111–119
Lehtilä K, Boalt E (2008) The use and usefulness of artificial herbivory in plant-herbivore studies. In: Weisser WW, Siemann E (eds) Insects and ecosystem function. Springer, Berlin, pp 257–275
Lu J, Robert CAM, Riemann M et al (2015) Induced jasmonate signaling leads to contrasting effects on root damage and herbivore performance. Plant Physiol 167:1100–1116
Mankin RW, Johnson SN, Grinev DV et al (2008) New experimental techniques for studying root herbivory. In: Johnson SN, Murray PJ (eds) Root feeders – an ecosystem perspective. CABI, Wallingford, pp 20–32
Masters GJ (1995) The impact of root herbivory on aphid performance – field and laboratory evidence. Acta Oecol 16:135–142
Masters GJ (2004) Below-ground herbivores and ecosystem processes. In: Weisser WW, Siemann E (eds) Insects and ecosystem function. Springer, Berlin, pp 94–112
McDougall BM (1970) Movement of ^{14}C-photosynthate into roots of wheat seedlings and exudation of ^{14}C from intact roots. New Phytol 69:37–46
McKenzie SW, Vanbergen AJ, Hails RS et al (2013) Reciprocal feeding facilitation by above- and below-ground herbivores. Biol Lett 9:20130341
Minagawa M, Wada E (1984) Stepwise enrichment of ^{15}N along food-chains – further evidence and the relation between $\delta^{15}N$ and animal age. Geochim Cosmochim Acta 48:1135–1140
Moujahed R, Frati F, Cusumano A et al (2014) Egg parasitoid attraction toward induced plant volatiles is disrupted by a non-host herbivore attacking above or belowground plant organs. Front Plant Sci 5:601

Murray PJ, Clements RO (1992) Studies on the feeding of *Sitona lineatus* L. (Coleoptera, Curculionidae) on white clover (*Trifolium repens* L.) seedlings. Ann Appl Biol 121:233–238

Murray PJ, Clements RO (1994) Investigations of the host feeding preferences of *Sitona* weevils found commonly on white clover (*Trifolium repens*) in the UK. Entomol Exp Appl 71:73–79

Murray PJ, Clegg CD, Crotty FV et al (2009) Dissipation of bacterially derived C and N through the meso- and macrofauna of a grassland soil. Soil Biol Biochem 41:1146–1150

Neveu N, Grandgirard J, Nenon JP et al (2002) Systemic release of herbivore-induced plant volatiles by turnips infested by concealed root-feeding larvae *Delia radicum* L. J Chem Ecol 28:1717–1732. https://doi.org/10.1023/A:1020500915728

Nielsen UN, Osler GHR, Campbell CD et al (2010a) The influence of vegetation type, soil properties and precipitation on the composition of soil mite and microbial communities at the landscape scale. J Biogeogr 37:1317–1328

Nielsen UN, Osler GHR, Campbell CD et al (2010b) The enigma of soil animal species diversity revisited: the role of small-scale heterogeneity. PLoS One 5(7):e11567

Orgiazzi A, Dunbar MB, Panagos P et al (2015) Soil biodiversity and DNA barcodes: opportunities and challenges. Soil Biol Biochem 80:244–250

Orians CM, Jones CG (2001) Plants as resource mosaics: a functional model for predicting patterns of within-plant resource heterogeneity to consumers based on vascular architecture and local environmental variability. Oikos 94:493–504

Orians CM, Ardón M, Mohammad BA (2002) Vascular architecture and patchy nutrient availability generate within-plant heterogeneity in plant traits important to herbivores. Am J Bot 89:270–278

Passioura JB (2006) The perils of pot experiments. Funct Plant Biol 33:1075–1079

Pedrotti L, Mueller MJ, Waller F (2013) *Piriformospora indica* root colonization triggers local and systemic root responses and inhibits secondary colonization of distal roots. PLoS One 8:e69352

Pollierer MM, Langel R, Körner C et al (2007) The underestimated importance of belowground carbon input for forest soil animal food webs. Ecol Lett 10:729–736

Ponsard S, Arditi R (2000) What can stable isotopes ($\delta\ ^{15}N$ and $\delta\ ^{13}C$) tell about the food web of soil macro-invertebrates? Ecology 81:852–864

Poorter H, Bühler J, van Dusschoten D et al (2012) Pot size matters: a meta-analysis of the effects of rooting volume on plant growth. Funct Plant Biol 39:839–850

Power SA, Barnett KL, Ochoa-Huesco R et al (2016) DRI-grass: a new experimental platform for addressing grassland ecosystem responses to future precipitation scenarios in South-East Australia. Front Plant Sci 7:1373

Price EAC, Hutchings MJ, Marshall C (1996) Causes and consequences of sectoriality in the clonal herb *Glechoma hederacea*. Vegetatio 127:41–54

Quinn MA, Hall MH (1992) Compensatory response of a legume root-nodule system to nodule herbivory by *Sitona hispidulus*. Entomol Exp Appl 64:167–176

Raghu S, Dhileepan K (2005) The value of simulating herbivory in selecting effective weed biological control agents. Biol Control 34:265–273

Rasmann S, Turlings TCJ (2007) Simultaneous feeding by aboveground and belowground herbivores attenuates plant-mediated attraction of their respective natural enemies. Ecol Lett 10:926–936

Rasmann S, Köllner TG, Degenhardt J et al (2005) Recruitment of entomopathogenic nematodes by insect-damaged maize roots. Nature 434:732–737

Rhea-Fournier D, González G (2017) Methodological considerations in the study of earthworms in forest ecosystems. In: Chakravarty S, Shukla G (eds) Forest ecology and conservation. InTech, Rijeka, pp 47–76

Ritz K (2011) Views of the underworld: in situ visualization of soil biota. In: Ritz K, Young IM (eds) Architecture and biology of soils: life in inner space. CABI, Wallingford, pp 1–12

Robert CAM, Erb M, Duployer M et al (2012a) Herbivore-induced plant volatiles mediate host selection by a root herbivore. New Phytol 194:1061–1069

Robert CAM, Erb M, Hibbard BE et al (2012b) A specialist root herbivore reduces plant resistance and uses an induced plant volatile to aggregate in a density-dependent manner. Funct Ecol 26:1429–1440

Rostas M, Cripps MG, Silcock P (2015) Aboveground endophyte affects root volatile emission and host plant selection of a belowground insect. Oecologia 177:487–497

Rushton SP, Luff ML (1984) A new electrical method for sampling earthworm populations. Pedobiologia 26:15–19

Russell EJ (1912) Soil conditions and plant growth. Longmans, Green and Co., London

Ryalls JMW (2016) The impacts of climate change and belowground herbivory on aphids via primary metabolites. PhD thesis, Western Sydney University, Sydney

Ryalls JMW, Moore BD, Riegler M et al (2015) Amino acid-mediated impacts of elevated carbon dioxide and simulated root herbivory on aphids are neutralized by increased air temperatures. J Exp Bot 66:613–623

Sapkota R, Nicolaisen M (2015) High-throughput sequencing of nematode communities from total soil DNA extractions. BMC Ecol 15:3

Schädler M, Jung G, Brandl R et al (2004) Secondary succession is influenced by belowground insect herbivory on a productive site. Oecologia 138:242–252

Scheu S, Falca M (2000) The soil food web of two beech forests (*Fagus sylvatica*) of contrasting humus type: stable isotope analysis of a macro- and a mesofauna-dominated community. Oecologia 123:285–296

Schmidt O, Scrimgeour CM, Handley LL (1997) Natural abundance of ^{15}N and ^{13}C in earthworms from a wheat and a wheat-clover field. Soil Biol Biochem 29:1301–1308

Schneider K, Migge S, Norton RA et al (2004) Trophic niche differentiation in soil microarthropods (Oribatida, Acari): evidence from stable isotope ratios ($^{15}N/^{14}N$). Soil Biol Biochem 36:1769–1774

Schöning C, Wurst S (2016) Positive effects of root-knot nematodes (*Meloidogyne incognita*) on nitrogen availability do not outweigh their negative effects on fitness in *Nicotiana attenuata*. Plant Soil 400:381–390

Setälä H, Aarnio T (2002) Vertical stratification and trophic interactions among organisms of a soil decomposer food web – a field experiment using ^{15}N as a tool. Eur J Soil Biol 38:29–34

Simon C, Daniel R (2011) Metagenomic analyses: past and future trends. Appl Environ Microb 77:1153–1161

Steinger T, Müller-Schärer H (1992) Physiological and growth responses of *Centaurea maculosa* (Asteraceae) to root herbivory under varying levels of interspecific plant competition and soil nitrogen availability. Oecologia 91:141–149

Subler S, Baranski CM, Edwards CA (1997) Earthworm additions increased short-term nitrogen availability and leaching in two grain-crop agroecosystems. Soil Biol Biochem 29:413–421

Taberlet P, Coissac E, Pompanon F et al (2012) Towards next-generation biodiversity assessment using DNA metabarcoding. Mol Ecol 21:2045–2050

Taina IA, Heck RJ, Elliot TR (2008) Application of X-ray computed tomography to soil science: a literature review. Can J Soil Sci 88:1–20

Tariq M, Wright DJ, Bruce TJA et al (2013) Drought and root herbivory interact to alter the response of above-ground parasitoids to aphid infested plants and associated plant volatile signals. PLoS One 8(7):e69013

Thibaud M-C, Arrighi J-F, Bayle V et al (2010) Dissection of local and systemic transcriptional responses to phosphate starvation in *Arabidopsis*. Plant J 64:775–789

Thorn AM, Orians CM (2011) Patchy nitrate promotes inter-sector flow and ^{15}N allocation in *Ocimum basilicum*: a model and an experiment. Funct Plant Biol 38:879–887

Tiunov AV (2007) Stable isotopes of carbon and nitrogen in soil ecological studies. Biol Bull 34:395–407

Torode MD, Barnett KL, Facey SL et al (2016) Altered precipitation impacts on above- and belowground grassland invertebrates: summer drought leads to outbreaks in spring. Front Plant Sci 7:1468

Traugott M, Pázmándi C, Kaufmann R et al (2007) Evaluating $^{15}N/^{14}N$ and C^{13}/C^{12} isotope ratio analysis to investigate trophic relationships of elaterid larvae (Coleoptera : Elateridae). Soil Biol Biochem 39:1023–1030

Turlings T, Davison A, Ricard I et al (2005) Above-and belowground olfactometers for high throughput bioassays. In: Noldus LPJJ, Grieco F, Loijens LWS, Zimmerman PH (eds) Proceedings of Measuring Behavior 2005, 5th international conference on methods and techniques in behavioral research. Wageningen, 30 August–2 September 2005. Noldus Information Technology, Wageningen, p 208

van Dam NM, Samudrala D, Harren FJM et al (2012) Real-time analysis of sulfur-containing volatiles in *Brassica* plants infested with root-feeding *Delia radicum* larvae using proton-transfer reaction mass spectrometry. AoB Plants 2012:pls021

Vandegehuchte ML, de la Peña E, Bonte D (2010) Interactions between root and shoot herbivores of *Ammophila arenaria* in the laboratory do not translate into correlated abundances in the field. Oikos 119:1011–1019

Vanderklift MA, Ponsard S (2003) Sources of variation in consumer-diet $\delta^{15}N$ enrichment: a meta-analysis. Oecologia 136:169–182

Vanholme B, De Meutter J, Tytgat T et al (2004) Secretions of plant-parasitic nematodes: a molecular update. Gene 332:13–27

Wäckers FL, Bezemer TM (2003) Root herbivory induces an above-ground indirect defence. Ecol Lett 6:9–12

Wade RN, Karley AJ, Johnson SN et al (2017) Impact of predicted precipitation scenarios on multitrophic interactions. Funct Ecol 31:1647–1658

Walling LL (2000) The myriad plant responses to herbivores. J Plant Growth Regul 19:195–216

Wallinger C, Juen A, Staudacher K et al (2012) Rapid plant identification using species- and group-specific primers targeting chloroplast DNA. PLoS One 7(1):e29473

Wallinger C, Staudacher K, Schallhart N et al (2013) The effect of plant identity and the level of plant decay on molecular gut content analysis in a herbivorous soil insect. Mol Ecol Resour 13:75–83

Walter DE, Kethley J, Moore JC (1987) A heptane flotation method for recovering microarthropods from semiarid soils, with comparison to the Merchant-Crossley high gradient extraction method and estimates of microarthropod biomass. Pedobiologia 30:221–232

War AR, Sharma HC, Paulraj MG et al (2011) Herbivore induced plant volatiles: their role in plant defense for pest management. Plant Signal Behav 6:1973–1978

Wilkinson DM (2008) Testate amoebae and nutrient cycling: peering into the black box of soil ecology. Trends Ecol Evol 23:596–599

Wilson K, Gunn A, Cherrett JM (1995) The application of a rhizotron to study the subterranean effects of pesticides. Pedobiologia 39:132–143

Zappala S, Helliwell JR, Tracy SR et al (2013) Effects of X-ray dose on rhizosphere studies using x-ray computed tomography. PLoS One 8(6):e67250

Zhang M, Crocker RL, Mankin RW et al (2003) Acoustic identification and measurement of activity patterns of white grubs in soil. J Econ Entomol 96:1704–1710

Zvereva EL, Kozlov MV (2012) Sources of variation in plant responses to belowground insect herbivory: a meta-analysis. Oecologia 169:441–452

Chapter 3
Modeling Aboveground–Belowground Interactions

Katrin M. Meyer

3.1 Introduction

Aboveground–belowground interactions are complex (Wardle 2002), because they include nonlinear relationships, interconnected feedbacks, scale-dependent dynamics, and an intermediate number of constituents. Large numbers of constituents can be averaged, small numbers can be addressed individually, but the study of intermediate numbers can be very challenging in practice (Schaffer 1981). Models can help to capture the complexity of aboveground–belowground interactions (van der Putten et al. 2009). They offer a systematic approach to explore consequences of assumptions where data are scarce, they provide mechanistic, bottom-up system information that might not be apparent from empirical top-down system samples, and they can be used to identify knowledge gaps and generate testable hypotheses. Nevertheless, models are purposeful simplifications of reality and thus always simplify and always need a specific purpose or question to guide the simplification. This means that models are not a universal remedy for aboveground–belowground complexity and need to be applied wisely and where possible in conjunction with empirical approaches to shed light on aboveground–belowground interactions.

Questions are thus at the heart of all modeling endeavors. Many of the questions addressed by aboveground–belowground models are not exclusive to modeling, but models can often include more factors, more interactions, more perspectives, and more types of data (e.g., qualitative data) than many empirical approaches (Meyer et al. 2009a). Hence, questions of aboveground–belowground models are typically

K. M. Meyer (✉)
Faculty of Forest Sciences and Forest Ecology, Department of Ecosystem Modelling, University of Göttingen, Göttingen, Germany
e-mail: kmeyer5@uni-goettingen.de

T. Ohgushi et al. (eds.), *Aboveground–Belowground Community Ecology*, Ecological Studies 234, https://doi.org/10.1007/978-3-319-91614-9_3

derived from the more complex realms of aboveground–belowground community ecology, for instance:

- How do all aboveground and belowground functional groups interact to influence plant performance (Schröter et al. 2004)?
- What is the relative importance of the different relationships in an aboveground–belowground community (Eisenhauer et al. 2012; Meyer et al. 2012)?
- Which aboveground–belowground effects are direct and which are indirect (Veen et al. 2010)?
- What are important thresholds in community dynamics given certain aboveground–belowground interactions (Fibich et al. 2010)?
- What are the underlying (physiological) mechanisms of aboveground–belowground interactions (Biondini 2001)?
- How does global change influence the interactions and feedbacks in aboveground–belowground communities (Sistla et al. 2014)?
- How are trophic interactions related to competitive interactions (Körner et al. 2014)?
- How do spatial relationships influence aboveground–belowground interactions (Levine et al. 2006)?
- How do individual properties of aboveground organisms affect plant and herbivore performance belowground and vice versa (Meyer et al. 2009b)?

These questions are reflected in the aims of the models.

Aims of models can generally be description, explanation, or prediction (Grimm and Railsback 2005, p. 36). Descriptive models identify and represent the characteristic properties of the system at hand. They can be used to demonstrate principles for the purposes of clarifying the corresponding relationships. In aboveground–belowground community ecology, descriptive models may, for instance, aim to identify and visualize aboveground and belowground functional types and their interactions (Schröter et al. 2004). Models that aim at explanation often implement hypotheses about a system to test them by model analysis. This can involve several model scenarios representing alternative hypotheses which are then compared by means of the respective model outputs. For example, such a model could be used to implement and test the hypothesis that aboveground trophic levels are more important than belowground trophic levels for plant performance (Meyer et al. 2009b). Models that aim to make predictions usually require the greatest level of detail and thus large amounts of data. Validation of model results against independent data is very important for predictive models, because only successfully validated models generate credible predictions. Predictive aboveground–belowground models are, for instance, common in crop sciences, where yields are predicted depending on pest species dynamics (Tixier et al. 2013).

Many model types are available to pursue the different questions and aims of community ecology studies (Fig. 3.1). Here, the typology of van der Putten et al. (2009) is used as a starting point and extended by two model types: statistical models and functional–structural models (Table 3.1). Conceptual models and graphical models describe and visualize relevant components of a system and their

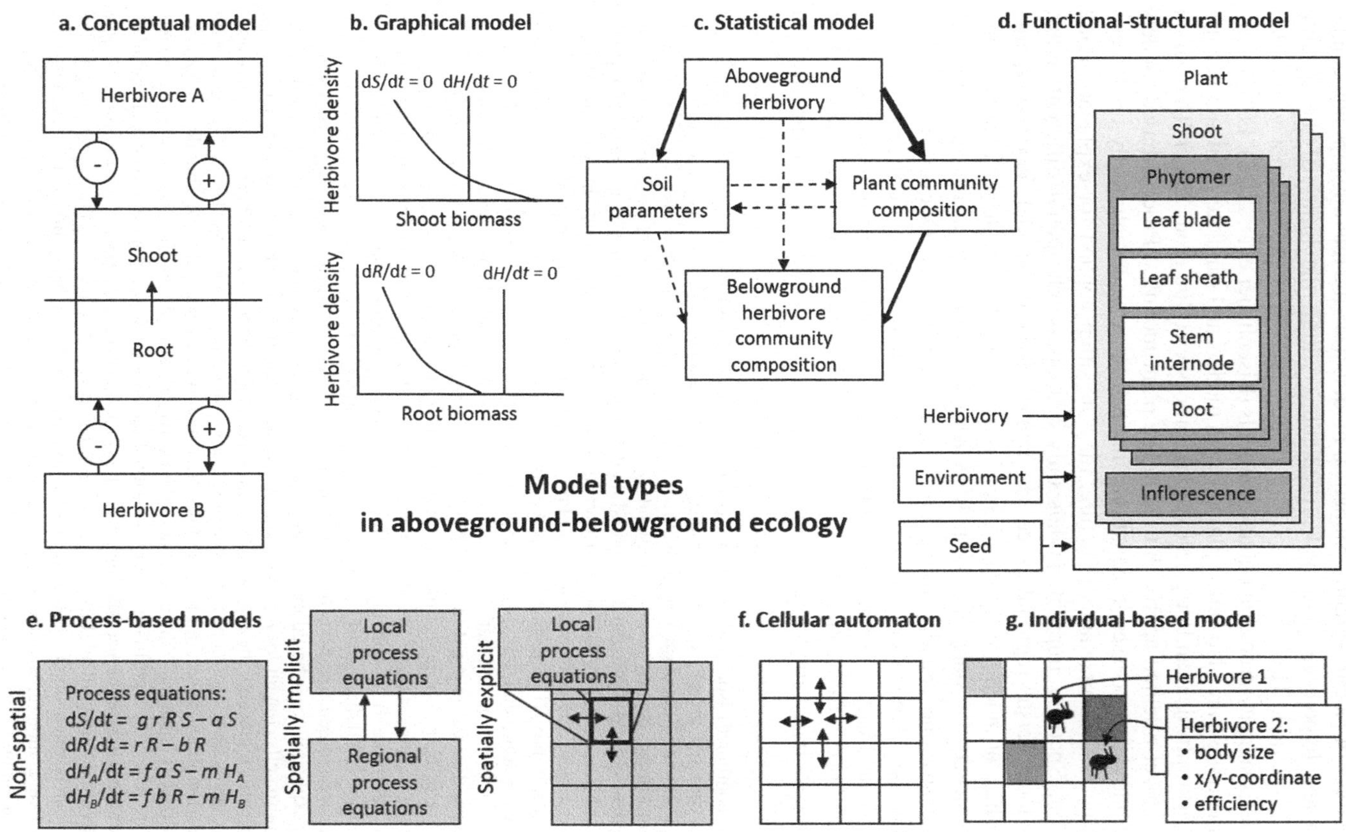

Fig. 3.1 Examples visualizing the different model types in aboveground–belowground ecology. (**a**) Conceptual model consisting of boxes for factors or actors or state variables and arrows for processes. Plus signs indicate positive effects and minus signs indicate negative effects. (**b**) Graphical model with zero-net-growth

relationships. Graphical models additionally rely on mathematical models and often take the form of isocline graphs. Statistical models describe relationships and can be used to test hypotheses and make predictions within the limits of the specific statistical method. This chapter only considers statistical models that go beyond simple tests such as structural equation models (Eisenhauer et al. 2015). Functional–structural models are also called plant growth models and simulate plant architecture arising from physical and/or physiological processes (Guo et al. 2011). They often take an object-oriented approach to simulate plant modules. Process-based models capture relevant processes in equations that can be solved analytically or numerically. Due to their mechanistic nature, they are often applied for explanation and prediction. Spatially implicit models only address rough spatial differentiations, such as local versus regional processes. Spatially explicit models consider spatial relationships at much greater detail. For instance, in cellular automata, neighboring cells influence the state of a cell. In other grid-based approaches, all cells can influence any one cell depending on distance and location (Meyer et al. 2010). Individual-based models or agent-based models simulate interactions between individual organisms giving rise to population and community dynamics (Grimm and Railsback 2005). This bottom-up approach is opposite to the top-down approach of most process-based models that impose population and community dynamics and

---

Fig. 3.1 (continued) isoclines that show combinations of x- and y-values for which the model produces zero net growth of root biomass R, shoot biomass S, or herbivore densities H. (**c**) Statistical structural equation model inspired by Veen et al. (2010). Bold arrows correspond to significant effects, dashed arrows to hypothesized but non-significant effects, and arrow width reflects effect size. (**d**) Functional–structural model inspired by Groot and Lantinga (2004). Boxes represent objects in the model, solid arrows indicate effects, dashed arrows indicate transformations. Each object can perform actions. (**e**) Process-based models formalize processes with process equations (grey boxes) corresponding to the arrows in conceptual models (see **a**). Process-based models can be nonspatial (left grey box), spatially implicit (centre grey boxes), and spatially explicit (right grey boxes). The nonspatial model consists of a possible set of equations for the processes in the conceptual model in panel (**a**); S: shoot biomass, R: root biomass, H_A: density of herbivore A, H_B: density of herbivore B, t: time, g: conversion rate of root growth into shoot growth, r: root growth rate, a: proportion of shoot biomass eaten by herbivore A, b: proportion of root biomass eaten by herbivore B, f: conversion efficiency of shoot or root biomass eaten into herbivore individuals, m: mortality rate of herbivore A and B. Spatially implicit process-based models couple a set of process-based equations for local dynamics with a set of equations for regional dynamics. Spatially explicit process-based models (including reaction–diffusion models) are usually based on a grid whose cells contain local process equations and equations on interactions with neighboring cells. The grid represents the regional dynamics. (**f**) Cellular automata are based on grids whose cells have states such as presence/absence of species or amounts of biomass. The cells interact with their direct neighbors according to predefined transition rules. These rules can (but do not have to) be formalized as equations. Thus, cellular automata include spatially explicit process-based models if these are based on a grid (see **e**). (**g**) Individual-based models or agent-based models simulate individual organisms as objects (Herbivores 1 and 2). Objects have properties whose values change from individual to individual. Individuals interact with each other and with their environment. The environment is often implemented as spatial grid cells that can have different properties, e.g., amount of vegetation (grey shades). In contrast to cellular automata, interactions can stretch beyond the direct neighbors and organisms are not only represented as states, but as objects with properties in individual-based models

Table 3.1 Overview of model types, basic model aims (description, explanation or prediction), selected examples from aboveground–belowground community ecology, and the modeled interactions in these examples[a]

Model types[b]	Model aims[c]	Examples	Modeled interactions[d]
Conceptual models	Description (explanation)	Cahill (1999)	PC
		Schröter et al. (2004)	PT_2^2M
Graphical models	Description (explanation)	Fibich et al. (2010)	PT_1C
Statistical models	Description (explanation) (prediction)	Lamb (2008), Lamb and Cahill Jr. (2008), Lamb et al. (2009)	PC PCF
		Eisenhauer et al. (2012)	PT_1CF
		Veen et al. (2010)	PT_1^1
Functional–structural models	Description (explanation) (prediction)	Drouet and Pagès (2003, 2007), Postma and Lynch (2011)	P
		Groot and Lantinga (2004)	PT^1
Process-based models	(Description) explanation prediction	Cheeseman (1993), van Wijk (2011), Feller et al. (2015), Eid et al. (2016)	P
		Asaeda et al. (2000, 2001), Lazzarotto et al. (2009)	PC
		Sistla et al. (2014), Biondini (2001), Ruget et al. (2002), Brisson et al. (2003), Scheiter and Higgins (2013)	PT_1
		Van Noordwijk and Lusiana (1998)	PT^1CF
		Bever (2003)	PT_1C
		Tixier et al. (2013)	PT^2C
		Goudard and Loreau (2008)	PT^2
		Jia et al. (2009)	PT_1^1
		Willocquet et al. (2008)	PT_1^1C
Spatially implicit process-based models	(Description) explanation prediction	Huston and DeAngelis (1994)	P
Spatially explicit reaction–diffusion models	(Description) explanation prediction	Barbier et al. (2008)	PCF
Spatially explicit cellular automaton models	(Description) explanation (prediction)	Levine et al. (2006)	PT_1
Individual-based models	Description explanation (prediction)	Zhang et al. (2013)	PCF
		Bonanomi et al. (2005)	PT_1
		May et al. (2009)	PT^1C
		Körner et al. (2014), Pfestorf et al. (2016)	PT_1^1C
		Meyer et al. (2009b, 2012)	PT_2^2M

[a]Only those examples are included that explicitly consider root–shoot interactions (and not only plant–resource interactions)
[b]Based on van der Putten et al. (2009) except for statistical models and functional–structural models
[c]Parentheses indicate less importance
[d]*P*—pure root–shoot interactions inside the plant, T_x^y—trophic interactions between plant and x belowground trophic levels and y aboveground trophic levels, *C*—competitive interactions among plants, *F*—facilitative interactions among plants, *M*—mutualistic interactions between plants and mutualists such as earthworms

simulate matter or energy fluxes instead of individual organisms (Schröter et al. 2004). Individual-based models are most often used for explanation, i.e., to demonstrate principles and analyze scenarios.

The origins of aboveground–belowground models are very diverse. Whereas many empirical aboveground–belowground studies have arisen from pure aboveground studies (Chap. 2), aboveground–belowground models are not necessarily extensions of aboveground approaches. Rather, it is the model type that influences the genesis of an aboveground–belowground model. For instance, functional–structural aboveground–belowground models have more commonly arisen from pure root architecture models than from pure aboveground models (Guo et al. 2011; e.g., Postma and Lynch 2011). Process-based aboveground–belowground models often originate from classical physiological models that early on have considered interactions between roots and shoots (e.g., Cheeseman 1993). In cases where aboveground–belowground process-based models arose from food-web models, it is more common that these food-web models capture only belowground interactions and no aboveground interactions (e.g., de Ruiter et al. 1995). Some process-based models (e.g., Bever 2003) and all graphical aboveground–belowground models (e.g., Fibich et al. 2010) are based on classic models from ecological theory such as the Lotka–Volterra model or the Rosenzweig–MacArthur model. These models are not specific to aboveground or belowground communities and neither are conceptual or statistical models. Individual-based aboveground–belowground models are an exception to this rule, because they more commonly originate from models that focus on aboveground interactions (e.g., May et al. 2009; as basis of Körner et al. 2014).

In the following, first, a review of model applications that link aboveground and belowground interactions will show that there are only relatively few approaches thus far. Then, promising avenues for future application of models in aboveground–belowground community ecology will be delineated. A practical guide on how to model aboveground–belowground interactions will conclude this chapter.

3.2 Models of Aboveground–Belowground Interactions

Application examples of models of aboveground–belowground interactions are not as manifold as the diversity of available model types might suggest. The examples compiled here (Table 3.1) are representative of all the available models, with a bias towards studies that involve trophic rather than non-trophic interactions with many rather than few trophic levels. This focus on trophic interactions reflects the aboveground–belowground perspective adopted in this book. In the following, first, simple shoot–root interaction models will be explained, because they are the core of all models compiled in this chapter. Then, models involving trophic interactions aboveground or belowground will be introduced. Finally, it will be explained how non-trophic interactions such as plant–plant and plant–mutualist interactions are integrated into aboveground–belowground models, either alone or in combination with trophic interactions.

3.2.1 *Shoot–Root Interactions*

Shoot–root interactions are the simplest form of aboveground–belowground interactions and are thus the target of most aboveground–belowground models ("P" in Table 3.1). These models simulate allocation of plant biomass to shoot and root at various levels of resolution. Most functional–structural models are restricted to pure shoot–root interactions. Accurate simulation of plant structure already requires great levels of detail, so computational capacities for additional trophic or non-trophic interactions are limited. Functional–structural models are often more highly resolved belowground than aboveground (e.g., Postma and Lynch 2011). However, most shoot–root allocation models are mechanistic process-based models. One of the classic shoot–root allocation models is SIMPLE (Cheeseman 1993), which simulates carbon and nitrogen fluxes between shoot and root compartments and can be used to test physiological hypotheses. Eid et al. (2016) modeled shoot and root growth as a function of photosynthesis, respiration, mortality, and translocation from shoots to roots. With his ecohydrological shoot–root-allocation model, van Wijk (2011) explains plant rooting strategies in dry ecosystems with the maximization of transpiration. Taking a systems biology approach, Feller et al. (2015) simulate aboveground–belowground interactions as exchange of sugars and phosphate between shoot and root compartments to explain the dynamic adaptation of shoot–root ratios in response to environmental conditions.

3.2.2 *Trophic Interactions*

Trophic interactions have been considered as implicit effects and as explicit interactions belowground and aboveground ("T_x^y" in Table 3.1). Trophic interactions are much more commonly addressed in the form of implicit effects than as explicit dynamic interactions in aboveground–belowground models. Implicit effects are given when, for instance, grazing or cutting aboveground biomass is simulated as biomass reduction without feedbacks to the herbivore organisms. Taking an object-oriented functional–structural modeling approach, Groot and Lantinga (2004) explored the effect of cutting on plant growth and digestibility. They simulated plant morphology as aggregation of objects such as leaf blade, leaf sheath, stem internode, and root and found good agreement between field experimental data and model outputs. Implicit belowground effects have been modeled in the form of plant–soil feedback effects (Bonanomi et al. 2005; Levine et al. 2006). Implicit aboveground and belowground effects have been taken into account in a process-based model of mowing and root fragmentation effects on an invasive weed species that aimed at identifying the management interval that maintains shoot biomass at a constant level (Jia et al. 2009). Explicit belowground interactions are part of the biogeochemical ecosystem model SCAMPS that simulates interactions between plant and microbial pools via soil organic matter (Sistla et al. 2014). When the study systems become more complex, direct and indirect trophic effects can be revealed with statistical structural equation models. For instance, Veen et al. (2010) differentiate between alternative hypotheses on direct and indirect effects of aboveground vertebrate herbivory on belowground nematode herbivores. This study also exemplifies how effects (here of aboveground herbivores) and responses (here of belowground herbivores) can be studied, linking aboveground and belowground interactions.

3.2.3 Plant–Plant Interactions

Non-trophic interactions between neighboring plants can have negative (competition) or positive (facilitation) effects on the interacting plants. In their simplest form, plant–plant interactions have been combined with shoot–root interactions without considering any trophic interactions ("C" for competitive interactions, "F" for facilitative interactions, and "CF" for both in Table 3.1). Conceptual models on competitive asymmetry (Cahill 1999) have inspired a statistical structural equation model that shows that root competition—even when it is very intense—is unimportant for plant community structure (Lamb and Cahill 2008). Rather, aboveground competition affects plant diversity (Lamb et al. 2009). A more mechanistic approach has been adopted in the process-based model PROGRASS (Lazzarotto et al. 2009). It simulates belowground competition of grass and clover for nitrogen and can be used to study effects of management such as fertilization and harvesting schemes. One of the rare aquatic examples in aboveground–belowground modeling captures the competition between macrophytes and different functional types of phytoplankton (Asaeda et al. 2001). The spatial nature of competitive and facilitative interactions is reflected in the zone-of-influence approach adopted to simulate facilitation and size-symmetric and size-asymmetric competition aboveground and belowground (e.g., Zhang et al. 2013). Another spatially explicit aboveground–belowground model (Barbier et al. 2008) uses kernels to reflect the variation in the decrease in competition strength around plant individuals of different plant species. Where zones-of-influence or kernels of different individuals overlap, the resources in the overlapping areas or volumes must be shared between individuals. The large number of available models involving plant–plant interactions may be due to three reasons. First, there might be more questions that require the consideration of plant–plant interactions than questions requiring trophic interactions. Second, trophic interactions may only recently have come into the focus of research. Finally, it might simply be easier to incorporate interactions among plants than interactions between plants and other trophic levels.

3.2.4 Plant–Plant and Trophic Interactions

Non-trophic interactions between plants have in some cases been combined with trophic interactions in aboveground–belowground models ("T_x^y" in combination with "C" or "F" or both in Table 3.1). Interactions between plants are usually competitive or facilitative, but can also be trophic when (hemi-) parasitic plants are involved. Along these lines, Fibich et al. (2010) implemented a classic Rosenzweig–MacArthur predator–prey model for the relationship between a plant and its root hemiparasite. The model was extended by including aboveground competition for light between the plant and its hemiparasite. This model was the first to successfully reproduce the reduction in hemiparasite numbers seen empirically at high environmental productivity. The statistical structural equation model by Eisenhauer et al. (2012) connects plant communities and decomposer communities, which can be thought of as implicit competitive, facilitative, and trophic interactions. Based on their model analysis, Eisenhauer et al. (2012) claim that the role of decomposers in the relationship between plant diversity and ecosystem functioning has been underappreciated. The WaNuLCAS model on tree–soil–crop interactions in agroforests treats aboveground trophic interactions implicitly (as harvest), but considers competition between plants explicitly. This model allows the inclusion and analysis of complex management scenarios such as alley cropping, contour hedgerows, parkland systems, and fallow–crop mosaics. Similarly, a series of individual-based models (May et al. 2009; Körner et al. 2014; Pfestorf et al. 2016) simulates herbivory implicitly, but competition explicitly via aboveground and belowground zones-of-influence. Whether or not an individual plant is affected by grazing depends on a weighted lottery based on the overall grazing probability, whereas belowground herbivory is modeled as percentage effect on the total available belowground biomass.

Competitive and trophic interactions have been considered more explicitly in the soil feedback model by Bever (2003). This model links two plants by competition and includes the effects of their soil communities on themselves and on each other as well as the effects of the plants on the soil communities. The difference between explicit and implicit inclusion lies in these two-way trophic interactions. By linking a crop model and a food web model, the banana growth model (Tixier et al. 2013) benefits from both (usually separate) worlds: explicit population dynamics of three trophic levels beyond the plant and explicit output of harvestable biomass of the banana crop and of a cover crop, including explicit competition for nitrogen between the two crops. Another example of an agrophysiological model that explicitly includes competitive and trophic interactions is the WHEATPEST model (Willocquet et al. 2008). Moreover, this model is very flexible in that it has been parameterized for an exceptionally broad range of aboveground and belowground pest species.

3.2.5 Mutualistic and Trophic Interactions

Non-trophic mutualistic interactions between plants and mutualists (often earthworms) have been combined with trophic interactions in a few aboveground–belowground models ("T_x^yM" in Table 3.1). Mutualistic interactions are difficult to include in classic equation-based models without generating positive feedback loops. However, mutualistic interactions and also other non-trophic interactions can be modeled as modifiers of trophic interactions in process-based food web models (Goudard and Loreau 2008). This means that trophic interactions would have a less detrimental effect on a plant when a mutualist is present. Mutualistic interactions have also been considered in the conceptual model for enhancing the biological realism of dynamic global vegetation models by Schröter et al. (2004). This conceptual model asks for several trophic levels aboveground and belowground on top of the mutualistic interactions. However, it might be difficult to implement all these demands in dynamic global vegetation models due to their already complex nature and large-scale applications.

If aboveground–belowground interactions of several trophic and non-trophic levels are to be implemented at a smaller scale such as food chains linked to one plant individual, individual-based models are very suitable for the task. An aboveground–belowground interactions model (Meyer et al. 2009b, 2012) involved three trophic levels aboveground, two trophic levels belowground, a plant with shoot and root compartments, and a mutualist. This model showed that belowground interactions can be more important for plant performance than aboveground interactions. Mutualistic interactions between earthworms and the plant had especially substantial effects on plant biomass (Meyer et al. 2009b). This model was also used to demonstrate that the paradox of enrichment can also apply in communities with multiple aboveground and belowground trophic and non-trophic levels. At high levels of enrichment at one end of a land-use gradient, there were abrupt destabilizations of the multitrophic aboveground–belowground system in the form of high plant mortalities (Meyer et al. 2012). Moreover, the effects of aboveground trophic levels on plant performance were more important under enriched conditions, whereas belowground effects were more important under more natural conditions.

3.2.6 Community Complexity

The complexity of a community is a real challenge for aboveground–belowground models (as it is for empirical approaches). Only very few of the existing aboveground–belowground models are community models in the sense that they incorporate more than three species and more than one type of interaction (e.g., Willocquet et al. 2008; Meyer et al. 2009b; Tixier et al. 2013; Körner et al. 2014; Pfestorf et al. 2016). Of course, complexity should not be pursued for complexity's sake. But modelers should take advantage of the added value that modeling approaches offer, especially in complex cases such as aboveground–belowground interactions in communities. For instance, where empirical data are scarce and the true value of a parameter is unknown, models can be used to explore the consequences of many different parameter values for community dynamics. Moreover, the relative importance of different interactions for community dynamics can be evaluated with a sensitivity analysis.

Different aspects of complexity pose different kinds of challenges to modeling. None of the models with more than one trophic level aboveground and belowground explicitly includes the spatial dimension of the community. Hence, addressing spatial relationships in aboveground–belowground interaction models seems to be a greater challenge than including more species or more interactions. However, spatial relationships are important for population and community dynamics (Durrett and Levin 1994). They determine whether a theoretically possible interaction really takes place or is absent or restricted due to spatial vegetation patterns, limited home ranges, differential movement speeds aboveground versus belowground, or the existence of refuges. However, explicitly including space (and time) in a model also raises questions of scaling.

Scaling in space and time poses a serious challenge to complex aboveground–belowground models. Sooner or later in the model development process, the question arises at which scale or scales the aboveground–belowground model should operate. For instance, accounting for different spatio-temporal scales has been pointed out as a problem of models that aim at optimizing ecosystem services (Tixier et al. 2013). Models that cover multiple spatial or temporal scales have been developed (e.g., Fig. 3.2), but they are rare. This is because the transfer of information between scales is technically challenging and virtually always involves loss of information. The technical challenge of scaling is due to the heterogeneities, nonlinearities, and feedbacks of the interactions to be modeled and scaled. More commonly, the modeled processes are integrated at one specific scale. This may also involve loss of information, but is technically easier once the specific scale has been identified. Mismatches in the spatial and temporal scales of the real processes underlying the model (Sayre and Vittorio 2009) aggravate the scaling problems in aboveground–belowground models, because spatial and temporal scales then have to be treated separately.

Another dimension of complexity is the fact that models always must be linked to empirical data for validation. This has been considered a serious bottleneck for crop models (Meine van Noordwijk 1996). Coupling a model to empirical data does not necessarily increase the complexity of the modeled system, but it does enhance the complexity of model analysis. This is especially true when empirical data are scarce and inverse modeling techniques (Grimm et al. 2005) have to be applied to determine parameter values. Inverse parameterization compares available empirical data to the outputs of several model versions that differ in the value of the parameter to be determined. The parameter value that produces the best match is then chosen. In spite of these difficulties, aboveground–belowground models and empirical data from greenhouse experiments or field studies have successfully been coupled in some cases (e.g., Cheeseman 1993; Meyer et al. 2009b; van der Putten et al. 2009; Jeltsch et al. 2013; Hol et al. 2016; Pfestorf et al. 2016).

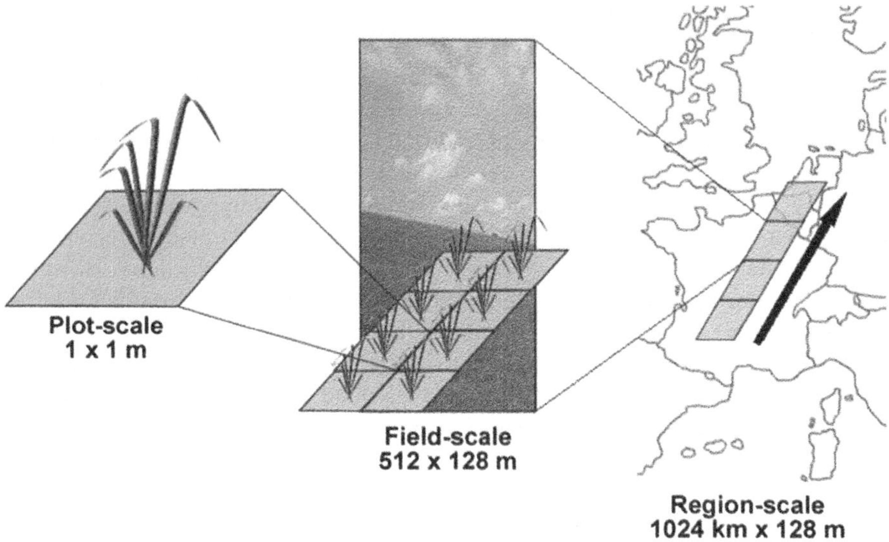

Fig. 3.2 Multi-scale concept of a model simulating northwards range expansion of Mediterranean plants that captures abiotic and biotic interactions with aboveground and belowground enemies. One plot accommodates one plant individual. Biotic interactions occur at the plot and field scale. The population-level output of the individual-based field-scale simulations is used as input to the regional-scale model. Based on the probability of a population to reach the northern end of a field, populations of plants and their aboveground and belowground enemies move northwards in the region-scale model. The field-scale submodel is run for several scenarios differing, for instance, in enemy identity or in temperature to provide the required input to the regional-scale model. Such a multi-scale model can be used to assess intracontinental range expansion speed or the enemy release hypothesis (Figure design by M. van Oorschot)

3.3 The Future of Aboveground–Belowground Models

Future modeling endeavors can advance aboveground–belowground community ecology threefold: first by providing more complete analyses, second by filling gaps in knowledge, and third by inspiring new perspectives. Examples of gaps in knowledge that models can help filling are:

- How do competitive, facilitative, trophic, and mutualistic interactions influence and modify each other in their effect on the involved organisms?
- Which roles do positive and negative feedbacks play for aboveground–belowground interactions?
- How important are interactions between plants and microorganisms relative to other aboveground–belowground interactions in a community?
- How do different types of aboveground–belowground interactions affect the invasibility of a community?
- What is the role of intraspecific interactions and intraspecific variability in the context of aboveground–belowground interactions?

- How do interactions of plants with the environment influence community dynamics relative to interactions with other organisms?

Technically, these advancements correspond to three main avenues for future aboveground–belowground modeling: first to improve and simplify current models, second to develop more complex models, and third to adopt entirely new modeling approaches. The following sections explore how these avenues look like and how we can advance aboveground–belowground community ecology when going along these avenues.

3.3.1 The First Avenue: Model Simplification

The first avenue is to improve current models by updating their data basis, implementing more efficient algorithms, and simplifying model structure where possible. This may involve reformulating the original research question and adapting model structure accordingly. Such continuous model updating is also known as the modeling cycle (Grimm and Railsback 2005) (Fig. 3.3), but is not very often put into practice because it requires extra investment in an apparently completed model project. However, the investment is worthwhile because newly available data may open up new topics for investigation with the model and more efficient model structure may make more complete analyses or other types of analyses possible. In the most extreme case, the original model and its outputs turn out to be wrong, which makes updates mandatory. This first avenue of improving current models also encourages further data collection targeted at filling gaps in the input parameters and reference data for the validation of aboveground–belowground models.

3.3.2 The Second Avenue: Model Expansion

The second avenue for future aboveground–belowground modeling is to take advantage of increasing computing power to expand research questions and models to include more factors. Model expansion can help to fill current gaps in knowledge. However, caution should be taken to not make a model unnecessarily complex with respect to the research question and the available empirical data for parameterization and validation. Unnecessary model complexity can be avoided by starting with the simplest possible model with the fewest variables, parameters, and processes, iterating through the modeling cycle (Fig. 3.3) and always considering to further simplify rather than adding model structure. Where question and data allow for model expansion, there are several factors that can be included in future aboveground–belowground models to fill knowledge gaps. These factors can be grouped into three directions of interaction: interactions between species,

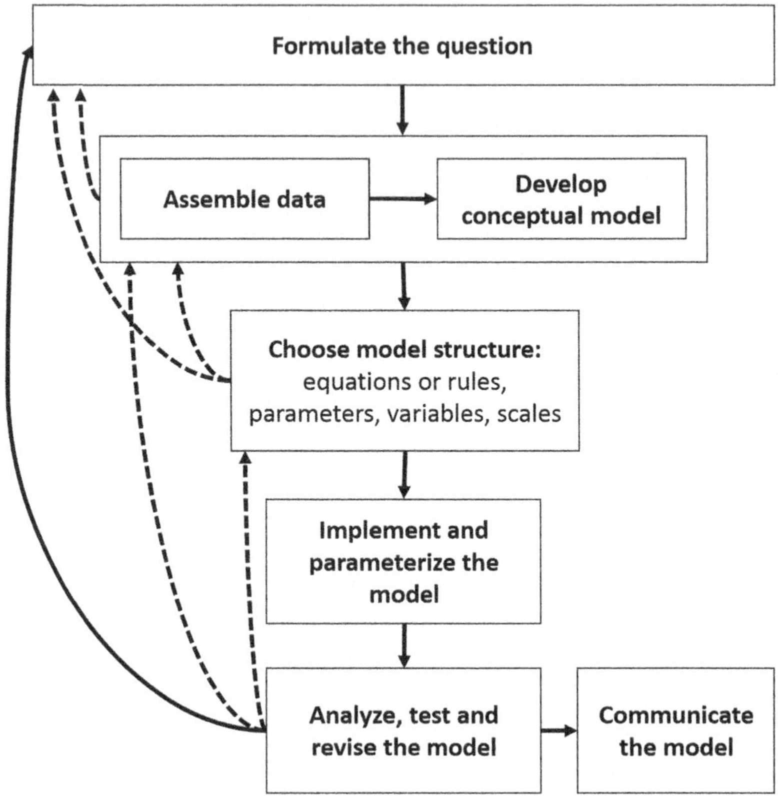

Fig. 3.3 The modeling cycle detailing the steps from question to model and back (modified based on Grimm and Railsback 2005)

intraspecific interactions and variability, and interactions between species and the environment.

In the first interaction direction, more different types of interactions such as trophic, competitive, facilitative, mutualistic, or other interspecific interactions could be combined to provide a more complete picture of a community. To this end, Tixier et al. (2013) suggested linking soil–plant models that stress plant–plant interactions with food web models that focus on trophic interactions. Moreover, negative and positive feedbacks should be considered explicitly (van der Putten et al. 2009), because they can enhance mechanistic understanding of aboveground–belowground interactions. It should also be explored under which conditions negative feedbacks become positive feedbacks and positive feedbacks become negative ones. Where applicable, special roles of selected functional groups in an interaction network should be implemented, as in the case of ecosystem engineers (Schröter et al. 2004). Some groups of organisms are underrepresented in aboveground–belowground models, among them microorganisms that colonize plant leaf and root surfaces (Meyer and Leveau 2012; Esser et al. 2015). Considering interactions with microorganisms may fill gaps in the understanding of the mechanisms of aboveground–belowground interactions. When these interactions are more fully reflected in future aboveground–belowground models, these models can also be used to investigate the invasibility of communities and to test interaction-related hypotheses such as the enemy release hypothesis (see also Fig. 3.2). Here, models have the great advantage over experimental approaches that they allow for countless manipulations and scenarios that can be run with very high replication (Meyer et al. 2009a).

In the second interaction direction, intraspecific interactions and variability may also be benefits of future aboveground–belowground models (Pfestorf et al. 2016). Intraspecific interactions mainly refer to intraspecific competition, i.e., density-dependent reproduction or mortality that give rise to positively or negatively density-dependent population growth. This is particularly important where coexistence of species is investigated, because classical ecological theory in the form of the Lotka–Volterra model predicts that intraspecific competition needs to be stronger than interspecific competition to promote stable coexistence of two species (see also Gause 1934). In terms of intraspecific variability, it is especially important that greater resolution of root traits is included in future approaches (May et al. 2009). Along similar lines, physiological and ecological aspects should be linked more tightly and different pathways of information flow should be considered, for instance in the form of volatiles (Meyer et al. 2012).

In the third interaction direction, interactions with the environment could be included more explicitly in future aboveground–belowground models. Such models with more detailed representations of the relevant environmental conditions such as temperature, pH, soil moisture, or nutrient availability could be used to study how context-dependent the outcomes of aboveground–belowground interactions are (Meyer et al. 2009b). These models will also allow agronomists to address the full complexity of agricultural questions, which requires models with great environmental detail (Tixier et al. 2013). One of these questions is, for instance, how large are the trade-offs between pesticide use and environmental risks (Tixier et al. 2006). Fortunately, there are also some types of models that require only a coarse spatial resolution and thus less detail, such as decision models in agricultural applications (Tixier et al. 2013). When environmental conditions are more explicitly covered, their changes can also be explicitly simulated to investigate the responses of aboveground–belowground interactions to, for instance, global climate change or to the abiotic components of land-use change. Of course, biotic components of land-use change should also be considered when land-use change is modeled. This involves replacing plant species identities by those of the species planted in the new land use and adapting the whole network of associated species aboveground and belowground. This will provide a more complete perspective on the relative importance of abiotic and biotic interactions for the implications of land-use change. Changes in environmental conditions also have consequences for ecosystem services. Hence, more aboveground–belowground model applications that produce output on ecosystem services are called for (Schröter et al. 2004; Tixier et al. 2013).

3.3.3 The Third Avenue: New Modeling Techniques

The third avenue towards future aboveground–belowground models is to explore entirely new technical approaches as bases for the modeling endeavors. These approaches also encompass empirical approaches that help to provide more complete or more finely resolved data on which to base parameterization or validation of the models. For example, molecular tools such as stable isotope analysis should be used more frequently in order to obtain full food web data for complex aboveground–belowground models (Tixier et al. 2013). For aboveground–belowground models that focus on plant morphogenesis and the genetic regulation of plant development, cell-based simulation approaches have been advocated to capture different types of plant cells (Dupuy et al. 2007). Game theory offers new techniques that have not yet been applied in aboveground–belowground modeling. These techniques can be used to model plant defense strategies or plant allocation responses to global change (McNickle and Dybzinski 2013). Similarly, static or dynamic Bayesian network models have not yet been frequently applied in aboveground–belowground modeling. However, they can be very useful where food web modelers want to include stochasticity (Tixier et al. 2013). Adopting new modeling techniques can thus also open up new ecological perspectives on aboveground–belowground community ecology.

3.4 Practical Guide to Modeling Aboveground–Belowground Interactions

How can you now make your own model of aboveground–belowground interactions? Whether you are an experienced modeler or a modeling novice, all your modeling endeavors (not only the ones related to aboveground–belowground interactions) will usually follow the same general principles that make up the modeling cycle (Fig. 3.3) (Grimm and Railsback 2005). These principles are derived from the definition of a model as a purposeful simplification of reality. Thus, the first step of the modeling cycle is to define the purpose of a model in form of the model question. This question needs to be as specific as possible to be useful as a decision tool for simplification (see the examples of model questions in the introduction to this chapter). Often, model questions arise either explicitly or implicitly from the exploration of empirical data. This is an explicit process when data give rise to questions on the underlying mechanisms, and these mechanisms are more amenable to modeling than empirical approaches. This can for instance be due to logistic constraints. Implicitly, the data that we process influence our preconceptions about the mechanisms that shape and drive community dynamics and thus also influence the questions that we ask about them.

The second step is to develop a conceptual model based on data assembly. The conceptual model is a collection of hypotheses about the factors, influences, and interactions that may be relevant for answering the model question. These hypotheses might only be mental hypotheses, but they are in fact often expressed in the form of graphics with factors in boxes and arrows representing influences or interactions. A conceptual model is a crucial milestone for a successful model of a system, because it makes explicit the knowledge, assumptions, and intuition about the system. Note that conceptual models can be a step in the modeling cycle or a self-contained model type (Fig. 3.1). The quality of the conceptual model strongly depends on the available data. Data are explicitly required for parameterization and validation of a model, but also more implicitly as a guideline during the definition of model structure. Data quality obviously varies, but any type and quality of data related to the model question is welcome at the conceptual modeling stage. This includes quantitative data obtained from literature sources or from experiments and observations carried out by the modelers themselves, as well as qualitative data in the form of expert knowledge or "guesstimates." If the model concept is transparent about the nature of the data and the estimated uncertainty in the data, even the most uncertain data can be useful in the modeling process if it is related to the model question. It is good practice go back and forth between question formation and data assembly/model conceptualization in several iterations to sharpen and synchronize question and concept until the next step can be taken.

In the third step, the model structure is fixed, i.e., concrete model equations, rules, and algorithms are devised, the set of parameters and variables is defined, and the temporal and spatial scales are determined. Parameters are the input values of a model that stay constant in any one model run, whereas variables express different states of the modeled system and vary accordingly. Scale is composed of grain and extent, i.e., the smallest and the largest spatial or temporal unit to be distinguished in the model. A careful definition of the model scales is especially important in aboveground–belowground interaction models, because the dynamics of aboveground–belowground interactions is often scale-dependent. Deciding to choose a particular model structure may have as a consequence that the conceptual model and sometimes even the model question have to be modified.

The fourth step is to implement and parameterize the model. This means that the model structure is turned into a set of spelled-out equations in the case of equation-based models such as process-based and reaction–diffusion models and into programming code in the case of rule-based models such as most cellular automata and individual-based models (Table 3.1). Additionally, standard parameter values and the initial values of the variables need to be defined based on data from the second step. This requires a certain degree of technical knowledge in mathematics or software design, which can of course also be outsourced to mathematicians or software experts. Literature on equation-based modeling in biology abounds; classic references are Adler (2005) and Edelstein-Keshet (2005). For rule-based models, the free software NetLogo (Wilensky 1999) is highly suitable for programming novices, because it offers ample self-learning material and an extensive library with ready-made models as starting points. This software is specialized on agent- or individual-based models, but can also be applied to implement any other rule-based model structure. Grimm and Railsback (2005) line out the theoretical basis of agent- and individual-based modeling in ecology, and Railsback and Grimm (2012) provide a practical introduction to agent-based modeling for beginners. Once all obvious mistakes have been eliminated from the set of equations or from the code and the model is running smoothly, the next step can be taken.

The fifth step involves model analysis, testing, and revision. For model analysis, scenarios need to be defined in the form of sets of parameter values that reflect different aspects of the model question. Parameter values can also be defined beyond the scope of the model question to test how the model responds to more extreme parameterizations. Testing should also include checking the plausibility of all model results and explicit model validation against reference data that were not used during parameterization. Once the model is successfully validated, model analysis can also include a sensitivity analysis to determine the relative importance of the input parameters with respect to model output (Cariboni et al. 2007). A sound model analysis usually takes ten times as much time as all previous steps taken together (Grimm and Railsback 2005). A particularly efficient and systematic way to test a rule-based model is pattern-oriented modeling (Grimm et al. 2005), which can also be used for inverse parametrization. Patterns for which reference data exist need to be discerned in the model output. These patterns can be spatial patterns, but can also be nonspatial patterns, e.g., size-frequency distributions or number and identity of species in an aboveground–belowground food web. The model is then run with several different model structures and sets of parameter values, and the resulting patterns are compared with the reference patterns from reality. The model version that produces the best fit between modeled and real patterns can then be considered as most realistic model structure and parameterization. Thereby, pattern-oriented modeling can also be used to identify unknown parameter values in an inverse parameterization where data for the unknown parameters are scarce, but data for reference patterns abound. During the process of model analysis, there is often cause for model revision, sometimes back to the modification of the original model question. This can either be because errors or uncertainty in important model parts have emerged and force the discarding of the current model and starting anew or because the results of the model analysis have inspired entirely new questions.

As the last step after successful model analysis, the model can be communicated in presentations and publications. This requires precise documentation of the model (e.g., Grimm et al. 2010, 2014). Ideally, documentation accompanies the whole model building process right from the start at the model question. Finally, spin-offs of the model can be considered, so that the modeling cycle starts again.

3.5 Conclusions

Models can capture parts of the complexity of aboveground–belowground interactions with the aim to describe, explain, and predict aboveground–belowground community patterns. Models come with the advantage of broad independence of logistic constraints, except, of course, for computing power limitations. Many of the existing aboveground–belowground models have been developed in applied fields such as the agricultural sciences, but classic community ecology and theory is underrepresented. Similarly, community models that address the full range of possible interactions are rare in aboveground–belowground ecology. One reason for this might be that in aboveground–belowground community ecology expertise from many different disciplines is required. Another reason is that models must simplify, by definition, so that at least initially not all factors or interactions should be included in a model. Once the simple initial models are understood, more factors and interactions can be added by iterating the modeling cycle. A broad range of model types is available to implement these models, but their full potential is yet to be exploited. This is particularly true for game theoretical approaches and Bayesian network models. One of the challenges that future modeling endeavors face is the appropriate representation of spatial and temporal scales including methods for scaling up and down between different scales. All these challenges require concerted efforts of modelers and empiricists in the future. Ideally, modelers and empiricists should work closely together right from the start of a project, defining question, data requirements, and methods together (Jeltsch et al. 2013). Along the same lines, greater emphasis should be put on targeted data collections to fill gaps in parameterization and validation of models. With these efforts, more facets of the complexity of aboveground–belowground interactions will be elucidated in future aboveground–belowground models.

Acknowledgements This chapter has immensely benefited from the insightful comments of Marleen Cobben.

References

Adler FR (2005) Modeling the dynamics of life: calculus and probability for life scientists, 2nd edn. Thomson Brooks/Cole, Belmont

Asaeda T, Trung VK, Manatunge J (2000) Modeling the effects of macrophyte growth and decomposition on the nutrient budget in shallow lakes. Aquat Bot 68:217–237. https://doi.org/10.1016/S0304-3770(00)00123-6

Asaeda T, Trung VK, Manatunge J et al (2001) Modelling macrophyte–nutrient–phytoplankton interactions in shallow eutrophic lakes and the evaluation of environmental impacts. Ecol Eng 16:341–357. https://doi.org/10.1016/S0925-8574(00)00120-8

Barbier N, Couteron P, Lefever R et al (2008) Spatial decoupling of facilitation and competition at the origin of gapped vegetation patterns. Ecology 89:1521–1531

Bever JD (2003) Soil community feedback and the coexistence of competitors: conceptual frameworks and empirical tests. New Phytol 157:465–473. https://doi.org/10.1046/j.1469-8137.2003.00714.x

Biondini M (2001) A three-dimensional spatial model for plant competition in an heterogeneous soil environment. Ecol Model 142:189–225. https://doi.org/10.1016/S0304-3800(01)00285-X

Bonanomi G, Giannino F, Mazzoleni S (2005) Negative plant-soil feedback and species coexistence. Oikos 111:311–321. https://doi.org/10.1111/j.0030-1299.2005.13975.x

Brisson N, Gary C, Justes E et al (2003) An overview of the crop model STICS. Eur J Agron 18: 309–332. https://doi.org/10.1016/S1161-0301(02)00110-7

Cahill JF (1999) Fertilization effects on interactions between above- and belowground competition in an old field. Ecology 80:466–480. https://doi.org/10.1890/0012-9658(1999)080[0466:FEOIBA]2.0.CO;2
Cariboni J, Gatelli D, Liska R et al (2007) The role of sensitivity analysis in ecological modelling. Ecol Model 203:167–182
Cheeseman JM (1993) Plant growth modelling without integrating mechanisms. Plant Cell Environ 16:137–147. https://doi.org/10.1111/j.1365-3040.1993.tb00855.x
de Ruiter PC, Neutel A-M, Moore JC (1995) energetics, patterns of interaction strengths, and stability in real ecosystems. Science 269:1257–1260. https://doi.org/10.1126/science.269.5228.1257
Drouet J-L, Pagès L (2003) GRAAL: a model of GRowth, Architecture and carbon ALlocation during the vegetative phase of the whole maize plant. Ecol Model 165:147–173. https://doi.org/10.1016/S0304-3800(03)00072-3
Drouet J-L, Pagès L (2007) GRAAL-CN: a model of GRowth, Architecture and ALlocation for Carbon and Nitrogen dynamics within whole plants formalised at the organ level. Ecol Model 206:231–249. https://doi.org/10.1016/j.ecolmodel.2007.03.036
Dupuy L, Mackenzie J, Rudge T et al (2007) A system for modelling cell-cell interactions during plant morphogenesis. Ann Bot 101:1255–1265. https://doi.org/10.1093/aob/mcm235
Durrett R, Levin S (1994) The importance of being discrete (and spatial). Theor Popul Biol 46:363–394
Edelstein-Keshet L (2005) Mathematical models in biology. Society for Industrial and Applied Mathematics, Philadelphia
Eid EM, Galal TM, Dakhil MA et al (2016) Modeling the growth dynamics of *Pistia stratiotes* L. populations along the water courses of south Nile Delta, Egypt. Rend Lincei Sci Fis Nat 27:375–382. https://doi.org/10.1007/s12210-015-0492-4
Eisenhauer N, Reich PB, Isbell F (2012) Decomposer diversity and identity influence plant diversity effects on ecosystem functioning. Ecology 93:2227–2240
Eisenhauer N, Bowker MA, Grace JB et al (2015) From patterns to causal understanding: structural equation modeling (SEM) in soil ecology. Pedobiologia 58:65–72. https://doi.org/10.1016/j.pedobi.2015.03.002
Esser DS, Leveau JHJ, Meyer KM (2015) Modeling microbial growth and dynamics. Appl Microbiol Biotechnol 99:8831–8846. https://doi.org/10.1007/s00253-015-6877-6
Feller C, Favre P, Janka A et al (2015) Mathematical modeling of the dynamics of shoot-root interactions and resource partitioning in plant growth. PLoS One 10:e0127905. https://doi.org/10.1371/journal.pone.0127905
Fibich P, Lepš J, Berec L (2010) Modelling the population dynamics of root hemiparasitic plants along a productivity gradient. Folia Geobot 45:425–442. https://doi.org/10.1007/s12224-010-9080-7
Gause GF (1934) The struggle for existence, by G. F. Gause. The Williams & Wilkins, Baltimore
Goudard A, Loreau M (2008) Nontrophic interactions, biodiversity, and ecosystem functioning: an interaction web model. Am Nat 171:91–106. https://doi.org/10.1086/523945
Grimm V, Railsback SF (2005) Individual-based modeling and ecology. Princeton University Press, Princeton
Grimm V, Revilla E, Berger U et al (2005) Pattern-oriented modeling of agent-based complex systems: lessons from ecology. Science 310:987–991
Grimm V, Berger U, DeAngelis DL et al (2010) The ODD protocol: a review and first update. Ecol Model 221:2760–2768. https://doi.org/10.1016/j.ecolmodel.2010.08.019
Grimm V, Augusiak J, Focks A et al (2014) Towards better modelling and decision support: documenting model development, testing, and analysis using TRACE. Ecol Model 280:129–139. https://doi.org/10.1016/j.ecolmodel.2014.01.018
Groot JC, Lantinga EA (2004) An object-oriented model of the morphological development and digestibility of perennial ryegrass. Ecol Model 177:297–312. https://doi.org/10.1016/j.ecolmodel.2004.02.008

Guo Y, Fourcaud T, Jaeger M et al (2011) Plant growth and architectural modelling and its applications. Ann Bot 107:723–727. https://doi.org/10.1093/aob/mcr073

Hol WHG, Raaijmakers CE, Mons I et al (2016) Root-lesion nematodes suppress cabbage aphid population development by reducing aphid daily reproduction. Front Plant Sci 7:111 https://doi.org/10.3389/fpls.2016.00111

Huston MA, DeAngelis DL (1994) Competition and coexistence: the effects of resource transport and supply rates. Am Nat 144:954–977

Jeltsch F, Blaum N, Brose U et al (2013) How can we bring together empiricists and modellers in functional biodiversity research? Basic Appl Ecol 14:93–101. https://doi.org/10.1016/j.baae.2013.01.001

Jia X, Pan XY, Li B et al (2009) Allometric growth, disturbance regime, and dilemmas of controlling invasive plants: a model analysis. Biol Invasions 11:743–752. https://doi.org/10.1007/s10530-008-9288-0

Körner K, Pfestorf H, May F et al (2014) Modelling the effect of belowground herbivory on grassland diversity. Ecol Model 273:79–85. https://doi.org/10.1016/j.ecolmodel.2013.10.025

Lamb EG (2008) Direct and indirect control of grassland community structure by litter, resources, and biomass. Ecology 89:216–225

Lamb EG, Cahill JF Jr (2008) When competition does not matter: grassland diversity and community composition. Am Nat 171:777–787. https://doi.org/10.1086/587528

Lamb EG, Kembel SW, Cahill JF (2009) Shoot, but not root, competition reduces community diversity in experimental mesocosms. J Ecol 97:155–163. https://doi.org/10.1111/j.1365-2745.2008.01454.x

Lazzarotto P, Calanca P, Fuhrer J (2009) Dynamics of grass–clover mixtures—an analysis of the response to management with the PROductive GRASsland Simulator (PROGRASS). Ecol Model 220:703–724. https://doi.org/10.1016/j.ecolmodel.2008.11.023

Levine JM, Pachepsky E, Kendall BE et al (2006) Plant-soil feedbacks and invasive spread: plant-soil feedbacks and invasive spread. Ecol Lett 9:1005–1014. https://doi.org/10.1111/j.1461-0248.2006.00949.x

May F, Grimm V, Jeltsch F (2009) Reversed effects of grazing on plant diversity: the role of belowground competition and size symmetry. Oikos 118:1830–1843. https://doi.org/10.1111/j.1600-0706.2009.17724.x

McNickle GG, Dybzinski R (2013) Game theory and plant ecology. Ecol Lett 16:545–555. https://doi.org/10.1111/ele.12071

Meyer KM, Leveau JHJ (2012) Microbiology of the phyllosphere: a playground for testing ecological concepts. Oecologia 168:621–629. https://doi.org/10.1007/s00442-011-2138-2

Meyer KM, Mooij WM, Vos M et al (2009a) The power of simulating experiments. Ecol Model 220:2594–2597. https://doi.org/10.1016/j.ecolmodel.2009.06.001

Meyer KM, Vos M, Mooij WM et al (2009b) Quantifying the impact of above-and belowground higher trophic levels on plant and herbivore performance by modeling. Oikos 118:981–990

Meyer KM, Wiegand K, Ward D (2010) Spatially explicit modelling of savanna processes. In: Hill MJ, Hanan NP (eds) Ecosystem function in savannas: measurement and modeling at landscape to global scales. Taylor & Francis, Boca Raton

Meyer KM, Vos M, Mooij WM et al (2012) Testing the paradox of enrichment along a land use gradient in a multitrophic aboveground and belowground community. PLoS One 7:e49034. https://doi.org/10.1371/journal.pone.0049034

Pfestorf H, Körner K, Sonnemann I et al (2016) Coupling experimental data with individual-based modelling reveals differential effects of root herbivory on grassland plant co-existence along a resource gradient. J Veg Sci 27:269–282. https://doi.org/10.1111/jvs.12357

Postma JA, Lynch JP (2011) Theoretical evidence for the functional benefit of root cortical aerenchyma in soils with low phosphorus availability. Ann Bot 107:829–841. https://doi.org/10.1093/aob/mcq199

Railsback SF, Grimm V (2012) Agent-based and individual-based modeling: a practical introduction. Princeton University Press, Princeton

Ruget F, Brisson N, Delécolle R et al (2002) Sensitivity analysis of a crop simulation model, STICS, in order to choose the main parameters to be estimated. Agronomie 22:133–158. https://doi.org/10.1051/agro:2002009
Sayre N, Vittorio AV (2009) Scale. In: Kitchin R, Thrift NJ (eds) International encyclopedia of human geography, 1st edn. Elsevier, Amsterdam, pp 19–28
Schaffer WM (1981) Ecological abstraction: the consequences of reduced dimensionality in ecological models. Ecol Monogr 51:383–401
Scheiter S, Higgins SI (2013) Intermediate coupling between aboveground and belowground biomass maximises the persistence of grasslands. PLoS One 8:e61149. https://doi.org/10.1371/journal.pone.0061149
Schröter D, Brussaard L, De Deyn G et al (2004) Trophic interactions in a changing world: modelling aboveground–belowground interactions. Basic Appl Ecol 5:515–528. https://doi.org/10.1016/j.baae.2004.09.006
Sistla SA, Rastetter EB, Schimel JP (2014) Responses of a tundra system to warming using SCAMPS: a stoichiometrically coupled, acclimating microbe–plant–soil model. Ecol Monogr 84:151–170. https://doi.org/10.1890/12-2119.1
Tixier P, Risède J-M, Dorel M et al (2006) Modelling population dynamics of banana plant-parasitic nematodes: a contribution to the design of sustainable cropping systems. Ecol Model 198:321–331. https://doi.org/10.1016/j.ecolmodel.2006.05.003
Tixier P, Peyrard N, Aubertot J-N et al (2013) Modelling interaction networks for enhanced ecosystem services in agroecosystems. Adv Ecol Res 49:437–480
van der Putten WH, Bardgett RD, de Ruiter PC et al (2009) Empirical and theoretical challenges in aboveground–belowground ecology. Oecologia 161:1–14. https://doi.org/10.1007/s00442-009-1351-8
Van Noordwijk M, Lusiana B (1998) WaNuLCAS, a model of water, nutrient and light capture in agroforestry systems. Agroforest Syst 43:217–242. https://doi.org/10.1023/A:1026417120254
Van Noordwijk M, Van de Geijn SC (1996) Root, shoot and soil parameters required for process-oriented models of crop growth limited by water or nutrients. Plant Soil 183:1–25
van Wijk MT (2011) Understanding plant rooting patterns in semi-arid systems: an integrated model analysis of climate, soil type and plant biomass: plant rooting strategies. Glob Ecol Biogeogr 20:331–342. https://doi.org/10.1111/j.1466-8238.2010.00601.x
Veen GF, Olff H, Duyts H et al (2010) Vertebrate herbivores influence soil nematodes by modifying plant communities. Ecology 91:828–835
Wardle DA (2002) Communities and ecosystems: linking the aboveground and belowground components. Princeton University Press, Princeton
Wilensky U (1999) Center for connected learning and computer-based modeling. Northwestern University. Evanston, IL. NetLogo.
Willocquet L, Aubertot JN, Lebard S et al (2008) Simulating multiple pest damage in varying winter wheat production situations. Field Crop Res 107:12–28. https://doi.org/10.1016/j.fcr.2007.12.013
Zhang W-P, Jia X, Damgaard C et al (2013) The interplay between above- and below-ground plant-plant interactions along an environmental gradient: insights from two-layer zone-of-influence models. Oikos 122:1147–1156. https://doi.org/10.1111/j.1600-0706.2012.20877.x

Chapter 4
Intraspecific Plant–Soil Feedbacks Link Ecosystem Ecology and Evolutionary Biology

Jennifer A. Schweitzer, Michael Van Nuland, and Joseph K. Bailey

4.1 Introduction

Beavers build lodges, prairie dogs build towns, and birds build nests, all of which feed back to reinforce the engineering behavior of the organism via the fitness consequences of building the structure (i.e., representing the extended phenotype; Dawkins 1982, 2004). Trees can also be considered "engineers" that build nests. "Nests" are built through the accumulation of leaf/root organic matter or root exudates that alter the soil nutrient environment; likewise microorganisms respond to variation in plant traits altering microbial diversity, activity, and function in soil. Variation in the interaction between plants and soil biota alters how energy flows through ecosystems and how nutrients are cycled. Moreover, the response of soil biota to plants and plant responses to soil biota have fitness and performance consequences (Hoeksema and Thompson 2007; Friesen et al. 2011; Lau and Lennon 2012; Hoeksema et al. 2012; Wagner et al. 2014; terHorst et al. 2014; Evans et al. 2016; Rúa et al. 2016; Van Nuland et al. 2016). Such above- and belowground linkages are often referred to as an "extended phenotype" and have a foundation in evolutionary theory associated with "indirect genetic effects" (indirect genetic effects occur when traits of one individual affect traits in other individuals; Moore et al. 1997; Odling-Smee et al. 2003; Whitham et al. 2006; Schweitzer et al. 2008,

J. A. Schweitzer (✉) · J. K. Bailey
Department of Ecology and Evolutionary Biology, University of Tennessee, Knoxville, TN, USA
e-mail: Jen.Schweitzer@utk.edu; Joe.Bailey@utk.edu

M. Van Nuland
Department of Ecology and Evolutionary Biology, University of Tennessee, Knoxville, TN, USA

Department of Biology, Stanford University, Stanford, CA, USA
e-mail: mvannula@vols.utk.edu

T. Ohgushi et al. (eds.), *Aboveground–Belowground Community Ecology*, Ecological Studies 234, https://doi.org/10.1007/978-3-319-91614-9_4

2012; terHorst and Zee 2016; Van Nuland et al. 2016). For example, plants possess genetically based traits that affect biotic interactions and that condition biotic and abiotic properties of soil (Madritch et al. 2009; Schweitzer et al. 2008, 2012; Bezemer et al. 2010; Waring et al. 2015). Heritable variation in plant susceptibility or resistance to pathogens and the quantity and quality of above- and belowground plant inputs to the soil influences the soil community.

Specifically, plants affect a diverse group of belowground biota, including mutualists, root herbivores, pathogens, and saprotrophs (Schweitzer et al. 2008; terHorst et al. 2014; Busby et al. 2014; Lamit et al. 2015), which can feed back to have fitness consequences of both (Pregitzer et al. 2010; Smith et al. 2012; Rúa et al. 2016; Van Nuland et al. 2016). In this way, plants can condition the environment their offspring and other plant species may encounter (Schweitzer et al. 2012, 2014; terHorst and Zee 2016; Van Nuland et al. 2016). Building on the growing literature showing the genetic basis to plant conditioning of soil via biotic and abiotic means (simplified to plant–soil for simplicity) and the role of plant–soil feedback (PSF) in determining plant performance and fitness, the goal of this chapter is to: (1) review evolutionary theory relevant to plant–soil interactions; (2) provide examples of genetically based variation in PSF; and (3) identify the evolutionary consequences of PSF.

4.2 Evolutionary Theory Relevant to Plant–Soil Interactions

The theory associated with the concepts of the *extended phenotype*, *community genetics*, and *niche construction* are fundamental to understanding the evolutionary consequences of plant–soil linkages and PSF (see Table 4.1 for glossary of terms). While the subtle definitions and evidence for these terms have been debated (Hunter 2009; Dawkins 2004; Odling-Smee et al. 2013; Laland et al. 2015), the terms share the underlying premise that ecological and evolutionary processes can influence one another (often described as "eco-evolutionary dynamics"; Hendry 2017). Importantly, these terms also share a common theoretical framework built on phenotypic models of indirect genetic effects (Moore et al. 1997; Wolf et al. 1998; Mutic and Wolf 2007; McGlothlin and Brodie 2009; Bailey 2012; Bailey et al. 2014). Indirect genetic effects (IGE) occur when the fitness and phenotype of one individual changes due to the genetic and phenotypic identity of interacting individuals (Moore et al. 1997; Wolf et al. 1998, 2011; Wolf 2003). In quantitative genetics, trait values or phenotypes of individuals are a consequence of both the genotype and the environment. When IGE occur, a portion of the environment term is due to interactions with another individual and can have effects that alter the direction and strength of response to selection (Bijma 2014). A growing literature is showing the importance and magnitude of IGE in agricultural and natural environments with models developing interaction coefficients and approaches that allow for the

Table 4.1 Glossary of ecological and evolutionary terms used in the chapter

Community genetics	Study of the genetic interactions that occur between species and their biotic and abiotic environment in complex communities
Ecosystem processes	Processes that transfer energy or materials from one pool to another
Extended phenotype	Concept whereby the genes of an organism can also be expressed beyond the individual
Geographic mosaic theory of co-evolution	The process by which variation in species interactions varies geographically, leading to evolutionary "cold-" and "hotspots" of coevolution
Indirect genetic effects	Interacting phenotype model whereby the genes in one individual of a species affect the expression of traits in another individual
Interspecific indirect genetic effects	Interacting phenotype model whereby the genes in one individual of a species affect the expression of traits of an individual of another species
Niche construction	The process whereby an organism's metabolism, activities, or behavior modify their own niche leading to feedbacks
Plant–soil feedback	Process by which plants alter the abiotic or biotic properties of soil that influences plant growth or fitness

estimation of IGE in the field (Bailey 2012; Bailey and Zuk 2012; Bijma and Wade 2008; Wolf et al. 2011; Costa e Silva et al. 2013; Bailey et al. 2014).

IGE theory has many important implications across levels of biological organization. At the individual and population level, IGE have been shown to be fundamental for social evolution and coadaptation, mate choice and sexual selection, evolution of aggression, and parental care (Schneider et al. 2016 and references therein). At the community level, interspecific indirect genetic effects (IIGE) represent the fundamental unit of the coevolutionary process and are important in plant–plant, plant–herbivore, plant–pollinator, microbe–microbe, and plant–microbe linkages and feedbacks (Crespi 2001; Shuster et al. 2006; Wolf et al. 1998, 2011; Allan et al. 2012; Costa e Silva et al. 2013; Genung et al. 2013). IIGE occur when genes that are expressed in the focal genotype of a species (e.g., a mycorrhizal fungal species) map to the phenotypes of an individual of another species (e.g., a tree genotype; Shuster et al. 2006; Wolf et al. 2011). For example, one of the first studies to map genotype–phenotype functions among co-occurring individual plants found a functional relationship between the genotypes and phenotypes of *Arabidopsis thaliana* and the expression of similar traits in their conspecific neighbors (Wolf et al. 2011). Specifically, using a combination of quantitative trait loci (QTL) and structural equation model (SEM) approaches, the authors found that a QTL of large effect (controlling plant size and development) in focal plants affected size and development phenotypes in neighbor plants. In a study using genotypic interactions of two species, *Solidago altissima* and *Solidago gigantea*, Genung et al. (2013) tested whether IIGE (i.e., the interactions of individual plant genotypes from two species grown together) that had affected living focal plants would also affect leaf litter decomposition rate, as well as nitrogen (N) and phosphorus (P) dynamics after the focal plant senesced. Genetic interactions of two co-occurring plant genotypes while the plants were living had afterlife consequences, which affected N and P immobilization and release. These results indicated that the effects of IGE were present at the ecosystem level and did not diminish across levels of organization.

Plant genotype interactions with diverse taxa in the soil also represent IIGE as plant genotype–soil biota interactions influence plant and biota resource acquisition, protection from other pathogens/herbivores and myriad other physiological effects that have fitness consequences for both the plant genotype and interacting soil biota. Thus, the interactions of plants and soil biota that lead to plant–soil conditioning have the same evolutionary potential as plant–plant interactions whereby interacting genotypes can impact plant phenotype. For example, when *Andropogon gerardii* ecotypes collected from phosphorus P- and N-limited grasslands were grown with all possible "home" and "'away" combinations of soils and mycorrhizal communities, local adaptation between plants and mycorrhizal communities was found that resulted in maximum exchange of the most limiting soil nutrient resource for each (Johnson et al. 2010). Reciprocal transplant experiments such as those used in Johnson et al. (2010) consistently indicate that varying selective pressures from soil microbes or nutrients lead to patterns of local adaptation and geographic mosaics of plant–microbe interactions that vary in strength (Hoeksema and Thompson 2007; Piculell et al. 2008; Pregitzer et al. 2010; Smith et al. 2012; Andonian et al. 2012; Rúa et al. 2016). Such IGE and IIGE are increasingly appreciated as common evolutionary mechanisms that provide a theoretical basis for the evolutionary consequences of species "social" interactions, such as those between plants and soil communities (Wagner et al. 2014; Panke-Buisse et al. 2014; Van Nuland et al. 2017). Additionally, a growing literature in Eco–Evo dynamics in a range of contexts is using the Price equation to integrate ecological and evolutionary theory that is also applicable to plant–soil interactions (Collins and Gardner 2009; Fox and Kerr 2012; Genung et al. 2011; Govaert et al. 2016). With this foundation in evolutionary theory, it is possible to show how genetic interactions among species lead to genetic divergence based on PSF that vary on the landscape.

4.3 Genetically Based Variation in PSF

Recent reviews show the genetic basis to plant–soil linkages that are critical to demonstrating evolution due to PSF (Schweitzer et al. 2012; Waring et al. 2015; terHorst and Zee 2016; Van Nuland et al. 2016). Plant–soil linkages at the level of plant species have been shown to result in feedbacks that influence many aspects of plant communities, including invasion by exotic species, plant competitive interactions, and successional dynamics (van der Putten et al. 1993, 2001, 2013, 2016; Reynolds et al. 2003; Bever 2003; Casper and Castelli 2003; Callaway et al. 2004; Kardol et al. 2006, 2013; Kulmatiski et al. 2008; de la Peña et al. 2010). For example, a classic example demonstrates the role of primary succession in plant species in facilitating late successional species (Kardol et al. 2006). Feedbacks among plants, biotic communities, and soil can also be found within species that can impact plant species persistence, local adaptation, range expansion, and community shifts in forests (Pregitzer et al. 2010; Felker-Quinn et al. 2011; Van Nuland et al. 2017; Semchenko et al. 2017; Pfennigwerth et al. 2017). For example, the net effects of diverse components of the soil community that interact with the living plant or plant detritus can result in either positive, neutral, or negative feedback to plant performance or persistence that can vary through time. Negative plant–soil feedbacks prevent genotypes or populations from persisting in certain locations or at high abundances while positive plant–soil feedbacks are mechanisms for persistence or local adaptation (Johnson et al. 2010; Pregitzer et al. 2010; Smith et al. 2012; Rúa et al. 2016). For example, positive feedbacks within populations of invasive *Ailanthus altissima* lead to persistence while negative feedbacks within some populations lead to populations of the invasive tree that may not persist over time (Felker-Quinn et al. 2011; see below). While our ecological understanding of PSF is growing, there are few examples focused on the genetic basis of plant–soil feedback or their evolutionary consequences.

When PSF within a species have been studied, positive, neutral, and negative feedbacks have been found and models have shown that this variation in PSF responses have different evolutionary consequences (Bever et al. 1997; Eppinga et al. 2011; Schweitzer et al. 2014). Schweitzer et al. (2014) developed a simple mathematical model of plant evolution to explore the relationship between the sign and magnitude of feedback and the divergence of plant traits and the ecosystem functions they support along a soil gradient. Additionally, an individual-based simulation model was constructed to study the conditions under which plant–soil feedbacks occurred, niche construction (i.e., soil conditioning) evolved, and plant traits diverged. These models addressed the relationship between positive and negative PSF with variation in plant conditioning of soils, the strength of selective gradients, and the relative importance of local adaptation. The models suggest that feedbacks (in any direction) between soils and plants may commonly result in divergence of plant traits and ecosystem function across environmental gradients due to a range of feedback responses resulting in positive fitness outcomes. The simulation model indicated that plant traits can diverge with niche construction, and traits can be selected for in response to niche construction. However, the magnitude of feedbacks and how strongly they evolve depends on the amount of gene flow and the strength of selective gradients over time. Specifically, the divergence of the trait responsible for the evolution of feedback was highest when the conditioning potential of plants on soil was high, plant gene flow was intermediate, and selection was strong.

4.3.1 Intraspecific Feedbacks

Understanding the causes and consequences of genetic variation in PSF is a major frontier for plant evolutionary biology. Schweitzer et al. (2014) mathematically showed that positive, neutral, and negative feedbacks within a species can all arise, dependent upon the degree of conditioning and the strength of environmental gradients. This directly contrasts with previous work that has largely focused on PSF at the species-level (Kulmatiski et al. 2008; Anacker et al. 2014; Bennett et al. 2017; Fitzpatrick et al. 2017), and recent empirical results support this model prediction. For example, using soils collected from beneath replicates ($n = 3$–5) of three female genotypes of 20-year-old *Populus angustifolia* trees grown in a common garden, half-sibling seed families collected from those maternal genotypes performed differently when they were grown in soils that contained either a live or sterilized 10 mL inoculation of soil conditioned by their mother compared to other mothers (soil inocula represented <5% of the total volume of soil to minimize differences in soil nutrient content resulting from sterilization, following methods in Pregitzer et al. 2010; Fig. 4.1). Results showed positive, neutral and negative feedbacks that were consistent with "parental care" by trees. Importantly, when these soils were autoclaved and sterilized, the strength and direction of the feedbacks changed. These results are notable because the soils were collected from replicated maternal genotypes that were randomly planted in a common garden. Overall, the results demonstrate: (1) variation in soil conditioning by maternal genotypes (i.e., trees build nests as in Schweitzer et al. 2008 which used the same replicate genotypes); (2) seed families collected from maternal genotypes perform differently in soils collected from their mothers relative to other maternal soils; and (3) variation in the response of seed families to maternal soils is likely related to the biotic response of the soil to variation in maternal genotypes because the inoculum was too small to impact soil nutrient conditions. While feedbacks such as this show variation in the level of "maternal care," only recently have studies begun probing whether PSF might vary among populations and divergent environments.

Variation in PSF across populations of a species is a valuable, but rarely considered, metric to assess how ecosystem ecology and evolution are connected. Felker-Quinn et al. (2011) found that population-level genetic variation of three geographically distinct populations of an invasive tree, *Ailanthus altissima*, grown in soil that had been conditioned by these populations resulted in genetic families that varied in growth patterns and traits. When seedlings from each half-sibling family and population were reciprocally grown in soils from each population, with both a live soil and sterile soil treatments, feedbacks from population origin differentially influenced the performance of seedlings. Two populations showed positive feedbacks, whereby seedling performance was enhanced in "home" soils rather than "away" soils, consistent with local adaptation of the populations to soil conditions. In contrast, one population performed better in "away" soils than "home" soils, indicating negative feedback effects. Moreover, the population genetic variance was lost when the seedlings were grown in a sterile control soil for a range of plant traits indicating that how we understand evolution of plants on the landscape is related to the soil microbiome. These results show variation in PSF responses among populations (i.e., both positive and negative PSF were found) and the expression of additive and population level genetic variation that depended upon the presence of the soil biotic community.

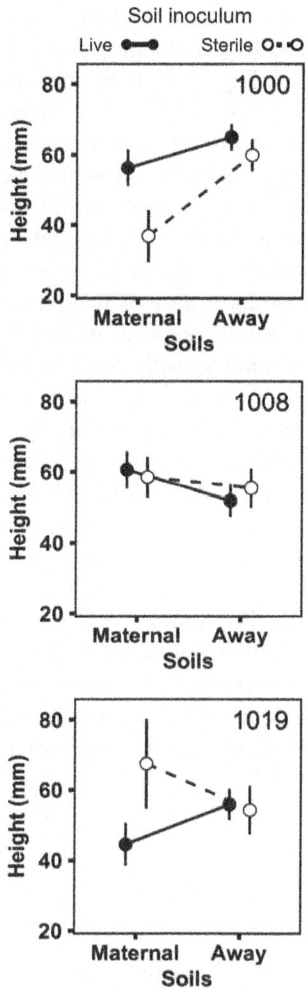

Fig. 4.1 A range of plant–soil feedback (PSF) responses can be found across individual genotypes (or populations) of a riparian species, *Populus angustifolia*, leading to different evolutionary consequences. Results from a replicated ($n = 3$–5) reciprocal transplant study of three maternal genotypes (with half-sibling seeds and their associated soils grown for 20 years in a common garden) show that positive feedbacks (i.e., local adaptation) as well as neutral and negative feedbacks occurred due to maternal conditioning of soils (i.e., conditioning by female genotypes). Symbols represent plant responses growing in either live (closed) or sterile (open) soils implicating the soil microbiome as the primary driver maintaining these feedbacks (Maternal soil × Tree Genotype × Soil Sterilization $p = 0.0455$, $F = 3.16$). Positive PSF, whereby plant fitness is higher in "home" soils relative to "away" soils, is indicative of local adaptation (genotype 1008). Neutral and negative PSF effects may indicate maladaptation that may lead to different evolutionary trajectories (genotypes 1000 and 1019, respectively). Soil and seed collection, inoculum application, and greenhouse growth conditions as in Pregitzer et al. (2010). Points represent mean height ±1 standard error

Lastly, positive, neutral, and negative feedbacks among populations in different environments show the potential for geographic mosaics in how plant–soil interactions respond to warming climates. Using seven field populations of *Populus angustifolia* along elevation gradients in the western USA, Van Nuland et al. (2017) examined the strength of plant–soil conditioning on soil chemistry and biotic communities as well as subsequent patterns in PSF, experimentally, across the interior, edge, and areas beyond current range boundaries. They found positive, neutral, and negative PSF effects when plants were reciprocally grown in each of the soil types such that trees growing in the interior portion of a populations range showed positive feedbacks, trees grown along the edge of the elevation distribution showed neutral feedbacks, and trees grown in soils beyond current range limits showed negative feedbacks. Moreover, the populations varied in their overall growth responses to soils that were collected beneath trees versus random locations away from trees. Three of the seven populations showed positive feedback effects (larger growth in "home" vs. "away" soils) that are likely related to soil biotic effects, while the remaining four populations showed neutral PSF. Collectively, these experimental results show that plant conditioning can lead to multiple feedback outcomes with implications for potential range shifts with climate change. From an evolutionary standpoint, the populations showing positive feedback at the center of their distributions are more likely to be locally adapted to their specific soils. In contrast, the populations at the edge of their distribution showing neutral feedback may show little co-evolutionary interactions. Overall, these examples support model predictions that genetically based positive, neutral, and negative feedbacks may all occur among genotypes within a population, among populations, and among populations along environmental gradients (see also Semchenko et al. 2017) (Fig. 4.2).

4.3.2 PSF and Plant Trait Divergence

The model from Schweitzer et al. (2014) also predicts that the conditioning effects of species interactions can themselves act as a selective agent driving divergence in plant traits and ultimately how ecosystems function. This parallels a previous model that revealed how plants can adaptively regulate their soil nutrient environment (Kylafis and Loreau 2008). It is this dynamic aspect of conditioning that is fundamental to integrating ecosystem ecology and evolutionary biology through PSF. Importantly, the Van Nuland et al. (2017) study, described above, provides some of the first empirical evidence that confirms these model predictions. Specifically, they related the strength of soil conditioning on soil N pools to the degree of PSF across seven populations in the western USA and found a positive relationship between the degree of soil N conditioning in the field and the direction and magnitude of feedbacks mediated by soil communities (Fig. 4.3). Importantly, differences in PSF were driven by among-population level differences in growth. These results indicate that plant traits that strongly impact soil nutrients will lead to more positive PSF's and stronger patterns of local adaptation, and specific members of the soil

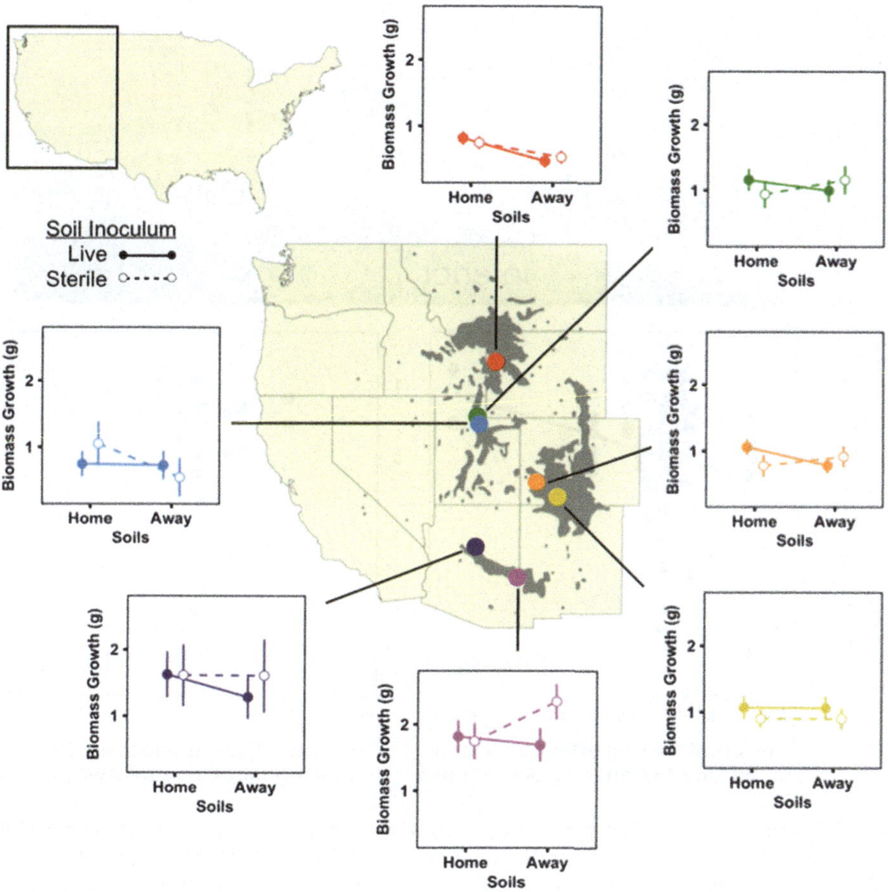

Fig. 4.2 Populations within a species can geographically vary in their response to conditioned soil biota across large environmental gradients. Results of plant growth responses to conditioned versus unconditioned soils, in combination with live (closed symbols; total $n = 286$) and sterile (open symbols; total $n = 284$) treatments, show that three of the seven populations perform better with their "home" population soil biota than random "away" soil biota. Such positive PSF suggest these populations may be locally adapted to their soil communities, whereas the remaining four populations showed neutral feedback effects and may be affected more strongly by other environmental factors. Figure modified from Van Nuland et al. (2017). Points represent mean biomass growth ±1 standard error

biotic community (i.e., those closely tied to N-cycling) may be responsible for mediating these outcomes. Though we highlight this example of a positive relationship, it is not difficult to see how this concept could extend to negative relationships. Pathogens (or other interactions such as reduced mycorrhizal function) may respond positively to soil conditioning by plants creating poor conditions for offspring and negative fitness effects locally, but increase plant fitness in "away" soils.

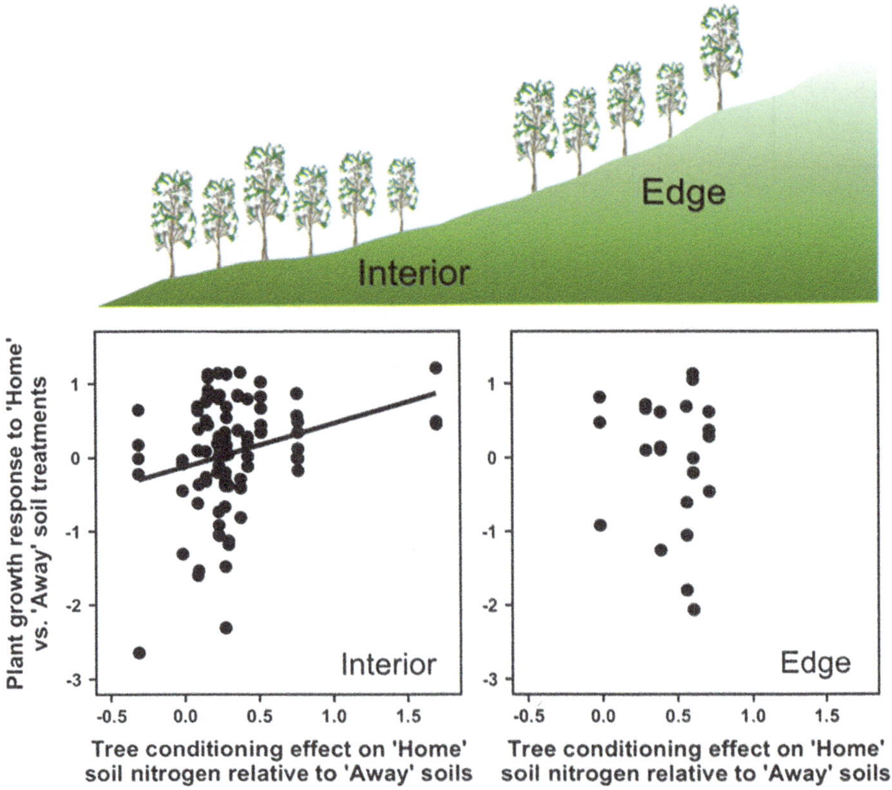

Fig. 4.3 The act of "nest" building is related to how plants respond to their nests. For trees within the interior of elevation ranges ($n = 104$), a positive relationship exists between the strength of plant–soil conditioning [calculated as the difference in "home" vs. "away" total soil nitrogen (N)] and the magnitude and direction of plant–soil feedback (PSF) (calculated as the growth response of cuttings to "home" vs. "away" live soil inoculum treatments). However, there was no correlation between soil N conditioning and feedbacks for trees at the range edge ($n = 34$). These results emphasize the consequences of geographic variation in plant–soil interactions, provide support for theoretical models (Kylafis and Loreau 2008; Schweitzer et al. 2014), and show how positive conditioning effects lead to stronger patterns of local adaptation (positive feedbacks). Figure modified from Van Nuland et al. (2017). Points represent mean biomass growth ± 1 standard error

Together these recent model advances, backed by empirical studies, show that intraspecific positive, neutral, and negative feedbacks can arise due to variation in soil conditioning that may vary geographically. These results indicate that both local adaptation and maladaptation are equally likely to occur which can drive divergence in plant traits and ecosystem function in different ways. The soil sterilization components of these studies show that the cumulative effect of the biotic community due to plant conditioning mediates this response.

4.4 Synthesis and Conclusions

Examination of the role of interspecific indirect genetic interactions among species and their ecosystem level effects and feedbacks allows for the examination of ecosystems within an evolutionary framework that could transform our understanding of the factors that affect the strength and direction of natural selection (Schoener 2011; Matthews et al. 2011, 2014). The data to date show all of the important elements required for plant–soil feedback to be an evolutionary force on the landscape: (1) heritable variation in plant traits that condition the physical and chemical properties of soils and alters soil biotic communities; (2) variation in the degree to which plants condition soils and change biotic communities and ecosystem properties (i.e., soil N or phenology, Schweitzer et al. 2008; Wagner et al. 2014; Panke-Buisse et al. 2014); and (3) intraspecific variation in the direction of feedback at the individual, within- and among-populations, across gradients that leads to positive fitness outcomes. These results are in contrast to recent assertions that specific groups of soil communities will lead to specific directions in PSF (e.g., that ectomycorrhizal communities lead to positive feedback and arbuscular mycorrhizal communities lead to negative feedback; Bennett et al. 2017) and demonstrate equivocally that plant soil linkages and feedbacks not only have ecological consequences but may also play under-appreciated and critical roles in evolutionary processes.

In support of model predictions, new advances in PSF (described above) within a plant species show that there is variation in the degree that plants condition soils, leading to differences in the magnitude and direction of PSF (Mangan et al. 2010; Van Nuland et al. 2017). These results are important as variation in degree of conditioning (i.e., niche construction) and the direction of feedback can lead to divergence over time such that the trait responsible for plant soil feedback can evolve (Schweitzer et al. 2014). The model, now supported by empirical data, shows that the magnitude and direction of PSF by a genotype is dependent upon the degree a genotype alters a soil, the strength of selection, and the mismatch between soils and the ecological traits of the "home" genotype. Variation in the ability to alter soils can alter the magnitude of PSF, depending on the strength of selection for the soil-altering trait, making it paramount to better understand specific traits or community members that determine conditioning. Over time this result can lead to the divergence of the niche construction trait, meaning that the ecosystem function may diverge among populations due to differences in feedbacks. Elucidating the trait responsible for the soil N conditioning would allow for specific examination of divergence with selection experiments or F_{ST}–Q_{ST} comparisons that allow for the determination of selection versus drift for this trait. This could be paired with next-generation sequencing to identify particular soil taxa that respond to conditioning and impact plant phenotypes.

The documentation of positive, neutral, and negative feedbacks within a species with the examples above provides evidence of strong selective gradients within and among populations that can lead to matches or mismatches in ecological traits and soils that can change the selective landscape. Matches can lead to local adaptation due to soil conditioning that can reinforce traits while mismatches represent maladaptation that could lead to reductions in population size or genetic diversity. These data also provide support for the role of geographic variation and environmental gradients on species interactions that can allow hot- and cold-spots of co-evolution in particular species interactions (Thompson 2005; Hoeksema and Thompson 2007; Hoeksema et al. 2012). Pathogens (or other interactions such as reduced mycorrhizal function) may respond positively to soil conditioning by plants creating poor conditions for offspring and negative fitness effects locally, but increase plant fitness in "away" soils.

Together, the genetic interactions among plant genotypes and soil biotic communities that lead to soil conditioning (IIGE) and PSF along environmental gradients provide the theoretical background, mechanisms, and selective environment that can lead to divergence on the landscape. Importantly, either positive or negative feedbacks can affect fitness leading to divergence; thus, the manner and magnitude of soil conditioning (i.e., "nest-building") can have legacy effects that impact the evolutionary trajectory of a population regardless of direction of the feedback. Ongoing and future work to understand the details of the genetic and genomic relationships among plants, their associated soil biotic communities, and soil properties is a critical frontier to better understand IIGE among plants, biotic communities, and soils (Long 1996; Baker et al. 1997; Herrera Paredes and Lebeis 2016). Likewise, understanding how PSF varies across gradients (Smith-Ramesh and Reynolds 2017; Pfennigwerth et al. 2017) and how PSF might reinforce environmental conditions that exist on the landscape remains an important frontier that will provide critical evidence of evolutionary processes at work in natural ecosystems. In conclusion, the "nests" that trees build represent important "extended phenotypes" that link ecosystem ecology to evolution via either positive or negative plant–soil feedbacks, connecting important fields of study that will greatly benefit from more overlap. Future work to continue elucidating this relationship has the potential to provide a predictive framework for understanding the evolutionary consequences of plant–soil ecological linkages in a changing world.

Acknowledgements Thanks for T. Ohgushi, S. Wurst, and S. Johnson for the opportunity to contribute to this volume. We are also grateful for the helpful comments and edits of Casey terHorst and an anonymous reviewer. Support for MVN was provided by a National Science Foundation Graduate Research Fellowship DGE-0929298.

References

Allan GJ, Shuster SM, Woolbright S et al (2012) Perspective: interspecific indirect genetic effects (IIGEs): linking genetics and genomics to community ecology and ecosystem processes. In: Ohgushi T, Schmitz OJ, Holt RD (eds) Trait-mediated indirect interactions: ecological and evolutionary perspectives. Cambridge University Press, New York, pp 295–323

Anacker BL, Klironomos JN, Maherali H et al (2014) Phylogenetic conservatism in plant-soil feedback and its implications for plant abundance. Ecol Lett 17:1613–1621
Andonian K, Hierro JL, Khetsuriani L et al (2012) Geographic mosaics of plant–soil microbe interactions in a global plant invasion. J Biogeogr 39:600–608
Bailey NW (2012) Evolutionary models of extended phenotypes. Trends Evol Ecol 27:561–569
Bailey NW, Zuk M (2012) Socially flexible female choice differs among populations of the Pacific field cricket: geographical variation in the interaction coefficient psi (C). Proc R Soc Lond B Biol Sci 279:3589–3596
Bailey JK, Genung MA, Ware I et al (2014) Indirect genetic effects: an evolutionary mechanism linking feedbacks, genotypic diversity and coadaptation in a climate change context. Funct Ecol 28:87–95
Baker B, Zambryski P, Staskawicz B et al (1997) Signaling in plant-microbe interactions. Science 276:726–733
Bennett JA, Maherali H, Reinhart KO et al (2017) Plant-soil feedbacks and mycorrhizal type influence temperate forest population dynamics. Science 355:181–184
Bever JD (2003) Soil community feedback and the coexistence of competitors: conceptual frameworks and empirical tests. New Phytol 157:465–473
Bever JD, Westover KM, Antonovics J (1997) Incorporating the soil community into plant population dynamics: the utility of the feedback approach. J Ecol 85:561–573
Bezemer TM, Fountain MT, Barea JM et al (2010) Divergent composition but similar function of soil food webs of individual plants: plant species and community effects. Ecology 91: 3027–3036
Bijma P (2014) The quantitative genetics of indirect genetic effects: a selective review of modeling issues. Heredity 112:61–69
Bijma P, Wade MJ (2008) The joint effects of kin, multilevel selection and indirect genetic effects on response to genetic selection. J Evol Biol 21:1175–1188
Busby PE, Newcombe G, Dirzo R et al (2014) Differentiating genetic and environmental drivers of plant-pathogen community interactions. J Ecol 102:1300–1309
Callaway RM, Thelen GC, Rodriguez A et al (2004) Soil biota and exotic plant invasion. Nature 427:731–733
Casper BB, Castelli JP (2003) Evaluating plant–soil feedback together with competition in a serpentine grassland. Ecol Lett 10:394–300
Collins S, Gardner A (2009) Integrating physiological, ecological and evolutionary change: a price equation approach. Ecol Lett 12:744–757
Costa e Silva J, Potts BM, Bijma P et al (2013) Genetic control of interactions amongst individuals: contrasting outcomes of indirect genetic effects arising from neighbour disease infection and competition in a forest tree. New Phytol 197:631–641
Crespi BJ (2001) The evolution of social behaviour in microorganisms. Trends Ecol Evol 16: 178–183
Dawkins R (1982) The extended phenotype. Oxford University Press, New York
Dawkins R (2004) Extended phenotype – but not too extended. A reply to Laland, Turner and Jablonka. Biol Philos 19:377–396
de la Peña E, de Clercq N, Bonte D et al (2010) Plant-soil feedback as a mechanism of invasion by *Carpobrotus edulis*. Biol Invasions 12:3637–3648
Eppinga MB, Kaproth MA, Collins AR et al (2011) Litter feedbacks, evolutionary change and exotic plant invasion. J Ecol 99:503–514
Evans JA, Lankau RA, Davis AS et al (2016) Eco-evolutionary feedbacks in the invasive plant *Alliaria petiolata*. Funct Ecol 30:1053–1061
Felker-Quinn E, Bailey JK, Schweitzer JA (2011) Geographic mosaics of plant-soil feedbacks in a highly invasive plant: implications for invasion success. Ecology 92:1208–1214
Fitzpatrick CR, Gehant L, Kotanen PM et al (2017) Phylogenetic relatedness, phenotypic similarity, and plant-soil feedbacks. J Ecol 105:786–800

Fox JW, Kerr B (2012) analyzing the effects of species gain and loss on ecosystem function using the extended price equation partition. Oikos 121:290–298

Friesen ML, Porter SS, Stark SC et al (2011) Microbially mediated plant functional traits. Annu Rev Ecol Evol Syst 42:23–46

Genung MA, Ubeda F, Fitzpatrick B et al (2011) Genetic variation and community change-selection, evolution and feedbacks. Funct Ecol 25:408–419

Genung MA, Bailey JK, Schweitzer JA (2013) The afterlife of interspecific indirect genetic effects: ecosystem processes as the gene-less products of genetically-based interactions. PLoS One 8: e53718

Govaert L, Pantel JH, De Meester L (2016) Eco-evolutionary partitioning metrics: assessing the importance of ecological and evolutionary contributions to population and community change. Ecol Lett 19:839–853

Hendry AP (2017) Eco-evolutionary dynamics. Princeton University Press, Princeton

Herrera Paredes S, Lebeis SL (2016) Giving back to the community: microbial mechanisms of plant-soil feedback. Funct Ecol 30:1043–1052

Hoeksema JD, Thompson JN (2007) Evolved geographic structure in a widespread plant-ectomycorrhizal interaction: pines and false truffles. J Evol Biol 20:1148–1163

Hoeksema JD, Hernandez JV, Rogers DR et al (2012) Geographic divergence in a species-rich symbiosis: interactions between monterey pines and ectomycorrhizal fungi. Ecology 93:2274–2285

Hunter P (2009) Extended phenotype redux. How far can the reach of genes extend in manipulating the environment of an organism? EMBO Rep 10:212–215

Johnson NC, Wilson GWT, Bowker MA et al (2010) Resource limitation is a driver of local adaptation in mycorrhizal symbioses. Proc Natl Acad Sci USA 107:2093–2098

Kardol P, Bezemer MT, van der Putten WH (2006) Temporal variation in plant-soil feedbacks controls succession. Ecol Lett 9:1080–1088

Kardol P, Deyn GB, Laliberté E et al (2013) Biotic plant–soil feedbacks across temporal scales. J Ecol 101:309–315. https://doi.org/10.1111/1365-2745.12046

Kulmatiski A, Beard KH, Stevens J et al (2008) Plant-soil feedbacks: a meta-analytical review. Ecol Lett 11:980–992

Kylafis G, Loreau M (2008) Ecological and evolutionary consequences of niche construction for its agent. Ecol Lett 11:1072–1081

Laland KN, Uller T, Feldman MW et al (2015) The extended evolutionary synthesis: its structure, assumptions and predictions. Proc Biol Sci 282:20151019

Lamit LJ, Holeski LM, Flores-Rentería L et al (2015) Tree genotype influences ectomycorrhizal fungal community structure: ecological and evolutionary implications. Funct Ecol 24:124–134

Lau JA, Lennon JT (2012) Rapid responses of soil microorganisms improve plant fitness in novel environments. Proc Natl Acad Sci USA 109:14058–14062

Long SR (1996) *Rhizobium* symbiosis: nod factors in perspective. Plant Cell 8:1885–1898

Madritch MD, Greene SL, Lindroth RL (2009) Genetic mosaics of ecosystem functioning across aspen-dominated landscapes. Oecologia 160:119–127

Mangan S, Schnitzer SA, Herre EA et al (2010) Negative plant-soil feedback predicts tree-species relative abundance in a tropical forest. Nature 466:752–755

Matthews B, Narwani A, Hausch S et al (2011) Toward an integration of evolutionary biology and ecosystem science. Ecol Lett 14:690–701

Matthews BL, De Meester L, Jones CG et al (2014) Under niche construction: a bridge between ecology, evolution and ecosystem science. Ecol Monogr 84:245–263

McGlothlin JW, Brodie ED (2009) How to measure indirect genetic effects: the congruence of trait-based and variance-partitioning approaches. Evolution 63:1785–1795

Moore AJ, Brodie ED, Wolf JB (1997) Interacting phenotypes and the evolutionary process. I. Direct and indirect genetic effects of social interactions. Evolution 51:1352–1362

Mutic JJ, Wolf JB (2007) Indirect genetic effects from ecological interactions in *Arabidopsis thaliana*. Mol Ecol 16:2371–2381

Odling-Smee J, Laland KN, Feldman MW (2003) Niche construction: the neglected process in evolution. Princeton University Press, Princeton
Odling-Smee J, Erwin DH, Palkovacs EP et al (2013) Niche construction theory: a practical guide for ecologists. Q Rev Biol 88:3–28
Panke-Buisse K, Poole AC, Goodrich JK et al (2014) Selection on plant microbiomes reveals reproducible impacts on plant function. ISME J 9:980–989
Pfennigwerth AA, Van Nuland M, Bailey JK et al (2017) Plant-soil feedbacks may mediate shrub expansion in declining forests, but only in the right light. J Ecol 106:179–194
Piculell BJ, Hoeksema JD, Thompson JN (2008) Interactions of biotic and abiotic environmental factors on an ectomycorrhizal symbiosis, and the potential for selection mosaics. BMC Biol 6:23
Pregitzer CP, Bailey JK, Hart SC et al (2010) Soils as agents of selection: feedbacks between plants and soils alter seedling survival and performance. Evol Ecol 24:1045–1049
Reynolds HL, Packer A, Bever JD et al (2003) Grassroots ecology: plant-microbe interactions as drivers of plant community structure and dynamics. Ecology 84:2281–2291
Rúa MA, Antonika A, Anunes PM et al (2016) Home field advantage? Evidence of local adaptation among plants, soil, and arbuscular mycorrhizal fungi through meta-analysis. BMC Evol Biol 16:122
Schneider J, Attalah J, Levine JD (2016) Social structure and indirect genetic effects: genetics of social behaviour. Biol Rev 92:1027–1038
Schoener TW (2011) The newest synthesis: understanding the interplay of evolutionary and ecological dynamics. Science 331:426–429
Schweitzer JA, Bailey JK, Fischer DG et al (2008) Soil microorganism-plant Interactions: heritable relationship between plant genotype and associated microorganisms. Ecology 89:773–781
Schweitzer JA, Bailey JK, Fischer DG et al (2012) Functional and heritable consequences of plant genotype on community composition and ecosystem processes. In: Ohgushi T, Schmitz O, Holt R (eds) Trait-mediated indirect interactions: ecological and evolutionary perspectives aspects of trait-mediated indirect interactions. British Ecological Society Ecological Reviews. Cambridge University Press, Cambridge, pp 371–390
Schweitzer JA, Juric I, van der Voorde T et al (2014) Are there evolutionary consequences of plant-soil feedback? Funct Ecol 28:55–64
Semchenko M, Saar S, Lepik A (2017) Intraspecific genetic diversity modulates plant soil feedback and nutrient cycling. New Phytol 216:90–98. https://doi.org/10.1111/nph.14653
Shuster SM, Lonsdorf EV, Wimp GM et al (2006) Community heritability measures the evolutionary consequences of indirect genetic effects on community structure. Evolution 60: 991–1003
Smith DS, Schweitzer JA, Turk P et al (2012) Soil-mediated local adaptation alters seedling survival and performance. Plant Soil 352:243–251
Smith-Ramesh L, Reynolds HL (2017) The next frontier in plant-soil feedback research: unraveling context dependence across biotic and abiotic gradients. J Veg Sci 28:484–494
terHorst CP, Zee PC (2016) Eco-evolutionary dynamics in plant-soil feedbacks. Funct Ecol 30: 1062–1072
terHorst CP, Lennon JT, Lau JA (2014) The relative importance of rapid evolution for plant-microbe interactions depends on ecological context. Proc R Soc Lond B Biol Sci 281:20140028
Thompson JN (2005) The geographic mosaic of co-evolution. University of Chicago Press, Chicago, IL
van der Putten WH, Van Dijk C, Peters BAM (1993) Plant–specific soil-borne diseases contribute to succession in foredune vegetation. Nature 362:53–56
van der Putten WH, Vet LEM, Harvey JA et al (2001) Linking above- and belowground multitrophic interactions of plants, herbivores, pathogens, and their antagonists. Trend Ecol Evol 16: 547–554
van der Putten WH, Bardgett RD, Bever JD et al (2013) Plant-soil feedbacks: the past, the present and future challenges. J Ecol 101:265–276

van der Putten WH, Bradford MA, Brinkman EP et al (2016) Where, when and how plant-soil feedback matters in a changing world. Funct Ecol 30:1109–1121

Van Nuland ME, Wooliver RC, Pfennigwerth A et al (2016) Plant-soil feedbacks: connecting ecosystem ecology and evolution. Funct Ecol 30:1032–1042

Van Nuland ME, Bailey JK, Schweitzer JA (2017) Divergent plant-soil feedbacks could alter future elevation ranges and ecosystem dynamics. Nat Ecol Evol 1:0150. https://doi.org/10.1038/s41559-017-0150

Wagner M, Lundberg D, Coleman-Derr D et al (2014) Natural soil microbes alter flowering phenology and the intensity of selection on flowering time in a wild *Arabidopsis* relative. Ecol Lett 17:717–726

Waring BG, Álvarez-Cansino L, Barry KE et al (2015) Pervasive and strong effects of plants on soil chemistry: a meta-analysis of individual plant 'Zinke' effects. Proc R Soc Lond B Biol Sci 282:1812

Whitham TG, Bailey JK, Schweitzer JA et al (2006) A framework for community and ecosystem genetics: from genes to ecosystems. Nat Rev Genet 7:510–523

Wolf JB (2003) Genetic architecture and evolutionary constraint when the environment contains genes. Proc Natl Acad Sci USA 100:4655–4660

Wolf JB, Brodie ED, Cheverud JM et al (1998) Evolutionary consequences of indirect genetic effects. Trends Ecol Evol 13:64–69

Wolf JB, Mutic JJ, Kover PX (2011) Functional genetics of intraspecific ecological interactions in *Arabidopsis thaliana*. Philos Trans R Soc B Biol Sci 366:1358–1367

Chapter 5
Fungal-Mediated Above–Belowground Interactions: The Community Approach, Stability, Evolution, Mechanisms, and Applications

Alison E. Bennett, Peter Orrell, Antonino Malacrino, and Maria José Pozo

5.1 Introduction

Within this chapter, we will focus on above–belowground interactions in which belowground organisms influence aboveground organisms (or vice versa) primarily via a shared host plant associated with a fungus (for ease throughout the chapter, we will refer to these as "fungal-mediated above–belowground interactions"). Almost exclusively, the fungi studied mediating these above–belowground interactions are belowground and commonly considered to be mutualistic (e.g. mycorrhizal fungi, plant-growth promoting fungi (PGPF), or endophytes), and the aboveground organisms are insects (e.g. herbivores and pollinators) visiting host plants. Arbuscular mycorrhizal (AM) fungi associate with over 80% of plant species, obligately depend on plants, and act as root mutualists whose hyphae exploit soil areas that plants cannot access themselves, and deliver nutrients (predominantly phosphorus, but also nitrogen and trace minerals) to plant roots in return for carbon. Ectomycorrhizal (ECM) fungi are facultative root mutualists that associate with root tips and whose

A. E. Bennett (✉)
Department of Evolution, Ecology, and Organismal Biology, Ohio State University, Columbus, OH, USA
e-mail: bennett.1242@osu.edu

P. Orrell
Ecological Sciences, James Hutton Institute, Dundee, UK
e-mail: Peter.Orrell@hutton.ac.uk

A. Malacrino
Department of Physics, Chemistry and Biology (IFM), Linköping University, Linköping, Sweden

M. J. Pozo
Department of Soil Microbiology and Symbiotic Systems, Estación Experimental del Zaidín (CSIC), Granada, Spain
e-mail: mjpozo@eez.csic.es

T. Ohgushi et al. (eds.), *Aboveground–Belowground Community Ecology*, Ecological Studies 234, https://doi.org/10.1007/978-3-319-91614-9_5

hyphae deliver predominantly nitrogen for their primarily coniferous hosts (Smith and Read 2008). PGPF consist of a number of fungal species, such as *Trichoderma* spp., which have been shown to promote plant growth and defence (Pieterse et al. 2014), and endophytes are organisms that live within host plant tissues without causing harm to the plant (Clay and Schardl 2002). Technically, AM fungi qualify as endophytes because they live inside host plant tissues. Very little research has been conducted on belowground fungal pathogen effects on aboveground insects (Shikano et al. 2017) despite our knowledge from aboveground systems that fungal pathogens can have equally strong influences on insects (Hauser et al. 2013), and that fungal–plant associations of any kind could prime plants for greater insect defence (see Sect. 5.5), both above- and belowground.

Fungi have been shown to have a wide array of surprising effects on host plant interactions with aboveground insects. A number of foliar endophytes produce toxic compounds (such as alkaloids) that can directly influence herbivores (Saikkonen et al. 2013), but this has so far not yet been demonstrated in root endophytes. Other endophytes, including root endophytes, have been shown to alter plant defences (Gill et al. 2016; Shikano et al. 2017), and AM fungi and ectomycorrhizal fungi (Gehring and Bennett 2009) and PGPF (Pieterse et al. 2014) have been shown to alter both direct and indirect plant defences. Direct plant defences involve changes in secondary chemistry or structures that can be constitutive (present early in development before herbivory) or induced (altered after herbivore feeding), and indirect plant defences (which can also be constitutive or induced) typically involve volatile organic compounds (VOCs) that recruit herbivore natural enemies (Rasmann et al. 2017). Alteration of direct plant defences appears to have the greatest effects on generalist chewing herbivores and beneficial or no effects on specialist sap sucking herbivores (Koricheva et al. 2009), but at least AM fungi have been shown to also alter VOCs and indirect defences in favour of host plants: the volatile blends released by plants associated with AM fungi (vs. no AM fungi) can be more attractive to natural herbivore enemies of specialist herbivores (Babikova et al. 2014a; Bennett et al. 2016; Bezemer et al. 2005; Gange et al. 2003; Guerrieri et al. 2004; Hempel et al. 2009; Hoffmann et al. 2011a, b; Schausberger et al. 2012), although it should be noted that, to date, fungal-mediated signalling to enemies of generalist herbivores has not been tested. AM fungi have also been shown to promote plant tolerance to herbivory (Borowicz 2013). Thus, belowground fungi can clearly influence aboveground herbivores.

Most beneficial fungi are also known to promote the growth and nutritional status of plants, but both growth and plant defence effects often depend on microbial species and plant species or genotype (Bennett et al. 2016; e.g. Kempel et al. 2010, 2013), plant age (Tomczak and Müller 2017), herbivore identity (Koricheva et al. 2009; Schweiger et al. 2014b), and the abiotic environment (Pineda et al. 2013). Beneficial fungi can also promote plant phenotypic plasticity in traits that improve resistance or tolerance to both abiotic and biotic stress (e.g. root and shoot architecture, flowering, and stress responses) (Goh et al. 2013; Millar and Bennett 2016), but this depends on the biotic and abiotic context (e.g. water, nutrient, and light availability; temperature; soil chemical composition; and biotic interactions).

Therefore, the effect of any fungus on herbivore performance will be due to a balance between a positive effect via the improved quality and quantity of the host plant and a negative effect due to changes in plant defences (Gehring and Bennett 2009; Koricheva et al. 2009) but will be modified by a variety of external factors.

Our goal within this chapter is to review the literature, but to also highlight what we feel are the biggest areas for future research within this field: the community approach, stability, evolution, mechanisms, and application of these interactions.

5.2 The Community Approach

Above–belowground ecology has shown that organisms living above and below the soil surface influence each other. These effects mainly occur via plants, which exist in both above (shoot) and belowground (roots) compartments and, by interacting with both communities, act as link between the two compartments. In the last ecology review published on the influence of mycorrhizal fungi on above–belowground interactions, Gehring and Bennett (2009) emphasized the need for a community approach to studies of fungal-mediated above–belowground interactions. The community approach examines multiple simultaneously interacting species and is referred to as the community approach because these interaction species form communities. In particular, they pushed for future studies incorporating a greater number of species and potential interactions and, most importantly, communities of organisms (fungi, plants, and insects). This approach is based on evidence that plants rarely interact with a single fungus or insect species, but rather interact with a community of organisms, and this greater diversity of interactions (via competition, facilitation, or predation) will likely reveal more varied outcomes that better match observed patterns in nature than when a smaller suite of species is considered.

With the arrival of new -omics techniques, which are a powerful tool to finely investigate above–belowground interactions using a culture-independent (e.g. molecular) approach (Zhang et al. 2010), we have the opportunity to truly use a community approach to understand fungal-mediated above–belowground interactions (Fig. 5.1). -Omics techniques can produce large amounts of complex data, and the next step after data collection is to analyse these complex data in a meaningful way. A growing number of researchers are combining meta-omics tools with network analyses (models of species interactions between organisms of different guilds) to analyse how shifts in above–belowground interactions occur (e.g. Toju et al. 2015). We expect these advanced techniques to increase the level of detail and rapidly expand the number of studies of fungi and other organisms within above–belowground interactions.

Within this section, we highlight current advances in fungal-mediated above–belowground interactions and focus on how -omics techniques could expand our knowledge of the influence of interactions along the trophic ladder.

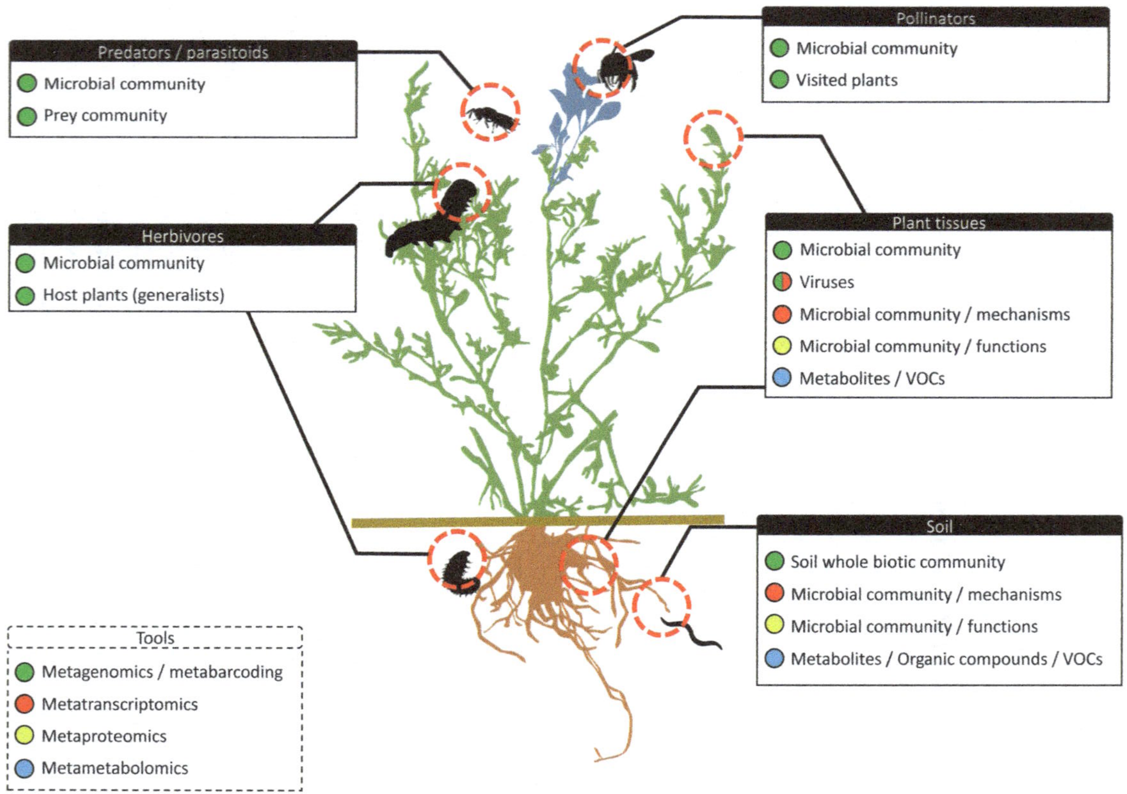

Fig. 5.1 An overview of the application of meta-omics tools that can be used in the community approach of fungal-mediated above–belowground interactions. Four types of meta-omics tools are considered: metagenomics (represented by green circles), metatranscriptomics (represented by red circles), metaproteomics (represented by yellow circles), and metametabolomics (represented by blue circles). The application of these techniques for understanding plant mediated interactions is broken down for all compartments and trophic levels: predators/parasitoids, herbivores, pollinators, plant tissues, and soil

5.2.1 Interactions Throughout the Trophic Ladder

While most studies of fungal-mediated above–belowground interactions have focused on fungal effects on insects or vice versa, there is growing awareness that aboveground fungi (e.g. endophytes) can influence belowground fungi (e.g. mycorrhizal fungi) and vice versa. Aboveground endophytes negatively impact AM fungal colonization in grasses (Mack and Rudgers 2008; Omacini et al. 2006) while the impact of AM fungi on endophytes is less clear. In one study, there was no reverse effect on foliar endophytes (Mack and Rudgers 2008), while in a second study the presence of AM fungi altered the foliar endophyte community composition (Eschen et al. 2010). In order to determine the impacts of these interactions within communities, we propose further studies to determine whether these changes in fungal abundance translate to changes in function or ecosystem services provided by these partners.

How do fungal-mediated above–belowground influences at one trophic level translate to the next trophic level? For example, if AM fungi or endophytes suppress insect herbivory, does this negatively impact parasitoids feeding on those herbivores? A growing number of studies are examining these questions in fungal-mediated and non-fungal-mediated above–belowground interactions (reviewed in van Dam and Heil 2011). AM fungi often appear to increase parasitism of herbivores feeding on host plants in several systems (Babikova et al. 2014a; Bennett et al. 2016; Gange et al. 2003; Guerrieri et al. 2004; Hempel et al. 2009; Hoffmann et al. 2011a, b; Schausberger et al. 2012; but see Bezemer et al. 2005), and endophytes have been shown to alter release of volatiles and herbivore enemy attraction (Lugtenberg et al. 2016). In contrast, fewer studies have examined the reverse: whether higher trophic level aboveground organisms can alter fungi belowground. In one set of studies, an endosymbiont within the gut of an herbivore reduced plant root allocation (Bennett et al. 2016; Hackett et al. 2013), which could possibly affect belowground communities. Can higher trophic level aboveground organisms (e.g. parasitoids) influence belowground fungal trophic interactions? We are not aware of any studies that have addressed this question in above–belowground interactions. Whilst the mechanisms that drive many of these interactions are still unknown, -omics tools now allow us to take a whole picture of our study systems at multiple spatial and temporal scales, allowing us to more clearly understand fungal-mediated above–belowground interactions—even in previously difficult to manipulate environments such as soil or inside plant tissues.

5.3 Stability

How stable are above–belowground interactions? Are they common interactions that occur repeatedly throughout systems or are they transient and occur under only the right circumstances? We know that systems can have multiple stable states influenced by a host of different factors (e.g. season and climate, presence or absence of organisms, etc.). Are above–belowground interactions common interactions occurring across most stable states or do they occur infrequently across stable states? These are major questions that are just beginning to be answered.

Throughout this section, we use the definition of stability proposed by Pimm (1984): whether, once disturbed, a system returns to its original equilibrium (or state) or is shifted to a new equilibrium (or state). As proposed by Pimm (1984), this definition has five aspects: asymptotic stability (whether, after disturbance, a system returns to its equilibrium following a predictable pattern), variability (how time and space influence the coefficient of variation of a trait in the system), persistence (how long a disturbance must be to shift a system away from its equilibrium), resistance (a measure of the change in a system before and after disturbance), and resilience (the rate at which a system returns to its equilibrium following disturbance) (reviewed in Donohue et al. 2016). Applied to fungal-mediated above–belowground interactions, stability refers to the likelihood that disturbances (natural or anthropogenic) will cause these interactions to dissolve. In other words, are above–belowground interactions likely to persist despite disturbance or do they only occur under specific environmental conditions?

To our knowledge, only a small proportion of the literature has tried to address this question despite earlier papers highlighting this gap (e.g. van der Putten et al. 2004). Most studies manipulated plant species or functional diversity and found plants and plant functional group play a big role in ecosystem stability, creating top down effects on belowground organisms (Eisenhauer et al. 2011; Kowalchuk et al. 2002; McElroy et al. 2012; Wardle et al. 2000) and bottom up effects on aboveground organisms (Eisenhauer et al. 2011; Wardle et al. 2000). In some cases, however, plant associations with both above- and belowground organisms can decouple spatial and temporal stability within plant communities because these organisms can influence the plant community in different ways (Eisenhauer et al. 2011). It is not surprising, given their approach, that the above examples found a primary influence of plants on system stability. By instead manipulating the belowground community, we see AM fungi can promote plant community temporal stability (Yang et al. 2014), but the bottom up influence of fungi on aboveground plant–insect interactions has yet to be explored. Thus, it seems likely that more balanced designs (manipulation of below- and aboveground organisms in addition to plants) will allow for a greater understanding of the forces that influence stability in fungal-mediated above–belowground interactions.

Several disturbance factors could alter fungal-mediated above–belowground interactions. Abiotic variables, such as temperature, nutrient availability, water availability, pH, and even sunlight, can influence the composition of soil fungal communities (e.g. Mulder and de Zwart 2003). These variables can directly influence fungal communities or influence them indirectly through changes in host plant composition. Many of these factors also lead to changes in fungal community composition (Gehring et al. 2014), suggesting that any disturbance may result in reshuffled fungal communities. Many of the same abiotic variables alter aboveground communities. For example, increasing nitrogen availability reduces insect diversity but increases insect abundance in grassland systems (Haddad et al. 2000) while pH can interact with plant diversity to alter arthropod diversity (Brandle et al. 2001; Choi 2015; but see Sanderson 1992). Thus, any disturbance that alters these variables (e.g. N and pH) would also alter above–belowground interactions, and disturbances (particularly due to environmental change) could threaten the stability of above–belowground interactions.

Thus, we have yet to sufficiently answer the question: How does the influence of environmental change alter the stability of above–belowground interactions? This is a major question because it influences whether we can continue to utilize above–belowground interactions in agriculture, against invasive species, and to promote conservation. Below we examine the influence of abiotic and biotic disturbance factors (many caused by environmental change) on the stability of fungal-mediated above–belowground interactions.

5.3.1 Influence of Abiotic Environmental Change on Stability

Given that most research on the influence of abiotic environmental change on stability has occurred in only aboveground or only belowground systems, in this section we explore conclusions from these separate compartments and draw predictions, based on the current limited literature, on how abiotic environmental change will alter combined above and belowground systems.

There is a growing literature focusing on how fungal interactions belowground will respond to abiotic environmental change. A meta-analysis focused on the influence of climate changes (elevated CO_2, drought, nitrogen, and temperature) on the function of fungal symbionts found that influences were dependent upon the symbiont and the climate change factor measured (Kivlin et al. 2013). For example, dark septate endophyte, but not AM fungi, promoted plant growth under elevated temperatures. Focusing on community stability (and not only plant growth), AM fungi have been shown to increase temporal stability even in the presence of increasing phosphorus availability (Yang et al. 2014). This research creates an interesting contrast to the work of Gehring et al. (2014) in which environmental changes (drought, competition, parasitism, and herbivory) altered community composition similarly, suggesting that possibly environmental change selects for more plastic fungal species that can alter their phenotype (and function) in response to multiple stresses.

Aboveground in non-fungal-mediated systems, there is evidence that management and abiotic change can have dramatic effects on insect communities—stronger effects than due to solely plant community diversity (Hudewenz et al. 2012). Pollinators have been shown to increase yield in agro-ecosystems faced with heat stress or warming (Bishop et al. 2016; Rader et al. 2013) and thereby contribute to ecosystem stability. There is more conflicting evidence about the influence of climate change factors on herbivores and their enemies (e.g. Pincebourde et al. 2017). Similar patterns to pollinators have been demonstrated for trophic chains (e.g. plants, herbivores, and their predators) exposed to warming in some systems—together they seem to buffer the influence of change and promote stability (Sentis et al. 2013). But in other systems increases in warming, water, and nitrogen availability have been shown to increase herbivore abundance but induce greater variability in herbivore enemy feeding and abundance (and even declines in abundance) thereby potentially reducing community stability (de Sassi and Tylianakis 2012; Haddad et al. 2000; Hentley and Wade 2017; Liman et al. 2017). Thus, it is difficult to draw general conclusions about the influence of climate changes aboveground on insects, particularly insects in higher trophic levels (Hentley and Wade 2017).

Here, we combine our knowledge of how separate below- and aboveground systems respond to abiotic changes to predict how stability of fungal-mediated above–belowground systems might be altered. For example, the variation in stability of aboveground trophic chains (plants–herbivores–herbivore enemies) could be driven by the influence of nutrient stoichiometry on herbivores and their enemies (Abbas et al. 2014)—and we know that belowground interactions can influence (Spohn 2016) and be influenced by nutrient stoichiometry as well (Johnson et al. 2015). Thus, the relative availability of nutrients in a system could play a large role in predicting the stability of several above–belowground interactions. In a second example involving elevated CO_2, significant previous research has identified that CO_2 can alter defence signalling pathways in host plants leading to changes in direct defences that influence herbivory (reviewed in Hentley and Wade 2017; Zavala et al. 2013). Fungal associations can also alter plant secondary chemistry (Gehring and Bennett 2009; Ludwig-Müller 2015; Lugtenberg et al. 2016) directly and indirectly. Thus, in the case of elevated CO_2, we may see no change (and thus stability) in fungal-mediated above–belowground interactions if CO_2 and fungi counteract each other's influence, or we could see great instability if their interaction has multiplicative effects. Alteration of defence signalling pathways by changes in other atmospheric gases (e.g. O_3, SO_2, and NO_2) (reviewed in Boullis et al. 2015) or fungi (reviewed in Rasmann et al. 2017) can also alter the production of secondary chemicals for indirect defence, which may further confuse results as recent research has shown that increases in temperature (often expected to co-occur with changes in atmospheric gases) can alter the perception of volatiles by insects (Sentis et al. 2015). Change in secondary compound perception could reduce system stability if it alters the ability of herbivore enemies to find hosts or maintain stability if reduced perception additively interacts with increased production of volatiles. Thus, while we can make predictions, many factors need to be tested before we can draw concrete conclusions about the influence of environmental change on the stability of fungal-mediated above–belowground interactions.

While immediate abiotic changes may alter interactions, the legacy of that change may also be important for later generations. Extreme climate events, such as drought, can reduce the function of soil communities [e.g. reduce decomposition and restoration (Bloor and Bardgett 2012)], but how does this reduction in function (or any other legacy effect) influence the next generation of organisms? Wurst and Ohgushi (2015) hypothesize that there are two factors likely to produce legacy effects:

changes in plant traits and changes in soil characteristics. They discuss both short-term (up to days) and long-term (weeks to years) influences of changes in plant and soil traits. For example, changes in defensive compounds within plants can alter the communities of herbivores feeding on plants both in the short- and long term, whereas in the short term changes in fungal activity in the soil could lead to changes in decomposition that lead to long term changes in nutrient availability. Limited research in this area has revealed that plant–soil feedbacks can alter plant–herbivore interactions in the next generation of plants (Kos et al. 2015; Morrien et al. 2011), although longer term feedbacks (i.e. with field collected soils) do not influence plant–herbivore interactions (Williams et al. 2014). Belowground, root herbivory can produce legacy effects for plant–AM fungal associations in the next generation in the same soil (Sonnemann et al. 2013). More experiments incorporating legacy effects will help predict the long-term stability of above–belowground interactions.

Could changing climates create phenological mismatches in above–belowground interactions? Climate changes are already influencing phenology of plants (Parmesan and Hanley 2015) and common organisms in above–belowground interactions (Cooper 2014), which could lead to phenological mismatches in above–belowground interactions. The change in interaction intensity between herbivores and herbivore enemies under increased temperature and precipitation in North America (Diehl et al. 2013) is a potential example of climate change driven phenological mismatch. Are phenological mismatches occurring belowground as well? Due to the difficult nature of observing interactions belowground, we are decades behind on collecting the needed phenological data to answer this question for most belowground organisms, including fungi (Miller-Rushing et al. 2010). As a result, abiotic change could drive phenological mismatches in above–belowground systems, yet we do not have the data to test for this question.

Alterations in abiotic factors associated with changing climates are also pushing changes in species ranges. This has been relatively well documented above ground (reviewed in Chuang and Peterson 2016), but, aside from invasive fungi (Dickie et al. 2016), is less well documented in belowground systems. There is evidence suggesting that AM fungi and endophytes vary in their temperature tolerances (Barrett et al. 2014; Bunn et al. 2009; Corbin et al. 2017; Heinemeyer and Fitter 2004; Klironomos et al. 2001), so increasing temperatures may exclude some fungal species from some habitats or may temporally shift when a fungal species is associated with plants. Some fungi disperse more slowly than plants (e.g. AM fungi), so loss of fungal symbionts may select for plants with less dependence on fungal partners at their range edge (van Grunsven et al. 2007). The effects of reduced fungal dependence (or reduced negative plant–soil feedbacks) appear to have an additive interaction with above ground herbivory (Morrien et al. 2011). Thus, changes in species ranges may allow above–belowground partners to become spatially or temporally separated leading to decreased stability in these interactions.

Aside from abiotic climate changes, habitat fragmentation and disturbance can also have strong influences on the prevalence of organisms in any environment. If disturbance and habitat fragmentation spatially or temporally separate fungi, insects, and their plant partners, we are likely to see similar influences on the stability of fungal-mediated above–belowground interactions as might be observed in the case of shifting species ranges and phenological mismatches. For example, belowground soil disturbance and habitat fragmentation alter the species composition and diversity of fungi (reviewed in Crowther et al. 2014; Lenoir et al. 2016; Uroz et al. 2016). Specifically, tillage is well known to reduce AM fungal composition and diversity (reviewed in Verbruggen and Kiers 2010). Aboveground both soil disturbance and habitat fragmentation have been shown to alter plant species diversity and composition (Kershaw and Mallik 2013; Paillet et al. 2010), and this influence alone can alter both belowground and aboveground interactions (e.g. Cahill et al. 2008). Independent of the influence on plant communities, we know that habitat fragmentation can also alter insect species diversity and abundance (reviewed in Joern and Laws 2013; Vasconcelos and Bruna 2012). As a result, it seems likely that habitat fragmentation will threaten the stability of fungal-mediated above–belowground interactions by removing key partners within interactions.

5.3.2 Influence of Biotic Environmental Change

One of the major factors influencing the stability of any interaction is the introduction of a novel species into the interaction. The vast majority of research into this area in above–belowground interactions has been focused on the influence of invasive plants, insects, and microbes (reviewed in Bennett 2013; van der Putten et al. 2007). Invasive species can lead to the loss of organisms important for stability (van der Putten 2012) and can strongly influence interaction strengths (Rodríguez-Echeverría and Traveset 2015). For example, invasive plants can influence both above and belowground systems, but the influence of invasive species on above and belowground systems differs and varies by habitat (McCary et al. 2016). Invasive plant species are well known to both interact differently with fungi (Inderjit and van der Putten 2010; Kulmatiski et al. 2008; Rout and Callaway 2012) and alter belowground fungal diversity (e.g. Vogelsang and Bever 2009). Invasive plants can reduce the abundance of many arthropods, particularly herbivores (Litt et al. 2014). The presence or absence of a belowground fungal partner has been hypothesized to promote the success of some invasive species (Levine et al. 2004; Menzel et al. 2017; Pringle et al. 2009; Reinhart and Callaway 2006; Richardson et al. 2000). As a result, introduced species are likely to dramatically alter above–belowground interactions involving fungi—but this has rarely been addressed quantitatively (van der Putten 2012).

Genetic variation within partners may be a biotic factor limiting stability within above–belowground interactions. This is particularly important for managed systems (e.g. agriculture) where the genetic make-up of partners may be well-defined and limited. For example, within potato aphids some but not all aphid holobionts (aphid genotype plus endosymbiotic bacterial genotype) have been shown to reduce allocation belowground (Bennett et al. 2016; Hackett et al. 2013), while belowground AM fungi have been shown to promote parasitism success of aboveground aphids, but this depends on potato genotype (Bennett et al. 2016). Genetic variation in AM fungi has also been shown to alter aboveground tobacco–herbivore interactions (Wooley and Paine 2007). Thus, genetic variation likely alters stability of above–belowground interactions, but how has yet to be clarified.

All of the above topics raise another important question for understanding the stability of above–belowground interactions: Is species identity or species function more important for maintaining fungal-mediated above–belowground interactions? In other words, if we replace species within interactions with species of the same function, is stability maintained? Plant functional group has been suggested to have a stronger influence on soil microbial diversity (Milcu et al. 2013) while above we suggest stress may select for more phenotypically plastic fungal partners. This suggests that plant and fungal function may be more important than species identity for stability of systems, so is plant and fungal functional diversity also more important than species identity for herbivores in above–belowground interactions? Only one study has tested this in a plant–soil feedback experiment and found that the plant functional group had a greater effect on aphids in the next generation than plant species identity (Kos et al. 2015). Thus, early indicators suggest that functional diversity may be more important than species diversity, but future studies in this area will provide a greater understanding of the importance of species identity and species function.

As a result, we are just beginning to address the factors that will lead to an understanding of stability in fungal-mediated above–belowground interactions.

5.4 Evolution

While we can be confident that natural selection is acting in fungal above–belowground systems, we have very little research documenting evolution in these systems. To date the only known specific publication on the topic is not data driven, but developed hypotheses for how AM fungi might alter plant responses to antagonists (such as insect herbivores) and in turn how that might alter the population dynamics of antagonists (Bennett et al. 2006). There have been broad brush evolution experiments in which soil communities and plants have been exposed to selection pressures and changes in response documented (e.g. Bonte et al. 2010; Lau and Lennon 2012; Panke-Buisse et al. 2015), but these studies have not looked at fungi themselves or documented how selection altered the organisms involved. However, they present a first step for documenting the effect of selection in fungal above–belowground interactions.

Here, we present three basic areas in which we expect selection to influence fungal above–belowground interactions: simple (one-way) evolutionary responses driven by fungi or aboveground organisms; evolutionary feedbacks and co-evolutionary arms races; and the relative strength of direct and indirect interactions on selection. We hope that these discussions promote future research in these areas that will further enhance our understanding of fungal above–belowground interactions.

5.4.1 Simple (One-Way) Evolutionary Responses

The simplest means of determining whether fungi can alter evolution within above–belowground communities is to assess whether the outcome of selection changes in the presence and absence of fungal partners by measuring changes in heritability or traits associated with interactions. For example, if a fungal partner alters plant chemistry, selection might favour herbivores with a greater tolerance for plant secondary chemicals. In tall fescue (*Festuca arundinacea* (syn., *Schedonorus arundinaceus*, and *Lolium arundinaceum*)), the production of alkaloids by the endophyte (*Neotyphodium coenophialum*) in the leaves negatively impacts herbivores allowing endophyte-colonized tall fescue to escape herbivory, outcompete uninfected neighbours, and become dominant in the environment (Clay 1988; Clay and Holah 1999). The influence of endophytes in tall fescue has knock-on effects on other interactions throughout the system (Finkes et al. 2006; Lemons et al. 2005; Mack and Rudgers 2008; Rudgers and Clay 2008). However, to date there are no tests of whether herbivores in endophyte-infested tall fescue systems have been selected for greater tolerance to alkaloids or higher fitness than naïve herbivores. In contrast, previous research has shown that increased herbivory can increase the heritability (h^2) of the endophyte in tall fescue (Clay et al. 2005), but we do not know if higher herbivore pressure leads to an increase in alkaloid production by the endophyte. "Simple" selection experiments like those described above require multiple generations of growth under experimental conditions using a combination of greenhouse and field approaches and present the first step to demonstrating patterns of selection within fungal-mediated above–belowground interactions.

5.4.2 Evolutionary Feedbacks, Co-evolutionary Arms Races, and Selection Mosaics

Selection at one end of the fungal-mediated above–belowground interaction "chain" influences selection at the other end. Returning to the above example, if endophyte colonization of fescue selects for herbivores with increased alkaloid tolerance, could increased herbivory by alkaloid tolerant herbivores promote selection for greater alkaloid production (or alternate alkaloid compounds) within the endophyte? Thus, do evolutionary feedbacks occur within above–belowground interactions? Also can fungi in above–belowground interactions alter co-evolutionary arms races? Most beneficial plant–fungal associations prime plant defence response pathways (see the Sect. 5.5), which has been suggested to increase constitutive or inducible plant defences. If fungi increase secondary chemicals in host plants, could this promote faster selection for tolerance within insects or overcome insect tolerance to these defences and reduce the opportunity for selection to promote tolerance within insects? This is a research area ripe for exploration. Many of the indirect selective influences on fungal-mediated above–belowground interactions are likely to produce diffuse patterns of co-evolution [an interaction that occurs with the addition of multiple organisms to a system which alters the pattern of selection on traits of focal species (Haloin and Strauss 2008)] and are thus likely to be more complicated than simple one-way evolutionary responses. Also, selection mosaics like those first described by Thompson (2005) likely also influence these interactions, but have never been studied. Both fungal partners (e.g. Wolfe et al. 2007) and herbivores (e.g. Thompson 1994) can be very patchy in the landscape, and this patchiness should produce patches where the feedbacks and arms races described above are stronger, weaker, or non-existent.

5.4.3 Direct or Indirect Influences

Here, we focus on whether fungal influences on selection within above–belowground interactions are driven predominantly by direct or indirect interactions. In the fescue–endophyte example, endophytes have a direct influence on herbivores via the production of alkaloids within host plant tissues (Clay 1988; Clay and Holah 1999). However, many other fungal influences on above–belowground interactions occur indirectly via changes in host plant phenotypes. Defence priming by beneficial microbes (see Sect. 5.5) can indirectly alter plant interactions aboveground. Thus, the nature of these interactions, be they indirect or direct, produces a number of questions: Do direct or indirect interactions more commonly influence, and which has the greatest influence, on selection in above–belowground systems? While we expect direct selection to be strong, we know that indirect interactions can also have strong, and sometimes unexpected, selective influences (Haloin and Strauss 2008). To test for direct versus indirect influences, we recommend comparing across generations the influence of plants hosting fungi that produce their own toxins versus those that alter plant chemistry on a common herbivore. Thus, we can clearly see the potential for fungi to alter selection within above–belowground interactions, but there are future opportunities to document that selection.

5.5 Mechanisms

The vast majority of research into mechanisms driving the indirect interactions observed via host plants in fungal-mediated above–belowground interactions has focused on aboveground pathogens and insects and belowground on Endophytes and AM fungi. As a result, we will focus on these two sets of organisms, and we highlight new areas of research needed to further describe the mechanisms leading to the influence of fungi in indirect above–belowground interactions.

5.5.1 What Are the Mechanisms That Influence Fungal Above–Belowground Interactions?

Soil fungal communities and insects can influence each other by affecting plant metabolic pathways thereby altering the phenotype of the shared host plant. Endophytic and AM fungi can influence aboveground insects via changes in the nutritional status, in the architecture, and in the accumulation of defence-related metabolites in plants. The leaf metabolome has been shown to be a product of plant species identity and fungal partners below- (Schweiger et al. 2014a; Schweiger and Müller 2015) and aboveground. Fungi can change plant phenotypes independent of interactions with insects, but in multiple systems belowground fungi have been shown to alter plant responses to aboveground herbivory by reallocating resources and activating inducible defence mechanisms (Gehring and Bennett 2009; Pineda et al. 2013).

Inducible defences allow plants to flexibly manage their resources towards defence or growth by triggering defences only when necessary, and priming, which reduces the lag time from defence activation to fully functional defences, is key for the efficiency of these defence responses (Martínez-Medina et al. 2016). Beneficial soil microbes commonly prime plant defences, increasing the readiness of induced defences through changes in the plant immune system (Selosse et al. 2014). The fitness benefits of priming have been shown to outweigh their costs under stress (Martínez-Medina et al. 2016). Specifically, the cost and underlying mechanisms have been well established in plant–pathogen interactions (Conrath et al. 2015; Mauch-Mani et al. 2017), but priming of antiherbivore defensive responses by beneficial fungi in plants has been less well studied (Bandoly et al. 2016; Frost et al. 2008; Kim and Felton 2013; Mauch-Mani et al. 2017; Fig. 5.2). Priming against herbivores has, however, been mostly studied at the molecular level, especially with regard to herbivore induced volatiles (VOCs), insect oviposition (reviewed in Hilker and Fatouros 2015), and treatment with plant defence elicitors (reviewed in Conrath et al. 2015).

The mechanisms regulating plant interactions with and priming by beneficial microbes are shared by taxonomically different groups of beneficial microbes (Partida-Martinez and Heil 2011; Pieterse et al. 2014; Van Wees et al. 2008). There are multiple examples of beneficial rhizobacterial or fungal primed defences in plants including endophytes such as *Piriformospora indica*, *Trichoderma* spp., and AM fungi (Pieterse et al. 2014; Pineda et al. 2013), and may be a consequence of the modulation of the plant immune system associated with root colonization and the establishment of the symbiosis that results in defence-related signalling (Selosse et al. 2014; Zamioudis and Pieterse 2012). A molecular dialogue between the symbionts and plants modulates host defences to favour the beneficial interaction, and, in the case of AM fungi, a symbiotic programme is triggered for mycorrhizal development (Genre and Russo 2016). We expect this modulation of the immune system to act in two ways. First, it should enhance local tolerance to endophytic fungi (like AM fungi) promoting colonization. For example, small secreted fungal peptides injected into root cells block specific regulators of plant defence signalling locally, resulting in a partial local desensitization that allows colonization (Plett and Martin 2015). Second, this modulation should prime plant defences placing the plant into a readiness state. During attack by soil pathogens, mycorrhizal fungal primed plants accumulate more pathogenesis-related proteins, callose, and phenolics compared to non-mycorrhizal plants, and this early and strong reaction is pivotal for successful defence (Jung et al. 2012). This primed state extends to the shoots and primes responses to foliar pathogens (Sanchez-Bel et al. 2016) and primes responses to herbivorous insects via differential transcriptional regulation of herbivore defence-related genes, accumulation of insect anti-feedant compounds (Babikova et al. 2014a, b; Bennett et al. 2009; Fontana et al. 2009; Song et al. 2010) and VOCs (reviewed in Rasmann et al. 2017).

Inducible plant defences and their coordination with growth and development is orchestrated by a complex signalling network in which plant phytohormones play a central role (Khan et al. 2015; Okada et al. 2015; Piasecka et al. 2015). Among them, jasmonates are key regulators of direct and indirect plant defences to wounding and insects, including the production of toxic and anti-feedant compounds and volatile signals (Okada et al. 2015). Typically, defence priming by beneficial microbes is dependent on jasmonate signalling pathways (Pieterse et al. 2014), as has been shown for *Trichoderma* spp. and AM fungi (Jung et al. 2012; Martínez-Medina et al. 2013; Sanchez-Bel et al. 2016; Song et al. 2014). Although the mechanisms of induced resistance against herbivores also involve priming of jasmonic acid (JA)-dependent responses (Song et al. 2014; Van Oosten et al. 2008), other mechanisms, still uncharacterized, are also likely operating (Pineda et al. 2013).

Mycorrhizal fungi tend to negatively affect generalist leaf chewers and have a neutral or positive effect on specialist leaf chewers and phloem feeders (Gehring and Whitham 2002; Koricheva et al. 2009), and this pattern correlates with the plant defence responses these herbivores trigger (Pozo and Azcón-Aguilar 2007). Attacker-specific signal signatures (determined partly by temporal and spatial patterns of mechanical tissue injury), intricate crosstalk among the different signalling pathways, and potential attacker-mediated suppression of host defences through effector proteins shape the plant response to herbivores (Howe and Jander 2008; Pieterse et al. 2014; Plett and Martin 2015). For example, induced resistance frequently consists of changes in secondary metabolites that are toxic for generalist herbivores but rarely affect specialist herbivores, because leaf chewers rupture cells thereby releasing and ingesting toxic metabolites while also triggering the plant jasmonate-dependent wound response. By contrast phloem feeders avoid cell damage and feed on phloem sap, which contains lower levels of toxic compounds. Thus, fungi which prime plants may also positively impact aphid growth and fecundity by making plants better-quality hosts through improved nutrition (Gehring and Bennett

2009; Koricheva et al. 2009). This priming effect can also be transmitted through common mycorrhizal mycelial networks (Babikova et al. 2013; Johnson and Gilbert 2015; Song et al. 2010, 2014) influencing jasmonate signalling that promotes direct defence against generalists (Song et al. 2014) and indirect defence against phloem feeders (Rasmann et al. 2017) in neighbouring plants. Thus, there is growing evidence that priming influences both direct and indirect defences, but that this priming influence may have different functions based on the herbivorous pest.

Remarkably, similar patterns have been also found in non-fungal-mediated above–belowground interactions involving rhizobacteria (e.g. *Pseudomonas fluorescens*) (Pineda et al. 2013). *P. fluorescens* root colonization induces systemic *Arabidopsis thaliana* resistance against the generalist caterpillar *Spodoptera exigua*, but not the specialist caterpillar *Pieris rapae* (Van Oosten et al. 2008). In this system, Induced Systemic Resistance (ISR) was associated with priming for an enhanced expression of genes regulated by the jasmonic acid/ethylene (JA/ET)-signalling pathway (Van Oosten et al. 2008). In contrast, rhizobacterial colonization did not affect the performance of the specialist aphid *Brevicoryne brassicae*, whereas it enhanced the intrinsic growth rate of the generalist aphid *Myzus persicae* despite evidence of priming for JA responses (Pineda et al. 2012). Thus, rhizobacteria influence direct defence priming aboveground in plants similarly to beneficial fungi, but research is still needed to confirm if fungi and rhizobacteria use the same mechanisms.

5.5.2 Context Dependency in Mechanisms of Fungal-Mediated Above–Belowground Interactions

The impact of belowground plant–fungal interactions on aboveground insects is highly context dependent, because plant responses to most abiotic and biotic factors are regulated to some extent by phytohormone signalling (Pozo et al. 2015). Plants can also often regulate their interaction with microbes using phytohormone signalling (Lakshmanan et al. 2014). For example, nearly all phytohormones studied to date appear to play a role in AM fungal formation and/or functioning (Boivin et al. 2016; Gutjahr 2014; Zwanenburg et al. 2016). Since phytohormones mediate plant responses to both environmental changes and the mycorrhizal symbiosis, the impact of the AM fungal symbiosis on plant–insect interactions will vary with environmental conditions. Thus, the final outcomes of fungal-mediated above–belowground interactions are highly context dependent (Pineda et al. 2013; Pozo et al. 2015).

Another source of context dependency is plant species or genotypic variation, as the degree of fungal priming in individual plant species or genotypes may depend on plant genetic variation in the JA pathway, the accumulation of related forms, or the degree of upregulation of JA biosynthetic or response genes. Testing for genetic variation in immunity priming will require the development of molecular markers associated with priming. Molecular markers associated to priming have been characterized in model plants (Conrath et al. 2015; Mauch-Mani et al. 2017) leading the way for equivalent markers to be developed in a wider array of plant species in the future.

Insect herbivores can also influence soil fungal communities through changes in carbon allocation (e.g. Frost and Hunter 2004), altering photosynthesis levels, or root exudation (Bardgett et al. 1998) which may combine with plant defence signalling to affect mycorrhizal colonization—but to date no one has demonstrated whether these changes influence plant defence or whether priming influences are irreversible following fungal colonization. Aboveground herbivory alters the allocation of carbon compounds to roots (Orians et al. 2011), and the activation of defence-related gene expression and accumulation of defence metabolites in roots has been reported within minutes of aboveground herbivory: the increase of jasmonate-related gene expression in roots has been shown to occur within 24 h after aboveground herbivores begin feeding on plants (Fragoso et al. 2014). Thus, aboveground herbivory effects on fungi depend on the interplay of opposite forces: increased allocation of carbon compounds should benefit soil fungal communities but defoliation reduces plant photosynthates (and, therefore, fungal symbiont resources) and activation of hormonal signalling pathways may also affect fungal endosymbionts. The influence of hormonal signalling will likely depend on the pathways activated (as each pathway has a different regulatory role on the colonization of roots by fungi) and on the intensity of herbivory. For example, abscisic acid (ABA) and JA signalling, commonly activated by chewing insects, are positive regulators of the AM fungal–plant symbiosis (Pozo et al. 2015). These signalling pathways are also activated in response to abiotic stress (reviewed in Meena et al. 2017), so that environmental conditions will also influence AM fungal root colonization (Pozo et al. 2015). However, no study has tested whether a negative impact of herbivory on fungal associations leads to changes in phytohormone signalling in attacked plants.

Thus, while there are still big gaps in our understanding, the central role of phytohormones is well established. Most of the mechanisms behind fungal-mediated above–belowground interactions are regulated by phytohormones, so that crosstalk in signalling in response to different cues explains much of the context dependency of interactions, and there are common patterns that may help explain fungal and plant responses to insects. Thus, understanding of the mechanisms behind fungal-mediated above–belowground interactions is still in its infancy, but, based on recent discoveries, is likely to develop rapidly in the future.

5.6 Applications

Here we explore if fungal-mediated above- and belowground interactions could be used as novel biotechnologies to improve agricultural production (Orrell and Bennett 2013), restore natural and degraded habitats, promote ecosystem services, and mitigate climate change impacts. We consider how these interactions can be utilized and controlled to promote positive effects. Potentially, within agro-ecosystems, these interactions could improve crop yields, reduce pest and pathogen effects, and improve pollination services, while in restoration they could improve the speed and stability of the restoration efforts and promote associated ecosystem services. Fungi can have both positive and negative impacts on aboveground systems. However, by altering the microbial species (Bennett et al. 2009; Gange et al. 2003; Hu et al. 2010) and genotype (Angelard et al. 2010) composition and/or changes in environmental management (Fitter 1991; Johnson et al. 2015), we can identify specific interactions with positive effects in a range of systems. Currently, direct evidence is largely unavailable for real world applications of fungal-mediated above–belowground interactions; however, here we propose a range of applications and review the technology that will make these applications possible (Fig. 5.2).

The easiest means of promoting fungal-mediated above–belowground interactions is adding fungal inocula to managed systems. In agricultural environments, fungal inocula can influence both the quantity and quality of crop yields (e.g. Ceballos et al. 2013) and could promote plant–pollinator interactions (reviewed in Barber and Gorden 2015), as well as promote defence against herbivores (Gehring and Bennett 2009; Pineda et al. 2013) and pathogens (reviewed in Borowicz 2001). However, current fungal inocula on the market are not typically designed for specific crops, growing systems, or environments (Middleton et al. 2015) nor do these inocula provide management advice to promote interactions or evidence of their effectiveness. Thus, for successful uptake of these inocula products, we need to collect data on their performance in commercial agriculture (e.g. Ceballos et al. 2013).

These same ecosystem services (nutrient delivery, pest and pathogen defence, and pollination) are also important in natural systems. For example, soil fungal community diversity and plant–soil feedbacks are hypothesized to be needed for rare plant establishment and persistence (Bothe et al. 2010). As a result, fungi are also important for restoration (Bever et al. 2003; McCain et al. 2011; Middleton et al. 2010), and invasion by non-native plant species (Suding et al. 2013). Thus, fungal inocula are increasingly promoted in restoration (Jayachandran and Fisher 2008; Quoreshi 2008) and are being applied through methods such as hydroseeding (e.g. Vogelsang and Bever 2010) or via fertilizer spreaders.

Promoting plant–fungal interactions should cascade through to aboveground plant interactions with insect communities (Bennett 2010). For example, native plant species promote the restoration of native aboveground plant–pollinator interaction networks (Menz et al. 2011; Morandin and Kremen 2013), and beneficial fungi can promote plant pollinator interactions through improvements in floral rewards, plant architecture, floral display, and volatile release (reviewed in Barber and Gorden 2015). Soil fungal communities may also influence pollination through changes in plant community composition (Cahill et al. 2008) which also alters the availability of floral rewards, plant architecture, floral display, and volatile release. Plant–fungal interactions also likely influence other ecosystem services such as attraction of herbivore enemies (Rasmann et al. 2017). Thus, manipulating soil fungal diversity in restoration by adding soil inocula (Bever et al. 2003; Middleton

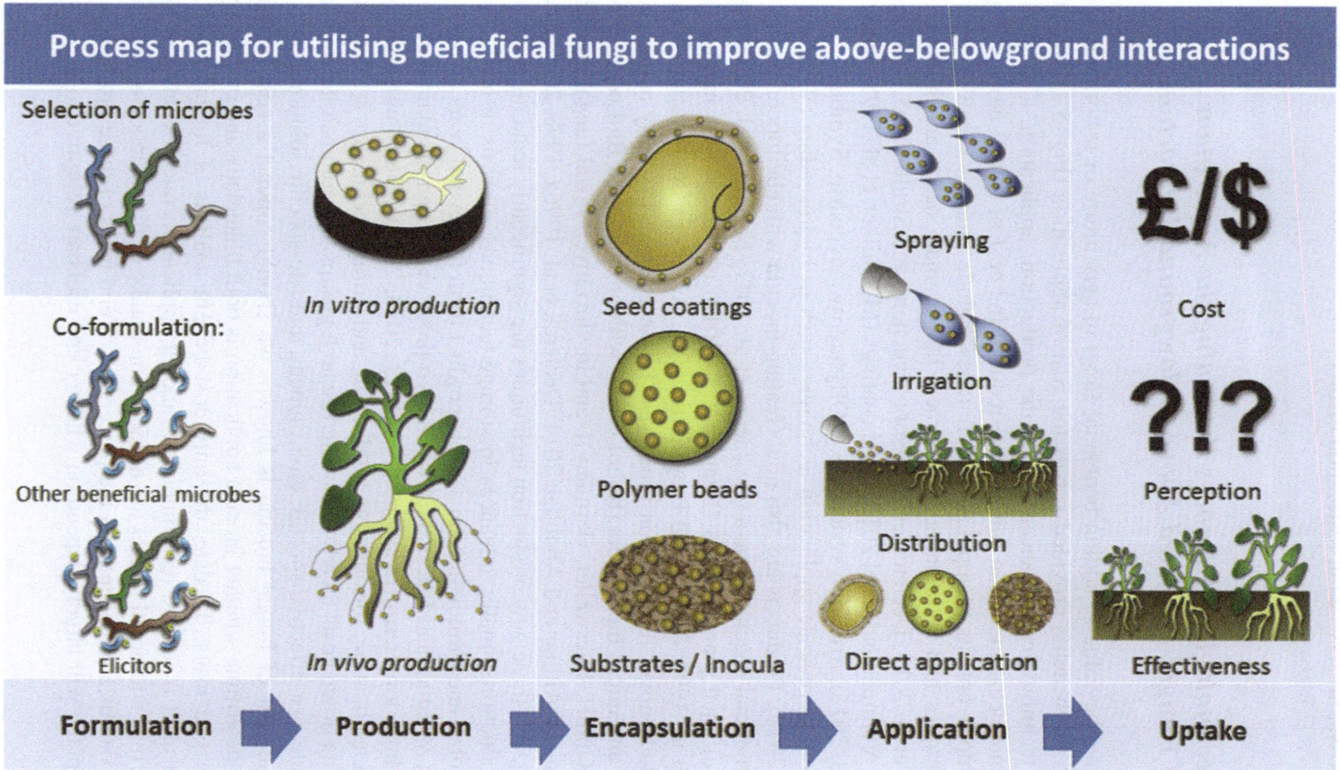

Fig. 5.2 A process map for the utilization of beneficial fungi to promote above–belowground interactions in managed systems. This map demonstrates that there are five steps to successfully applying fungi in managed systems: formulation (picking the appropriate fungi for use), production of fungi, encapsulation of fungi to allow easy application, application of fungi, and uptake of fungal inocula by end users

et al. 2010) or fungicide (McCain et al. 2011) can cascade up the trophic ladder to promote ecosystem services in restoration efforts.

5.6.1 Technological Advances Promoting the Application of Fungal-Mediated Above–Belowground Interactions

To determine the value of adding beneficial fungi in agriculture or restoration, we must weigh the benefits provided versus the cost of application (Fig. 5.2). Recent significant technological advancements in the production, application, and use of beneficial microbes (including fungi) have reduced the cost of fungal inocula. For example, improvements in application techniques include multiple novel encapsulation techniques and entrapment in polymers like alginate beads and hydrogels (De Jaeger et al. 2011; Declerck et al. 1996b; Plenchette and Strullu 2003; Vassilev et al. 2005; Vemmer and Patel 2013), seed coatings (McQuilken et al. 1998), sprays (Lohse et al. 2015), irrigation systems (e.g. Offyougrow Mycodrip, Symbiom s.r.o., Czechia), and large-scale application techniques such as hydroseeding (Vogelsang and Bever 2010). Producers are also now creating inocula with elicitors or multiple microbe beneficial species or taxa (De Jaeger et al. 2011; Nzanza et al. 2012) that provide additional plant benefits (Nzanza et al. 2012) or promote the interaction (Garbaye et al. 1992; Tarkka and Frey-Klett 2008). Advances in production techniques include novel fermentation processes (Lohse et al. 2015), fully automated large-scale and micro-bioreactors [RoboLector® with an integrated BioLector® Pro (m2p-labs GmbH, Baesweiler, Germany)], and in vitro mycorrhizal fungal production techniques (Akhtar and Abdullah 2014; Declerck et al. 1996a, 1998, 2005; Ijdo et al. 2011), and these new production techniques are significantly reducing costs of inocula production. Finally, advances in storage techniques allow us to now store target microbes and formulations in a stable product form for several years (Vassilev et al. 2005; Vemmer and Patel 2013) while cryo-preservation can maintain individual fungal lines long term (Lalaymia et al. 2012, 2013, 2014). The combination of these novel application, formulation, production, and storage techniques are dramatically driving down costs thereby allowing research groups and commercial companies to create cost-effective species, environment, and site-specific solutions.

Once the cost of fungal inocula is known, we should conduct a cost–benefit analysis to determine whether inocula application is both financially and practically feasible for end users. To date, the only cost–benefit analysis of fungal inocula application used twice the recommended inocula application rate (Ceballos et al. 2013), and thus additional studies are needed. The uptake of fungal inocula usage in managed systems also depends on the perception of end users of these products, as practical uptake will largely depend on whether end users are receptive to this new technology.

5.6.2 *Overcoming Limitations to the Application of Fungal-Mediated Above–Belowground Interactions*

We also need to test the application of these inocula for different plant species, habitats, and stresses to determine their reliability in the face of mitigating factors such as climate change, alterations in management practices, or invasive species. In addition, we can use data from these tests to provide information for land managers on how to manage their systems to promote these interactions [e.g. through a reduction in nutrient inputs (Johnson et al. 2015), and careful application of agro-chemicals (reviewed in Baar 2008)].

Competition between fungi and other microbes in the rhizosphere may also suppress the potential benefits provided by fungal inocula. For example, above-ground endophytes that promote herbivore defence have been shown to be negatively correlated with AM fungi in roots (Mack and Rudgers 2008; Omacini et al. 2006) and surrounding soils (Chuchou et al. 1992). The PGPF *Trichoderma harzianum* has been shown to mycoparasitize AM fungi, decreasing AM fungal hyphal viability (De Jaeger et al. 2010). More competitive AM fungi can also compete for root space to the detriment of plant growth (Bennett and Bever 2009) and potentially herbivore defence (Bennett et al. 2009). Thus, no one inoculum is likely to work in every environment due to these interactions, so a suite of products will need to be developed for a range of environments with different competitors.

Currently, there is little research on the use of fungal inocula to promote fungal-mediated above–belowground interactions, cost–benefit analyses of inocula use, or the potential uptake of these inocula by and perception of end users. Thus, further research is needed on potential fungal inocula to develop evidence-based advice on their usage. Further advances in production, storage, and application techniques are also needed to deliver cost-effective solutions that can be applied to a wide range of situations. We also recommend breeding crop species to further take advantage of these interactions and using genetic markers to match combinations of fungi and plant species that result in positive interactions. While agricultural intensification, natural habitat loss and degradation, land use change, and climate change will not be remedied by fungal-mediated above–belowground fungal interactions alone, promotion of fungal-mediated above–belowground interactions could form part of a vital toolset of biotechnologies to mitigate these pressures and promote sustainable agricultural intensification, restoration of natural habitats, and mitigation of climate change.

Thus, our knowledge of fungal-mediated above–belowground interactions is expanding revealing large areas still in need of exploration (communities, stability, evolution, and mechanisms), but also identifying application of these interactions in managed systems.

Acknowledgements We would like to acknowledge the financial support of Scottish Government Rural and Environment Science and Analytical Services Division 2016–2020 Work packages 1.3.1 and 2.3.8, and COST Action FA1405. In addition, we would like to thank Jenny Slater for contributing to early discussions on the chapter and Alison Karley and Philip Smith for comments that improved the chapter.

References

Abbas M, Klein AM, Ebeling A et al (2014) Plant diversity effects on pollinating and herbivorous insects can be linked to plant stoichiometry. Basic Appl Ecol 15:169–178. https://doi.org/10.1016/j.baae.2014.02.001

Akhtar MS, Abdullah SNA (2014) Mass production techniques of arbuscular mycorrhizal fungi: major advantages and disadvantages: a review. Biosci Biotechnol Res Asia 11:1199–1204

Angelard C, Colard A, Niculita-Hirzel H et al (2010) Segregation in a mycorrhizal fungus alters rice growth and symbiosis-specific gene transcription. Curr Biol 20:1216–1221. https://doi.org/10.1016/j.cub.2010.05.031

Baar J (2008) From production to application of arbuscular mycorrhizal fungi in agricultural systems: requirements and needs. In: Varma A (ed) Mycorrhiza: state of the art, genetics and molecular biology, eco-function, biotechnology, eco-physiology, structure and systematics. Springer, Berlin, pp 361–373. https://doi.org/10.1007/978-3-540-78826-3_18

Babikova Z, Gilbert L, Bruce TJA et al (2013) Underground signals carried through common mycelial networks warn neighbouring plants of aphid attack. Ecol Lett 16:835–843. https://doi.org/10.1111/ele.12115

Babikova Z, Gilbert L, Bruce T et al (2014a) Arbuscular mycorrhizal fungi and aphids interact by changing host plant quality and volatile emission. Funct Ecol 28:375–385

Babikova Z, Gilbert L, Randall KC et al (2014b) Increasing phosphorus supply is not the mechanism by which arbuscular mycorrhiza increase attractiveness of bean (*Vicia faba*) to aphids. J Exp Bot 65:5231–5241. https://doi.org/10.1093/jxb/eru283

Bandoly M, Grichnik R, Hilker M et al (2016) Priming of anti-herbivore defence in *Nicotiana attenuata* by insect oviposition: herbivore-specific effects. Plant Cell Environ 39:848–859. https://doi.org/10.1111/pce.12677

Barber NA, Gorden NLS (2015) How do belowground organisms influence plant-pollinator interactions? J Plant Ecol 8:1–11. https://doi.org/10.1093/jpe/rtu012

Bardgett RD, Wardle DA, Yeates GW (1998) Linking above-ground and below-ground interactions: how plant responses to foliar herbivory influence soil organisms. Soil Biol Biochem 30:1867–1878. https://doi.org/10.1016/s0038-0717(98)00069-8

Barrett G, Campbell CD, Hodge A (2014) The direct response of the external mycelium of arbuscular mycorrhizal fungi to temperature and the implications for nutrient transfer. Soil Biol Biochem 78:109–117

Bennett AE (2010) The role of soil community biodiversity in insect biodiversity. Insect Conserv Divers 3:157–171

Bennett AE (2013) Can plant–microbe–insect interactions enhance or inhibit the spread of invasive species? Funct Ecol 27:661–671

Bennett AE, Bever JD (2009) Trade-offs between arbuscular mycorrhizal fungal competitive ability and host growth promotion in *Plantago lanceolata*. Oecologia 160:807–816

Bennett AE, Alers-Garcia J, Bever JD (2006) Three-way interactions among mutualistic mycorrhizal fungi, plants, and plant enemies: hypotheses and synthesis. Am Nat 167:141–152

Bennett AE, Bever JD, Bowers MD (2009) Arbuscular mycorrhizal fungal species suppress inducible plant responses and alter defensive strategies following herbivory. Oecologia 160:771–779

Bennett AE, Millar NS, Gedrovics E et al (2016) Plant and insect microbial symbionts alter the outcome of plant-herbivore-parasitoid interactions: implications for invaded, agricultural and natural systems. J Ecol 104:1734–1744. https://doi.org/10.1111/1365-2745.12620

Bever JD, Schultz PA, Miller RM et al (2003) Prairie mycorrhizal fungi inoculant may increase native plant diversity on restored sites (Illinois). Ecol Restor 21:311–312

Bezemer TM, De Deyn GB, Bossinga TM et al (2005) Soil community composition drives aboveground plant-herbivore-parasitoid interactions. Ecol Lett 8:652–661. https://doi.org/10.1111/j1461-0248.2005.00762.x

Bishop J, Jones HE, Lukac M et al (2016) Insect pollination reduces yield loss following heat stress in faba bean (*Vicia faba* L.). Agric Ecosyst Environ 220:89–96. https://doi.org/10.1016/j.agee.2015.12.007

Bloor JMG, Bardgett RD (2012) Stability of above-ground and below-ground processes to extreme drought in model grassland ecosystems: interactions with plant species diversity and soil nitrogen availability. Perspect Plant Ecol 14:193–204. https://doi.org/10.1016/j.ppees.2011.12.001

Boivin S, Fonouni-Farde C, Frugier F (2016) How auxin and cytokinin phytohormones modulate root microbe interactions. Front Plant Sci 7. https://doi.org/10.3389/fpls.2016.01240

Bonte D, De Roissart A, Vandegehuchte ML et al (2010) Local adaptation of aboveground herbivores towards plant phenotypes induced by soil biota. PLoS One 5:e11174. https://doi.org/10.1371/journal.pone.0011174

Borowicz VA (2001) Do arbuscular mycorrhizal fungi alter plant-pathogen relations? Ecology 82:3057–3068

Borowicz VA (2013) The impact of arbuscular mycorrhizal fungi on plant growth following herbivory: a search for pattern. Acta Oecol 52:1–9. https://doi.org/10.1016/j.actao.2013.06.004

Bothe H, Turnau K, Regvar M (2010) The potential role of arbuscular mycorrhizal fungi in protecting endangered plants and habitats. Mycorrhiza 20:445–457. https://doi.org/10.1007/s00572-010-0332-4

Boullis A, Francis F, Verheggen FJ (2015) Climate change and tritrophic interactions: will modifications to greenhouse gas emissions increase the vulnerability of herbivorous insects to natural enemies? Environ Entomol 44:277–286. https://doi.org/10.1093/ee/nvu019

Brandle M, Amarell U, Auge H et al (2001) Plant and insect diversity along a pollution gradient: understanding species richness across trophic levels. Biodivers Conserv 10:1497–1511. https://doi.org/10.1023/a:1011815325503

Bunn R, Lekberg Y, Zabinski C (2009) Arbuscular mycorrhizal fungi ameliorate temperature stress in thermophilic plants. Ecology 90:1378–1388. https://doi.org/10.1890/07-2080.1

Cahill JFJ, Elle E, Smith GR et al (2008) Disruption of a belowground mutualism alters interactions between plants and their floral visitors. Ecology 89:1791–1801

Ceballos I, Ruiz M, Fernández C et al (2013) The *in vitro* mass-produced model mycorrhizal fungus, *Rhizophagus irregularis*, significantly increases yields of the globally important food security crop Cassava. PLoS One 8:e70633. https://doi.org/10.1371/journal.pone.0070633

Choi SW (2015) Bottom-up impact of soils on the network of soil, plants, and moths (Lepidoptera) in a South Korean temperate forest. Can Entomol 147:405–418. https://doi.org/10.4039/tce.2014.57

Chuang A, Peterson CR (2016) Expanding population edges: theories, traits, and trade-offs. Glob Change Biol 22:494–512. https://doi.org/10.1111/gcb.13107

Chu-Chou M, Guo B, An Z-Q et al (1992) Suppression of mycorrhizal fungi in fescue by the *Acremonium coenophialum* endophyte. Soil Biol Biochem 24:633–637. https://doi.org/10.1016/0038-0717(92)90041-u

Clay K (1988) Fungal endophytes of grasses: a defensive mutualism between plants and fungi. Ecology 69:10–16

Clay K, Holah J (1999) Fungal endophyte symbiosis and plant diversity in successional fields. Science 285:1742–1744

Clay K, Schardl C (2002) Evolutionary origins and ecological consequences of endophyte symbiosis with grasses. Am Nat 160:S99–S127

Clay K, Holah J, Rudgers JA (2005) Herbivores cause a rapid increase in hereditary symbiosis and alter plant community composition. Proc Natl Acad Sci USA 102:12465–12470. https://doi.org/10.1073/pnas.0503059102

Conrath U, Beckers GJM, Langenbach CJG et al (2015) Priming for enhanced defense. Annu Rev Phytopathol 53:97–119. https://doi.org/10.1146/annurev-phyto-080614-120132

Cooper EJ (2014) Warmer shorter winters disrupt arctic terrestrial ecosystems. Annu Rev Ecol Evol Syst 45:271–295. https://doi.org/10.1146/annurev-ecolsys-120213-091620

Corbin C, Heyworth ER, Ferrari J et al (2017) Heritable symbionts in a world of varying temperature. Heredity 118:10–20. https://doi.org/10.1038/hdy.2016.71

Crowther TW, Maynard DS, Crowther TR et al (2014) Untangling the fungal niche: the trait-based approach. Front Microbiol 5:579. https://doi.org/10.3389/fmicb.2014.00579

De Jaeger N, Declerck S, de la Providencia IE (2010) Mycoparasitism of arbuscular mycorrhizal fungi: a pathway for the entry of saprotrophic fungi into roots. FEMS Microbiol Ecol 73:312–322. https://doi.org/10.1111/j.1574-6941.2010.00903.x

De Jaeger N, de la Providencia IE, Rouhier H et al (2011) Co-entrapment of *Trichoderma harzianum* and *Glomus* sp within alginate beads: impact on the arbuscular mycorrhizal fungi life cycle. J Appl Microbiol 111:125–135. https://doi.org/10.1111/j.1365-2672.2011.05035.x

de Sassi C, Tylianakis JM (2012) Climate change disproportionately increases herbivore over plant or parasitoid biomass. PLoS One 7. https://doi.org/10.1371/journal.pone.0040557

Declerck S, Strullu DG, Plenchette C (1996a) In vitro mass-production of the arbuscular mycorrhizal fungus, *Glomus versiforme*, associated with Ri T-DNA transformed carrot roots. Mycol Res 100:1237–1242

Declerck S, Strullu DG, Plenchette C et al (1996b) Entrapment of in vitro produced spores of *Glomus versiforme* in alginate beads: in vitro and in vivo inoculum potentials. J Biotechnol 48:51–57. https://doi.org/10.1016/0168-1656(96)01396-x

Declerck S, Strullu DG, Plenchette C (1998) Monoxenic culture of the intraradical forms of *Glomus* sp. isolated from a tropical ecosystem: a proposed methodology for germplasm collection. Mycologia 90:579–585. https://doi.org/10.2307/3761216

Declerck S, Strullu D-G, Fortin A (eds) (2005) In vitro culture of mycorrhizas. Springer, Berlin. https://doi.org/10.1007/b138925

Dickie IA, Nunez MA, Pringle A et al (2016) Towards management of invasive ectomycorrhizal fungi. Biol Invasions 18:3383–3395. https://doi.org/10.1007/s10530-016-1243-x

Diehl E, Sereda E, Wolters V et al (2013) Effects of predator specialization, host plant and climate on biological control of aphids by natural enemies: a meta-analysis. J Appl Ecol 50:262–270. https://doi.org/10.1111/1365-2664.12032

Donohue I, Hillebrand H, Montoya JM et al (2016) Navigating the complexity of ecological stability. Ecol Lett 19:1172–1185. https://doi.org/10.1111/ele.12648

Eisenhauer N, Milcu A, Allan E et al (2011) Impact of above- and below-ground invertebrates on temporal and spatial stability of grassland of different diversity. J Ecol 99:572–582. https://doi.org/10.1111/j.1365-2745.2010.01783.x

Eschen R, Hunt S, Mykura C et al (2010) The foliar endophytic fungal community composition in *Cirsium arvense* is affected by mycorrhizal colonization and soil nutrient content. Fungal Biol 114:991–998. https://doi.org/10.1016/j.funbio.2010.09.009

Finkes LK, Cady AB, Mulroy JC et al (2006) Plant-fungus mutualism affects spider composition in successional fields. Ecol Lett 9:344–353

Fitter A (1991) Costs and benefits of mycorrhizas: implications for functioning under natural conditions. Cell Mol Life Sci 47:350–355

Fontana A, Reichelt M, Hempel S et al (2009) The effects of arbuscular mycorrhizal fungi on direct and indirect defense metabolites of *Plantago lanceolata* L. J Chem Ecol 35:833–843. https://doi.org/10.1007/s10886-009-9654-0

Fragoso V, Rothe E, Baldwin IT et al (2014) Root jasmonic acid synthesis and perception regulate folivore-induced shoot metabolites and increase *Nicotiana attenuata* resistance. New Phytol 202:1335–1345. https://doi.org/10.1111/nph.12747

Frost CJ, Hunter MD (2004) Insect canopy herbivory and frass deposition affect soil nutrient dynamics and export in oak mesocosms. Ecology 85:3335–3347. https://doi.org/10.1890/04-0003

Frost CJ, Mescher MC, Carlson JE et al (2008) Plant defense priming against herbivores: getting ready for a different battle. Plant Physiol 146:818–824. https://doi.org/10.1104/pp.107.113027

Gange AC, Brown VK, Aplin DM (2003) Multitrophic links between arbuscular mycorrhizal fungi and insect parasitoids. Ecol Lett 6:1051–1055

Garbaye J, Churin JL, Duponnois R (1992) Effects of substrate sterilization, fungicide treatment, and mycorrhization helper bacteria on ectomycorrhizal formation of pedunculate oak (*Quercus robur*) inoculated with *Laccaria laccata* in two peat bare-root nurseries. Biol Fertil Soils 13:55–57. https://doi.org/10.1007/bf00337239

Gehring C, Bennett A (2009) Mycorrhizal fungal–plant–insect interactions: the importance of a community approach. Environ Entomol 38:93–102

Gehring CA, Whitham TG (2002) Mycorrhizae-herbivore interactions: population and community consequences. In: van der Heijden MGA, Sanders IR (eds) Mycorrhizal ecology, Ecological studies, vol 157. Springer, Berlin, pp 295–320

Gehring CA, Mueller RC, Haskins KE et al (2014) Convergence in mycorrhizal fungal communities due to drought, plant competition, parasitism, and susceptibility to herbivory: consequences for fungi and host plants. Front Microbiol 5:306. https://doi.org/10.3389/fmicb.2014.00306

Genre A, Russo G (2016) Does a common pathway transduce symbiotic signals in plant-microbe interactions? Front Plant Sci 7. https://doi.org/10.3389/fpls.2016.00096

Gill SS, Gill R, Trivedi DK et al (2016) *Piriformospora indica*: potential and significance in plant stress tolerance. Front Microbiol 7. https://doi.org/10.3389/fmicb.2016.00332

GmbH mp-l (2017) RoboLector automated fermentation. http://www.m2p-labs.com/bioreactors/robolector/

Goh CH, Vallejos DFV, Nicotra AB et al (2013) The impact of beneficial plant-associated microbes on plant phenotypic plasticity. J Chem Ecol 39:826–839. https://doi.org/10.1007/s10886-013-0326-8

Guerrieri E, Lingua G, Digilio MC et al (2004) Do interactions between plant roots and the rhizosphere affect parasitoid behaviour? Ecol Entomol 29:753–756

Gutjahr C (2014) Phytohormone signaling in arbuscular mycorhiza development. Curr Opin Plant Biol 20:26–34. https://doi.org/10.1016/j.pbi.2014.04.003

Hackett SC, Karley AJ, Bennett AE (2013) Unpredicted impacts of insect endosymbionts on interactions between soil organisms, plants and aphids. Proc R Soc Lond B Biol Sci 280:20131275

Haddad NM, Haarstad J, Tilman D (2000) The effects of long-term nitrogen loading on grassland insect communities. Oecologia 124:73–84. https://doi.org/10.1007/s004420050026

Haloin JR, Strauss SY (2008) Interplay between ecological communities and evolution: Review of feedbacks from microevolutionary to macroevolutionary scales. Ann NY Acad Sci 1133:87–125

Hauser TP, Christensen S, Heimes C et al (2013) Combined effects of arthropod herbivores and phytopathogens on plant performance. Funct Ecol 27:623–632. https://doi.org/10.1111/1365-2435.12053

Heinemeyer A, Fitter AH (2004) Impact of temperature on the arbuscular mycorrhizal (AM) symbiosis: growth responses of the host plant and its AM fungal partner. J Exp Bot 55:525–534

Hempel S, Stein C, Unsicker SB et al (2009) Specific bottom-up effects of arbuscular mycorrhizal fungi across a plant-herbivore-parasitoid system. Oecologia 160:267–277. https://doi.org/10.1007/s00442-009-1294-0

Hentley WT, Wade RN (2017) Global change, herbivores and their natural enemies. In: Johnson SN, Jones TH (eds) Global climate change and terrestrial invertebrates. Wiley, Chichester, pp 177–200

Hilker M, Fatouros NE (2015) Plant responses to insect egg deposition. Annu Rev Entomol 60:493–515. https://doi.org/10.1146/annurev-ento-010814-020620
Hoffmann D, Vierheilig H, Schausberger P (2011a) Arbuscular mycorrhiza enhances preference of ovipositing predatory mites for direct prey-related cues. Physiol Entomol 36:90–95. https://doi.org/10.1111/j.1365-3032.2010.00751.x
Hoffmann D, Vierheilig H, Schausberger P (2011b) Mycorrhiza-induced trophic cascade enhances fitness and population growth of an acarine predator. Oecologia 166:141–149. https://doi.org/10.1007/s00442-010-1821-z
Howe GA, Jander G (2008) Plant immunity to insect herbivores. Annu Rev Plant Biol 59:41–66. https://doi.org/10.1146/annurev.arplant.59.032607.092825
Hu JL, Lin XG, Wang JH et al (2010) Arbuscular mycorrhizal fungal inoculation enhances suppression of cucumber fusarium wilt in greenhouse soils. Pedosphere 20:586–593
Hudewenz A, Klein A-M, Scherber C et al (2012) Herbivore and pollinator responses to grassland management intensity along experimental changes in plant species richness. Biol Conserv 150:42–52. https://doi.org/10.1016/j.biocon.2012.02.024
Ijdo M, Cranenbrouck S, Declerck S (2011) Methods for large-scale production of AM fungi: past, present, and future. Mycorrhiza 21:1–16
Inderjit, van der Putten WH (2010) Impacts of soil microbial communities on exotic plant invasions. Trends Ecol Evol 25:512–519. https://doi.org/10.1016/j.tree.2010.06.006
Jayachandran K, Fisher J (2008) Arbuscular mycorrhizae and their role in plant restoration in native ecosystems. In: Siddiqui ZA, Akhtar MS, Futai K (eds) Mycorrhizae: sustainable agriculture and forestry. Springer, Dordrecht, pp 195–209
Joern A, Laws AN (2013) Ecological mechanisms underlying arthropod species diversity in grasslands. Annu Rev Entomol 58:19–36. https://doi.org/10.1146/annurev-ento-120811-153540
Johnson D, Gilbert L (2015) Interplant signalling through hyphal networks. New Phytol 205:1448–1453. https://doi.org/10.1111/nph.13115
Johnson NC, Wilson GWT, Wilson JA et al (2015) Mycorrhizal phenotypes and the law of the minimum. New Phytol 205:1473–1484. https://doi.org/10.1111/nph.13172
Jung SC, Martínez-Medina A, Lopez-Raez JA et al (2012) Mycorrhiza-induced resistance and priming of plant defenses. J Chem Ecol 38:651–664. https://doi.org/10.1007/s10886-012-0134-6
Kempel A, Schmidt AK, Brandl R et al (2010) Support from the underground: induced plant resistance depends on arbuscular mycorrhizal fungi. Funct Ecol 24:293–300. https://doi.org/10.1111/j.1365-2435.2009.01647.x
Kempel A, Nater P, Fischer M et al (2013) Plant-microbe-herbivore interactions in invasive and non-invasive alien plant species. Funct Ecol 27:498–508. https://doi.org/10.1111/1365-2435.12056
Kershaw HM, Mallik AU (2013) Predicting plant diversity response to disturbance: applicability of the intermediate disturbance hypothesis and mass ratio hypothesis. Crit Rev Plant Sci 32:383–395. https://doi.org/10.1080/07352689.2013.791501
Khan MIR, Fatma M, Per TS et al (2015) Salicylic acid-induced abiotic stress tolerance and underlying mechanisms in plants. Front Plant Sci 6:462. https://doi.org/10.3389/fpls.2015.00462
Kim J, Felton GW (2013) Priming of antiherbivore defensive responses in plants. Insect Sci 20:273–285. https://doi.org/10.1111/j.1744-7917.2012.01584.x
Kivlin SN, Emery SM, Rudgers JA (2013) Fungal symbionts alter plant responses to global change. Am J Bot 100:1445–1457. https://doi.org/10.3732/ajb.1200558
Klironomos JN, Hart MM, Gurney JE et al (2001) Interspecific differences in the tolerance of arbuscular mycorrhizal fungi to freezing and drying. Can J Bot 79:1161–1166
Koricheva J, Gange AC, Jones T (2009) Effects of mycorrhizal fungi on insect herbivores: a meta-analysis. Ecology 90:2088–2097
Kos M, Tuijl MAB, de Roo J et al (2015) Species-specific plant-soil feedback effects on above-ground plant-insect interactions. J Ecol 103:904–914. https://doi.org/10.1111/1365-2745.12402

Kowalchuk GA, Buma DS, de Boer W et al (2002) Effects of above-ground plant species composition and diversity on the diversity of soil-borne microorganisms. Anton Leeuw Int J G 81:509–520

Kulmatiski A, Beard KH, Stevens JR et al (2008) Plant–soil feedbacks: a meta–analytical review. Ecol Lett 11:980–992. https://doi.org/10.1111/j.1461-0248.2008.01209.x

Lakshmanan V, Selvaraj G, Bais HP (2014) Functional soil microbiome: belowground solutions to an aboveground problem. Plant Physiol 166:689–700. https://doi.org/10.1104/pp.114.245811

Lalaymia I, Cranenbrouck S, Draye X et al (2012) Preservation at ultra-low temperature of in vitro cultured arbuscular mycorrhizal fungi via encapsulation-drying. Fungal Biol 116:1032–1041. https://doi.org/10.1016/j.funbio.2012.07.007

Lalaymia I, Declerck S, Cranenbrouck S (2013) Cryopreservation of in vitro-produced *Rhizophagus* species has minor effects on their morphology, physiology, and genetic stability. Mycorrhiza 23:675–682. https://doi.org/10.1007/s00572-013-0506-y

Lalaymia I, Cranenbrouck S, Declerck S (2014) Maintenance and preservation of ectomycorrhizal and arbuscular mycorrhizal fungi. Mycorrhiza 24:323–337. https://doi.org/10.1007/s00572-013-0541-8

Lau JA, Lennon JT (2012) Rapid responses of soil microorganisms improve plant fitness in novel environments. Proc Natl Acad Sci USA 109:14058–14062. https://doi.org/10.1073/pnas.1202319109

Lemons A, Clay K, Rudgers JA (2005) Connecting plant-microbial interactions above and belowground: a fungal endophyte affects decomposition. Oecologia 145:595–604. https://doi.org/10.1007/s00442-005-0163-8

Lenoir I, Fontaine J, Sahraoui ALH (2016) Arbuscular mycorrhizal fungal responses to abiotic stresses: a review. Phytochemistry 123:4–15. https://doi.org/10.1016/j.phytochem.2016.01.002

Levine JM, Adler PB, Yelenik SG (2004) A meta-analysis of biotic resistance to exotic plant invasions. Ecol Lett 7:975–989. https://doi.org/10.1111/j.1461-0248.2004.00657.x

Liman AS, Dalin P, Bjorkman C (2017) Enhanced leaf nitrogen status stabilizes omnivore population density. Oecologia 183:57–65. https://doi.org/10.1007/s00442-016-3742-y

Litt AR, Cord EE, Fulbright TE et al (2014) Effects of invasive plants on arthropods. Conserv Biol 28:1532–1549. https://doi.org/10.1111/cobi.12350

Lohse R, Jakobs-Schonwandt D, Vidal S et al (2015) Evaluation of new fermentation and formulation strategies for a high endophytic establishment of *Beauveria bassiana* in oilseed rape plants. Biol Control 88:26–36. https://doi.org/10.1016/j.biocontrol.2015.05.002

Ludwig-Müller J (2015) Plants and endophytes: equal partners in secondary metabolite production? Biotechnol Lett 37:1325–1334. https://doi.org/10.1007/s10529-015-1814-4

Lugtenberg BJJ, Caradus JR, Johnson LJ (2016) Fungal endophytes for sustainable crop production. FEMS Microbiol Ecol 92:fiw194. https://doi.org/10.1093/femsec/fiw194

Mack KML, Rudgers JA (2008) Balancing multiple mutualists: asymmetric interactions among plants, arbuscular mycorrhizal fungi, and fungal endophytes. Oikos 117:310–320. https://doi.org/10.1111/j.2007.0030-1299.15973.x

Martínez-Medina A, Fernández I, Sánchez-Guzmán MJ et al (2013) Deciphering the hormonal signaling network behind the systemic resistance induced by *Trichoderma harzianum* in tomato. Front Plant Sci 4. https://doi.org/10.3389/fpls.2013.00206

Martínez-Medina A, Flors V, Heil M et al (2016) Recognizing plant defense priming. Trends Plant Sci 21:818–822. https://doi.org/10.1016/j.tplants.2016.07.009

Mauch-Mani B, Baccelli I, Luna E et al (2017) Defense priming: an adaptive part of induced resistance. Annu Rev Plant Biol 68:485–512. https://doi.org/10.1146/annurev-arplant-042916-041132

McCain KNS, Wilson GWT, Blair JM (2011) Mycorrhizal suppression alters plant productivity and forb establishment in a grass-dominated prairie restoration. Plant Ecol 212:1675–1685. https://doi.org/10.1007/s11258-011-9940-0

McCary MA, Mores R, Farfan MA et al (2016) Invasive plants have different effects on trophic structure of green and brown food webs in terrestrial ecosystems: a meta-analysis. Ecol Lett 19:328–335. https://doi.org/10.1111/ele.12562

McElroy MS, Papadopoulos YA, Adl MS (2012) Complexity and composition of pasture swards affect plant productivity and soil organisms. Can J Plant Sci 92:687–697. https://doi.org/10.4141/cjps2011-147

McQuilken MP, Halmer P, Rhodes D (1998) Application of microorganisms to seeds. In: Burges HD (ed) Formulation of microbial biopesticides. Springer, Dordrecht, pp 255–285

Meena KK, Sorty AM, Bitla UM et al (2017) Abiotic stress responses and microbe-mediated mitigation in plants: the omics strategies. Front Plant Sci 8:172. https://doi.org/10.3389/fpls.2017.00172

Menz MHM, Phillips RD, Winfree R et al (2011) Reconnecting plants and pollinators: challenges in the restoration of pollination mutualisms. Trends Plant Sci 16:4–12. https://doi.org/10.1016/j.tplants.2010.09.006

Menzel A, Hempel S, Klotz S et al (2017) Mycorrhizal status helps explain invasion success of alien plant species. Ecology 98:92–102. https://doi.org/10.1002/ecy.1621

Middleton EL, Bever JD, Schultz PA (2010) The effect of restoration methods on the quality of the restoration and resistance to invasion by exotics. Restor Ecol 18:181–187. https://doi.org/10.1111/j.1526-100X.2008.00501.x

Middleton EL, Richardson S, Koziol L et al (2015) Locally adapted arbuscular mycorrhizal fungi improve vigor and resistance to herbivory of native prairie plant species. Ecosphere 6:276. https://doi.org/10.1890/es15-00152.1

Milcu, Allan E, Roscher CA et al (2013) Functionally and phylogenetically diverse plant communities key to soil biota. Ecology 94:1878–1885. https://doi.org/10.1890/12-1936.1

Millar NS, Bennett AE (2016) Stressed out symbiotes: hypotheses for the influence of abiotic stress on arbuscular mycorrhizal fungi. Oecologia 182:625–641. https://doi.org/10.1007/s00442-016-3673-7

Miller-Rushing AJ, Hoye TT, Inouye DW et al (2010) The effects of phenological mismatches on demography. Philos Trans R Soc B 365:3177–3186. https://doi.org/10.1098/rstb.2010.0148

Morandin LA, Kremen C (2013) Hedgerow restoration promotes pollinator populations and exports native bees to adjacent fields. Ecol Appl 23:829–839

Morrien E, Engelkes T, van der Putten WH (2011) Additive effects of aboveground polyphagous herbivores and soil feedback in native and range-expanding exotic plants. Ecology 92:1344–1352. https://doi.org/10.1890/10-1937.1

Mulder C, de Zwart D (2003) Assessing fungal species sensitivity to environmental gradients by the Ellenberg indicator values of above-ground vegetation. Basic Appl Ecol 4:557–568. https://doi.org/10.1078/1439-1791-00141

Nzanza B, Marais D, Soundy P (2012) Yield and nutrient content of tomato (*Solanum lycopersicum* L.) as influenced by *Trichoderma harzianum* and *Glomus mosseae* inoculation. Sci Hortic 144:55–59. https://doi.org/10.1016/j.scienta.2012.06.005

Okada K, Abe H, Arimura G (2015) Jasmonates induce both defense responses and communication in Monocotyledonous and Dicotyledonous plants. Plant Cell Physiol 56:16–27. https://doi.org/10.1093/pcp/pcu158

Omacini M, Eggers T, Bonkowski M et al (2006) Leaf endophytes affect mycorrhizal status and growth of co-infected and neighbouring plants. Funct Ecol 20:226–232. https://doi.org/10.1111/j.1365-2435.2006.01099.x

Orians CM, Thorn A, Gómez S (2011) Herbivore-induced resource sequestration in plants: why bother? Oecologia 167:1–9. https://doi.org/10.1007/s00442-011-1968-2

Orrell P, Bennett AE (2013) How can we exploit above–belowground interactions to assist in addressing the challenges of food security? Front Plant Sci 4:1–11. https://doi.org/10.3389/fpls.2013.00432

Paillet Y, Bergès L, Hjältén J et al (2010) Biodiversity differences between managed and unmanaged forests: meta-analysis of species richness in Europe. Conserv Biol 24:101–112. https://doi.org/10.1111/j.1523-1739.2009.01399.x

Panke-Buisse K, Poole AC, Goodrich JK et al (2015) Selection on soil microbiomes reveals reproducible impacts on plant function. ISME J 9:980–989. https://doi.org/10.1038/ismej.2014.196

Parmesan C, Hanley ME (2015) Plants and climate change: complexities and surprises. Ann Bot 116:849–864. https://doi.org/10.1093/aob/mcv169

Partida-Martinez LP, Heil M (2011) The microbe-free plant: fact or artifact? Front Plant Sci 2:100. https://doi.org/10.3389/fpls.2011.00100

Piasecka A, Jedrzejczak-Rey N, Bednarek P (2015) Secondary metabolites in plant innate immunity: conserved function of divergent chemicals. New Phytol 206:948–964. https://doi.org/10.1111/nph.13325

Pieterse CMJ, Zamioudis C, Berendsen RL et al (2014) Induced systemic resistance by beneficial microbes. Annu Rev Phytopathol 52:347–375. https://doi.org/10.1146/annurev-phyto-082712-102340

Pimm SL (1984) The complexity and stability of ecosystems. Nature 307:321–326. https://doi.org/10.1038/307321a0

Pincebourde S, van Baaren J, Rasmann S et al (2017) Plant-insect interactions in a changing world. In: Sauvion N, Thiery D, Calatayud PA (eds) Insect-plant interactions in a crop protection perspective, Advances in botanical research, vol 81. Academic/Elsevier, London, pp 289–332. https://doi.org/10.1016/bs.abr.2016.09.009

Pineda A, Zheng SJ, van Loon JJA et al (2012) Rhizobacteria modify plant-aphid interactions: a case of induced systemic susceptibility. Plant Biol 14:83–90. https://doi.org/10.1111/j.1438-8677.2011.00549.x

Pineda A, Dicke M, Pieterse CMJ et al (2013) Beneficial microbes in a changing environment: are they always helping plants to deal with insects? Funct Ecol 27:574–586. https://doi.org/10.1111/1365-2435.12050

Plenchette C, Strullu DG (2003) Long-term viability and infectivity of intraradical forms of *Glomus intraradices* vesicles encapsulated in alginate beads. Mycol Res 107:614–616. https://doi.org/10.1017/s0953756203007482

Plett JM, Martin F (2015) Reconsidering mutualistic plant-fungal interactions through the lens of effector biology. Curr Opin Plant Biol 26:45–50. https://doi.org/10.1016/j.pbi.2015.06.001

Pozo MJ, Azcón-Aguilar C (2007) Unraveling mycorrhiza-induced resistance. Curr Opin Plant Biol 10:393–398

Pozo MJ, López-Ráez JA, Azcón-Aguilar C et al (2015) Phytohormones as integrators of environmental signals in the regulation of mycorrhizal symbioses. New Phytol 205:1431–1436. https://doi.org/10.1111/nph.13252

Pringle A, Bever JD, Gardes M et al (2009) Mycorrhizal symbioses and plant invasions. Annu Rev Ecol Evol Syst 40:699–715. https://doi.org/10.1146/annurev.ecolsys.39.110707.173454

Quoreshi AM (2008) The use of mycorrhizal biotechnology in restoration of disturbed ecosystem. In: Siddiqui ZA et al (eds) Mycorrhizae: sustainable agriculture and forestry. Springer, New York, pp 303–320

Rader R, Reilly J, Bartomeus I et al (2013) Native bees buffer the negative impact of climate warming on honey bee pollination of watermelon crops. Glob Change Biol 19:3103–3110. https://doi.org/10.1111/gcb.12264

Rasmann S, Bennett AE, Biere A et al (2017) Root symbionts: powerful drivers of plant above- and belowground indirect defences. Insect Sci. https://doi.org/10.1111/1744-7917.12464

Reinhart KO, Callaway RM (2006) Soil biota and invasive plants. New Phytol 170:445–457

Richardson DM, Allsopp N, D'Antonio CM et al (2000) Plant invasions—the role of mutualisms. Biol Rev 75:65–93

Rodríguez-Echeverría S, Traveset A (2015) Putative linkages between below- and aboveground mutualisms during alien plant invasions. AoB Plants 7. https://doi.org/10.1093/aobpla/plv062

Rout ME, Callaway RM (2012) Interactions between exotic invasive plants and soil microbes in the rhizosphere suggest that 'everything is not everywhere'. Ann Bot 110:213–222. https://doi.org/10.1093/aob/mcs061

Rudgers JA, Clay K (2008) An invasive plant–fungal mutualism reduces arthropod diversity. Ecol Lett 11:831–840. https://doi.org/10.1111/j.1461-0248.2008.01201.x

s.r.o. S (2012) Offyougrow mycodrip. Symbiom s.r.o. https://www.symbiom.cz/en/p-17-offyougrow-mycodrip/

Saikkonen K, Gundel PE, Helander M (2013) Chemical ecology mediated by fungal endophytes in grasses. J Chem Ecol 39:962–968. https://doi.org/10.1007/s10886-013-0310-3

Sanchez-Bel P, Troncho P, Gamir J et al (2016) The nitrogen availability interferes with mycorrhiza-induced resistance against *Botrytis cinerea* in tomato. Front Microbiol 7:1598. https://doi.org/10.3389/fmicb.2016.01598

Sanderson RA (1992) Diversity and evenness of Hemiptera communities on naturally vegetated derelict land in NW England. Ecography 15:154–160. https://doi.org/10.1111/j.1600-0587.1992.tb00018.x

Schausberger P, Peneder S, Jurschik S et al (2012) Mycorrhiza changes plant volatiles to attract spider mite enemies. Funct Ecol 26:441–449. https://doi.org/10.1111/j.1365-2435.2011.01947.x

Schweiger R, Müller C (2015) Leaf metabolome in arbuscular mycorrhizal symbiosis. Curr Opin Plant Biol 26:120–126. https://doi.org/10.1016/j.pbi.2015.06.009

Schweiger R, Baier MC, Persicke M et al (2014a) High specificity in plant leaf metabolic responses to arbuscular mycorrhiza. Nat Commun 5. https://doi.org/10.1038/ncomms4886

Schweiger R, Heise AM, Persicke M et al (2014b) Interactions between the jasmonic and salicylic acid pathway modulate the plant metabolome and affect herbivores of different feeding types. Plant Cell Environ 37:1574–1585. https://doi.org/10.1111/pce.12257

Selosse MA, Bessis A, Pozo MJ (2014) Microbial priming of plant and animal immunity: symbionts as developmental signals. Trends Microbiol 22:607–613. https://doi.org/10.1016/j.tim.2014.07.003

Sentis A, Hemptinne JL, Brodeur J (2013) Effects of simulated heat waves on an experimental plant-herbivore-predator food chain. Glob Change Biol 19:833–842. https://doi.org/10.1016/j.tim.2014.07.003

Sentis A, Ramon-Portugal F, Brodeur J et al (2015) The smell of change: warming affects species interactions mediated by chemical information. Glob Change Biol 21:3586–3594. https://doi.org/10.1111/gcb.12932

Shikano I, Rosa C, Tan CW et al (2017) Tritrophic interactions: microbe-mediated plant effects on insect herbivores. Annu Rev Phytopathol 55:313–331. https://doi.org/10.1146/annurev-phyto-080516-035319

Smith SE, Read DJ (2008) Mycorrhizal symbiosis, 3rd edn. Academic, New York

Song YY, Zeng RS, Xu JF, Li J et al (2010) Interplant communication of tomato plants through underground common mycorrhizal networks. PLoS One 5:e13324. https://doi.org/10.1371/journal.pone.0013324

Song YY, Ye M, Li C et al (2014) Hijacking common mycorrhizal networks for herbivore-induced defence signal transfer between tomato plants. Sci Rep 4. https://doi.org/10.1038/srep03915

Sonnemann I, Hempel S, Beutel M et al (2013) The root herbivore history of the soil affects the productivity of a grassland plant community and determines plant response to new root herbivore attack. PLoS One 8:e56524. https://doi.org/10.1371/journal.pone.0056524

Spohn M (2016) Element cycling as driven by stoichiometric homeostasis of soil microorganisms. Basic Appl Ecol 17:471–478. https://doi.org/10.1016/j.baae.2016.05.003

Suding KN, Stanley Harpole W, Fukami T et al (2013) Consequences of plant–soil feedbacks in invasion. J Ecol 101:298–308. https://doi.org/10.1111/1365-2745.12057

Tarkka MT, Frey-Klett P (2008) Mycorrhiza helper bacteria. In: Varma A (ed) Mycorrhiza. Springer, New York, pp 113–132

Thompson JN (1994) The coevolutionary process. University of Chicago Press, Chicago

Thompson JN (2005) The geographic mosaic of coevolution. University of Chicago Press, Chicago
Toju H, Guimarães PR, Olesen JM et al (2015) Below-ground plant–fungus network topology is not congruent with above-ground plant–animal network topology. Sci Adv 1:e1500291. https://doi.org/10.1126/sciadv.1500291
Tomczak VV, Müller C (2017) Influence of arbuscular mycorrhizal stage and plant age on the performance of a generalist aphid. J Insect Physiol 98:258–266. https://doi.org/10.1016/j.jinsphys.2017.01.016
Uroz S, Buee M, Deveau A et al (2016) Ecology of the forest microbiome: highlights of temperate and boreal ecosystems. Soil Biol Biochem 103:471–488. https://doi.org/10.1016/j.soilbio.2016.09.006
van Dam NM, Heil M (2011) Multitrophic interactions below and above ground: en route to the next level. J Ecol 99:77–88. https://doi.org/10.1111/j.1365-2745.2010.01761.x
van der Putten WH (2012) Climate change, aboveground-belowground interactions, and species' range shifts. Annu Rev Ecol Evol Syst 43:365–383. https://doi.org/10.1146/annurev-ecolsys-110411-160423
van der Putten WH, de Ruiter PC, Bezemer TM et al (2004) Trophic interactions in a changing world. Basic Appl Ecol 5:487–494. https://doi.org/10.1016/j.baae.2004.09.003
van der Putten WH, Klironomos JN, Wardle DA (2007) Microbial ecology of biological invasions. ISME J 1:28–37. https://doi.org/10.1038/ismej.2007.9
van Grunsven RHA, van der Putten WH, Bezemer TM et al (2007) Reduced plant-soil feedback of plant species expanding their range as compared to natives. J Ecol 95:1050–1057
Van Oosten VR, Bodenhausen N, Reymond P et al (2008) Differential effectiveness of microbially induced resistance against herbivorous insects in Arabidopsis. Mol Plant Microbe Interact 21:919–930. https://doi.org/10.1094/mpmi-21-7-0919
Van Wees SCM, Van der Ent S, Pieterse CMJ (2008) Plant immune responses triggered by beneficial microbes. Curr Opin Plant Biol 11:443–448. https://doi.org/10.1016/j.pbi.2008.05.005
Vasconcelos HL, Bruna EM (2012) Arthropod responses to the experimental isolation of Amazonian forest fragments. Zoologia 29:515–530
Vassilev N, Nikolaeva I, Vassileva M (2005) Polymer-based preparation of soil inoculants: applications to arbuscular mycorrhizal fungi. Rev Environ Sci Biotechnol 4:235–243. https://doi.org/10.1007/s11157-005-2098-2
Vemmer M, Patel AV (2013) Review of encapsulation methods suitable for microbial biological control agents. Biol Control 67:380–389
Verbruggen E, Kiers ET (2010) Evolutionary ecology of mycorrhizal functional diversity in agricultural systems. Evol Appl 3:547–560. https://doi.org/10.1111/j.1752-4571.2010.00145.x
Vogelsang KM, Bever JD (2009) Mycorrhizal densities decline in association with nonnative plants and contribute to plant invasion. Ecology 90:399–407
Vogelsang KM, Bever JD (2010) The use of native plants and mycorrhizal fungi for slope stabilization and topsoil management. Indiana University, Bloomington
Wardle DA, Bonner KI, Barker GM (2000) Stability of ecosystem properties in response to above-ground functional group richness and composition. Oikos 89:11–23. https://doi.org/10.1034/j.1600-0706.2000.890102.x
Williams A, Birkhofer K, Hedlund K (2014) Above- and below-ground interactions with agricultural management: effects of soil microbial communities on barley and aphids. Pedobiologia 57:67–74. https://doi.org/10.1016/j.pedobi.2014.01.004
Wolfe BE, Mummey DL, Rillig MC et al (2007) Small-scale spatial heterogeneity of arbuscular mycorrhizal fungal abundance and community composition in a wetland plant community. Mycorrhiza 17:175–183. https://doi.org/10.1007/s00572-006-0089-y
Wooley SC, Paine TD (2007) Can intra-specific genetic variation in arbuscular mycorrhizal fungi (*Glomus etunicatum*) affect a mesophyll-feeding herbivore (*Tupiocoris notatus* Distant)? Ecol Entomol 32:428–434. https://doi.org/10.1111/j.1365-2311.2007.00883.x

Wurst S, Ohgushi T (2015) Do plant- and soil-mediated legacy effects impact future biotic interactions? Funct Ecol 29:1373–1382. https://doi.org/10.1111/1365-2435.12456

Yang GW, Liu N, Lu WJ et al (2014) The interaction between arbuscular mycorrhizal fungi and soil phosphorus availability influences plant community productivity and ecosystem stability. J Ecol 102:1072–1082. https://doi.org/10.1111/1365-2745.12249

Zamioudis C, Pieterse CMJ (2012) Modulation of host immunity by beneficial microbes. Mol Plant Microbe Interact 25:139–150. https://doi.org/10.1094/mpmi-06-11-0179

Zavala JA, Nabity PD, DeLucia EH (2013) An emerging understanding of mechanisms governing insect herbivory under elevated CO_2. Annu Rev Entomol 58:79–97. https://doi.org/10.1146/annurev-ento-120811-153544

Zhang WW, Li F, Nie L (2010) Integrating multiple 'omics' analysis for microbial biology: application and methodologies. Microbiology 156:287–301. https://doi.org/10.1099/mic.0.034793-0

Zwanenburg B, Pospisil T, Zeljkovic SC (2016) Strigolactones: new plant hormones in action. Planta 243:1311–1326. https://doi.org/10.1007/s00425-015-2455-5

Chapter 6
Interactions Involving Rhizobacteria and Foliar-Feeding Insects

Kiran R. Gadhave and Alan C. Gange

6.1 Introduction

The term "Rhizosphere" was coined and defined by Lorenz Hiltner as a soil compartment influenced by plant roots (Hiltner 1904). It is the largest habitat in the nature (Hinsinger et al. 2009). The coexistence of a wide variety of microorganisms in the rhizosphere facilitates numerous multitrophic interactions, below- as well as aboveground (Mendes et al. 2013). Rhizobacteria comprise the predominant group that influences soil health and plant biology to a significant extent. Here, we focus on two beneficial parts of the plant-associated bacteria group: rhizobia and Plant Growth-Promoting Rhizobacteria (PGPR or "Rhizobacteria"). Rhizobia have long been known to provide growth benefits to legumes, through their N-fixing abilities. PGPR is a term that covers a disparate set of bacterial species that exist in the rhizosphere and which provide growth benefits to plants via a variety of mechanisms. The majority of these reported benefits occur in major food crop systems and have been the subject of several reviews (Johri et al. 2003; Lugtenberg and Kamilova 2009; Parray et al. 2016). However, a feature of these reviews is that the interactions between rhizobacteria and foliar-feeding insects are rarely mentioned (but see Pineda et al. 2013a, 2015, 2017 for an exception). An examination of the literature shows that this is a situation in which ecology lags behind plant biochemistry and molecular biology (Gange et al. 2012). While the effects of PGPR on plant biochemical

K. R. Gadhave (✉)
School of Biological Sciences, Royal Holloway, University of London, Egham, Surrey, UK

Department of Entomology and Plant Pathology, North Carolina State University, Raleigh, NC, USA
e-mail: krgadhave@ncsu.edu

A. C. Gange
School of Biological Sciences, Royal Holloway, University of London, Egham, Surrey, UK
e-mail: A.Gange@rhul.ac.uk

T. Ohgushi et al. (eds.), *Aboveground–Belowground Community Ecology*, Ecological Studies 234, https://doi.org/10.1007/978-3-319-91614-9_6

pathways and signaling is well documented (particularly in *Arabidopsis*), the ecological consequences of these interactions are poorly known and their effects at the community level unexplored. Our aim here is to document some of the pertinent interaction studies and their mechanisms and to show that many more of them are needed to develop the broader understanding of these interactions in an ecological and evolutionary context. We review whether and how PGPR specifically interact with insect herbivores from chemical, molecular, and ecological standpoint, beyond their conventional, well-documented applications as biofertilizer, phytostimulator, and plant stress controller. More specifically, we dissect these effects on the basis of diverse herbivore-feeding guilds (generalist vs. specialist, phloem feeders vs. chewers); constitutive and induced plant defenses; plant, insect, and rhizobacterial species; lab versus greenhouse studies; and the outcome of effects (positive, negative, or neutral). An array of these aspects form the focus of this chapter.

6.2 Rhizobia and Foliar-Feeding Insects

Given the importance of rhizobia in agricultural production and crops such as soybean (*Glycine max*), it is perhaps surprising that the first demonstration of rhizobia affecting foliar insect herbivores was performed with *Trifolium repens* and *Lotus corniculatus*. Insect herbivores can broadly be categorized into (1) chewing versus phloem feeders based on the feeding mechanism, (2) generalist versus specialist feeders based on host plant range, and (3) foliage versus root feeders based on the site of feeding. Kempel et al. (2009) found that rhizobia increased the growth of the generalist *Spodoptera littoralis* larvae in acyanogenic strains of *T. repens*, but not in cyanogenic strains, suggesting that the extra nitrogen provided by the rhizobia may be used for the production of N-based defense compounds. Enhanced production of cyanogenic defense compounds caused by rhizobia and leading to reduced chewing herbivore performance was also shown by Thamer et al. (2011) using Mexican bean beetle (*Epilachna varivestis*) feeding on lima bean (*Phaseolus lunatus*). Rhizobia were also reported to trigger plant defense via increase in jasmonic acid-induced volatiles in lima bean–Mexican bean beetle system (Ballhorn et al. 2013). Unlike mycorrhizal fungi, showing negative effects on generalist leaf-chewing herbivores while positive effects on specialist sap-feeders, effects of Rhizobia on insect herbivores appear to be context-dependent. Furthermore, *Rhizobium*, despite being one of the most extensively studied genera with regards to nitrogen cycling, has been sparsely explored with regard to its bottom-up effects on plant herbivores (van der Heijden et al. 2008). A few studies in this regard suggest that *Rhizobium* mediated bottom-up effects on insect herbivores with no clear and coherent effects on generalist versus specialist feeders or chewers versus phloem feeders. However, as with many above–below ground interactions involving microbes and insects, the outcome is often context-dependent, particularly affected by soil nitrogen levels (Dean et al. 2014). The symbiosis of Rhizobia with legumes can directly as well as indirectly influence the plant–herbivore interactions (Friesen et al. 2011; Lau and Lennon 2012). The context-dependency in these interactions is apparent, as the outcome may be influenced by the strain of rhizobia employed (Dean et al. 2009), the presence of other (non N-fixing) bacteria in the rhizosphere (Brunner et al. 2015), and feeding guild of herbivore involved (Dean et al. 2014). For instance, *Bradyrhizobium japonicum* conferred resistance to aphids feeding on soybean thus demonstrating direct manipulation of plant defenses by Rhizobia. Although such direct manipulation is less frequent, indirect effects of Rhizobia, in which Rhizobia trigger physiological changes in plant herbivores, have been reported. For instance, Whitaker et al. (2014) found that aphids (*Aphis glycines*) feeding on a nodulating strain of soybean produced honeydew with almost twice as much total sugar content than those feeding on a non-nodulating strain. Such differences in honeydew composition could have important feedback consequences for nutrient cycling in natural systems.

Katayama et al. (2014) showed that aphid honeydew added significant amounts of carbon to a soil, resulting in increased microbial growth and immobilization of soil N. The role that such interactions play in natural plant and insect communities is currently unknown, but one study (again with soybean) suggests that it may be very influential. Katayama et al. (2011) grew nodulating and non-nodulating strains of *G. max* in a garden experiment and found that the species diversity and abundance of herbivorous arthropods was higher on the nodulated plants. Furthermore, predatory insects were also more abundant on these plants. Overall, the community composition of foliar feeders differed greatly between the nodulated and non-nodulating plants. If such effects are repeated in natural plant communities, then it is likely that rhizobia play a significant role in the structure of insect communities. It is known that the presence of rhizobia and arbuscular mycorrhizal fungi (AMF) lead to increased plant diversity and productivity (Van der Heijden et al. 2016), which even if no direct (i.e., via plant chemistry) effects occurred, would still be expected to alter the composition of the associated insect communities (Siemann 1998).

6.3 PGPR and Insect Herbivores

Only a few rhizobacterial species have been studied for their effects on insect herbivores (Gange et al. 2012). Most of these studies seem to show negative effects of PGPR on pests in different crops and involve either individual *Bacillus* or *Pseudomonas* species or their combination with other microbial species (Fig. 6.1). The majority of studies have taken place in controlled conditions and have involved a limited range of insect orders, principally Lepidoptera and Hemiptera (mainly aphids) (Fig. 6.1). For instance, the growth, consumption rates, and digestive ability of cotton bollworm, *Helicoverpa armigera*, was reduced in *Pseudomonas gladioli*-treated cotton plants (Qingwen et al. 1998). The different strains of fluorescent *P. fluorescens* increased accumulation of defense molecules chitinase and proteinase inhibitors in rice and effectively suppressed leaf-folder *Cnaphalocrocis medinalis* infestation in field conditions (Saravanakumar et al. 2007). Furthermore, *P. fluorescens* treatment to *Arabidopsis* triggered induced systemic resistance

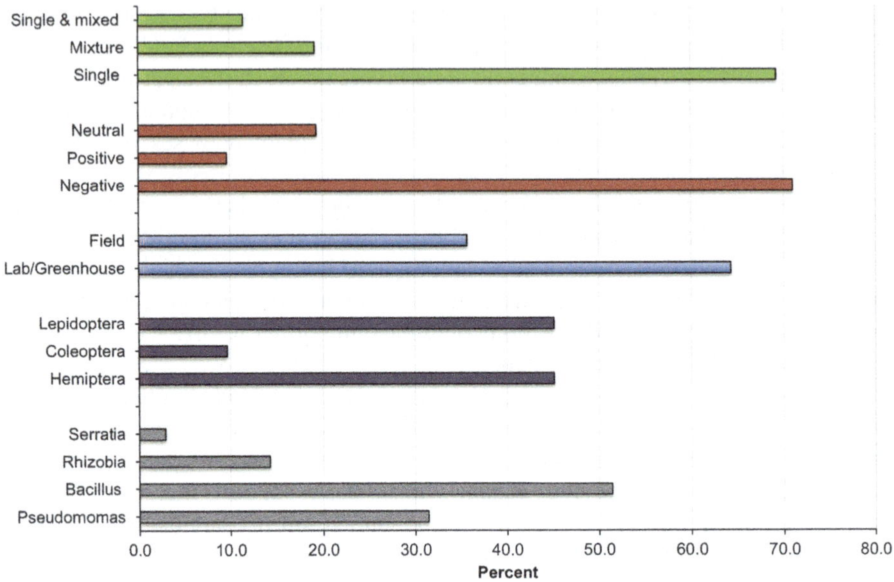

Fig. 6.1 The proportion of all published studies involving interactions between PGPR and foliar-feeding insects, summarized by inoculation method, outcomes for the insect, location of study, insect order used, and PGPR species employed

(ISR, see below) against different herbivores in *Arabidopsis* (Van der Ent et al. 2009). *Pseudomonas* spp. when co-inoculated with other microbial species were also found to be highly effective in reducing pest outbreaks. A PGPR mixture (*Pseudomonas putida* 89B-61, *Serratia marcescens* 90–166, *Flavimonas oryzihabitans* INR-5, and *B. pumilus* INR-7) significantly reduced cucumber beetle populations in field experiments in two separate seasons (Zehnder et al. 1997). Moreover, a combined inoculum of *P. fluorescens* and the entomopathogenic fungus, *Beauveria bassiana*, when applied as a seed and seedling treatment, produced a significant reduction in the incidence of larvae of two Lepidoptera, the leaf miner *Aproaerema modicella* in groundnut (Senthilraja et al. 2010) and a leaf-folder *Cnaphalocrocis medinalis* in rice (Karthiba et al. 2010). Both experiments were performed in glasshouse and field conditions, suggesting a degree of consistency in the treatments.

Apart from *Pseudomonas*, various species of plant growth-promoting *Bacillus* spp. were also found to negatively affect the growth and development of generalist and specialist insect herbivores (Vijayasamundeeswari et al. 2009; Valenzuela-Soto et al. 2010; Gadhave and Gange 2016). In cotton and tomato, bioformulations of *B. subtilis* reduced the growth of *H. armigera* larvae (Vijayasamundeeswari et al. 2009) and *Bemisia tabaci* (Valenzuela-Soto et al. 2010), respectively. However, multiple *Bacillus* species seem to exert less of an effect on plant herbivores. For instance, *B. subtilis* and *B. amyloliquefaciens* together failed to suppress green peach aphid (*Myzus persicae*) populations (Herman et al. 2008). In a more recent study, Gadhave et al. (2016a) showed that a mixture of *Bacillus* spp., *B. cereus*, *B. subtilis*, and *B. amyloliquefaciens* was not as effective as the individual species in reducing field infestations of the cabbage aphid *Brevicoryne brassicae*. The apparent lack of effects with combinations of PGPR may be why the majority of published studies involve single species inoculations of PGPR (Fig. 6.1). Several factors, including competition between the microbes, nature and diversity of plant microbiome, feeding guild of herbivore, plant species, and differential manipulation of plant biochemistry, are likely to be important in shaping the outcome of plant–herbivore interaction. Interspecies competition between microbes, in particular, is likely to be the single most important factor determining the subsequent bottom up effects on plants and consequently on herbivores as it lies at the bottom of the interaction ladder. Furthermore, the highly dynamic nature of the microbial community with regard to soil, crop, and environmental conditions makes it difficult to predict the immediate effects of microbial species on plants and herbivores.

In this respect, PGPR seem to be similar to AM fungi, in that addition of multiple species to plants may have far less effect on foliar herbivore performance than the single species present in the mixture (Gadhave et al. 2016a). Furthermore, as with AM fungi, these effects are context-specific and depend on the identity of the plant and insect species and degree of insect specialism (Pineda et al. 2010; Gadhave et al. 2016b). Ecologists are fortunate in that there is a rich literature on the effects of PGPR on plant defenses and signaling, at the biochemical and molecular levels. Reference to this literature should be very helpful in understanding the outcome of the community-level experiments, which need to be done, but are currently lacking.

6.4 Chemical Cues Determining PGPR–Plant–Insect Interactions

Rhizobacteria promote plant growth through increased nutrient uptake and biotic and abiotic stress tolerance (Gange et al. 2012). The increased nutritional quality of plants may directly benefit insects in terms of increased growth and development (Schoonhoven et al. 2005; Bukovinszky et al. 2009). Conversely, increased plant nutrient status may also encourage increased stress tolerance. For instance, rhizobacteria-mediated nutrient uptake can compensate for plant biomass loss after herbivory (Kempel et al. 2009) and facilitate the synthesis of plant defensive secondary metabolites (Aziz et al. 2016).

Plants employ an array of constitutive and induced defenses to deal with antagonists (Pineda et al. 2017). Rhizobacteria induce systemic resistance in plants that effectively suppress invading pathogens and insect herbivores (Van Oosten et al. 2008). An array of proteins and secondary metabolites, from both rhizobacteria and plants, are involved in the elicitation of ISR. These include (1) bacterial traits like lipopolysaccharides (LPS) and siderophores (Van Loon et al. 1998; Ramamoorthy et al. 2001); (2) airborne organic volatile compounds, 2,3-butanediol and acetoin (Ping and Boland 2004; Ryu et al. 2004); and (3) individual bacterial compounds such as cyclic lipopeptides (Ongena et al. 2007), antifungal factor,

2,4-diacetylphloroglucinol (Phl) (Iavicoli et al. 2003) and *N*-acyl homoserine lactones (AHLs) (Schuhegger et al. 2006).

The majority of volatiles released by plants, in response to herbivory and other stresses, belong to either of the following groups: terpenoids, phenylpropanoids/benzenoids, fatty acid derivatives, and amino acid derivatives (War et al. 2012). These constitute about 1% of plant secondary metabolites, with about 1700 compounds from 90 plant families (Dudareva et al. 2006). It appears that the chemical cues triggering plant–microbe and plant–herbivore interactions are relatively well characterized. However, whether and how major chemical cues are shared directly or indirectly when PGPR, plant, and insect interact remains poorly studied.

To date, only a few studies have explored these multi-trophic interactions in the context of deciphering linked biochemical cues. Most of these studies include *Bacillus* and *Pseudomonas* species and involve the manipulation of a variety of constitutive (e.g., bark, waxy cuticle, spines, etc.) and induced (e.g., HIPVs) plant molecules specific to plant and insect species. For instance, *Bacillus* spp.-treated cotton plants showed increased levels of gossypol, a key secondary metabolite that reduced the fitness of *Spodoptera exigua* (Zebelo et al. 2016). *B. amyloliquefaciens* (GB03) enhanced sulfur assimilation and aliphatic and indolic glucosinolates in response to *S. exigua* herbivory in *Arabidopsis* (Aziz et al. 2016). Another *Bacillus* spp., *B. pumilus* INR7, altered the cucurbitacin profiles in cucumber plants, resulting in the growth of cucumber beetles being reduced (Zehnder et al. 1997).

Two independent studies with the common rhizobacterium *P. fluorescens* on plant volatile emissions showed similar results in two different model systems. In the first study, *P. fluorescens* WCS417r suppressed the emission of the terpene (E)-alpha-bergamotene, methyl salicylate, and lilial in response to feeding by larvae of the generalist Lepidopteran *Mamestra brassicae*. This triggered indirect plant defense through an increased attraction of the parasitoid *Microplitis mediator* to caterpillar-infested plants (Pangesti et al. 2015). In the second study, *P. fluorescens* altered the herbivore-induced plant volatile profiles of *Arabidopsis thaliana* Col-0 and interfered with the attraction of the parasitoid *Diaeretiella rapae* (Pineda et al. 2013b). The olfactory preference behavior and oviposition of this wasp was reduced on rhizobacteria-treated aphid (*Myzus persicae*)-infested plants. These effects were caused by modification of the jasmonic acid signaling pathway, leading to altered herbivore-induced plant volatiles. All of the above-reported studies suggest that rhizobacteria benefit plants, by modulating their biochemistry, to counter the herbivore attack. However, studies involving more rhizobacterial species and extending the plants to those in non-crop situations would enable us to draw more robust conclusions about the role of PGPR in natural communities.

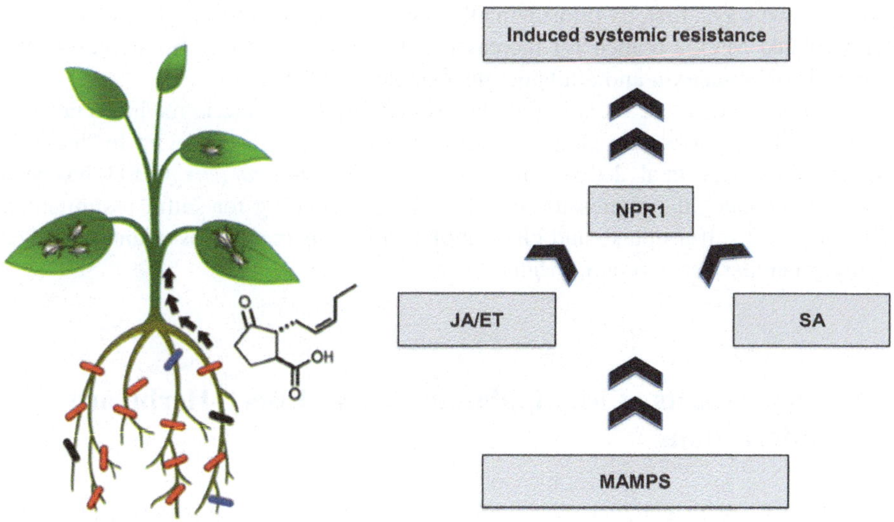

Fig. 6.2 A simplified model depicting a series of events associated with the PGPR-mediated induction of systemic resistance following an herbivore attack (modified from Niu et al. 2011). At the outset, plants recognize microbe-associated molecular patterns (MAMPs, shown by red, black, and blue rods). As a result, plant transcription factor genes (e.g., *MYB72* in *Arabidopsis*) that trigger ISR signals in roots are activated. These signals are further mobilized from roots to shoot with transporting factors (shown by upward arrows), which leads to concurrent activation of the jasmonic acid (JA), ethylene (ET), and salicylic acid (SA) plant signaling pathways with the help of regulatory protein NPR1. Consequently, induced systemic resistance is fine-tuned against an invading herbivore

6.5 PGPR and Plant Signaling

Priming is a rapid physiological response to pathogen and insect attacks. This response is orchestrated by plant signaling hormones through individual and cross-talk of different plant signaling pathways (Fig. 6.2) (Van der Ent et al. 2009). Priming against insects has been studied in a few model systems. Most of these systems highlight the role of jasmonic acid (JA)-dependent and JA-independent responses in *Arabidopsis* (Valenzuela-Soto et al. 2010). For instance, *P. fluorescens* inoculation to *Arabidopsis* and rice exhibited differential expression of transcripts encoding metabolites between treated versus untreated plants leading to increased defense responses against herbivores (Valenzuela-Soto et al. 2010; Saveetha et al. 2010). *Bacillus* spp.-treated cotton plants showed increased transcript levels of jasmonic acid responsive genes and gossypol (Zebelo et al. 2016). These biochemical and molecular changes led to reduced larval feeding and development of *Spodoptera exigua* on PGPR-treated plants. Furthermore, *P. fluorescens* WCS417r enhanced herbivore defense-related genes *PDF1.2* and *HEL* in *Arabidopsis* against *S. exigua* (Van Oosten et al. 2008). More recently, Pangesti et al. (2016) showed that rhizobacteria differentially modulate plant immune systems to counter *M. brassicae* attack in two ways, first, by prioritizing the JA/ET-regulated ORA59-branch over the JA-regulated MYC2-branch for increased defense and secondly by increasing the synthesis of camalexin and aliphatic glucosinolates (GLS).

Apart from the direct alteration of plant signaling, rhizobacteria-mediated nutrient assimilation in plants and parasitoid attraction have also been found to fine-tune plant defense (Aziz et al. 2016). For instance, *B. amyloliquefaciens* (GB03) has been shown to enhance the expression of GB03 genes encoding for sulfur assimilation. The increased sulfur uptake and glucosinolate levels increased host plant resistance to the generalist herbivore, *S. exigua*.

6.6 Key Aspects of Manipulating PGPR–Insect–Herbivore Interactions

If ecologists are to conduct meaningful community-level experiments with rhizobacteria and foliar-feeding insects, it is essential that the manipulations performed are realistic. Microbe-free plants do not exist in nature and so appropriate controls must be carefully considered (Partida-Martinez and Heil 2011). It is tempting to use mixtures of commercially produced bacterial species that are now widely available (Tailor and Joshi 2014), but there are important lessons to learn from their use in agriculture. One of the key issues with using microbial inoculants in agriculture and horticulture has been their inconsistency in performance at spatial and temporal scales (Herrmann and Lesueur 2013). The inconsistency is a result of context-specific effects of PGPR on herbivores (described above) and the confounding effects of extraneous factors such as species-specificity and environment (Pineda et al. 2010). For instance, different microbial species (*Pseudomonas* vs. *Bacillus*) can have different effects on a given crop (cereal vs. legume) in manipulating resistance against different herbivores (generalist vs. specialist; chewer vs. phloem feeder). Failure to take these factors into account leads to discrepancies in results in laboratory/glasshouse versus field conditions. Indeed, microbial inoculants often show promising results in controlled conditions, but produce inconsistent results in the field (Herrmann and Lesueur 2013). In natural communities, plants appear to recruit a consortium of rhizobacteria and to develop opportunistic plant–bacteria mutualisms that solve context-dependent ecological problems (Berendsen et al. 2012). The interdependence of PGPR and the plant microbiome adds another layer of complexity to variation in the efficacy of microbial inoculants. Below, we provide examples of some key factors that explain inconsistency in microbial inoculant performance in various conditions and which need to be taken into account in ecological experiments.

6.6.1 Differences in the Plant Microbiome

The effects of microbial inoculants containing rhizobacteria on the rhizosphere microbiome are largely unknown. A few recent studies analyzing rhizobacterial communities through high-throughput 16S rRNA amplicon sequencing suggest the predominance of the bacterial phyla *Proteobacteria*, *Firmicutes*, *Actinobacteria*, and *Bacteroidetes* in the rhizo- and endospheres of different plants (Bulgarelli et al. 2012; de Campos et al. 2013; Jin et al. 2014; Kröber et al. 2014). Although the association between rhizobacteria, the plant-associated bacterial community, and herbivory is poorly studied, we speculate that context-specificity in these interactions will vary within and among the prevalent bacterial groups (Humphrey et al. 2014; Chaudhry et al. 2016). There is a distinct possibility that a variety of external factors may cumulatively determine the outcome of these interactions. A recent study showed that *Pedobacter* spp. and *P. fluorescens* infections were negatively associated with *Scaptomyza nigrita* (Diptera: Drosophilidae) herbivory in bittercress (*Cardamine cordifolia*) in the field, but positively associated in controlled conditions (Humphrey et al. 2014). In the same study, bittercress phyllosphere bacterial diversity and abundance was higher in insect-damaged versus undamaged leaves. This suggests the possible intervention of other predominant bacterial groups, which are more likely to be prevalent in the field conditions, in differential shaping of plant–insect interactions.

A high degree of colonization competitiveness, an ability of a microorganism to successfully colonize roots, and a buildup of a sufficiently large population in the rhizosphere are two prerequisites for any rhizobacterial species to have a plant-mediated effect on the resident endophytic bacterial community (Bashan 1998). Once such species build sufficiently large populations (e.g., by quorum sensing and biofilm formation), their chances of successful establishment increase. This could be achieved by the early seed inoculation coupled with optimized repetitive field applications with a stable carrier based formulation. The microbiome-mediated effects of rhizobacteria on plant insect interactions are further complicated by the inoculation of (1) single versus multiple and (2) preexisting versus "exotic" bacterial species to plants. A proposed model by Gadhave et al. (2016b) suggests that the inoculation of single preexisting microbial species is likely to prime the plant for systemic defense against insect herbivores. Increased relative abundance and colonization competitiveness of single species is likely to be antagonistic to other less beneficial bacteria in the rhizosphere. Conversely, the inoculation of preexisting multiple, exotic single, and exotic multiple bacterial species is unlikely to trigger any chemical changes in plants or to significantly influence insect infestation. For example, *B. amyloliquefaciens* FZB42 has been shown to last in the rhizosphere over 5 weeks (Kröber et al. 2014), alter foliar glucosinolate levels, and suppress insect populations (Gadhave 2015). Furthermore, Conn and Franco (2004) and Trabelsi and Mhamdi (2013) suggested that multiple rhizobacterial species adversely affected endophytic bacterial communities in two different model systems, whereas single inoculation of those species did not. Our recent study (Gadhave et al. 2018) reported similar results in that the effects of individual *B. cereus*, *B. subtilis*, and *B. amyloliquefaciens* on endophytic bacterial community were more robust than those of a mixture of the species. A similar pattern was identified in identical but different study in which cabbage aphid growth and infestation was severely reduced in individual *Bacillus* spp. treatments than the mixed one (Gadhave et al. 2016a). Thus, it is possible that the extended effects of *Bacillus* spp. inoculation on cabbage aphids were manifested through endophytic community composition.

Adding "exotic" bacteria can cause a shift within the root bacterial community structure and determine the fate of these interactions (de Campos et al. 2013; Schmidt et al. 2014). The prime reason why mixed bacterial species fail to be as effective as single species could be the competition within added bacterial species for nutrients and niches. Furthermore, if the added mixture of bacteria is nonnative to the soil or crop, the failure of these exotic species to either acclimatize to the rhizosphere conditions or to compete with more aggressive root colonizers in the rhizosphere could produce either inconsistent and undesirable effects on plants and herbivores.

6.6.2 Competition

The disparity in results in field versus controlled conditions may also be explained by priority effects whereby indigenous microbial species outcompete all introduced species (Werner and Kiers 2014; Gadhave et al. 2016b). The inclusion of multiple microbial species in commercial inoculants is a common feature (Mayer et al. 2010). The prime (economic, not ecological) reason is that if some of the species fail to perform, others may come to the rescue to attain the desirable results on crops and herbivores. However, the use of multiple species in microbial inoculants may not necessarily be a good thing, especially when these are being used for pest management (Gadhave et al. 2016b). Multiple species in an inoculant may be compatible in static conditions, but can compete with each other for nutrients and niche in the rhizo- and endospheres (Adesemoye and Kloepper 2009). This competition may affect their initial population density (Elsas et al. 1986), root colonization ability, and the associated multitrophic bottom-up effects.

6.7 PGPR in Community Ecology

Currently, with the exception of *Arabidopsis*, the literature concerning PGPR interactions with foliar-feeding insects is dominated by controlled or glasshouse studies of pests attacking crop plants (Fig. 6.1). Indeed, experiments involving "model plants" that are widely used in ecological studies of arbuscular mycorrhizal–insect interactions, such as *Asclepias* spp., *Plantago* spp., or *Senecio* (*Jacobaea*) spp. (Hartley and Gange 2009), are absent. Field experiments are rarer, but suggest that PGPR can have negative effects on foliar-feeding insects, when inoculated as seed treatments and/or as drenches to the roots (Sudhakar et al. 2011; Gadhave et al. 2016a). If the interactions outlined above with species such as tomato and various *Brassica* spp. are repeated in natural situations, then PGPR could have profound effects in natural plant and insect community structure. To date, the only rhizobacterial–insect community study is that of Katayama et al. (2011) described above and involving rhizobia. Eisenhauer (2012) describes how interactions between soil microbes can affect plant growth and diversity, but to date no experiments involving AM fungi, PGPR (and/or rhizobia), and foliar-feeding insects have been performed in controlled or natural situations. These interactions should not just involve the herbivores, but also the natural enemies of those herbivores (Boutard-Hunt et al. 2009; Gadhave et al. 2016a). Evidence suggests that PGPR and AM fungi will have synergistic effects on insects that may well be negative, given that such interactions have been seen with plant–parasitic nematodes (Akhtar and Panwar 2013).

Van der Heijden et al. (2016) have shown that the functional diversity of microbes in the rhizosphere is important and that AM fungi and rhizobia complement each other, resulting in enhanced plant diversity and productivity. Such effects are highly likely to influence insect diversity and populations, through a variety of mechanisms that include plant chemistry, architecture, species richness, and abundance. We suggest that PGPR are a neglected part of plant–insect community ecology and deserve further attention, both in natural situations and in an applied aspect, through their potential role as "biofertilizers" and "bioprotectants."

6.8 The -Omics Approach and Future Prospects

The -omics approaches enable qualitative and quantitative characterization of biomolecules that translate into the structure and function of an individual or species (Rotroff and Motsinger-Reif 2016). These include genomics, transcriptomics, proteomics, and metabolomics and processes that vary from sequencing to in silico modeling of signaling pathways (Seaver et al. 2012). These approaches can potentially generate extensive databases that are capable of dissecting rhizobacteria–plant–insect herbivore interactions both at the molecular and community level. Currently, rhizobacteria–plant and plant–herbivore interactions appear to be better characterized than plant-associated bacteria–insect herbivore and plant-associated bacteria–plant–insect herbivore interactions (Fig. 6.3). Thus far, only a few integrative databases exploring these interactions at the molecular level have been set up. The pathogen–host interaction database (PHI-base: http://www.phi-base.org) is one such example that maintains the catalogs of genes that determine the outcome of host–pathogen interactions (Winnenburg et al. 2008). Furthermore, the PHI-base allows the in-depth characterization of plant endophytes and economically important pathogens, their properties, dynamics, and their interactions with plant hosts (Winnenburg et al. 2008; Urban et al. 2015). Such information is already being used in the manipulation of plant–*Agrobacterium* interactions (Pitzschke 2007), and

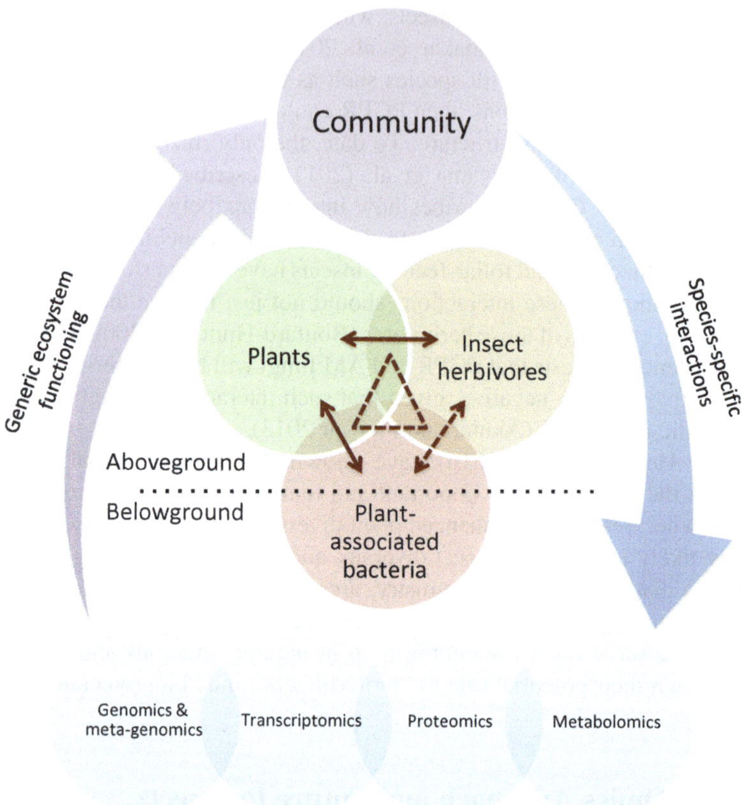

Fig. 6.3 A model depicting bottom-up and top-down approaches to exploring plant-associated bacteria–plant–insect interactions from molecular to community level and back. In particular, -omics tools can unravel micro-processes (e.g., gene–gene, protein–protein interactions) that govern species and community level interactions (generic ecosystem functioning). On the contrary, generic trends in community interactions can be used to decipher species-specific interactions at organismal and molecular level. Solid red arrows indicate relatively well-characterized PB-plant–insect herbivore interactions using -omics approaches, whereas broken red arrow and triangle indicate relatively less well studied interactions using these approaches

its use can be further extended to study other patho-systems. The information on plant–herbivore interactions is being gathered through insect genome projects such as i5K (http://i5k.github.io/genomes) with the aim of assembling genome sequences of 5000 economically and ecologically important insects (Sills et al. 2011). Such a substantial insect genome meta-data resource can help propel new integrative strategies for studying not just insect biology, but the dynamics of insect interactions with the insect microbiome, plants, and the environment. Future efforts should focus on gathering information on different study systems, using multiple variables at spatiotemporal scales through -omics approaches and assembling this information to understand the micro-processes in ecosystem functioning.

6.9 Concluding Remarks

Rhizobia and PGPR can modulate aboveground interactions and thereby extend their effects on foliar-feeding insects and their parasitoids. Individual studies suggest that the effects of both of these key belowground players on foliage feeders are mostly negative and are mediated through alteration of plant biochemistry, signaling, and the plant microbiome. However, these effects have been barely explored both in the field studies and at the community level. Recent studies have started to describe the mechanisms underpinning these interactions, mostly in controlled conditions, that ecologists can draw conclusions from. The current rhizobacteria–plant–insect study systems need to be more realistic in terms of addressing the issues associated with controlled versus field studies and involving spatial and temporal scales, while taking multiple rhizobacteria-, plant-, and herbivore-associated variables into account. No microbial group exists in isolation, but there are serious challenges to designing sensible studies that include each of these players in varying conditions. Newly emerging -omics approaches including metagenomics, metatranscriptomics, and metaproteomics are proving to be effective in exploring the taxonomic and functional variability in the plant microbiome. These modern approaches are imperative to advance the current understanding of rhizobacterial–plant interactions at the molecular level and to determine how these translate into ecological functions. Furthermore, they can be effectively used to generate meta-data that can encompass minute details on each of the interactions individually and in a broader context.

Acknowledgments We thank Royal Holloway, University of London, for funding some of our experimental work, Rajagopalbabu Srinivasan and Bhabesh Dutta for helpful discussions, and reviewers for constructive comments that greatly improved the manuscript.

References

Adesemoye AO, Kloepper JW (2009) Plant–microbes interactions in enhanced fertilizer-use efficiency. Appl Microbiol Biotechnol 85:1–12

Akhtar MS, Panwar J (2013) Efficacy of root-associated fungi and PGPR on the growth of *Pisum sativum* (cv. Arkil) and reproduction of the root-knot nematode *Meloidogyne incognita*. J Basic Microbiol 53:318–326

Aziz M, Nadipalli RK, Xie X et al (2016) Augmenting sulphur metabolism and herbivore defence in *Arabidopsis* by bacterial volatile signaling. Front Plant Sci. https://doi.org/10.3389/fpls.2016.00458

Ballhorn DJ, Kautz S, Schädler M (2013) Induced plant defense via volatile production is dependent on rhizobial symbiosis. Oecologia 172:833–846

Bashan Y (1998) Inoculants of plant growth-promoting bacteria for use in agriculture. Biotechnol Adv 16:729–770

Berendsen RL, Pieterse CM, Bakker PA (2012) The rhizosphere microbiome and plant health. Trends Plant Sci 17:478–486

Boutard-Hunt C, Smart CD, Thaler J et al (2009) Impact of plant growth-promoting rhizobacteria and natural enemies on *Myzus persicae* (Hemiptera: Aphididae) infestations in pepper. J Econ Entomol 102:2183–2191

Brunner SM, Goos RJ, Swenson SJ et al (2015) Impact of nitrogen fixing and plant growth-promoting bacteria on a phloem-feeding soybean herbivore. Appl Soil Ecol 86:71–81

Bukovinszky T, Poelman EH, Gols R et al (2009) Consequences of constitutive and induced variation in plant nutritional quality for immune defence of a herbivore against parasitism. Oecologia 160:299–308

Bulgarelli D, Rott M, Schlaeppi K et al (2012) Revealing structure and assembly cues for *Arabidopsis* root-inhabiting bacterial microbiota. Nature 488:91–95

Chaudhry V, Sharma S, Bansal K et al (2016) Glimpse into the genomes of rice endophytic bacteria: diversity and distribution of Firmicutes. Front Microbiol 7:2115. https://doi.org/10.3389/fmicb.2016.02115

Conn VM, Franco CM (2004) Effect of microbial inoculants on the indigenous actinobacterial endophyte population in the roots of wheat as determined by terminal restriction fragment length polymorphism. Appl Environ Microbiol 70:6407–6413

de Campos SB, Youn JW, Farina R et al (2013) Changes in root bacterial communities associated to two different development stages of canola (*Brassica napus* L. var oleifera) evaluated through next-generation sequencing technology. Microb Ecol 65:593–601

Dean JM, Mescher MC, De Moraes CM (2009) Plant-rhizobia mutualism influences aphid abundance on soybean. Plant Soil 323:187–196

Dean JM, Mescher MC, De Moraes CM (2014) Plant dependence on rhizobia for nitrogen influences induced plant defences and herbivore performance. Int J Mol Sci 15:1466–1480

Dudareva N, Negre F, Nagegowda DA et al (2006) Plant volatiles: recent advances and future perspectives. Crit Rev Plant Sci 25:417–440

Eisenhauer N (2012) Aboveground-belowground interactions as a source of complementarity effects in biodiversity experiments. Plant Soil 351:1–22

Elsas J, Dijkstra A, Govaert J et al (1986) Survival of *Pseudomonas fluorescens* and *Bacillus subtilis* introduced into two soils of different texture in field microplots. FEMS Microbiol Lett 38:151–160

Friesen ML, Porter SS, Stark SC et al (2011) Microbially mediated plant functional traits. Annu Rev Ecol Evol Syst 42:23–46. https://doi.org/10.1146/annurev-ecolsys-102710-145039

Gadhave KR (2015) Interactions between plant growth promoting rhizobacteria, foliar-feeding insects and higher trophic levels. PhD Thesis, Royal Holloway, University of London

Gadhave KR, Gange AC (2016) Plant-associated *Bacillus* spp. alter life-history traits of the specialist insect *Brevicoryne brassicae* L. Agric For Entomol 18:35–42

Gadhave KR, Finch P, Gibson TM et al (2016a) Plant growth-promoting *Bacillus* suppress *Brevicoryne brassicae* field infestation and trigger density-dependent and density-independent natural enemy responses. J Pest Sci 89:985–992

Gadhave KR, Hourston JE, Gange AC (2016b) Developing soil microbial inoculants for pest management: can one have too much of a good thing? J Chem Ecol 42:348–356

Gadhave KR, Devlin PF, Ebertz A et al (2018) Soil inoculation with *Bacillus* spp. modifies root endophytic bacterial diversity, evenness and community composition in a context specific manner. Microb Ecol. https://doi.org/10.1007/s00248-018-1160-x

Gange AC, Eschen R, Schroeder V (2012) The soil microbial community and plant foliar defences against insects. In: Iason GR, Dicke M, Hartley SE (eds) The ecology of plant secondary metabolites: from genes to global processes. Cambridge University Press, Cambridge, pp 170–188

Hartley SE, Gange AC (2009) Impacts of plant symbiotic fungi on insect herbivores: mutualism in a multitrophic context. Annu Rev Entomol 54:323–342

Herman MAB, Nault BA, Smart CD (2008) Effects of plant growth-promoting rhizobacteria on bell pepper production and green peach aphid infestations in New York. Crop Prot 27:996–1002

Herrmann L, Lesueur D (2013) Challenges of formulation and quality of biofertilizers for successful inoculation. Appl Microbiol Biotechnol 97:8859–8873

Hiltner L (1904) Uber neuere Erfahrungen und Probleme auf dem Gebiete der Bodenbakteriologie unter besonderden berucksichtigung und Brache. Arb Dtsch Landwirtsch Gesellschaft 98:59–78

Hinsinger P, Bengough AG, Vetterlein D et al (2009) Rhizosphere: biophysics, biogeochemistry and ecological relevance. Plant Soil 321:117–152

Humphrey PT, Nguyen TT, Villalobos MM et al (2014) Diversity and abundance of phyllosphere bacteria are linked to insect herbivory. Mol Ecol 23:1497–1515

Iavicoli A, Boutet E, Buchala A et al (2003) Induced systemic resistance in *Arabidopsis thaliana* in response to root inoculation with *Pseudomonas fluorescens* CHA0. Mol Plant Microbe Interact 16:851–858

Jin H, Yang XY, Yan ZQ et al (2014) Characterization of rhizosphere and endophytic bacterial communities from leaves, stems and roots of medicinal *Stellera chamaejasme* L. Syst Appl Microbiol 37:376–385

Johri B, Sharma A, Virdi J (2003) Rhizobacterial diversity in India and its influence on soil and plant health. Biotechnology 84:49–89

Karthiba L, Saveetha K, Suresh S (2010) PGPR and entomopathogenic fungus bioformulation for the synchronous management of leaffolder pest and sheath blight disease of rice. Pest Manag Sci 66:555–564

Katayama N, Zhang ZQ, Ohgushi T (2011) Community-wide effects of below-ground rhizobia on above-ground arthropods. Ecol Entomol 36:43–51

Katayama N, Silva AO, Kishida O et al (2014) Herbivorous insect decreases plant nutrient uptake: the role of soil nutrient availability and association of below-ground symbionts. Ecol Entomol 39: 511–518

Kempel A, Brandl R, Schädler M (2009) Symbiotic soil microorganisms as players in aboveground plant–herbivore interactions – the role of rhizobia. Oikos 118:634–640

Kröber M, Wibberg D, Grosch R et al (2014) Effect of the strain *Bacillus amyloliquefaciens* FZB42 on the microbial community in the rhizosphere of lettuce under field conditions analyzed by whole metagenome sequencing. Front Microbiol. https://doi.org/10.3389/fmicb.2014.00252

Lau JA, Lennon JT (2012) Rapid responses of soil microorganisms improve plant fitness in novel environments. Proc Natl Acad Sci USA 109:14058–14062

Lugtenberg B, Kamilova F (2009) Plant-growth-promoting rhizobacteria. Annu Rev Microbiol 63:541–556

Mayer J, Scheid S, Widmer F et al (2010) How effective are 'effective microorganisms®' (EM) results from a field study in temperate climate. Appl Soil Ecol 46:230–239

Mendes R, Garbeva P, Raaijmakers JM (2013) The rhizosphere microbiome: significance of plant beneficial, plant pathogenic, and human pathogenic microorganisms. FEMS Microbiol Rev 37:634–663

Niu DD, Liu HX, Jiang CH et al (2011) The plant growth–promoting rhizobacterium *Bacillus cereus* AR156 induces systemic resistance in *Arabidopsis thaliana* by simultaneously activating salicylate- and jasmonate/ethylene-dependent signaling pathways. Mol Plant Microbe Interact 24:533–542

Ongena M, Jourdan E, Adam A et al (2007) Surfactin and fengycin lipopeptides of *Bacillus subtilis* as elicitors of induced systemic resistance in plants. Environ Microbiol 9:1084–1090

Pangesti N, Weldegergis BT, Langendorf B et al (2015) Rhizobacterial colonization of roots modulates plant volatile emission and enhances the attraction of a parasitoid wasp to host-infested plants. Oecologia 178:1169–1180

Pangesti N, Reichelt M, Van de Mortel JE et al (2016) Jasmonic acid and ethylene signaling pathways regulate glucosinolate levels in plants during rhizobacteria-induced systemic resistance against a leaf-chewing herbivore. J Chem Ecol 42:1212–1225

Parray JA, Jan S, Kamili AN et al (2016) Current perspectives on plant growth-promoting rhizobacteria. Plant Growth Regul 35:877–902

Partida-Martinez LP, Heil M (2011) The microbe-free plant: fact or artifact? Front Plant Sci 2:100. https://doi.org/10.3389/fpls.2011.00100

Pineda A, Zheng SJ, Van Loon JJA et al (2010) Helping plants to deal with insects: the role of beneficial soil-borne microbes. Trends Plant Sci 15:507–514. https://doi.org/10.1016/j.tplants.2010.05.007

Pineda A, Dicke M, Pieterse CMJ et al (2013a) Beneficial microbes in a changing environment: are they always helping plants to deal with insects? Funct Ecol 27:574–586

Pineda A, Soler R, Weldegergis BT et al (2013b) Non-pathogenic rhizobacteria interfere with the attraction of parasitoids to aphid-induced plant volatiles via jasmonic acid signalling. Plant Cell Environ 36:393–404

Pineda A, Soler R, Pozo MJ et al (2015) Editorial: above-belowground interactions involving plants, microbes and insects. Front Plant Sci 6:318. https://doi.org/10.3389/fpls.2015.00318

Pineda A, Kaplan I, Bezemer TM (2017) Steering soil microbiomes to suppress aboveground insect pests. Trends Plant Sci 22:770–778

Ping L, Boland W (2004) Signals from the underground: bacterial volatiles promote growth in *Arabidopsis*. Trends Plant Sci 9:263–266

Pitzschke A (2007) Agrobacterium infection and plant defence—transformation success hangs by a thread. Front Plant Sci 4:115–126

Qingwen Z, Ping L, Gang W et al (1998) On the biochemical mechanism of induced resistance of cotton to cotton bollworm by cutting young seedling at plumular axis. Acta Phytophylacica Sin 25:209–212

Ramamoorthy V, Viswanathan R, Raguchander T et al (2001) Induction of systemic resistance by plant growth promoting rhizobacteria in crop plants against pests and diseases. Crop Prot 20: 1–11

Rotroff DM, Motsinger-Reif AA (2016) Embracing integrative multiomics approaches. Int J Genomics 2016:1715985. https://doi.org/10.1155/2016/1715985

Ryu CM, Farag MA, Hu CH et al (2004) Bacterial volatiles induce systemic resistance in *Arabidopsis*. Plant Physiol 134:1017–1026

Saravanakumar D, Muthumeena K, Lavanya N et al (2007) *Pseudomonas* induced defence molecules in rice plants against leaffolder (*Cnaphalocrocis medinalis*) pest. Pest Manag Sci 63: 714–721

Saveetha K, Karthiba L, Raveendran M et al (2010) Transcriptional analysis of molecular interactions between *Pseudomonas fluorescens* strain TDK1, *Oryza sativa* and *Cnaphalocrocis medinalis*. J Appl Entomol 134:762–773

Schmidt R, Köberl M, Mostafa A et al (2014) Effects of bacterial inoculants on the indigenous microbiome and secondary metabolites of chamomile plants. Front Microbiol. https://doi.org/10.3389/fmicb.2014.00064

Schoonhoven LM, Van Loon JJ, Dicke M (2005) Insect-plant biology. Oxford University Press, Oxford

Schuhegger R, Ihring A, Gantner S et al (2006) Induction of systemic resistance in tomato by *N*-acyl-L-homoserine lactone-producing rhizosphere bacteria. Plant Cell Environ 29:909–918

Seaver SM, Henry CS, Hanson AD (2012) Frontiers in metabolic reconstruction and modeling of plant genomes. J Exp Bot 63:2247–2258

Senthilraja G, Anand T, Durairaj C (2010) A new microbial consortia containing entomopathogenic fungus, *Beauveria bassiana* and plant growth promoting rhizobacteria, *Pseudomonas fluorescens* for simultaneous management of leafminers and collar rot disease in groundnut. Biocontrol Sci Technol 20:449–464

Siemann E (1998) Experimental tests of effects of plant productivity and diversity on grassland arthropod diversity. Ecology 79:2057–2070

Sills J, Robinson GE, Hackett KJ et al (2011) Creating a buzz about insect genomes. Science 331: 1386–1387

Sudhakar N, Thajuddin N, Murugesan K et al (2011) Plant growth-promoting rhizobacterial mediated protection of tomato in the field against cucumber mosaic virus and its vector *Aphis gossypii*. Biocontrol Sci Technol 21:367–386

Tailor A, Joshi BH (2014) Harnessing plant growth promoting rhizobacteria beyond nature: a review. J Plant Nutr 37:1534–1571

Thamer S, Schadler M, Bonte D et al (2011) Dual benefit from a belowground symbiosis: nitrogen fixing rhizobia promote growth and defence against a specialist herbivore in a cyanogenic plant. Plant Soil 341:209–219

Trabelsi D, Mhamdi R (2013) Microbial inoculants and their impact on soil microbial communities: a review. Biomed Res Int. https://doi.org/10.1155/2013/863240

Urban M, Irvine AG, Cuzick A et al (2015) Using the pathogen-host interactions database (PHI-base) to investigate plant pathogen genomes and genes implicated in virulence. Front Plant Sci 6:605

Valenzuela-Soto JH, Estrada-Hernandez MG, Ibarra-Laclette E et al (2010) Inoculation of tomato plants (*Solanum lycopersicum*) with growth-promoting *Bacillus subtilis* retards whitefly *Bemisia tabaci* development. Planta 231:397–410

Van der Ent S, Van Wees S, Pieterse CMJ (2009) Jasmonate signaling in plant interactions with resistance-inducing beneficial microbes. Phytochemistry 70:1581–1588

Van der Heijden MGA, Bardgett RD, Van Straalen NM (2008) The unseen majority: soil microbes as drivers of plant diversity and productivity in terrestrial ecosystems. Ecol Lett 11:296–310

Van der Heijden MGA, de Bruin S, Luckerhoff L et al (2016) A widespread plant-fungal-bacterial symbiosis promotes plant biodiversity, plant nutrition and seedling recruitment. ISME J 10:389–399

Van Loon LC, Bakker P, Pieterse CMJ (1998) Systemic resistance induced by rhizosphere bacteria. Annu Rev Phytopathol 36:453–483

Van Oosten VR, Bodenhausen N, Reymond P et al (2008) Differential effectiveness of microbially induced resistance against herbivorous insects in *Arabidopsis*. Mol Plant Microbe Interact 21: 919–930

Vijayasamundeeswari A, Ladhalakshmi D, Sankaralingam A et al (2009) Plant growth promoting rhizobacteria of cotton affecting the developmental stages of *Helicoverpa armigera*. J Plant Prot Res 49:239–243

War AR, Paulraj MG, Ahmad T (2012) Mechanisms of plant defence against insect herbivores. Plant Signal Behav 7:1306–1320

Werner GDA, Kiers ET (2014) Order of arrival structures arbuscular mycorrhizal colonization of plants. New Phytol. https://doi.org/10.1111/nph.13092

Whitaker MRL, Katayama N, Ohgushi T (2014) Plant-rhizobia interactions alter aphid honeydew composition. Arthropod Plant Interact 8:213–220

Winnenburg R, Urban M, Beacham A et al (2008) PHI-base update: additions to the pathogen–host interaction database. Nucleic Acids Res 36:572–576

Zebelo S, Song Y, Kloepper JW et al (2016) Rhizobacteria activates (+)-δ-cadinene synthase genes and induces systemic resistance in cotton against beet armyworm (*Spodoptera exigua*). Plant Cell Environ 39:935–943

Zehnder G, Kloepper J, Yao C et al (1997) Induction of systemic resistance in cucumber against cucumber beetles (Coleoptera: Chrysomelidae) by plant growth-promoting rhizobacteria. J Econ Entomol 90:391–396

Chapter 7
Belowground–Aboveground Interactions Between Pathogens and Herbivores

Bastien Castagneyrol, Pilar Fernandez-Conradi, Pil U. Rasmussen, Cécile Robin, and Ayco J. M. Tack

7.1 Introduction

In this chapter, the main focus will be on belowground–aboveground (BG–AG) interactions between plant-associated pathogens and herbivores. Most of herbivores studied in this context are arthropods. Hence, unless stated otherwise, the term *herbivore* will be used to refer to plant-feeding arthropods. It is now largely accepted that plant pathogens can interact strongly with herbivores when co-occurring on aboveground plant parts. Pathogens generally reduce the preference and performance of herbivores (Fernandez-Conradi et al. 2018), which can have cascading effects on the structure of insect communities found on terrestrial plants (Tack et al. 2012; Tack and Dicke 2013). The reciprocal effect of herbivores on pathogens has also been addressed, but there is no consensus yet on how pathogens respond to herbivore attack on the shared plant: studies have reported either positive, neutral, or negative effects of herbivores on pathogens (Hatcher 1995).

Cross-compartment interactions between herbivores and pathogens differ from interactions involving BG beneficial microbes and AG herbivores as they involve two plant antagonists that compete for a shared, limited, and defended resource. Likewise, cross-compartment pathogen–herbivore interactions differ from pathogen–herbivore interactions within the same compartment: while within-compartment pathogen–herbivore interactions may be both direct (e.g., herbivores acquiring

B. Castagneyrol · P. Fernandez-Conradi · C. Robin
UMR BIOGECO, INRA, University of Bordeaux, Cestas, France
e-mail: bastien.castagneyrol@inra.fr; cecile.robin@inra.fr

P. U. Rasmussen · A. J. M. Tack (✉)
Department of Ecology, Environment and Plant Sciences, Stockholm University, Stockholm, Sweden
e-mail: pil.rasmussen@su.se; ayco.tack@su.se

T. Ohgushi et al. (eds.), *Aboveground–Belowground Community Ecology*, Ecological Studies 234, https://doi.org/10.1007/978-3-319-91614-9_7

supplementary nutrients from pathogens or being exposed to their toxins) and indirect (e.g., plant-mediated interactions), cross-compartment interactions among pathogens and herbivores are inevitably indirect. Indirect interactions may involve changes in primary and secondary metabolites within the shared host plant. It is also important to stress that a cross-compartment interaction between a pathogen and an herbivore is just one type of indirect interaction between two organisms using the same resource. In real life, such species interactions are embedded within highly diverse plant-based food webs: for example, BG pathogens and herbivores could interact with a range of organisms aboveground, like herbivores, pathogens, endophytes, as well as their natural enemies; likewise, AG pathogens and herbivores could interact with the entire belowground community and not only with organisms tightly associated with their host's roots.

Here, we explore whether pathogen–herbivore interactions may also play an important role when the organisms are separated by the soil surface. These interactions have received little attention as compared to interactions between BG beneficial microorganisms and AG herbivores, a discrepancy that may be explained by the focus on BG beneficial organisms in studies for biocontrol development. However, as the outcome of plant attack by multiple attackers is not necessarily additive, recent studies in both community ecology and agroecology have increasingly focused on the outcome of tripartite interactions between plants, pathogens, and herbivores and its consequences for community dynamics and plant yield. In this chapter, we will explore the scant available literature on belowground–aboveground interactions between herbivores and pathogens and outline promising areas for future research. Throughout, given the scarcity of published studies on BG–AG interactions between herbivores and pathogens, we draw partly on findings, ideas, and insights from three related research areas that are accompanied by a wealth of published articles: (1) the study of interactions between BG and AG herbivores, (2) the study of interactions among pathogens and herbivores that both attack AG parts of the plant (Fernandez-Conradi et al. 2018; Tack and Dicke 2013), and (3) interactions between BG mutualistic microbes and AG herbivores.

In this chapter, we aim to: (1) provide a comprehensive overview of the literature on BG–AG interactions among herbivores and pathogens, by tabulating the available studies and discussing the patterns, (2) explore the sources of variation in the strength, direction, and symmetry of these interactions, including the role of the abiotic environment and the life history of both pathogens and herbivores, and therefore assess whether cross-compartment interactions among herbivores and pathogens are predictable, and (3) address the consequences of cross-compartment interactions on the ecology and evolution of plant-based communities.

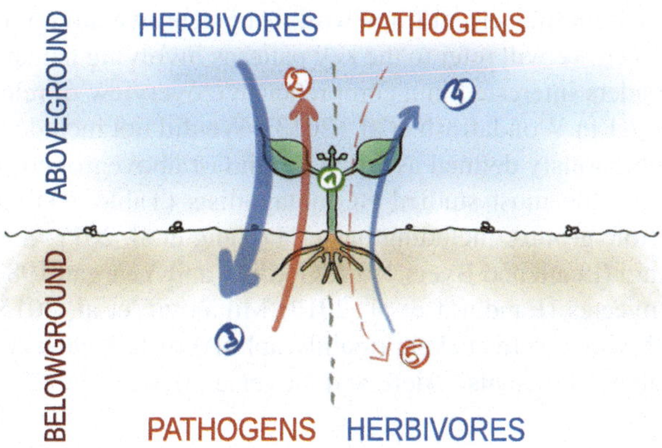

Fig. 7.1 Cross-compartment interactions between pathogens and herbivores. Blue arrows represent the effect of herbivores on pathogens. The reciprocal effect of pathogens on herbivores is shown by red arrows. (1) consequences of dual-attack on the host plant; (2) effects of BG pathogens on AG herbivores; (3) effects of AG herbivores on BG pathogens; (4) effects of BG herbivores on AG pathogens; (5) effects of AG pathogens on BG herbivores. Arrow width is proportional to the number of studies specifically addressing corresponding interactions

7.2 Consequences of Plant–Pathogen–Herbivore Interactions for Three Players: A Review of Patterns

To review patterns of cross-compartment interactions involving BG pathogens and AG herbivores or BG herbivores and AG pathogens (Fig. 7.1), we first screened reference lists of the recent reviews discussing plant–microbe–herbivore interactions (Wondafrash et al. 2013; Tack and Dicke 2013; Hauser et al. 2013; Biere and Goverse 2016). We further searched additional references in the Web of Science database (January 23, 2017) using the following combination of keywords: "aboveground" AND "belowground" AND "plant" AND "pathogen" AND "herbivor*" AND "insect."

Plant parasitic nematodes are by far the most studied BG antagonists of plants in the context of BG–AG pathogen–herbivore interactions (reviewed by Wondafrash et al. 2013). However, classifying nematodes as herbivores or pathogens is debatable. Plant parasitic nematodes include ectoparasites, which live outside the plant and puncture cell walls feeding on cell material using their stylet, as well as migratory endoparasites, which penetrate the root and continuously move through the root cells while feeding through the puncturing of cell walls. These plant parasitic nematodes can cause cell death, similar to what is seen in some leafhoppers (e.g., Hunter and Backus 1989). In contrast, sedentary endoparasites penetrate roots and induce permanent giant feeding cells within the plants, the most typical being the root-knot and cyst nematodes. By inducing permanent feeding cells that are not killed, such nematodes are more similar to gall-forming pathogens and galling herbivores, such as the ovary smut fungus *Ustilago maydis* on maize or oak gall wasps. As interactions involving nematodes have been thoroughly reviewed, and their placement within the current framework is unclear, we did not include them in Table 7.1. However, we will refer to the key patterns involving nematodes within the text and refer readers interested in a comprehensive overview of relevant nematode studies to Table 7.1 in Wondafrash et al. (2013). We did not include viruses, as they cannot be unambiguously defined as belowground or aboveground pathogens.

Pathogens were the most studied BG antagonists (Table 7.1) and belonged to different taxonomic groups, including bacteria (Yang et al. 2011; Song et al. 2015), necrotrophic fungi (Leath and Byers 1977; Godfrey and Yeargan 1987; McNee et al. 2003), and oomycetes (Landgraf et al. 2012; Milanović et al. 2015). Most of the studied AG herbivores were suckers (mainly aphids) or leaf-chewers (Lepidoptera, Coleoptera) (Table 7.1, see also Biere and Goverse 2016).

7.2.1 The Impact of Single and Dual Attack on Plant Performance

Both pathogens and herbivores are, on their own, harmful to plants. Harmful effects of dual attack by BG herbivores and AG pathogens (or, inversely, BG pathogens and AG herbivores) attacking distant compartments have been reported for plant growth (Alexander et al. 1981), biomass production (De Roissart et al. 2013; Saravesi et al. 2015), survival (Leath and Byers 1977), reproductive output (Barber et al. 2015), and crop yield (Godfrey and Yeargan 1987). Yet, this general tendency hides an important variability in plant–pathogen–herbivore interactions, with examples of antagonistic (i.e., the plant being less damaged than expected based on single attacks, Godfrey and Yeargan 1987; Yang et al. 2011), synergistic (i.e., the plant being more damaged than expected based on single attacks, Leath and Byers 1977), and additive effects of dual attack on plant performance (reviewed in Hauser et al. 2013).

Current knowledge on plant–pathogen–herbivore interactions in general suggests that most of dual attacks result in additive effects on plant performances, with surprisingly little evidence for synergistic effects of pathogens and herbivores on plant performance (Hauser et al. 2013). To predict the particular effect of dual attack by BG and AG attackers on plant performance, which currently seems like a distant future, we probably need an accurate knowledge of the reciprocal impact of BG and AG attackers on each other's performance. If one attacker has a positive effect on the second one, dual attack is likely to have a stronger negative impact on plant fitness than single attacks. If there is a negative effect of one attacker on the other, dual attack is likely to be less harmful than single attack. Finally, if the attackers do not affect each other's fitness, dual attack may simply have an additive impact on their host plant.

7.2.2 Interactions Between BG Pathogens and AG Herbivores

7.2.2.1 Effect of BG Pathogens on AG Herbivores

The effects of BG pathogens on AG herbivores were shown to be negative (McNee et al. 2003; Hong et al. 2011; McCarville et al. 2012; Kammerhofer et al. 2015), positive (De Roissart et al. 2013; Milanović et al. 2015; Kammerhofer et al. 2015), or neutral (Godfrey and Yeargan 1989) (Table 7.1). Wondafrash et al. (2013) extensively reviewed cross-compartment interactions between BG nematodes and AG herbivores and showed that the outcome of their distant interactions was contingent upon the feeding strategies of both nematodes (migratory vs. sedentary) and herbivores (leaf chewers vs. phloem feeders), with migratory nematodes principally reducing performance of phloem feeders such as aphids (Bezemer et al. 2005; Wurst and van der Putten 2007), while the impact of root infection by sedentary nematodes on AG herbivores is more variable (Wondafrash et al. 2013). Most of available literature addresses the effect of BG nematodes on AG herbivores. The few studies focusing on fungus pathogens are highlighted in Table 7.1. While evidence is lacking, current knowledge on plant–pathogen–herbivore interactions within the same compartment (Lazebnik et al. 2014) suggests that the direction and strength of the effect of BG pathogens on AG herbivores may depend on the pathogen lifestyle (necrotrophic vs. biotrophic). Likewise, it may depend on herbivore feeding guild, with different responses by chewers and sap-sucking insects. These differences and corresponding predictions will be detailed in Sect. 7.3.1.1 (Fig. 7.2, panels g and h).

7.2.2.2 Effect of AG Herbivores on BG Pathogens

AG herbivores may both positively and negatively affect BG pathogens. Several studies show that AG herbivores facilitate root colonization by BG pathogens and are associated with greater pathogen severity (Leath and Byers 1977; Alexander et al. 1981; Burrill et al. 1999; Saravesi et al. 2015; Kammerhofer et al. 2015). For instance, in a study by Leath and Byers (1977), it was found that root rot caused by the necrotrophic fungus *Fusarium roseum* was more severe when the plant was simultaneously attacked by aboveground aphids. Similar positive effects of AG herbivores on BG pathogens have been found for the southern pine beetle *Dendroctonus frontalis*, which increased colonization levels of the BG necrotrophic fungus *Heterobasidion annosum* on *Pinus taeda* roots (Alexander et al. 1981). On the other hand, AG herbivores may trigger systemic defenses effective in roots that may act against BG pathogens (Yang et al. 2011; Landgraf et al. 2012), thus reducing their incidence and severity (Song et al. 2015). The same variability in

Table 7.1 Overview of BG-AG plant–pathogen–herbivore interactions

	The study			The plant player		The pathogen player			
References	Methodological approach	System type (natural vs. applied)	Compartment attacked first	Plant species	Plant type	Pathogen species	Pathogen taxonomic group	Pathogen lifestyle	Pathogen compartment
Godfrey and Yeargan (1989)	Experimental	Applied	NA (timing not manipulated)	*Medicago sativa*	Crop	*Rhizoctonia* spp., *Fusarium* spp., *Pythium* spp.	Fungi	Necrotrophic	BG
Leath and Byers (1977)	Experimental	Natural	BG ↔ AG	*Trifolium pratense*	Herb	*Fusarium roseum*	Fungus	Necrotrophic	BG
Leath and Byers (1977)	Experimental	Natural	BG ↔ AG	*Trifolium repens*	Herb	*Fusarium roseum*	Fungus	Necrotrophic	BG
Leath and Byers (1977)	Experimental	Natural	BG ↔ AG	*Medicago sativa*	Herb	*Fusarium roseum*	Fungus	Necrotrophic	BG
Milanović et al. (2015)	Semi-experimental	Natural	BG → AG	*Quercus rubra*	Tree	*Phytophthora plurivora*	Oomycota	Hemibiotrophic	BG
Alexander et al. (1981)	Observational	Natural	BG → AG	*Pinus taeda*	Tree	*Heterobasidium annosum*	Fungus	Necrotrophic	BG
Lee et al. (2012)	Experimental	Applied	AG → BG	*Capsicum annuum*	Crop	*Bacillus subtilis GB03 (beneficial), Pseudomonas fluorescens Pf-5 (beneficial), Ralstonia solanacearum SL1931 (pathogenic)*	Rhizosphere bacteria	Mix of beneficial and biotrophic pathogen	BG

Saravesi et al. (2015)	Observational	Natural	AG → BG	*Betula pubescens*	Tree	Unidentified	Fungus	Beneficial, saprotrophic	BG
Song et al. (2015)	Experimental	Applied	AG → BG	*Nicotiana benthamiana*	Crop	*Agrobacterium tumefasciens*	Bacteria	Biotrophic	BG
Yang et al. (2011)	Experimental	Applied	AG → BG	*Capsicum annuum*	Crop	*Ralstonia solanacearum*	Bacteria	Biotrophic	BG
Barber et al. (2015)	Experimental	Applied	BG → AG	*Cucumis sativus*	Crop	*Pseudoperonospora cubensis*	Oomycota	Biotrophic	AG
Erb et al. (2009)	Experimental	Applied	BG → AG	*Zea mays*	Crop	*Setosphaeria turcica*	Fungus	Necrotrophic	AG
Damicone and Manning (1987)	Experimental	Applied	BG → AG	*Asparagus officinalis*	Crop	*Fusarium oxysporum*	Fungus	Necrotrophic	BG
Godfrey and Yeargan (1987)	Experimental	Applied	NA (timing not manipulated)	*Medicago sativa*	Crop	*Rhizoctonia* spp., *Fusarium* spp., *Pythium* spp.	Fungi	Necrotrophic	BG
Kostenko et al. (2012)	Observational	Natural	BG → AG AG → BG BG ↔ AG	*Jacobaea vulgaris*	Herb	Unidentified	Fungus	NA	BG
Landgraf et al. (2012)	Experimental	Applied	BG → AG	*Medicago truncatula*	Herb	*Aphanomyces euteiches*	Oomycota	Biotrophic	BG
Leath and Byers (1977)	Experimental	Natural	BG ↔ AG	*Medicago sativa*	Herb	*Fusarium roseum*	Fungus	Necrotrophic	BG
Lee et al. (2012)	Experimental	Applied	NA	*Capsicum annuum*	Crop	*Xanthomonas axonopodis pv. vesicatoria (Xav)*	Bacteria	Biotrophic	AG

(continued)

Table 7.1 (continued)

References	The study			The plant player		The pathogen player			
	Methodological approach	System type (natural vs. applied)	Compartment attacked first	Plant species	Plant type	Pathogen species	Pathogen taxonomic group	Pathogen lifestyle	Pathogen compartment
Lee et al. (2012)	Experimental	Applied	NA	*Capsicum annuum*	Crop	*Xanthomonas axonopodis* pv. *glycines (Xag)*	Bacteria	Biotrophic	AG
Marçais and Bréda (2006)	Experimental	Natural	BG → AG	*Quercus robur*	Tree	*Armillaria gallica*	Fungus	Necrotrophic	BG
McCarville et al. (2012)	Experimental	Applied	BG ↔ AG	*Glycine max*	Crop	*Heterodera glycines*	Nematode	Biotrophic	BG
McCarville et al. (2012)	Experimental	Applied	BG ↔ AG	*Glycine max*	Crop	*Cadophora gregata*	Fungus	Necrotrophic	BG (+AG)
McNee et al. (2003)	Experimental	Applied	BG → AG	*Pinus ponderosa*	Tree	*Heterobasidium annosum*	Fungus	Necrotrophic	BG
Burrill et al. (1999)	Observational	Natural	NA (timing not manipulated)	Various *Quercus* species	Tree	*Armillaria gallica*	Fungus	Necrotrophic	BG

References	The insect player				Consequences of dual attack			
	Herbivore species	Herbivore taxonomic group	Herbivore guild	Insect compartment	On the plant player	On the pathogen player	On the herbivore player	Additional remarks
Godfrey and Yeargan (1989)	*Hypera postica* (alfalfa weevil)	Coleoptera (weevil)	Chewing	AG	Coinfection ↓ plant density Coinfection ↓ plant longevity in a nonadditive	No effect	No effect	BG-herbivore favored BG pathogens

					way, with less reduction than the sum of single attacks by BG pathogen and AG herbivore			
Leath and Byers (1977)	*Acyrthosiphon pisum*	Homoptera (aphid)	Sucking	AG	↓ plant growth and survival in a nonadditive (synergistic) manner	↑ severity	↓ aphid population size	
Leath and Byers (1977)	*Acyrthosiphon pisum*	Homoptera (aphid)	Sucking	AG	↓ plant growth in a nonadditive (synergistic) manner	↑ severity	↓ aphid population size	
Leath and Byers (1977)	*Acyrthosiphon pisum*	Homoptera (aphid)	Sucking	AG	↓ plant growth and survival in a nonadditive (synergistic) manner	↑ severity	↓ aphid population size	
Milanović et al. (2015)	*Lymantria dispar*	Lepidoptera	Chewing	AG	Pathogen modifies leaf water and nutrient content, resulting in ↑ leaf digestibility When given the choice, herbivores consume four times more leaves of pathogen-infected	NA	Herbivore feeding on pathogen-infected trees have higher growth rate and are more efficient at converting ingested food into biomass	

(continued)

Table 7.1 (continued)

	The insect player				Consequences of dual attack			
References	Herbivore species	Herbivore taxonomic group	Herbivore guild	Insect compartment	On the plant player	On the pathogen player	On the herbivore player	Additional remarks
					trees, but consumption is similar in no-choice tests			
Alexander et al. (1981)	*Dendroctonus frontalis*	Coleoptera (bark beetle)	Chewing	AG	Growth rate ↓ in presence of herbivore	Colonization of the root system by the pathogen ↑ in plants attacked by the herbivore	NA	
Lee et al. (2012)	*Myzus persicae*	Homoptera (aphid)	Sucking	AG	Changes in expression of defense priming genes. Recruitment of beneficial soil bacteria	↓ abundance of soil pathogenic bacteria	NA	
Saravesi et al. (2015)	Geometrid	Lepidoptera	Chewing	AG	Defoliation ↓ fine root biomass	Defoliation ↑ the diversity of saprotrophic fungi	NA	
Song et al. (2015)	*Bemisia tabaci*	Hemiptera	Sucking	AG	Dual attack induces SA-mediated plant systemic defenses	↓ *A. tumefaciens* transformation efficiency	NA	

Yang et al. (2011)	*Bemisia tabaci*	Hemiptera	Sucking	AG	Dual attack induces SA-mediated plant systemic defences	Changes in root bacterial communities	NA	
Barber et al. (2015)	*Acalymma vitattum*	Coleoptera (Chrysomelidae)	Chewing	BG (larva) AG (adult)	BG herbivore causes: Pollinator visitation ↓ Leaf production ↓ Lower and seed production ↓ No effect of BG herbivore on AG herbivory	BG herbivore causes: infestation rate by AG pathogen ↓ Host resistance to the pathogen ↑	NA	
Erb et al. (2009)	*Diabrotica virgifera virgifera*	Coleoptera (Chrysomelidae)	Root herbivore	BG	Production of defensive compounds ↑ with coinfection Translocation of secondary metabolites from roots to shoots	Lesions and development ↓ in presence of BG herbivore	NA	Priming of AG defenses is mediated by abscisic acid (ABA) ABA is induced by BG herbivore feeding, not by artificial root removal
Damicone and Manning (1987)	*Ophiomyia simplex*	Diptera	Miner	AG	Mortality ↑ with coinfection	NA	NA	Effect of coinfection possibly contingent on interaction with competitors: significant three

(continued)

Table 7.1 (continued)

	The insect player				Consequences of dual attack			
References	Herbivore species	Herbivore taxonomic group	Herbivore guild	Insect compartment	On the plant player	On the pathogen player	On the herbivore player	Additional remarks
								way interaction between fungicide, insecticide and herbicide treatments No effect of BG pathogen on AG pathogen (other *Fusarium* species)
Godfrey and Yeargan (1987)	*Hypera postica* (alfalfa weevil)	Coleoptera (weevil)	Chewing	AG	Coinfection ↓ crop yield Large spatio-temporal variation in the strength of coinfection effect	NA	NA	
Kostenko et al. (2012)	*Agriotes lineatus* (BG) *Mamestra brassicae* (AG)	Coleoptera (Elateridae) Lepidoptera	Chewing	BG AG	Damage year n has legacy effects on soil microbial community and plant performance year $n + 1$	Not directly measured, but suspected: ↓ secondary metabolite concentration, with possible negative impact on BG pathogens and plant-soil feedback	NA	

Landgraf et al. (2012)	Mechanical leaf wounding	NA			Leaf wounding causes: – local chlorosis – ↑JA concentration in shoots – growth of stunted shoots – growth of stunted roots	Wounding – ↓ frequency pathogen infection – ↓ infection intensity		
Leath and Byers (1977)	*Empoasca fabae*	Hemiptera	Sucking	AG	Only AG herbivore ↓ yield ↓ plant winter survival in a non-additive (synergistic) manner	NA	NA	
Lee et al. (2012)	*Myzus persicae*	Homoptera (aphid)	Sucking	AG	AG herbivore triggers immunity, with no symptoms of BG pathogen	↓ symptoms	↓ AG herbivore density (by 45%)	Treatment with benzothiadiazole had the same effects on plant than aphids
Lee et al. (2012)	*Myzus persicae*	Homoptera (aphid)	Sucking	AG	↓ of hypersensitivity response compared to control	↓ and delayed plant hypersensitivity response to B pathogen	↓ AG herbivore density (by 45%)	Treatment with benzothiadiazole had the same effects on plant than aphids
Marçais and Bréda (2006)	Unidentified	NA	NA	AG	Pathogen affects plant ability to cope with acute stress such as defoliation	NA	NA	

(continued)

Table 7.1 (continued)

	The insect player				Consequences of dual attack			
References	Herbivore species	Herbivore taxonomic group	Herbivore guild	Insect compartment	On the plant player	On the pathogen player	On the herbivore player	Additional remarks
McCarville et al. (2012)	*Aphis glycines*	Homoptera (aphid)	Sucking	AG	NA	Coinfection by fungi and herbivores ↑ nematode fecundity	Coinfection by nematodes and fungi ↓ herbivore population growth	
McCarville et al. (2012)	*Aphis glycines*	Homoptera (aphid)	Sucking	AG	NA	Coinfection by nematodes and herbivores ↓ disease rate	Coinfection by nematodes and fungi ↓ herbivore population growth	
McNee et al. (2003)	*Ips paraconfusus*	Coleoptera (bark beetle)	Chewing	AG	NA	NA	↓ phloem consumption in leaving trees and phloem extracts in laboratory feeding trials	By reviewing the literature, the authors indicate that trees infected by the BG pathogen are more likely to be colonized by the AG herbivore

Burrill et al. (1999)	Unidentified	NA	Chewing	AG	NA	Colonization of the root system by the pathogen ↑ in defoliated stands. No effect of defoliation when sanitation thinning was done before the defoliation	NA	Observations reported at the stand level rather than the tree level Defoliation was not directly attributed to the AG herbivore *Lymantria dispar* but the study was carried out during an outbreak of this pest

The table includes key features of each primary study, taxonomic and functional identity of BG and AG attackers and consequences of BG–AG interactions for the plant, the pathogen, and the herbivore. "Compartment attacked first" indicates the order of the attack: the first attack was belowground (BG → AG), the first attack was aboveground (AG → BG), or below- and aboveground attack took place at the same time (BG ↔ AG). References are organized to mirror numbers in Fig. 7.1. *NA* indicates that the information could not be extracted from the study.

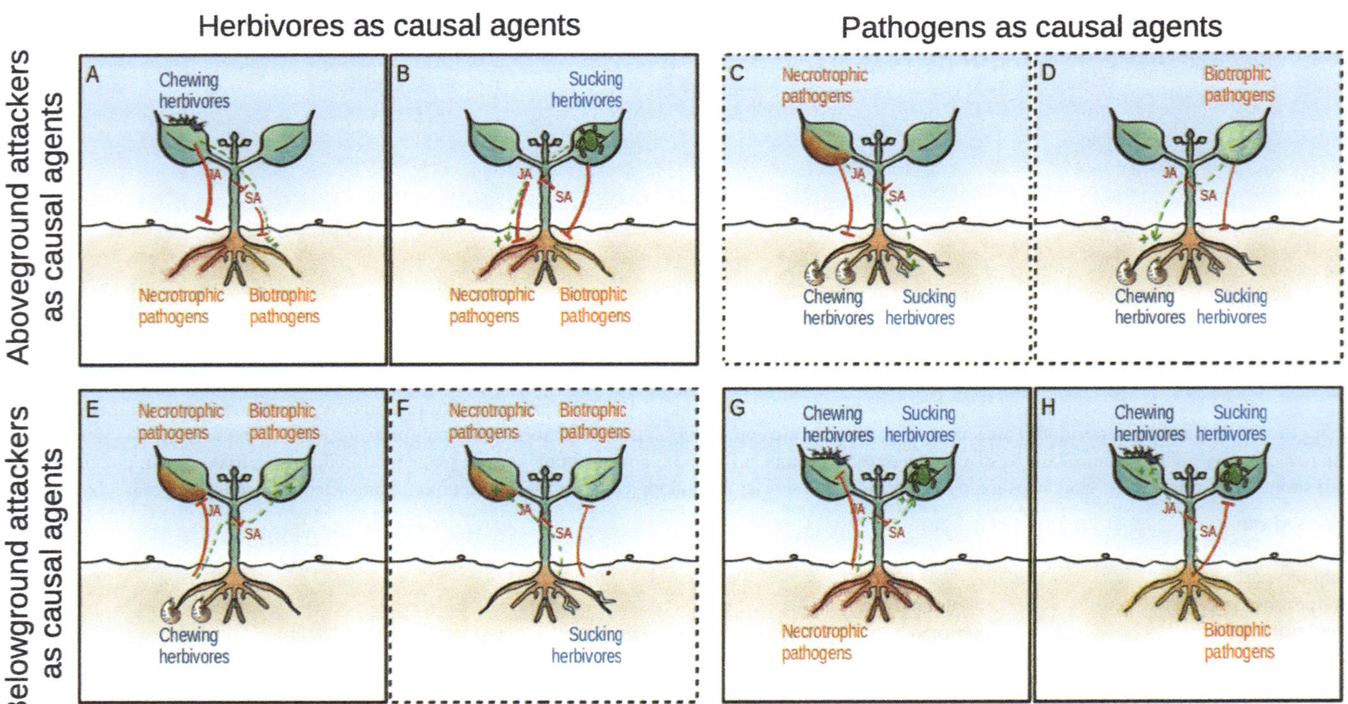

Fig. 7.2 Summary of hormonal pathways and cross-talk antagonisms involving BG and AG herbivores and pathogens. Arrows originating from one compartment indicate the causal effect of the corresponding attacker on the second attacker in the other compartment. Red and green arrows are for predicted negative and positive effects, respectively. Panels (**c**), (**d**), and (**f**) are bordered by a dashed line to indicate scenarios that are likely but for which no specific case study was retrieved

the response to AG herbivores is also reported for nematodes, where AG herbivores have been found to lower the number of plant nematodes (Kutyniok and Müller 2013) or, on the contrary, to make roots more attractive to nematodes (Kammerhofer et al. 2015).

7.2.3 Interactions Between AG Pathogens and BG Herbivores

Very little is known on the interaction between AG pathogens and BG herbivores. In fact, we could not find a single example of the impact of AG pathogens on BG herbivores. Hence, all examples of interactions between AG pathogens and BG herbivores presented in Table 7.1 refer to the effect of BG herbivores on AG pathogens. For instance, root damage by BG larvae of the chrysomelid *Diabrotica virgifera* was shown to induce defenses in maize leaves against the necrotrophic pathogen *Setosphaeria turcica* (Erb et al. 2009). Likewise, root herbivory by the specialist herbivore *Acalymma vitattum* was shown to increase cucumber leaf resistance to downy mildew *Pseudoperonospora cubensis* (Barber et al. 2015). This effect was stronger with higher herbivore abundance. Although these examples are consistent with previous studies reporting negative effects of root herbivory on AG herbivory (Erb et al. 2008), it is clearly premature to draw any generalizations at this stage.

7.2.4 Symmetry of Cross-Compartment Pathogen–Herbivore Interactions

An important question is whether species interactions between BG and AG antagonists are symmetric. Symmetry could take two forms: (1) species A negatively affects species B and species B also negatively affects species A; (2) species A positively affects species B and species B positively affects species A. Unfortunately, few studies simultaneously addressed the effect of a BG attacker on an AG attacker and the reciprocal effect of an AG attacker on a BG attacker (see Table 7.1: Godfrey and Yeargan 1989; McCarville et al. 2012; Lee et al. 2012; Kammerhofer et al. 2015). As one example, Leath and Byers (1977) reported an increased severity of the BG fungal pathogen *Fusarium roseum* when the host plant was simultaneously colonized by aphids, whereas aphid population size decreased on *Fusarium*-infected plants. McCarville et al. (2012) found that dual attack by aphids and the fungus *Cadophora gregata* increased plant infestation by root nematodes, whereas coinfection by nematodes and the fungal pathogen reduced aphid population growth.

7.2.5 General Patterns

Our understanding of BG–AG plant–pathogen–herbivore interactions is still in its infancy, and current evidence reveals a large diversity of interaction outcomes, with both positive, neutral, and negative effects reported on each of the three players. Given the varying responses, identifying general patterns and the factors that modify the direction and strength of the effect will require a large(r) number of studies. Beyond patterns, we may also change our focus to the mechanisms at play and develop a predictive framework. This will be the focus of the next section.

7.3 Mechanisms Shaping BG–AG Interactions Between Pathogens and Herbivores

Few studies in Table 7.1 explore the mechanisms underlying BG–AG interactions between herbivores and pathogens. However, given that a plant's response to herbivores, pathogens, and other organisms involves common signaling pathways and secondary compounds, we may assume that (1) interactions between pathogens (Blodgett et al. 2007), (2) interactions between herbivores (Erb et al. 2008; Johnson et al. 2012), (3) interactions between herbivores and mutualists (Koricheva et al. 2009), and (4) within-compartment interactions between pathogens and herbivores (Fernandez-Conradi et al. 2018) can help to predict the outcome of cross-compartment interactions between pathogens and herbivores (Van der Putten et al. 2001; van Dam and Heil 2011; Biere and Goverse 2016). Notably, while changes in defense-related hormonal pathways received a massive interest, other mechanisms like changes in plant quality, the possible interplay between biotic attackers and abiotic stressors, as well as the ecological and evolutionary consequences of dual attacks are relatively poorly addressed.

In this section, we recapitulate the recognized and putative mechanisms linking pathogens and herbivores across BG and AG compartments. However, as this topic has been extensively reviewed, we aim to be brief, and we refer readers interested in the fine hormonal and physiological mechanisms to the recent and extensive reviews on this topic (e.g., Wondafrash et al. 2013; Lazebnik et al. 2014; Biere and Goverse 2016). Importantly, while this section focuses on mechanisms related to primary and secondary chemistry, interactions may equally likely be mediated by changes in the quantity of the shared resource (the host plant) or, as discussed in Sect. 7.4, by interactions mediated by other members of the plant-associated food web.

7.3.1 Plant-Mediated AG–BG Interactions

7.3.1.1 Effects Mediated by Shared Defenses and Cross-Compartment Signaling

Transportation, induction, and priming—There are three nonexclusive mechanisms by which herbivory or pathogen infection in one compartment can make the other compartment increase its defense or readiness for attack: transportation of defensive secondary compounds, defense induction, and defense priming. Induction is the increase in concentration of secondary metabolites involved in defenses immediately following attack. Defense priming is the pre-activation of mechanisms that make plants able to better or more rapidly mount defense responses against attackers (Prime-A-Plant Group et al. 2006; Martinez-Medina et al. 2016). While translocation and induction directly result in an increase of basal defense levels, priming does not and may go unnoticed if only defensive compounds are targeted. If cross-compartment interactions rely—based on their spatial separation—more on defense priming than within-compartment interactions, BG–AG cross-compartment interactions may have been underestimated because of methodological issues (i.e., a focus on increased levels of compounds).

Several defensive compounds such as nicotine (an alkaloid) are exclusively produced in the roots but are effective against foliar herbivores and can migrate through long-distance transportation to AG parts (Dawson 1941; Kaplan et al. 2008; Bezemer et al. 2013). In tobacco plants, Kaplan et al. (2008) showed that the concentration of alkaloids decreased in shoots after plants were attacked BG by the root-knot nematode *Meloidogyne incognita*, whereas concentrations of chemical compounds synthesized in the shoots increased. From the literature addressing cross-compartment interactions between BG and AG herbivores, it is clear that root herbivory is commonly followed by an increase in basal levels of defenses in shoots, even in the absence of AG damage (reviewed by Erb et al. 2008), which can result from translocation, induction, or both.

The opposite, increase of basal defenses in roots following attacks in shoots, is also possible but more variable in terms of direction and intensity (Erb et al. 2008). AG herbivores and pathogens can induce the production and storage of defensive compounds in roots (which is common for alkaloids such as nicotine, Kaplan et al. 2008) or activate defense-related pathways resulting in the priming or induction of defenses in BG organs (Yang et al. 2011; Landgraf et al. 2012). For example, AG herbivory by the whitefly *Bemisia tabaci* activates the SA-dependent signaling in AG and BG organs, eliciting induced resistance of pepper plants to the soil-borne pathogen *Ralstonia solanacearum* (Yang et al. 2011).

Hormone signaling and shoot–root integration—Plant BG and AG parts are tightly interconnected by the plant vascular system, allowing long-distance communication between roots and shoots. Although plants respond locally to herbivore attack or pathogen infection, plant-level resistance to both pathogens and herbivores requires a complex integration at the plant scale, including root-to-shoot-to-root or shoot-to-root-to-shoot communication loops (reviewed by Biere and Goverse 2016). Such compartments' share of defenses involve uni- or bidirectional exchanges of molecules (e.g., RNA, peptides, phytohormones, or alkaloids) through xylem and phloem vessels (Lucas et al. 2013).

The induction of systemic resistance to herbivores and pathogens is mainly based on chemical defense pathways involving three key hormones acting as major players: salicylic acid (SA), jasmonic acid (JA), and ethylene (ET). A certain specificity in their induction by, and effectiveness against, different groups of herbivores and pathogens has long been assumed. For example, the SA pathway is usually induced by, and efficient against, biotrophic pathogens and sucking herbivores, whereas the JA pathway is principally activated by, and effective against, necrotrophic pathogens and leaf-chewers (Spoel et al. 2007; Ali and Agrawal 2012; Thaler et al. 2012; Lazebnik et al. 2014). In addition, there is a reciprocal cross-talk consisting of an antagonism between SA and JA signaling pathways in several systems (Thaler et al. 2012).

When such cross-talk exists, the impact of dual attack may result in either negative or positive interactions between herbivores and pathogens, where the direction of the interaction is predicted to depend on the specific combination of herbivore feeding guild and pathogen lifestyle (Fig. 7.2). For instance, it has been postulated that plant attack by a BG or AG chewing herbivore may activate the JA-pathway, thereby suppressing SA production, which may be detrimental to necrotrophic pathogens and beneficial to biotrophs in the other compartment (Fig. 7.2, panels a and e). On the contrary, plant attack by sucking herbivores may increase SA levels, and decrease levels of JA, which would benefit necrotrophs and be detrimental to biotrophs (Fig. 7.2, panels b and f). Similarly, plant infection by an AG or BG necrotrophic pathogen may increase JA levels and reduce SA levels, which may benefit piercing-sucking herbivores but be detrimental to chewing herbivores (Fig. 7.2, panels c and g). Finally, infection by an AG or BG biotrophic pathogen may upregulate the SA-pathway and downregulate the JA-pathway, which would be beneficial to chewers and detrimental to piercing-sucking herbivores (Fig. 7.2, panels d and h).

7.3.1.2 Effects Mediated by Altered Plant Nutritional Quality and the Abiotic Environment

Changes in plant nutritional quality and defenses can hardly be teased apart (Van der Putten et al. 2001), both concurring to shape defense syndromes (Agrawal and Fishbein 2006). Indeed, nutrient uptake by the roots does not only affect plant quality, but frequently affects both direct and indirect defenses (i.e., involving a third trophic level). As one example, the density of trichomes, which act as physical barriers against herbivores, as well as volatile compounds, which may be used for parasitoid recruitment, increase with nitrogen uptake (Bernays 1994; Van der Putten et al. 2001). As a consequence, changes in nutrient uptake resulting from root herbivory, infection by pathogens, or changes in abiotic conditions due to

N fertilization may have important consequences in terms of both host plant quality and subsequent defense production.

Effects of BG Damage and Abiotic Factors on AG Tissues and AG Organisms (Fig. 7.3a)—BG attackers can have multiple effects on AG plant quality. These include both changes in primary and secondary metabolites (Hatcher 1995; Van der Putten et al. 2001; Cipollini et al. 2002) and alteration of plant growth pattern and architecture (Bernays 1994; Van der Putten et al. 2001). Yet, these mechanisms can be triggered both by BG attackers, abiotic stresses, or a combination of both. It is, therefore, critical to acknowledge that spatial and temporal variation in the abiotic environment can impact the outcome of cross-compartment plant–pathogen–herbivore interactions.

BG attackers often cause nutrient or water stress (Fig. 7.3a). As such, they can mimic the well-known effects of both abiotic stresses on AG plant parts. For instance, water stress and root infection by *Phytophthora cinnamomi* have similar effects on stomatal conductance and the concentration of abscisic acid, a hormone involved in plant response to drought, in the xylem of chestnut (Maurel et al. 2004). Similarly, Erb et al. (2011) showed that the root herbivore *Diabrotica virgifera* induced changes in the quality of AG tissues that were mediated by the production and translocation of abscisic acid.

Stress-like effects of BG attackers on AG plant parts may cascade on AG herbivores and pathogens. The plant stress hypothesis (White 1974, 2009) predicts an increase in herbivore performances on drought-stressed plants (Gange and Brown 1989). However, whether these stress-induced changes are beneficial or detrimental to AG attackers may depend on their feeding habits (Huberty and Denno 2004) and in particular on whether they target foliage or wood, and healthy or declining trees (Jactel et al. 2012). BG herbivores and pathogens may, similar to water stress (White 1974, 2009), have contrasting effects on AG herbivores depending on the type of tissues they feed on: for example, AG herbivores feeding on young and actively growing leaves (i.e., flush-feeders, *sensu* White 2009) may be more hampered by BG attackers than AG herbivores that feed on older, senescent organs (i.e., senescence-feeders). Indeed, the latter herbivores may even benefit from regulatory mechanisms resulting in the release of soluble sugars and free amino acids in cells (Gutbrodt et al. 2011; Ximénez-Embún et al. 2016).

Abiotic stresses and BG herbivores and pathogens can interactively shape plant–pathogen and plant–herbivore interactions in AG plant parts. For instance, the strength of BG–AG interactions between the nematode *Heterodera schachtii* and aphids was found to be dependent on N-fertilization (Kutyniok and Müller 2013; Kutyniok et al. 2014): in low N-soil, nematodes had no effect on *Brevicoryne brassicae* aphids, whereas aphids increased nematode abundance in roots; in contrast, under high N, aphids reduced nematode abundance and cyst formation (Kutyniok and Müller 2013). On the contrary, the presence of nematodes decreased the abundance of the shoot-infesting aphid *Myzus persicae* only when N supply was low (Kutyniok et al. 2014). A direct consequence of such an interaction between the effects of BG attackers and abiotic factors on AG attackers is that the direction,

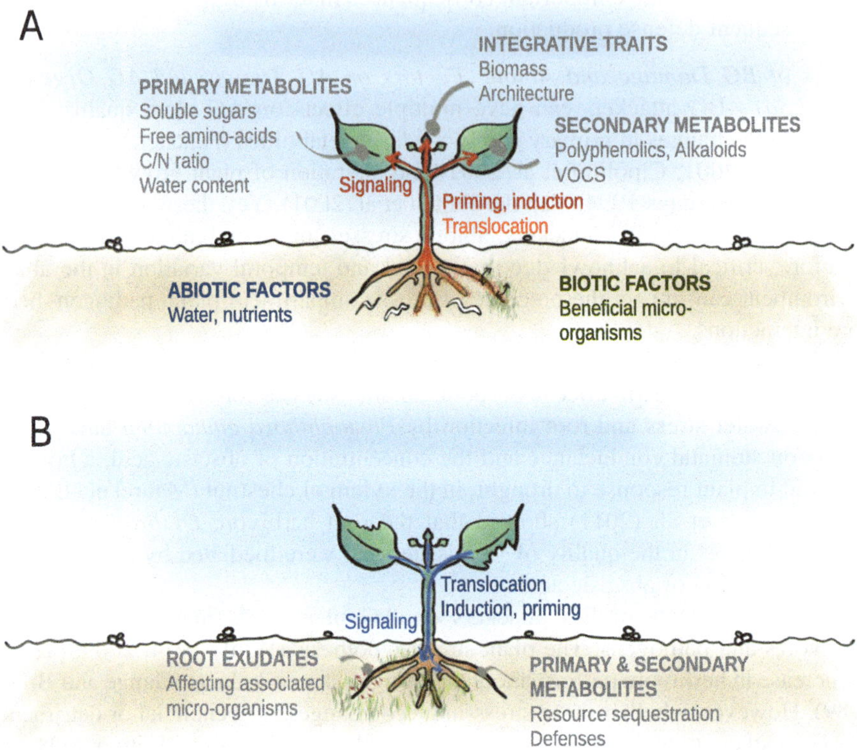

Fig. 7.3 Summary of mechanisms involved in cross-compartment interactions between herbivores and pathogens. (**a**) AG response to BG damage and (**b**) BG response to AG damage

strength, and underlying mechanisms of cross-compartment pathogen–herbivore interactions are expected to vary along abiotic gradients.

Altogether, the presence of spatial and temporal variation in the abiotic environment weakens our ability to infer general patterns on plant–pathogen–herbivore interactions. Yet, this role of the environment in mediating species interactions may have massive implications in agricultural systems where such biotic and abiotic stresses may or may not be controlled (through irrigation, fertilization, and pesticides). It, therefore, appears urgent to better address how abiotic factors can mediate BG–AG plant–pathogen–herbivore interactions.

Effects of AG Damage on the Quality of BG Tissues and BG Organisms (Fig. 7.3b)—AG herbivores and pathogens can strongly affect carbon dynamics and alter carbon allocation to AG and BG compartments (Orians et al. 2011). Foliar herbivory commonly increases resource allocation to roots, thereby reducing its availability to AG plant attackers, which is referred to as *induced resource sequestration* (Orians et al. 2011). Although this strategy may be seen as a way to secure resources and make them unreachable to AG herbivores and pathogens, induced resource sequestration may also have indirect effects on BG pathogens and BG herbivores. These indirect effects can range from positive, when roots act as a sink for photoassimilates, to negative, when these resources are invested and stored in roots as defensive compounds (see Sect. 7.4.1.1 and Biere and Goverse 2016). For instance, AG herbivory was found to increase levels of defensive secondary metabolites in roots, which can reduce plant quality to root herbivores and nematodes (Van Dam et al. 2005).

AG–BG interactions involving changes in plant nutritional quality are, generally, asymmetrical. BG herbivores and pathogens consume or destroy root tissues, which directly reduces the plant's ability to take up water and nutrients. The effects of root consumption propagate through the plant to AG parts, resulting in changes in the nutritional quality of AG plant tissues (e.g., changes in water content or concentration of free amino acids and soluble sugars). While AG herbivores and AG pathogens have also been shown to affect root quality, their systemic effect is generally weaker (Kaplan et al. 2008). Bezemer and van Dam (2005) proposed that such an asymmetry may further result from roots being exposed to herbivores early in the season before leaves are available to herbivores, making the plant ready to face AG herbivores and pathogens before they attack.

7.3.2 Intensity and Timing of Damage

7.3.2.1 The Intensity of Damage and Pathogen Infection

The consequences of BG or AG damage on plant quality, and hence on AG or BG attackers, depend on the amount of damage. However, very few studies manipulated, or even clearly reported, the amount of herbivory or the intensity of the infection (Marçais and Bréda 2006). This seems surprising, as herbivory can range from a few percent to full defoliation, and infections can range from a few lesions, which may increase plant quality due to the mobilization of nutrients, to entirely necrotic foliage or rotting roots (Agrios 2005). In the extreme case, the plants may die, which will dramatically affect the performance or the survival of other organisms feeding on the same plant, with a shift from biotrophic toward necrotrophic (i.e., hemi-biotrophs) and then saprotrophic species. For example, while Cardoza et al. (2003) found a positive effect of the necrotrophic fungus *Sclerotium rolfsii* on the development of *Spodoptera exigua* caterpillars when developing on fungus-infected peanut plants, this pathogen will ultimately kill its host, and the positive effect of infection may then reverse with increased inoculation density and disease progression. As one example, the effect of birch defoliation by geometrid moths on the birch fungal root community differed with the intensity and frequency of the attacks (Saravesi et al. 2015).

7.3.2.2 Does It Matter Who Comes First?

The strength and direction of BG–AG interactions between herbivores and pathogens may also be affected by the relative timing of attacks: herbivores and pathogens may attack the plant at the same time, or one of the species may arrive before the other (Mouttet et al. 2013). Indeed, in the most extreme case, one of the attackers may already be gone from the plant before the other attacker arrives. This naturally excludes any reciprocal effect and leaves us to probe the impact of the first on the second attacker. While this sounds trivial, we stress that this may be rather common in nature, where herbivores may move around and pathogens often have a restricted growing season. Indeed, early-season herbivores are known to have a pronounced impact on herbivore preference, performance, and community structure later in the season, where "later" can be hours, days, weeks, months, or even years (Van Zandt and Agrawal 2004; Stam et al. 2014). Importantly, the plant responses linking the first attacker to the second attacker may take place at different time scales: while induced defenses may take minutes to hours, changes in plant quality and quantity may take longer. Thus, even when the attackers are separated in time, it may be important to take into account the amount of time that has passed between the attack by the first and second attacker.

However, the majority of studies on species interactions focuses on cases where the timing of the two attackers at least partly overlaps. Here, the meta-analysis by Johnson et al. (2012) reported that AG herbivores had strong negative effects on BG herbivores when they attacked first in laboratory studies. In contrast, primary attacks by BG herbivores had only moderately positive and non-significant effects on AG herbivores.

It is critical to acknowledge that the effects of BG attackers on AG attackers, and vice versa, may vary non-linearly with both the intensity of damage and with time. While the hormonal signaling may be relatively fast (Sect. 7.3.1.1), the impact of damage on the quality of root and aerial tissues, or changes mediated by the composition of the other plant-associated biota, may take longer to establish and may last long after the initial damage was caused (Sect. 7.3.1.2). It is, therefore, not only the identity of the first attacker and the attacked compartment that matters for the second player but also the type of changes it induced in the host plant by the time it arrives (e.g., Li et al. 2016).

7.3.3 Annual Versus Perennial Plants: Does It Matter If Interactions Are Reset Every Year?

During their lifetime, perennial plants are exposed to a greater abundance and diversity of pathogens and herbivores than annual plants. Moreover, they experience profound ontogenetic changes in constitutive and induced defenses against different attackers (Boege and Marquis 2005; Barton and Koricheva 2010). They may also be more difficult to study, or, at least, there might be a bias toward more observational field studies for perennial plants such as trees (Marçais and Bréda 2006; Saravesi et al. 2015) and short-term, highly controlled studies for annual plants, including crops (Table 7.1).

The timing and diversity of attackers may differ strongly between annual and perennial plants, and results from short-term highly controlled studies may, therefore, lack relevance for perennial plants. After emergence, the first attacker of annual plants may have a large impact on how the plant will respond to future attacks in the same and opposite compartment (see Sect. 7.3.2.2). In contrast, the response of perennial plants to the first attack of the season may be weakened by a legacy of attacks by pathogens and herbivores in the previous year. Likewise, as large perennials like trees are attacked by a diverse community of herbivores and pathogens, the attack by a single herbivore or pathogen may leave a very weak imprint. In such cases, it seems hard to extrapolate the outcome and effect sizes of short-term and highly controlled laboratory experiments on annual or crop plants to the diversity and complexity of interactions occurring on long-living plants. Notably, there may also be intergenerational legacy effects in annual plants: induced changes in defenses in year $t - 1$ may affect the composition of soil microbial communities, which indirectly affects the next generation of the plant growing within the same soil (Kostenko et al. 2012).

Despite the scarcity of studies documenting cross-compartment interactions among trees, pathogens, and herbivores, forests ecologists have long recognized the importance of dual attacks for tree health. They defined primary pests as those pathogens and herbivores being able to successfully develop and reproduce on healthy trees (Wainhouse 2005). In contrast, secondary pests can only exploit trees that are first weakened by attack from primary pests or by an abiotic stress. For instance, severe defoliation by the Gypsy moth *Lymantria dispar* was shown to alter root chemistry and facilitate root colonization by *Armillaria* spp., a taxon that includes several secondary fungal pathogens and causes root rot (Burrill et al. 1999; Young and Giese 2003; Marçais and Bréda 2006).

7.4 Upscaling Plant–Pathogen–Herbivore Interactions: From Individuals to Communities and Ecosystems

In the previous section, we saw that spatial and temporal variation in the *abiotic* environment affects cross-compartment pathogen–herbivore interactions. This section will focus on how the *biotic* environment affects cross-compartment pathogen–herbivore interactions, and, vice versa, how cross-compartment pathogen–herbivore interactions affect the biotic environment. At the same time, we raise questions about the importance of BG–AG interactions among herbivores and pathogens within a community context.

7.4.1 How Do BG–AG Interactions Among Herbivores and Pathogens Compare to Other Types of Interactions?

As evidenced by Table 7.1, the majority of controlled greenhouse and field studies have demonstrated that herbivores and pathogens can strongly interact with each other, despite the spatial (and in some instances temporal) separation between the herbivore and the pathogen. However, a demonstration in the lab does not automatically translate into relevance in a natural setting. As may be evident, the data available to date does not allow to unambiguously answer the question raised in the section header. Nonetheless, we here make a first attempt to explore the relevance of these BG–AG interactions between herbivores and pathogens in understanding the dynamics of communities in the natural environment.

7.4.1.1 Comparing Within- and Between-Compartment Interactions Between Pathogens and Herbivores (Fig. 7.4a)

The BG and AG plant parts are frequently attacked by a diverse set of pathogens and herbivores, and reviews have highlighted that pathogen–pathogen interactions, herbivore–herbivore interactions, and interactions between herbivores and pathogens within the same compartment can have a major impact on plant-associated community structure (Kaplan and Denno 2007; Tack and Dicke 2013). But if plants are already attacked by a diverse set of herbivores and pathogens within the same compartment, how important and how different are cross-compartment interactions between herbivores and pathogens?

To answer this question, we can compare the relevance (effect size) of BG–AG interactions between pathogens and herbivores with the relevance (effect size) of interactions between pathogens and herbivores within the same compartment. Ideally, we would carry out a meta-analysis and compare studies within and between compartments: for instance, we can investigate whether the effect of BG pathogen infection similarly affects BG and AG herbivores. However, while there is a considerable number of studies on pathogen–herbivore interactions that can be compared through meta-analyses (Fernandez-Conradi et al. 2018), most focus on within compartment interactions. Among the very few studies dedicated to cross-compartment interactions, results are conflicting. For instance, the root necrotrophic pathogen *Heterobasidion annosum* produces phloem metabolites that negatively impact the bark beetle *Ips paraconfusus* (McNee et al. 2003). Interestingly, the effect size of this cross-compartment interaction is -1.22 (SD: ± 0.31), which is slightly stronger than the overall effect size for within-compartment interactions [mean $\pm$ 95%-CI: -0.42 (-0.64, -0.20)]. In contrast, Milanović et al. (2015) found that the performance of Gypsy moth larvae (*Lymantria dispar*) was higher when fed leaves from *Phytophthora*-infected trees than when fed leaves from healthy red oaks.

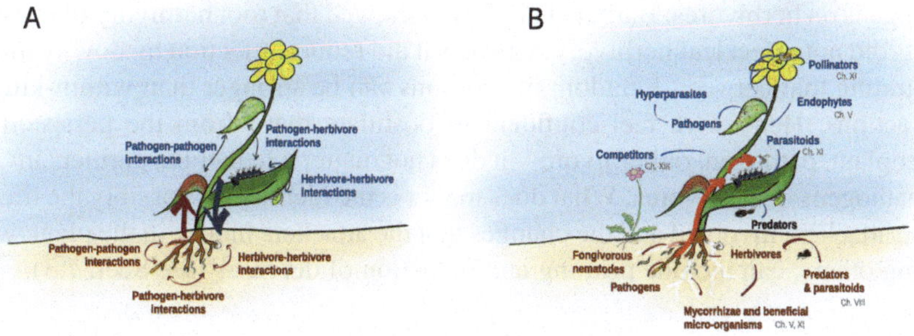

Fig. 7.4 Upscaling plant–pathogen–herbivore interactions to the community level. In panel (**a**) are shown both the interactions among pathogens and herbivores within the same compartment (thin arrows) as well as the interactions among compartments (thick arrows). A major challenge will be to assess the relative importance of within- versus between-compartment interactions and within- versus between-kingdom interactions. In other words: which types of interactions are most important within a community context? Panel (b) illustrates the complex web of multitrophic interactions within which belowground–aboveground interactions are embedded. The red arrows illustrate one possible interaction cascade, where a belowground pathogen affects the preference and density of an aboveground herbivore, which in turn affects the rate of attack by the parasitoid. The response of the parasitoid may be density-mediated (i.e., in response to changes in density of the herbivore) or trait-mediated (e.g., due to changes in the volatile composition of the plant or changes in behavior of the herbivore). In panel (**b**), the roman numerals (in grey font) refer to other chapters within this book

Given the diversity of mechanisms shaping BG–AG interactions among pathogens and herbivores, and their dependency on abiotic factors, it is obvious that these two studies need to be backed up by further research.

7.4.1.2 Comparing Within- and Between-Kingdom Interactions

Studies of interkingdom interactions between herbivores and pathogens are relatively few, as a common approach in entomology and pathology has been to isolate the effect of the focal organism group (insects or pathogens) by the use of insecticides, fungicides, or enclosures (Tack and Dicke 2013). This may be due to the fact that competition for resources has been assumed to increase with species similarity, which precluded much enthusiasm for studies of interactions among species with widely different lifestyles.

While relatively few studies exist on interactions between plant pathogens (Marçais et al. 2011; Kemen 2014), there is a wealth of literature on the interactions between insect herbivores (Denno et al. 1995; Kaplan and Denno 2007). Interestingly, the review by Kaplan and Denno (2007) has demonstrated that interactions among herbivores are highly variable, are similar in magnitude within and among feeding guilds (e.g., sap-sucking herbivores and chewers), and can range from negative to positive. Importantly, the effect sizes reported for these herbivore–herbivore interactions (e.g., Fig. 7.3 in Kaplan and Denno 2007) are within the same range as those reported by Fernandez-Conradi et al. (2017) for the effect of pathogens on herbivores. Barber et al. (2015) showed that root herbivory of cucumber plants did not affect leaf herbivory, whereas it did reduce infection by downy mildew, illustrating that between-kingdom interactions can be stronger than within-kingdom interactions. Hence, we feel confident to postulate that—from the perspective of either plant, pathogen, or herbivore—it does not matter whether the partners involved are pathogens or herbivores. What does matter is the identity, or possibly the lifestyle, of the attackers involved and the changes that the attackers induce in the plant, which, among others, can include priming and induction of defenses (see Sect. 7.3).

7.4.2 *How Do BG–AG Interactions Among Herbivores and Pathogens Affect Other Members of the Community, and Vice Versa?*

Plants are associated with a diverse plant-based community of organisms belonging to different trophic levels (Fig. 7.4b). Moreover, plants are not growing alone, but are embedded within plant communities. This community context may strongly mediate the interactions between pathogens and herbivores. At the same time, interactions between pathogens and herbivores will shape the surrounding community. Clearly, we need a community perspective. So how does the community context affect BG–AG interactions among pathogens and herbivores? And, vice versa, how do BG–AG interactions among pathogens and herbivores affect the surrounding community?

7.4.2.1 Plant Community Structure

The plant community surrounding the focal plant may affect the outcome of single and dual attack by herbivores and pathogens (Fig. 7.4b). For instance, Damicone et al. (1987) reported a significant interaction between fungicide, insecticide, and herbicide treatments, such that yield and survival of *Asparagus officinale* was strongly (and non-additively) reduced by dual-attack of the AG herbivore and BG pathogen in the absence of competitors, whereas dual attack resulted in additive effects on asparagus yield in the presence of competitors (Damicone et al. 1987). This study then suggests that the consequences of single and dual attack by pathogens and herbivores can be modified by the presence of competitors of the host plant. Moreover, the surrounding plant community can affect the likelihood and severity of BG–AG interactions among pathogens and herbivores: the risk of attack by both herbivores and pathogens on a given plant can be lower (i.e., associational resistance) or higher (i.e., associational susceptibility) in the presence of heterospecific neighbors (Underwood 2010; Hantsch et al. 2014; Nguyen et al. 2016). One fascinating direction would also be to focus on plant competitors that are attacked by the same pathogen and herbivore species.

The outcome of single and dual attacks by herbivores and pathogens may also affect the competitive ability of plants relative to conspecifics or heterospecifics and thereby affect the structure of the plant community (Fig. 7.4b). In one example, Godfrey and Yeargan (1987) showed how interactions of early season pests and pathogens changed the density of the surrounding plant community ("weed density") within alfalfa fields. Hopefully, future studies will target natural systems to explore whether BG–AG interactions among pathogens and herbivores result in changes in natural plant communities.

7.4.2.2 Herbivore Community Structure

When herbivore species respond differently to attack by a pathogen in the other compartment, this will result in a change in herbivore community structure. This may be either due to differences in herbivore preference or performance. As an example from an aboveground study, Tack et al. (2012) showed that the community structure of leaf miners and gallers associated with the oak *Quercus robur* changed with the intensity of infection by the powdery mildew *Erysiphe alphitoides* (a foliar biotrophic pathogen).

Given the highly variable response of herbivores to the presence of a pathogen within the same compartment (Tack and Dicke 2013), we can a priori expect that pathogen infection will differentially affect some members of the herbivore community in the opposite compartment, and infection will thereby result in changes in the herbivore community structure.

Lifestyle of the pathogen and herbivore may underlie some of the differences among herbivores in their response to infection (Fig. 7.2). In addition, part of the variability in the cross-talk between AG and BG plant parts could be explained by the degree of herbivore specialization (Kaplan et al. 2008; Ali and Agrawal 2012). As generalist and specialist herbivores differ in their effect on, and response to, qualitative and quantitative defenses (Ali and Agrawal 2012), the nature of changes in foliar quality induced by BG specialists and generalists may profoundly influence the nature of the response of AG specialists and generalists, and vice versa.

No studies have yet addressed how herbivore community structure would affect BG–AG interactions between pathogens and herbivores.

7.4.2.3 Pathogen Community Structure

No studies in Table 7.1 have measured the response of multiple pathogen species to herbivore attack. However, as explained in detail in Sect. 7.3, we may expect differences in response of pathogens to herbivore attack to be affected by the lifestyle of the pathogen (e.g., necrotrophic vs. biotrophic pathogens). Similar to the herbivores, we therefore expect pathogens to respond differently to attack by herbivores within the other compartment.

No studies have yet addressed how pathogen community structure would affect BG–AG interactions between pathogens and herbivores.

7.4.2.4 Microbial Community Structure

Soil biota may mediate the interactions between BG and AG attackers. For instance, root herbivores can affect root colonization by mycorrhizal fungi (reviewed by Johnson and Rasmann 2015), with consequences for plant nutrition and defense (Gange 2000). Such changes in the BG community of plant-associated beneficial organism may provide an indirect link between BG and AG attackers. However, the direction and strength of the effect of BG herbivores and pathogens on mycorrhizal fungi was reported to range from negative (Bennett et al. 2013) to neutral (Gange 2001) or even positive (Currie et al. 2006). Hence, predicting the strength and direction of mycorrhiza-mediated effects of BG herbivores and pathogens on AG attackers may be difficult (Chap. 5).

Likewise, AG herbivores and pathogens may change the quality and defense of BG plant tissue, with consequences for the soil biota (Gehring and Bennett 2009; Heath and Lau 2011). For example, defoliation of mountain birches by geometrid moths caused subsequent changes in taxonomic and functional composition of root fungal communities (Saravesi et al. 2015), and resource sequestration in roots following herbivore damage in AG organs was shown to influence root exudation by the grass *Poa pratensis* (Hamilton et al. 2008), which in turn may impact associated soil microorganisms (Kostenko et al. 2012). In an interesting study, Barber et al. (2015) assessed the impact of root herbivory on both root colonization by arbuscular mycorrhizal fungi and leaf infection by downy mildew; while both response variables were affected by root herbivory, it seems unlikely that arbuscular mycorrhizal fungi mediated the response of the pathogen to the root herbivore: root colonization was lowest at intermediate herbivory, whereas foliar infection was highest in the absence of herbivory.

We predict that the aboveground microbial community, including bacterial and fungal endophytes and epiphytes, may act as the aboveground equivalent of the soil biota and play an equally important role in mediating interactions between the BG and AG compartments (Jaber and Vidal 2010; Menjivar et al. 2012; Vacher et al. 2016). Overall, the role of microbes in mediating the response of the plant to BG and AG attack would be a promising avenue for future research.

7.4.2.5 Higher Trophic Levels

BG–AG interactions among pathogens and herbivores may also affect higher trophic levels (Bezemer et al. 2005). The attack of roots by pathogens and herbivores may induce volatile organic compounds (VOCs) that attract the natural enemies of herbivores, like parasitoids or entomopathogenic nematodes (Rasmann et al. 2005). Root exudates may play a similar signaling role within the belowground compartment. Notably, the impact of AG pathogens and herbivores on belowground natural enemies may involve both BG parasitoids and parasitic nematodes, as the latter play a particularly important role in the soil community (Strong et al. 1999). However, most of the existing studies taking into account natural enemies focused on within-compartment interactions between pathogens and herbivores (e.g., Cardoza et al. 2003; Tack et al. 2012) or cross-compartment interactions between herbivores (Soler et al. 2005, 2007).

We hope that future studies will address the impact of cross-compartment interactions between pathogens and herbivores on both natural enemy attack and the multitrophic community structure. Likewise, future studies may investigate whether induced changes in plant quality, VOCs, and root exudates also affect the performance of natural enemies of pathogens (e.g., fungal hyperparasites and snails). To our knowledge, no study has been dedicated to this topic.

7.4.3 *How Do BG–AG Interactions Among Herbivores and Pathogens Affect Ecosystem Dynamics?*

Given the strong impact of BG–AG interactions among herbivores and pathogens on plant performance and community composition, it seems likely that such interactions will also affect ecosystem processes like carbon dynamics, water dynamics, and decomposition in the litter layer. Alternatively, even a strong pathogen–herbivore interaction may leave only a weak imprint at the ecosystem level. We eagerly await studies that explore this.

7.5 Further Avenues for Future Research

As stated above, the available literature specifically addressing cross-compartment interactions between pathogens and herbivores is very small (Table 7.1). Hence, despite the speculations and predictions provided in this chapter, we think that no general patterns can be safely inferred, and we refrain from a final synthesis. In the previous section, we already discussed the need to assess the relative importance of plant–pathogen–herbivore interactions within a community and ecosystem context. Here, we hope to stimulate future research by outlining additional open questions with tentative predictions.

7.5.1 Are the Outcomes of Short-Term Laboratory Experiments and Observational Studies in Natural and Agricultural Systems Comparable?

Prediction The evidence seems inconsistent. Focusing on herbivores, Johnson et al. (2012) reported stronger cross-compartment interactions in short-term experiments than in observational studies. In contrast, Fernandez-Conradi et al. (2018) reported similar effect sizes for the impact of plant pathogens on herbivores when studies were conducted under highly controlled experiments or in the field. Overall, we expect that short-term experiments will be reflected to some degree in the field, although the effect sizes may generally be lower: short-term experiments do not take into account all variation or complexity (e.g., neighboring plants, abiotic, and biotic variation in the environment) and thereby are sometimes informative and sometimes not.

7.5.2 Can We Predict the Outcome of BG–AG Interactions Between Pathogens and Herbivores?

Prediction Yes, but only to some degree. Interactions may vary predictably as based on the pathogen and herbivore lifestyle (Fig. 7.2; Thaler et al. 2012; Biere and Goverse 2016) and specialization (Ali and Agrawal 2012; Thaler et al. 2012; Biere and Goverse 2016). Superimposed on this are the idiosyncrasies of the study system and variation in the outcome due to the abiotic and biotic environment (Sects. 7.3 and 7.4).

Suggestion for Future Studies To improve our understanding of the generality and mechanisms at play, we recommend studies to consider multiple herbivores or pathogens within the same study system (e.g., Kaplan et al. 2008; McCarville et al. 2012; Barber et al. 2015). To facilitate meta-analyses, we ask authors to systematically report detailed information on the biology of the studied attackers (notably their degree of specialization and the plant organs they damage) and include the sample size, the mean, and the variability for each experimental result, even for differences that are not statistically significant among treatments.

7.5.3 What About Other Animals, Like Viruses and Mammalian Herbivores?

Prediction Other organisms, like viruses and mammalian herbivores, are of major importance. In particular, there is an extensive literature on vector-transmitted viruses, which we ignored in this book chapter for two reasons. First, the distinction between BG and AG viruses is frequently unclear (admittedly, bacterial and fungal pathogens can also become systemic). Second, many viruses are transmitted by vectors, and the interaction between viruses and herbivores becomes very complex. Hence, we think that viruses are best treated separately. We did not find any studies on BG–AG interactions among pathogens and mammalian herbivores. But as both BG mammalian herbivores (like meadow voles eating roots) and AG mammalian herbivores (like grazers) play an important role in plant performance, we do think that BG–AG interactions between pathogens and mammalian herbivores are worth exploring. The strong impact of grass endophytes on grazers provides one example of the potential role of microbes on grazers; conversely, mammals may facilitate the entrance of pathogens into their plant host.

7.5.4 Are BG–AG Interactions Between Pathogens and Herbivores Symmetric?

Prediction We predict the absence of a general pattern of symmetry in BG–AG interactions among pathogens and herbivores. Symmetry in the direction of the effect may depend on the lifestyle of the pathogen and herbivore (Fig. 7.2). Symmetry in the strength of the effect (e.g., effect size) has not been studied for pathogen–herbivore interactions, but was notably absent for herbivore–herbivore interactions (Kaplan and Denno 2007). Because BG–AG interactions between herbivores and pathogens partially involve the same signaling pathways and may have comparable effects on the shared host plant, we expect symmetry in the strength of the effect to be absent for pathogen–herbivore interactions too. However, we note that the different metrics of herbivore and pathogen performance makes a quantitative comparison more difficult.

Most studies investigate unidirectional effects. While this is logical for studies where the first attacker is gone before the arrival of the second attacker, it seems more surprising for cases where attack by the herbivore and pathogen (partly) overlap.

7.5.5 Are There AG–BG–AG or BG–AG–BG Feedbacks?

Prediction Feedbacks are—within the context of BG–AG interactions among pathogens and herbivores—*terra incognita*. It would be fascinating to explore whether, for example, an early-season root herbivore can affect a foliar pathogen later in the season, which in turn affects BG herbivory.

7.5.6 What Is the Role of the Abiotic and Biotic Environment in Mediating BG–AG Interactions Among Pathogens and Herbivores? Can We Predict the Impact of Climate Change?

Prediction There are several studies showing that not only the strength, but also the sign, of species interactions can change with the abiotic or biotic environment (Chamberlain et al. 2014). This may be due to the fact that the hormonal signaling pathways involved in responses to herbivores and pathogens such as SA, JA, or even ET are often also involved in, and show cross-talk with hormones involved in responses to abiotic stresses such as ABA (Pieterse et al. 2012). As such, we predict that BG–AG interactions among pathogens and herbivores are variable in space and time. However, we feel it is too early to postulate in what context, and what way, the environment matters. It would be interesting to explore the relative importance of the abiotic environment and the biotic environment (and their interactions) on cross-compartment plant–pathogen–herbivore interactions. While it is to be expected that BG–AG plant–pathogen–herbivore interactions will be modified by climate change, we have no explicit predictions for what may happen.

7.5.7 What Are the Evolutionary Consequences of BG–AG Interactions Among Herbivores and Pathogens?

The outcome of BG–AG interactions among pathogens and herbivores may be affected by genetic variation within both the plant, pathogen, and herbivore (Biere and Tack 2013). However, few studies on pathogen–herbivore interactions have used multiple genotypes. McCarville et al. (2012) used six cultivars of soybean *Glycine max* that varied in their resistance to the soybean cyst nematode *Heterodera glycines* and showed that the interaction between the AG herbivore *Aphis glycines* and the BG pathogen *Cadophora gregata* varied between resistant vs. sensitive cultivars.

The non-additivity of single and dual attack on plant performance may affect selection on plant resistance (Biere and Tack 2013). As a hypothetical example, the negative impact of a common BG plant pathogen on plant performance may turn neutral, or even positive, in the presence of an AG herbivore. If so, the plant would not undergo selection for increased resistance to the pathogen in the presence of the AG herbivore. Moreover, negative effects of herbivore and pathogen attack on plant performance may be offset by beneficial indirect effects on other community members. However, in a study on the effects of root herbivory on the associated community of cucumber, Barber et al. (2015) showed that direct negative interactions on plant fitness were more important than indirect interactions with other community members: direct damage inflicted by a root herbivore was not compensated by indirect effects on mycorrhizal colonization, pollination, or foliar infection rates. The impact of single and dual attack by pathogens and herbivores on the evolution of plant resistance and tolerance would be an interesting research direction.

Likewise, BG–AG interactions among herbivores and pathogens may affect selection on both the pathogen and the herbivore (Biere and Tack 2013). As an empirical example, the selection pressure exerted by the presence of root-feeding nematodes on the common bean *Phaseolus vulgaris* resulted in spider mites (*Tetranychus urticae*) adapted to perform better on nematode-infected than nematode-free plants within the time span of ten mite generations (Bonte et al. 2010). As another example of soil-mediated selection, the perennial herb *Plantago lanceolata* showed higher resistance against its specialist powdery mildew *Podosphaera plantaginis* when growing in association with its local soil biota (Mursinoff and Tack 2017).

References

Agrawal AA, Fishbein M (2006) Plant defense syndromes. Ecology 87:S132–S149

Agrios GN (2005) Plant pathology, 5th edn. Elsevier Academic, Amsterdam

Alexander SA, Skelly JM, Webb RS (1981) Effects of *Heterobasidion annosum* on radial growth in Southern pine beetle-infested loblolly pine. Phytopathology 71:479–481

Ali JG, Agrawal AA (2012) Specialist versus generalist insect herbivores and plant defense. Trends Plant Sci 17:293–302. https://doi.org/10.1016/j.tplants.2012.02.006

Barber NA, Milano NJ, Kiers ET et al (2015) Root herbivory indirectly affects above- and belowground community members and directly reduces plant performance. J Ecol 103:1509–1518. https://doi.org/10.1111/1365-2745.12464

Barton KE, Koricheva J (2010) The ontogeny of plant defense and herbivory: characterizing general patterns using meta-analysis. Am Nat 175:481–493. https://doi.org/10.1086/650722

Bennett AE, Macrae AM, Moore BD et al (2013) Early root herbivory impairs arbuscular mycorrhizal fungal colonization and shifts defence allocation in establishing *Plantago lanceolata*. PLoS One 8:e66053. https://doi.org/10.1371/journal.pone.0066053

Bernays EA (ed) (1994) Insect-plant interactions. CRC Press, Boca Raton

Bezemer TM, van Dam NM (2005) Linking aboveground and belowg ground interactions via induced plant defenses. Trends Ecol Evol 20(11):617–624

Bezemer TM, De Deyn GB, Bossinga TM et al (2005) Soil community composition drives aboveground plant-herbivore-parasitoid interactions: plant-herbivore-parasitoid interactions. Ecol Lett 8:652–661. https://doi.org/10.1111/j.1461-0248.2005.00762.x

Bezemer TM, van der Putten WH, Martens H et al (2013) Above- and below-ground herbivory effects on below-ground plant-fungus interactions and plant-soil feedback responses. J Ecol 101:325–333. https://doi.org/10.1111/1365-2745.12045

Biere A, Goverse A (2016) Plant-mediated systemic interactions between pathogens, parasitic nematodes, and herbivores above- and belowground. Annu Rev Phytopathol 54: 499–527. https://doi.org/10.1146/annurev-phyto-080615-100245

Biere A, Tack AJM (2013) Evolutionary adaptation in three-way interactions between plants, microbes and arthropods. Funct Ecol 27:646–660. https://doi.org/10.1111/1365-2435.12096

Blodgett JT, Eyles A, Bonello P (2007) Organ-dependent induction of systemic resistance and systemic susceptibility in *Pinus nigra* inoculated with *Sphaeropsis sapinea* and *Diplodia scrobiculata*. Tree Physiol 27:511–517. https://doi.org/10.1093/treephys/27.4.511

Boege K, Marquis RJ (2005) Facing herbivory as you grow up: the ontogeny of resistance in plants. Trends Ecol Evol 20:441–448. https://doi.org/10.1016/j.tree.2005.05.001

Bonte D, Roissart AD, Vandegehuchte ML et al (2010) Local adaptation of aboveground herbivores towards plant phenotypes induced by soil biota. PLoS One 5:e11174. https://doi.org/10.1371/journal.pone.0011174

Burrill EA, Worrall JJ, Wargo PM et al (1999) Effects of defoliation and cutting in eastern oak forests on *Armillaria spp.* and a competitor, *Megacollybia platyphylla*. Can J For Res 29: 347–355. https://doi.org/10.1139/x99-001

Cardoza YJ, Lait CG, Schmelz EA et al (2003) Fungus-induced biochemical changes in peanut plants and their effect on development of beet armyworm, *Spodoptera exigua* Hübner (Lepidoptera: Noctuidae) larvae. Environ Entomol 32:220–228

Chamberlain SA, Bronstein JL, Rudgers JA (2014) How context dependent are species interactions? Ecol Lett 17:881–890. https://doi.org/10.1111/ele.12279

Cipollini ML, Paulk E, Cipollini DF (2002) Effect of nitrogen and water treatment on leaf chemistry in horsenettle (*Solanum carolinense*), and relationship to herbivory by flea beetles (*Epitrix spp.*) and tobacco hornworm (*Manduca sexta*). J Chem Ecol 28:2377–2398

Currie AF, Murray PJ, Gange AC (2006) Root herbivory by *Tipula paludosa* larvae increases colonization of *Agrostis capillaris* by arbuscular mycorrhizal fungi. Soil BiolBiochem 38: 1994–1997. https://doi.org/10.1016/j.soilbio.2005.11.015

Damicone JP, Manning WJ (1987) Influence of management practices on severity of stem and crown rot, incidence of Asparagus miner, and yield of Asparagus grown from transplants. Plant Dis 71(1):81–84

Damicone JP, Manning WJ, Ferro DN (1987) Influence of management practices on severity of stem and crown rot, incidence of asparagus miner, and yield of asparagus grown from transplants. Plant Dis 71:81–84

Dawson RF (1941) The localization of nicotine synthetic mechanism in the tobacco plant. Science 94:396–397. https://doi.org/10.1126/science.94.2443.396

De Roissart A, de la Peña E, Van Oyen L et al (2013) The presence of root-feeding nematodes – Not AMF – affects an herbivore dispersal strategy. Acta Oecol 52:38–44. https://doi.org/10.1016/j.actao.2013.07.005

Denno RF, McClure MS, Ott JR (1995) Interspecific interactions in phytophagous insects: competition reexamined and resurrected. Annu Rev Entomol 40:297–331. https://doi.org/10.1146/annurev.en.40.010195.001501

Erb M, Flors V, Karlen D et al (2009) Signal signature of aboveground-induced resistance upon belowground herbivory in maize. Plant J 59:292–302. https://doi.org/10.1111/j.1365-313X.2009.03868.x

Erb M, Köllner TG, Degenhardt J et al (2011) The role of abscisic acid and water stress in root herbivore-induced leaf resistance. New Phytol 189:308–320. https://doi.org/10.1111/j.1469-8137.2010.03450.x

Erb M, Ton J, Degenhardt J et al (2008) Interactions between arthropod-induced aboveground and belowground defenses in plants. Plant Physiol 146:867–874. https://doi.org/10.1104/pp.107.112169

Fernandez-Conradi P, Jactel H, Robin C et al (2018) Fungi reduce preference and performance of insect herbivores on challenged plants. Ecology 99:300–311

Gange A (2000) Arbuscular mycorrhizal fungi, Collembola and plant growth. Trends Ecol Evol 15: 369–372. https://doi.org/10.1016/S0169-5347(00)01940-6

Gange AC (2001) Species-specific responses of a root- and shoot-feeding insect to arbuscular mycorrhizal colonization of its host plant. New Phytol 150:611–618. https://doi.org/10.1046/j.1469-8137.2001.00137.x

Gange AC, Brown VK (1989) Effects of root herbivory by an insect on a foliar-feeding species, mediated through changes in the host plant. Oecologia 81:38–42

Gehring C, Bennett A (2009) Mycorrhizal fungal-plant-insect interactions: the importance of a community approach. Environ Entomol 38:93–102. https://doi.org/10.1603/022.038.0111

Godfrey LD, Yeargan KV (1987) Effects and interactions of early season pests on alfalfa yield in Kentucky. J Econ Entomol 80:248–256

Godfrey LD, Yeargan KV (1989) Effects of clover root curculio, alfalfa weevil (Coleoptera: Curculionidae), and soil-borne fungi on alfalfa stand density and longevity in Kentucky. J Econ Entomol 82:1749–1756

Gutbrodt B, Mody K, Dorn S (2011) Drought changes plant chemistry and causes contrasting responses in lepidopteran herbivores. Oikos 120:1732–1740. https://doi.org/10.1111/j.1600-0706.2011.19558.x

Hamilton EW, Frank DA, Hinchey PM et al (2008) Defoliation induces root exudation and triggers positive rhizospheric feedbacks in a temperate grassland. Soil Biol Biochem 40: 2865–2873. https://doi.org/10.1016/j.soilbio.2008.08.007

Hantsch L, Bien S, Radatz S et al (2014) Tree diversity and the role of non-host neighbour tree species in reducing fungal pathogen infestation. J Ecol 102:1673–1687. https://doi.org/10.1111/1365-2745.12317

Hatcher PE (1995) Three-way interactions between plant-pathogenic fungi, herbivorous insects and their host plants. Biol Rev 70:639–694. https://doi.org/10.1111/j.1469-185X.1995.tb01655.x

Hauser TP, Christensen S, Heimes C et al (2013) Combined effects of arthropod herbivores and phytopathogens on plant performance. Funct Ecol 27:623–632. https://doi.org/10.1111/1365-2435.12053

Heath KD, Lau JA (2011) Herbivores alter the fitness benefits of a plant–rhizobium mutualism. Acta Oecol 37:87–92. https://doi.org/10.1016/j.actao.2010.12.002

Hong SC, MacGuidwin A, Gratton C (2011) Soybean aphid and soybean cyst nematode interactions in the field and effects on soybean yield. J Econ Entomol 104:1568–1574. https://doi.org/10.1603/EC11084

Huberty AF, Denno RF (2004) Plant water stress and its consequences for herbivorous insects: a new synthesis. Ecology 85:1383–1398. https://doi.org/10.1890/03-0352

Hunter WB, Backus EA (1989) Mesophyll-feeding by the potato leafhopper, *Empoasca fabae* (Homoptera: Cicadellidae): results from electronic monitoring and thin-layer chromatography. Environ Entomol 18:465–472. https://doi.org/10.1093/ee/18.3.465

Jaber LR, Vidal S (2010) Fungal endophyte negative effects on herbivory are enhanced on intact plants and maintained in a subsequent generation. Ecol Entomol 35:25–36. https://doi.org/10.1111/j.1365-2311.2009.01152.x

Jactel H, Petit J, Desprez-Loustau M-L et al (2012) Drought effects on damage by forest insects and pathogens: a meta-analysis. Glob Chang Biol 18:267–276. https://doi.org/10.1111/j.1365-2486.2011.02512.x

Johnson SN, Clark KE, Hartley SE et al (2012) Aboveground–belowground herbivore interactions: a meta-analysis. Ecology 93:2208–2215. https://doi.org/10.1890/11-2272.1

Johnson SN, Rasmann S (2015) Root-feeding insects and their interactions with organisms in the rhizosphere. Annu Rev Entomol 60:517–535. https://doi.org/10.1146/annurev-ento-010814-020608

Kammerhofer N, Egger B, Dobrev P et al (2015) Systemic above- and belowground cross talk: hormone-based responses triggered by *Heterodera schachtii* and shoot herbivores in *Arabidopsis thaliana*. J Exp Bot 66:7005–7017. https://doi.org/10.1093/jxb/erv398

Kaplan I, Denno RF (2007) Interspecific interactions in phytophagous insects revisited: a quantitative assessment of competition theory. Ecol Lett 10:977–994. https://doi.org/10.1111/j.1461-0248.2007.01093.x

Kaplan I, Halitschke R, Kessler A et al (2008) Constitutive and induced defenses to herbivory in above- and belowground plant tissues. Ecology 89:392–406. https://doi.org/10.1890/07-0471.1

Kemen E (2014) Microbe–microbe interactions determine oomycete and fungal host colonization. Curr Opin Plant Biol 20:75–81. https://doi.org/10.1016/j.pbi.2014.04.005

Koricheva J, Gange AC, Jones T (2009) Effects of mycorrhizal fungi on insect herbivores: a meta-analysis. Ecology 90:2088–2097. https://doi.org/10.1890/08-1555.1

Kostenko O, van de Voorde TFJ, Mulder PPJ et al (2012) Legacy effects of aboveground-belowground interactions. Ecol Lett 15:813–821. https://doi.org/10.1111/j.1461-0248.2012.01801.x

Kutyniok M, Müller C (2013) Plant-mediated interactions between shoot-feeding aphids and root-feeding nematodes depend on nitrate fertilization. Oecologia 173: 1367–1377. https://doi.org/10.1007/s00442-013-2712-x

Kutyniok M, Persicke M, Müller C (2014) Effects of root herbivory by nematodes on the performance and preference of a leaf-infesting generalist aphid depend on nitrate fertilization. J Chem Ecol 40:118–127. https://doi.org/10.1007/s10886-014-0387-3

Landgraf R, Schaarschmidt S, Hause B (2012) Repeated leaf wounding alters the colonization of Medicago truncatula roots by beneficial and pathogenic microorganisms: effects of leaf wounding on root colonization. Plant Cell Environ 35:1344–1357. https://doi.org/10.1111/j.1365-3040.2012.02495.x

Lazebnik J, Frago E, Dicke M et al (2014) Phytohormone mediation of interactions between herbivores and plant pathogens. J Chem Ecol 40:730–741. https://doi.org/10.1007/s10886-014-0480-7

Leath KT, Byers RA (1977) Interaction of Fusarium root rot with pea aphid and potato leafhopper feeding on forage legumes. Phytopathology 67:226–229

Lee B, Lee S, Ryu C-M (2012) Foliar aphid feeding recruits rhizosphere bacteria and primes plant immunity against pathogenic and non-pathogenic bacteria in pepper. Ann Bot 110:281–290. https://doi.org/10.1093/aob/mcs055

Li Y, Stam JM, Poelman EH et al (2016) Community structure and abundance of insects in response to early-season aphid infestation in wild cabbage populations. Ecol Entomol 41:378–388. https://doi.org/10.1111/een.12308

Lucas WJ, Groover A, Lichtenberger R et al (2013) The plant vascular system: evolution, development and functions. J Integr Plant Biol 55:294–388. https://doi.org/10.1111/jipb.12041

Marçais B, Bréda N (2006) Role of an opportunistic pathogen in the decline of stressed oak trees. J Ecol 94:1214–1223. https://doi.org/10.1111/j.1365-2745.2006.01173.x

Marçais B, Caël O, Delatour C (2011) Interaction between root rot basidiomycetes and *Phytophthora* species on pedunculate oak. Plant Pathol 60:296–303. https://doi.org/10.1111/j.1365-3059.2010.02378.x

Martinez-Medina A, Flors V, Heil M et al (2016) Recognizing plant defense priming. Trends Plant Sci 21:818–822. https://doi.org/10.1016/j.tplants.2016.07.009978-0-471-33145-2

Maurel M, Robin C, Simonneau T et al (2004) Stomatal conductance and root-to-shoot signalling in chestnut saplings exposed to *Phytophthora cinnamomi* or partial soil drying. Funct Plant Biol 31:41–51

McCarville MT, O'Neal M, Tylka GL et al (2012) A nematode, fungus, and aphid interact via a shared host plant: implications for soybean management: interactions among a nematode, fungus, and aphid. Entomol Exp Appl 143:55–66. https://doi.org/10.1111/j.1570-7458.2012.01227.x

McNee WR, Bonello P, Storer AJ et al (2003) Feeding response of *Ips paraconfusus* to phloem and phloem metabolites of *Heterobasidion annosum*–inoculated ponderosa pine, *Pinus ponderosa*. J Chem Ecol 29:1183–1202

Menjivar RD, Cabrera JA, Kranz J et al (2012) Induction of metabolite organic compounds by mutualistic endophytic fungi to reduce the greenhouse whitefly *Trialeurodes vaporariorum* (Westwood) infection on tomato. Plant and Soil 352:233–241. https://doi.org/10.1007/s11104-011-0991-8

Milanović S, Lazarević J, Karadžić D et al (2015) Belowground infections of the invasive *Phytophthora plurivora* pathogen enhance the suitability of red oak leaves to the generalist herbivore *Lymantria dispar*: Phytophthora infections benefit *Lymantria dispar*. Ecol Entomol 40:479–482. https://doi.org/10.1111/een.12193

Mouttet R, Kaplan I, Bearez P et al (2013) Spatiotemporal patterns of induced resistance and susceptibility linking diverse plant parasites. Oecologia 173:1379–1386. https://doi.org/10.1007/s00442-013-2716-6

Mursinoff S, Tack AJM (2017) Spatial variation in soil biota mediates plant adaptation to a foliar pathogen. New Phytol 214:644–654. https://doi.org/10.1111/nph.14402

Nguyen D, Castagneyrol B, Bruelheide H et al (2016) Fungal disease incidence along tree diversity gradients depends on latitude in European forests. Ecol Evol 6:2426–2438. https://doi.org/10.1002/ece3.2056

Orians CM, Thorn A, Gómez S (2011) Herbivore-induced resource sequestration in plants: why bother? Oecologia 167(1). https://doi.org/10.1007/s00442-011-1968-2

Pieterse CMJ, Van der Does D, Zamioudis C et al (2012) Hormonal modulation of plant immunity. Annu Rev Cell Dev Biol 28:489–521. https://doi.org/10.1146/annurev-cellbio-092910-154055

Prime-A-Plant Group, Conrath U, Beckers GJM et al (2006) Priming: getting ready for battle. Mol Plant Microbe Interact 19:1062–1071. https://doi.org/10.1094/MPMI-19-1062

Rasmann S, Köllner TG, Degenhardt J et al (2005) Recruitment of entomopathogenic nematodes by insect-damaged maize roots. Nature 434:732–737. https://doi.org/10.1038/nature03451

Saravesi K, Aikio S, Wäli PR et al (2015) Moth outbreaks alter root-associated fungal communities in subarctic mountain birch forests. Microb Ecol 69:788–797. https://doi.org/10.1007/s00248-015-0577-8

Soler R, Bezemer TM, Van Der Putten WH et al (2005) Root herbivore effects on above-ground herbivore, parasitoid and hyperparasitoid performance via changes in plant quality. J Anim Ecol 74:1121–1130. https://doi.org/10.1111/j.1365-2656.2005.01006.x

Soler R, Harvey JA, Kamp AFD et al (2007) Root herbivores influence the behaviour of an above-ground parasitoid through changes in plant-volatile signals. Oikos 116:367–376. https://doi.org/10.1111/j.0030-1299.2007.15501.x

Song GC, Lee S, Hong J et al (2015) Aboveground insect infestation attenuates belowground *Agrobacterium*- mediated genetic transformation. New Phytol 207:148–158. https://doi.org/10.1111/nph.13324

Spoel SH, Johnson JS, Dong X (2007) Regulation of tradeoffs between plant defenses against pathogens with different lifestyles. Proc Natl Acad Sci USA 104:18842–18847. https://doi.org/10.1073/pnas.0708139104

Stam JM, Kroes A, Li Y et al (2014) Plant interactions with multiple insect herbivores: from community to genes. Annu Rev Plant Biol 65:689–713. https://doi.org/10.1146/annurev-arplant-050213-035937

Strong DR, Whipple AV, Child AL et al (1999) Model selection for a subterranean trophic cascade: root-feeding caterpillars and entomopathogenic nematodes. Ecology 80:2750–2761

Tack AJM, Dicke M (2013) Plant pathogens structure arthropod communities across multiple spatial and temporal scales. Funct Ecol 27:633–645. https://doi.org/10.1111/1365-2435.12087

Tack AJM, Gripenberg S, Roslin T (2012) Cross-kingdom interactions matter: fungal-mediated interactions structure an insect community on oak: fungus-plant-insect interactions. Ecol Lett 15:177–185. https://doi.org/10.1111/j.1461-0248.2011.01724.x

Thaler JS, Humphrey PT, Whiteman NK (2012) Evolution of jasmonate and salicylate signal cross-talk. Trends Plant Sci 17:260–270. https://doi.org/10.1016/j.tplants.2012.02.010

Underwood N (2010) Density dependence in insect performance within individual plants: induced resistance to *Spodoptera exigua* in tomato. Oikos 119:1993–1999. https://doi.org/10.1111/j.1600-0706.2010.18578.x

Vacher C, Hampe A, Porté AJ et al (2016) The phyllosphere: microbial jungle at the plant–climate interface. Annu Rev Ecol Syst 47:1–24. https://doi.org/10.1146/annurev-ecolsys-121415-032238

van Dam NM, Heil M (2011) Multitrophic interactions below and above ground: en route to the next level: below-ground-above-ground interactions. J Ecol 99:77–88. https://doi.org/10.1111/j.1365-2745.2010.01761.x

Van Dam NM, Raaijmakers CE, Van Der Putten WH (2005) Root herbivory reduces growth and survival of the shoot feeding specialist *Pieris rapae* on *Brassica nigra*. Entomol Exp Appl 115:161–170. https://doi.org/10.1111/j.1570-7458.2005.00241.x

Van der Putten WH, Vet LE, Harvey JA et al (2001) Linking above-and belowground multitrophic interactions of plants, herbivores, pathogens, and their antagonists. Trends Ecol Evol 16:547–554

Van Zandt PA, Agrawal AA (2004) Community-wide impacts of herbivore-induced plant responses in milkweed (*Asclepias syriaca*). Ecology 85:2616–2629. https://doi.org/10.1890/03-0622

Wainhouse D (2005) Ecological methods in forest pest management. Oxford University Press, Oxford

White TCR (1974) A hypothesis to explain outbreaks of looper caterpillars, with special reference to populations of *Selidosema suavis* in a plantation of *Pinus radiata* in New Zealand. Oecologia 16:279–301

White TCR (2009) Plant vigour versus plant stress: a false dichotomy. Oikos 118:807–808. https://doi.org/10.1111/j.1600-0706.2009.17495.x

Wondafrash M, Van Dam NM, Tytgat TOG (2013) Plant systemic induced responses mediate interactions between root parasitic nematodes and aboveground herbivorous insects. Front Plant Sci 4:87. https://doi.org/10.3389/fpls.2013.00087

Wurst S, van der Putten WH (2007) Root herbivore identity matters in plant-mediated interactions between root and shoot herbivores. Basic Appl Ecol 8:491–499. https://doi.org/10.1016/j.baae.2006.09.015

Ximénez-Embún MG, Ortego F, Castañera P (2016) Drought-stressed tomato plants trigger bottom–up effects on the invasive *Tetranychus evansi*. PLoS One 11:e0145275. https://doi.org/10.1371/journal.pone.0145275

Yang JW, Yi H-S, Kim H et al (2011) Whitefly infestation of pepper plants elicits defence responses against bacterial pathogens in leaves and roots and changes the below-ground microflora: leaf insects reshuffle root microflora and defences. J Ecol 99:46–56. https://doi.org/10.1111/j.1365-2745.2010.01756.x

Young RA, Giese RL (eds) (2003) Introduction to forest ecosystem science and management, 3rd edn. Wiley, Hoboken

Chapter 8
Soil Macro-Invertebrates: Their Impact on Plants and Associated Aboveground Communities in Temperate Regions

Susanne Wurst, Ilja Sonnemann, and Johann G. Zaller

8.1 Introduction

Compared to conspicuously diverse habitats such as tropical rain forests or coral reefs, the rich fauna and flora of the soil habitat is poorly described (Groombridge 1992). It supports a wide array of diverse animals, with representatives from every major phylum in the animal kingdom except the coelenterates and the echinoderms (Wallwork 1976). While plants regulate the quantity and quality of resources available for soil biota, belowground communities regulate plant growth and community composition (Wardle et al. 2004). This chapter focuses on soil macro-invertebrates visible to the naked eye and their soil- and plant-mediated impacts on aboveground arthropod communities in temperate regions. Changes in soil characteristics and plant (community) parameters are the crucial link between soil macro-invertebrates and aboveground communities. Although those interactions are seldom studied, they might play profound roles in terrestrial ecosystems, because of the ubiquity of soil macro-invertebrates and their strong impact on physical, chemical, and biological soil characteristics with consequences for plant performance. By affecting plant traits, the influence of soil macro-invertebrates may cascade up to higher trophic levels such as shoot herbivores and their antagonists and affect aboveground ecosystem functions such as plant productivity and resistance against pest organisms (Fig. 8.1). First, the present chapter will introduce the most important taxa of soil macro-invertebrates of temperate zones, belonging to different functional groups. Then their different effects

S. Wurst (✉) · I. Sonnemann
Functional Biodiversity, Institute of Biology, Freie Universität, Berlin, Germany
e-mail: s.wurst@fu-berlin.de; i.sonnemann@fu-berlin.de

J. G. Zaller
Institute of Zoology, University of Natural Resources and Life Sciences Vienna (BOKU), Vienna, Austria
e-mail: johann.zaller@boku.ac.at

T. Ohgushi et al. (eds.), *Aboveground–Belowground Community Ecology*, Ecological Studies 234, https://doi.org/10.1007/978-3-319-91614-9_8

Fig. 8.1 Soil macrofauna effects on (A) soil physics, (B) soil chemistry, and (C) other soil biota impact plant performance with consequences for aboveground arthropod communities. Examples: (A) Ecosystem engineers such as earthworms affect the soil structure and influence root growth (I); (B) root-feeding insect larvae change soil chemistry by inducing plant defense and root exudation (II), and dead macrofauna by releasing organic compounds and nutrients into the soil (III); (C) predatory beetle larvae feed on other soil biota (IV) with potential impact on their functions, and decaying macrofauna affect microbial communities (III). All these interactions may feedback on plants and indirectly on aboveground communities

on soil physics, soil chemistry, and other soil biota are summarized. Afterwards, we will provide an overview over the impacts of soil macro-invertebrates on individual plants and plant communities. Besides looking at the impact on plants, i.e., the first trophic level in terrestrial ecosystems, we also consider their influence on higher trophic levels above the ground such as herbivores and their antagonists. Based on state-of-the-art knowledge, plant-mediated links between soil macro-invertebrates and aboveground arthropod communities will be discussed. Feedback effects of aboveground communities and plants on the abundance and diversity of soil macro-invertebrates will not be considered in this chapter. This chapter will not provide a complete overview over the existing literature on soil macro-invertebrate effects on soil characteristics and plants, but rather focuses on soil-mediated impacts on plant performance and aboveground arthropod communities. In general, our chapter highlights the important role of soil macro-invertebrates for soil functions and plant performance with documented and potential cascading effects on aboveground communities. Future research directions are proposed to include soil macro-invertebrates in basic and applied studies on above–belowground community ecology.

8.2 Soil Macrofauna

Soil macro-invertebrates, also known as soil macrofauna, are invertebrates that are large enough to disrupt the soil by burrowing and feeding (Stork and Eggleton 1992). According to different definitions, they include all soil invertebrates that have body length >1 cm (Dunger 1964), body width >2 mm (Swift et al. 1979), and are visible to the naked eye (Kevan 1968). We limit our discussion to soil macrofauna in temperate regions, excluding termites because most are found in subtropics and tropics, and focus on soil-dwelling organisms rather than those that live at the soil surface. Soil macrofauna include different taxa such as Annelids (earthworms), Insects (adults and larvae), Myriapoda (millipedes, centipedes), Isopoda (woodlice), Araneae (spiders, scorpions), and Gastropoda (slugs, snails). In terms of abundance, biomass, and impact on the soil environment, earthworms and ants are considered as the most important macrofauna components of temperate soils. They are also called "ecosystem engineers" (Jones et al. 1997) because of their strong impact on the physical state of soil which affects the availability of resources and the habitats for other soil biota. Generally, the ecological function of ecosystem engineers is somewhat unique. While all macrofauna are big enough to burrow and move soil around, those identified as ecosystem engineers do so to a greater extent than the others, and their populations are large enough that their soil moving/manipulation is more noticeable and measurable.

Many representatives of the soil macrofauna play crucial roles in decomposition processes. As detritivores they live both in the litter layer and the soil, feed on organic debris, and promote decomposition processes. Earthworms, millipedes, woodlice, slugs, snails, and certain insect larvae (e.g., Bibionidae) are part of the functional group of detritivores. Other insect larvae belonging to different families such as Elateridae (click beetles), Curculionidae (weevils), or Tipulidae (crane flies) are well known as root feeders, and some also have pest status in agricultural systems. Other members of the soil macrofauna are predators, such as spiders, beetles, and centipedes. Several taxa are not restricted to a certain trophic level and fulfill various functions in soil. Earthworms, for instance, are known both as ecosystem engineers and as detrivores. Besides these well acknowledged roles, there is still uncertainty regarding the degree to which earthworms also feed on living root material (Bouché and Kretzschmar 1974; Baylis et al. 1986; Gunn and Cherrett 1993; Arnone and Zaller 2014). Ants are known as predators, microbial- and plant feeders, and opportunistic omnivores (Bardgett 2005). Both affect several ecosystem functions such as nutrient cycling, litter decomposition, and water infiltration.

In Sect. 8.3, we examine how the activities of different soil macrofauna alter the soil environment in terms of soil physics, soil chemistry, and soil biology.

8.3 Macrofaunal Effects on the Soil Environment

For the sake of clarity, we divide effects of soil macrofauna in physical, chemical, and biological components (Table 8.1). However, it is important to note that in reality, these processes are closely interlinked, and physical effects simultaneously have chemical and biological consequences and vice versa.

8.3.1 Soil Physics

8.3.1.1 Ecosystem Engineers

Representatives of soil macrofauna considered as ecosystem engineers are by definition those that have the most impact on soil physics. Most of what we know about soil fauna effects on physical properties in temperate soils is derived from studies involving earthworms, with pioneering works from the late nineteenth century (Hensen 1877; Darwin 1881). Several comprehensive reviews addressed this topic (Lavelle 1988, 2000; Bottinelli et al. 2015). Although earthworms are not able to change the soil texture, i.e., the relative proportions of the mineral particles sand, silt, and clay within the soil matrix, they can change the arrangement of soil organo-minerals (Butenschoen et al. 2009). This influences the water holding capacity, the capacity to retain plant nutrients, and the ability of roots to develop and grow through the soil. Burrowing of earthworms (Fig. 8.1A) has been shown to increase water infiltration and the pore volume in soil (Ernst et al. 2009; Spurgeon et al. 2013; Zaller et al. 2014). For anecic and endogeic earthworms, impressive burrow networks of up to 8900 km ha^{-1} with an overall internal surface of 12 ha ha^{-1} were found in a pasture (Kretzschmar 1982). Soils worked by earthworms also tend to have more water stable aggregates than soils without earthworms (Lee 1985;

Table 8.1 Documented influence of soil macrofauna on soil physical, chemical, and biological parameters that have potential consequences for plant growth and performance

Soil characteristics	Soil macrofauna activity and its consequences
Soil physics	
Soil structure	Feeding, burrowing, and mixing of mineral and organic material change the arrangement of soil organo-minerals
Soil aeration and porosity	Creation of biopores, redistribution, and mixing of organic and mineral particles alter the arrangement of solid soil particles and pore space
Aggregate stability	Mixing of mineral and organic particles and casting create stable aggregates
Soil water and infiltration	Burrowing, comminuting of organic matter, and nest structures create soil aggregates with high water holding capacity and improved water infiltration
Soil chemistry	
Decomposition and carbon cycling	Fragmenting of plant and animal residues; stimulation of microbial activity leading to accelerated decomposition and soil organic matter turnover
Nutrient cycling and bioavailability	Faunal excrements are hotspots of plant nutrients; macrofaunal activities in the rhizosphere and bulk soil can alter root exudation and nutrient availability for plants and other soil organisms
Soil pH	No general trend of soil macrofaunal effects
Soil redox conditions	Soil macrofauna creates an oxygen demand in the soil
Soil biology	
Soil microorganisms (prokaryotes, fungi)	Various (multidirectional) effects on the abundance, diversity, and activity of soil microorganisms
Plant symbioses	Feeding and redistribution of mycorrhizal spores; alteration of root colonization by mycorrhiza or rhizobia
Suppression of plant pests	Tendency of reducing plant-feeding nematodes
Soil meso- and macrofauna	Multiple interactions between different faunal components in soils; ecosystem engineers create habitats for microarthropods; predators control prey organisms
Root growth control	Ecosystem engineers, detritivores, and root herbivores affect root growth depending on their densities; predators may indirectly affect plant performance by influencing ecosystem engineers or herbivores

Marinissen and Hillenaar 1997). It has been stated that up to 50% of surface layer soil aggregates in temperate pastures consist of earthworm casts (Ponomareva 1950; van de Westeringh 1972). These casts are considered the quantitatively most important direct contribution of soil invertebrates to soil organic matter turnover (Wolters 2000). Soil erodibility is considered to be reduced by earthworm activity; however, it depends on soil types and the organic matter content in soils (Blanchart et al. 2004).

Similarly to earthworms, ants have an important impact on soil physical properties by creating macro-voids, galleries, chambers, and organo-mineral soil aggregates (Lobry de Bruyn and Conacher 1990; Lee and Foster 1991; MacMahon et al. 2000; Cammeraat and Risch 2008). Several researchers report clay enrichment of mounds compared with undisturbed soil nearby, probably through a vertical transport of soil from horizons rich in clay (Baxter and Hole 1967; Wiken et al. 1976). Soil bulk density can be decreased (Rogers 1972) or increased by ant activity (Drager et al. 2016). Ants may facilitate the flow of water through the soil by their channels and their effects on soil structure (James et al. 2008). On the other hand, ants expose bare soil around their burrows which can impede water infiltration and encourage soil erosion, especially when rainfall intensity is greater than the infiltration capacity of the macropores (Cerda and Jurgensen 2011). It appears that the effects on soil physical and hydrological properties strongly depend on the ant species involved, the type of soil, and the ecosystem inhabited as well as on the initial soil conditions (Cammeraat and Risch 2008).

8.3.1.2 Detritivores, Root Herbivores and Predators

Millipedes are often the dominant detritivores in habitats where earthworms are absent (Hopkin and Read 1992); however, little is known about their effects on soil physics. Larvae of millipedes have been documented to increase the development of soil aggregates >2 mm (Fujimaki et al. 2010). Beetles and their larvae can also play an important role in regulating physical soil functions (Nichols et al. 2008; Brown et al. 2010; Badorreck et al. 2012). From the few studies available, it can be concluded that burrowing Scarabidae and millipede species (Nichols et al. 2008; Snyder et al. 2009) positively influence soil hydrological properties by increasing water infiltration and porosity, and reducing surface water runoff, while their effects on soil aggregation dynamics appear to be little. By their burrowing activity, insect root herbivores (Bibionid or Tipulid larvae) have been found to loosen the soil around grass plants, thereby potentially exposing the plant roots to desiccation and frost damage (D'Arcy-Burt and Blackshaw 1991). Overall, there is a great gap of knowledge regarding the effects of detritivores, root herbivores, and predators on soil physics.

8.3.2 Soil Chemistry

8.3.2.1 Ecosystem Engineers

A great body of literature deals with the influence of ecosystem engineers on soil chemistry (Table 8.1). Earthworms alter soil chemistry by collecting and transporting soil organic matter (SOM) within the soil profile and by producing nutrient-rich castings on the soil surface and belowground. Ecosystem carbon (C) stocks are affected by earthworms by an initial fast decomposition associated with the newly formed casts (Martin 1991) and a longer-term stabilization of C within microaggregates formed within the casts. Therefore, soil organic matter (SOM) turnover is commonly accelerated due to earthworm activity (Gilot-Villenave et al.

1996). The availability of nitrogen in forms that plants can use is a basic determinant of soil fertility; the role of soil macrofauna in facilitating the nitrogen cycle is, therefore, of great importance. In soils with limited C supply, the presence of earthworms can strongly increase nitrogen (N) leaching, especially in absence of plants. When C availability is increased, earthworms are more effective in mobilizing N (Scheu 1993; Tiunov and Scheu 2004). However, the effect of earthworms on nitrogen mineralization depends on soil type and was calculated to range between 8 and 23 kg N ha^{-1} $year^{-1}$ for different ecosystems (Scheu 1994). Additionally, a small amount of N is contributed to N-cycling by earthworms via mucus and excretion (Curry et al. 1995). Besides N, several other bioavailable nutrients are enriched in earthworm casts, e.g., potassium, phosphorus (Lavelle 1988). Soil pH influences nutrient availability to plants, however to what extent earthworms alter soil pH is not well investigated.

Ants affect soil chemistry by collecting and transporting live and dead animal and plant materials to their nest structures and by the addition of salivary secretions and excreta in nest construction. The majority of ants increase carbon and nutrient levels, especially N, phosphorus, and potassium, as well as exchangeable magnesium and calcium (James et al. 2008). Chemical changes will be related to the physical alterations which have taken place and to accumulation of organic matter and decomposition processes which occur in and around the nest. Therefore, ant nests increase the spatial heterogeneity of soil properties and create unique micro-sites within ecosystems (Azcarate and Peco 2007). Alteration of soil pH at the nest site in comparison with the surrounding soil seems to be moderate (Lobry de Bruyn and Conacher 1990; Cammeraat and Risch 2008).

8.3.2.2 Detritivores, Root Herbivores, and Predators

While the contribution of earthworms and ants to soil chemistry is well recognized, few studies have examined the influence of other detritivores. Among the best studied detritivores are millipedes consuming organic detritus and microbial biomass living on this material (Hopkin and Read 1992). Undigested C is deposited in fecal pellets where it is subject to increased microbial activity and C loss through respiration (Maraun and Scheu 1996), but remaining C is subsequently protected in fecal pellet-derived aggregates (Toyota et al. 2006). It has been estimated that these millipede fecal pellets can account for up to 39% of organic soil layers (Dangerfield and Milner 1996). Because of a lack of systematic experiments involving millipedes, it is difficult to judge their effect on nitrogen cycling. Larvae of a millipede species (*Parafontaria laminata*) have been found to promote N mineralization and nitrification; however, this did not result in changes in the total amounts of C and N in the soil (Fujimaki et al. 2010). When a plant or animal dies, soil (micro)organisms break up the complex proteins, polypeptides, and nucleic acids of the decaying corpses and produce plant available ammonium, ions, nitrates, and nitrites (Fig. 8.1B,C). It was shown that movement of N from dead earthworm tissue into microbial biomass and plant tissue happens within a few days (Whalen et al.

1999). The impact of other detritivorous soil macrofauna such as woodlice, slugs, snails, and certain insect larvae (e.g., Bibionidae) on the cycling of C or other soil nutrients is not well studied.

Root feeding insect larvae may affect soil chemistry by enhancing or modifying root exudation (Fig. 8.1B), as damaged roots are thought to leak more than intact roots (Treonis et al. 2005). The effects of root-feeding larvae may be quite important for carbon cycling, since up to 80% of the plant photosynthates can be partitioned to root biomass and exudates, depending on environmental conditions (Li et al. 1994). Root herbivory by cranefly larvae (*Tipula paludosa*) have been shown to mainly enhance the content of organic C and carbohydrates in the rhizosphere solution of grasses and legumes (Treonis et al. 2005). Additionally, similar to aboveground plant parts, roots emit volatile compounds upon the attack by insect root herbivores. For maize plants attacked by larvae of the western corn rootworm (*Diabrotica virgifera virgifera*), these volatiles have been identified as (E)-beta-caryophyllene and α-humulene (Rasmann et al. 2005; Robert et al. 2012). In contrast, enhanced content of ammonium in the rhizosphere in the presence of root feeding insect larvae has been attributed to the degradation of larval waste (Treonis et al. 2005). Additionally, roots detached due to the feeding activity of insect root feeders may mineralize faster than intact roots, thus making more N available for subsequent uptake by plants (Murray and Clements 1998).

Whether and to what extent predacious soil macrofauna directly or indirectly affects soil chemistry has not been addressed in experimental studies to the best of our knowledge.

8.3.3 Macrofaunal Effects on Soil Biota

8.3.3.1 Ecosystem Engineers

Ecosystem engineers affect other soil biota by directly or indirectly modulating the availability of resources and by creating macropores, galleries, and caverns. These structures provide habitats for other organisms creating a high habitat diversity, which may in turn increase species diversity (Lavelle and Spain 2001). Further, earthworm casts are hotspots of microbial-driven processes, such as nutrient release or nutrient immobilization and decomposition (Aira et al. 2005). The presence of earthworms increases C limitation of soil microorganisms, due to increased availability of N and P in earthworm casts or a direct depletion of easily available C resources by earthworms. The role of earthworms as vectors of soil bacteria and their capacity to influence the population dynamics and impact of plant-associated microorganisms has been the focus of several studies. For example, earthworms' interactions with the symbiotic bacterium *Rhizobium* is ambiguous, leading to increased root colonization (Doube et al. 1994; Stephens et al. 1994; Kim et al. 2017), but also to a reduced survival of *Rhizobium* in the soil (Stephens et al. 1994). By spreading biocontrol agents, earthworms have been shown to reduce a fungal root disease in wheat (Doube et al. 1994; Thorpe et al. 1996). Among the best-studied interactions between earthworms and plant-associated soil biota are those involving arbuscular mycorrhizal fungi (AMF). Earthworms have been shown to selectively feed on fungal mycelia (Bonkowski et al. 2000), disperse AMF spores (Reddell and Spain 1991; Gange 1993; Lee et al. 1996), increase AMF biomass in the soil (Gormsen et al. 2004), and either increase (Ortiz-Ceballos et al. 2007) or do not affect root AMF colonization (Wurst et al. 2004a). Consequences of earthworm–AMF interactions on plant performance are commonly species-specific and vary from an increased plant nutrient uptake and productivity (Yu et al. 2005; Ma et al. 2006; Zaller et al. 2011a, b) to no interactive effects (Wurst et al. 2004a, b; Eisenhauer et al. 2009). Interactions with other soil biota have rarely been addressed. Earthworms tend to reduce the number of soil nematodes (Yeates 1981; Alphei et al. 1996; Räty and Huhta 2003) and decrease Collembola abundance (Zaller et al. 2016). Anecic earthworms increased the density and biomass of protozoa in burrow walls (Tiunov et al. 2001) but did not affect protozoa outside burrow walls (Zaller et al. 2016).

Ant nests have been shown to contain high amounts of microfauna and -flora (Wagner et al. 1997; MacMahon et al. 2000; Boulton et al. 2003; Holec and Frouz 2006; Amador and Gorres 2007). Greater microbial activities and different microbial assemblages of bacteria, fungi, and nitrogen-fixing bacteria within nests compared to reference soils have been reported (Holec and Frouz 2006; Boots and Clipson 2013). The soil biota activity in ant nests has been shown to be 160 times greater than in adjacent non-nest soils (Fernandez et al. 2014). However, functional diversity of soil microorganisms in ant mounds were considerably lower than that in earthworm casts (Amador and Gorres 2007). Ants also increased the colonization of plant roots by AMF, by modifying biotic and abiotic soil properties (Snyder et al. 2002; Dauber et al. 2008); however, effects of ants on microbial assemblages seem to be species-specific (Boots et al. 2012). Nest soils were reported to differ from surrounding soils in respect to densities of protozoa and microarthropods (Wagner et al. 1997; Boulton et al. 2003). In general, the results demonstrate that ant nests may constitute a significant source of spatial heterogeneity in soil biota. Ant nests are stable resource patches that can be utilized by organisms living in association with ants (myrmecophiles) such as mites (Campbell et al. 2013). To the best of our knowledge, no study investigated possible effects of ants on the distribution and fate of rhizobia in soil.

8.3.3.2 Detritivores, Root Herbivores, and Predators

Generally, the detritivore group, consisting of insect larvae, isopoda, and molluscs, crush and mix plant litter and process it for further decomposition by smaller fauna and microorganisms (Lavelle 1996). However, this has rarely been studied experimentally. Periodically swarming train millipedes (*Parafontaria laminata*) have been shown to increase soil microbial biomass, but mainly at high adult densities but not at larval stages (Toyota et al. 2006). We are not aware of studies, where interactions between soil detritivores (other than earthworms or ants) and other soil biota and their consequences on plant performances have been examined.

Changes in soil biota due to the activity of root feeding insect larvae have mainly been studied for microorganisms that are closely associated to plant roots. One mechanism by which insect root herbivores can affect root associated microorganisms is mechanical disruption or facilitation. For instance, larvae of the genus *Sitona* are specialized to feed on the root nodules of legume plants (Johnson and Rasmann 2015), thereby destroying the physical requirements of symbiotic interactions of legumes and N-fixing rhizobacteria. Independent of root nodules, root herbivory by insect root feeders is known to generally include the risk of subsequent colonization of roots by, mainly fungal, pathogens (reviewed by Brown and Gange 1990; Mortimer et al. 1999). The root feeders can transport root diseases from infected to uninfected roots, and pathogen entry into roots may be facilitated at herbivore-induced root lesions. Mycorrhizal hyphae may be ingested by insect root feeders, which thereby mechanically disrupt the hyphal network (Johnson and Rasmann 2015). Insect root herbivore effects on root colonization by mycorrhizal fungi are assumed to also be mediated by changes in the viability of mycorrhizal spores (Sonnemann et al. 2013) and in the quantity and quality of root exudates that stimulate mycorrhizal spore germination and hyphal growth (Currie et al. 2006). Reduced as well as increased mycorrhizal root colonization due to the presence of insect root herbivores has been reported, with most studies not finding any effect. Similar to root colonization by mycorrhizal fungi, insect root herbivore impacts on the soil bacterial community are suggested to be mediated by changes in soil chemistry (Treonis et al. 2005). Insect root feeders have been found to increase the number of *Pseudomonas* spp. and the bacterial utilization of sugars, amino acids, and carboxylic acids in grassland systems (Dawson et al. 2004; Grayston et al. 2001). Consequently, insect root herbivore-induced shifts in the soil microbial community are likely to impact nutrient cycling and availability (Grayston et al. 2001). Additional to their impact on soil bacteria and fungi, insect root feeders have been shown to influence larger organisms. Volatile compounds emitted by the plant upon insect root herbivore attack strongly attract entomopathogenic nematodes, which can infect the insect larvae and thus defend the plant against the root feeder (Rasmann et al. 2005). Also, the emission of root volatiles was found to be dependent on root herbivore density and was used by conspecific insect larvae to orientate towards intermediately infested plants, which benefit their development (Robert et al. 2012).

In general, predacious soil macrofauna such as beetles, spiders, and centipedes may affect the abundance and performance of other soil biota (Fig. 8.1C) with impact on their preys' functional roles (e.g., decomposition or root herbivory). As predators, ants were the most apparent invertebrate scavengers observed foraging on entomopathogenic nematode-killed insects (i.e., insect cadavers containing entomopathogenic nematodes and their symbiotic bacteria; Baur et al. 1998). Others demonstrated that the symbiotic bacteria of some species of entomopathogenic nematodes produce a compound that deters scavengers such as ants and, thus, could protect nematodes from being eaten during reproduction within insect cadavers (Zhou et al. 2002).

8.4 Macrofaunal Effects on Plants (Individual and Communities)

8.4.1 Ecosystem Engineers

Earthworms are known to affect plant growth by their effects on the physical, chemical, and biotic soil environment. In a review on earthworm effects on plant growth (Scheu 2003), 79% of the examined 67 studies reported significant increases of shoot biomass in the presence of earthworms, while 9% reported reductions. The reviewed studies have mainly focused on agricultural plants, while wild plants received less attention. Scheu proposed that earthworms affect plant productivity and communities by both direct and indirect effects. Direct effects include, for example, root feeding and transposal of seeds, while indirect effects are mediated by impacts on soil structure, mineralization processes, dispersal of microorganisms, and hormone-like effects. A recent meta-analysis (van Groenigen et al. 2014) documented that earthworm presence in agroecosystems leads to a 25% increase in crop yield and a 23% increase in aboveground biomass. Because the effects become larger when more crop residues are returned and disappear when soil nutrient availability is high, the authors suggest that earthworms stimulate plant growth mainly by releasing nitrogen locked in residues and soil organic matter. A less well investigated aspect is the influence of earthworms on plant communities via seed translocation (Zaller and Saxler 2007; Drouin et al. 2014; Eisenhauer et al. 2010; Laossi et al. 2010) and/or creating small-scale nutrient heterogeneity (Wurst et al. 2004a, b; Milcu et al. 2006; Zaller et al. 2013).

Also ants can affect plant growth, community structure, and diversity by creating soil heterogeneity (Jouquet et al. 2006). Ants also play an important role in seed dispersal. Seed dispersal by ants (myrmecochory) mediated by lipid-rich seed appendages (elaiosomes) as rewards has been suggested to occur in at least 11,000 plant species (Lengyel et al. 2009). Since ants predate on herbivores, they can also affect plant performance indirectly by reducing herbivory. A meta-analysis (Rosumek et al. 2009) documented that plants in an ant-free environment had 50% more herbivores and suffered almost twice as much herbivore damage than plants growing with ants. Consequently, plants without ants showed a reduction in biomass and reproduction by about 24%. However, the effects were much stronger in tropical than temperate regions, and especially strong in plants that provide shelter for ants (myrmecophytes). Another meta-analysis also showed that the ant–plant protection mutualism is generally positive for the plant and only occasionally neutral (Chamberlain and Holland 2009). But ants can also indirectly depress plant growth by affecting dung decomposer communities through territorial interference, with negative consequences for dung removal rates and soil nitrogen concentrations (Zhao et al. 2014).

8.4.2 Detritivores

Detrivorous macrofauna such as woodlice and millipedes may impact plants directly, by mainly feeding on seeds and seedlings, and indirectly via their effects on decomposition processes. Different species of woodlice (Isopoda) feed on seeds and seedlings (Saska 2008; Honek et al. 2009). Interestingly, herbivory by woodlice (*Porcellio scaber* and *Armadillidium vulgare*) differed between the *Arabidopsis thaliana* wild type and a jasmonate-deficient mutant (Farmer and Dubugnon 2009), suggesting a deterrence of woodlice herbivory by jasmonate. In a millipede (*Cylindroiulus caeruleocinctus,* Julidae), seed consumption has been documented (Koprdova et al. 2010). Indirect effects of detritivorous soil macrofauna on plant performance via impacts on litter decomposition are also plausible, but studies do not often take feedbacks on plants into account. It has been documented that litter diversity interacts with woodlice feeding on decomposition (Vos et al. 2011), but the consequences for plant performance was not assessed. Millipede compost has been compared with vermicompost (Antunes et al. 2016), and results indicate that both composts are equally efficient as substrate for the production of lettuce seedlings. In general, there is a considerable lack of studies on the impact of detritivorous soil macrofauna (except earthworms) on plant performance, although their important role in decomposition processes is widely acknowledged.

8.4.3 Root Herbivores

Macrofaunal impact on plants is known in the form of root herbivory by herbivorous soil living insects. Herbivorous soil living insects are found in the orders Orthoptera, Isoptera, Hemiptera, Thysanoptera, Neuroptera, Coleoptera, Diptera, and Lepidoptera (see reviews by Brown and Gange 1990; Mortimer et al. 1999), with the latter three being the most important. Still, within these orders, root feeding is restricted to few families and is done mostly by larval stages. Most root feeding insects have chewing mouthparts, while there are only few known sap suckers. Root feeding insects cover a wide range of host specificity, from highly specialized to generalist species. Which part of the root they feed on depends on their size and their mouthpart characteristics.

Literature on the impact of insect root herbivory on individual plants and plant communities has been reviewed by several authors (e.g., Brown and Gange 1990; Mortimer et al. 1999; Blossey and Hunt-Joshi 2003; Hunter 2001). Primarily, root damage by insect root herbivores impairs water and nutrient uptake as well as storage of resources in root tissues. It can disrupt vascular connection between roots and shoots and can reduce plant stability. Additionally, insect root herbivores have been shown to influence plant interactions with root colonizing microorganisms like the endorhiza bacterial community (Dematheis et al. 2013), AMF (Currie et al. 2006; Bennett et al. 2013), fungal pathogens, or symbiotic rhizobacteria.

Consequences of insect root herbivory for the plant depend on plant tolerance towards root herbivory, which is species-specific but also influenced by environmental factors. Drought and nutrient limitation, for example, can aggravate root herbivory effects. Plant tolerance towards root herbivory involves the reallocation of resources and the regrowth of lost tissues. Occasionally, even over-compensatory regrowth, below- as well as aboveground, occurs. Cases of intolerable insect root herbivore attack, however, result in reduced vegetative and reproductive plant growth. Besides tolerance, plant roots employ several other mechanisms to deal with (insect) root herbivore attack, namely constitutive structural and chemical defenses, inducible chemical defense as well as avoidance (van Dam 2009).

Insect root herbivores can be pests in agri- or horticultural plantings, forests, orchards, or turf sites, and considerably reducing quality and quantity of yields. In diverse natural plant communities, insect root herbivores are thought to influence plant community composition due to their impact on establishment, competitive ability, and reproduction of individual plants. In recent years, modeling approaches by Körner et al. (2014) and Pfestorf et al. (2016) revealed that conflicting results of empirical studies regarding the strength and direction of the root herbivore effect in plant communities may be explained by different root herbivore feeding modes (from unselective to highly plant trait specific) and interactions with resource availability and aboveground grazing.

8.4.4 Predators

Spiders, beetles, centipedes, and ants predate on other soil biota and may indirectly affect plant performance. For the centipedes *Lithobius curtipes* and *L. crassipes*, it has been documented that freshly fixed C by plants is found back in their tissues, suggesting that they also feed on roots or mycorrhizal hyphae (Pollierer et al. 2007; Goncharov et al. 2016). This shows that the classical classification of soil macrofauna in feeding groups is challenged by novel results involving isotope tracer techniques. Spiders mainly predate on the ground and may control herbivores that have life stages on the plant and ground and in soil (Rendon et al. 2016). In general, spiders exert top-down effects on soil arthropods and herbivores and may reduce leaf damage and increase the density of soybean (Rypstra and Marshall 2005). Interestingly, spiders may also affect litter decomposition by affecting the physiology of prey which is stressed by the spider presence. Hawlena et al. (2012) showed that the spider *Pisuarina mira* slowed litter decomposition, because its prey, the grasshopper (*Melanoplus femurrubrum*), changed its body C-to-N ratio in response to the spider presence, leading to equal decomposition of the grasshoppers, but legacy effects on grass litter decomposition by changes in belowground community function. For the predatory beetle *Agonum impressum*, indirect positive effects on plant performance have been shown, mediated by its effect on earthworm behavior (Zhao et al. 2013). Under predation by the beetle, the earthworms moved to deeper soil layers leading to positive effects on plant growth due to the resulting ecosystem engineering effects.

The indirect effects of macrofauna predators on plant performance are thus mediated by their effects on the behavior, the physiology, and the population size of their prey. However, there are only few studies investigating the feedbacks of macrofauna predators on plant productivity or other ecosystem functions. To what extent these interactions alter the diversity of plant communities is unclear.

8.5 Macrofaunal Effects on Aboveground Higher Trophic Levels

8.5.1 Ecosystem Engineers

The majority of studies on effects of macrofaunal ecosystem engineers on aboveground higher trophic levels dealt with earthworms. Earthworms were shown to affect aboveground herbivore performance indirectly by changes in plant traits. Two reviews exist so far (Wurst 2010, 2013) that document that the plant-mediated effects of earthworms on aboveground herbivores range from positive to negative and are likely mediated by altered resource uptake and/or changes in soil microbial communities affecting plant traits. Also changes in plant secondary compounds (Wurst et al. 2004a, b, 2006; Lohmann et al. 2009) and the expression of stress-responsive genes were reported in the presence of earthworms (Blouin et al. 2005; Jana et al. 2010; Puga-Freitas et al. 2012). The majority of the reviewed studies were done under controlled conditions in the greenhouse, using endogeic earthworm species and aphids (mainly *Myzus persicae* and *Rhopalosiphum padi*) as aboveground herbivores. The genetic variation of a host plant (*Vicia faba*) and aphid species identity (*Aphis fabae* and *Acyrthosiphon pisum*) was reported to mediate the impact of earthworms (*Eisenia veneta*) on aboveground aphid populations (Singh et al. 2014). Thus genotype- and species-specific effects may explain part of the variation in the results reported so far. Besides using aphids as aboveground herbivores, snails and slugs were considered in studies on plant communities (e.g., Wurst and Rillig 2011; Zaller et al. 2013; Trouvé et al. 2014). For instance, Zaller et al. (2013) reported that damage of grassland plant communities by the invasive slug *Arion vulgaris* was reduced when earthworms were present, but the earthworm effects on slug herbivory partly depended on the number of plant species grown in the community. In a similar study, slug (*A. vulgaris*) herbivory was only reduced by earthworms in AMF-inoculated plant communities (Trouvé et al. 2014). No earthworm effect on *A. vulgaris* fresh weight was detected. In general, the studies conducted so far show a great deal of context dependency. The feeding mode of the herbivores may matter, since the few studies involving aboveground chewing herbivores (mainly Gastropoda) did not detect impacts of earthworms on herbivore numbers or biomasses. For phloem-feeding aphids, inconsistent results have been reported: positive, negative, and neutral effects on aphid performance (numbers, reproduction) were documented. The mechanisms proposed for the plant-mediated impacts of earthworms on aboveground herbivores range from changes in nutrient availability to the induction of chemical defenses. Few studies (Wurst and Jones 2003; Poveda et al. 2005; Johnson et al. 2011) followed the effects of earthworms up to the third level, i.e., the antagonists of herbivores above the ground. It has been shown that parasitoids of aphids responded to aphid abundance. Negative or positive effects of earthworms on aphid abundances may thus cascade up to higher trophic levels such as aphid antagonists. Earthworms were recently shown to affect the visitation length of flowers by pollinators (Guo et al. 2016), but besides their documented positive impacts on flower numbers (e.g., Poveda et al. 2005; Guo et al. 2016) their effects on flower traits such as nectar quality have not been studied so far.

Ant effects on higher trophic levels aboveground have been very rarely studied. The one and, as far as we are aware, only exception is the study by Sanders and van Veen (2011). They studied the influence of colonies of two common European ant species (*Myrmica rubra* and *Lasius niger*) on an arthropod community via both the predation and ecosystem engineering pathways. The presence of one ant colony had positive effects on densities of decomposers (Collembola), herbivores (mainly hemipterans), and parasitoids. The authors concluded that ecosystem engineering effects rather than predation by ants explain the observed positive effects. However, the presence of two colonies showed neutral and negative effects on herbivores and parasitoids, likely by an increase of predation pressure. The study further documented that soil from ant nests (without ants) had not the same positive effects as intact nests. This study is among the first ones documenting ant effects on aboveground arthropod food webs. Former studies only focused on ant effects on soil biota such as bacteria, nematodes, mites, and collembolan (e.g., Boulton and Amberman 2006).

8.5.2 Detritivores

Detritivorous macrofauna (except earthworms) were not or very rarely considered in regard to their impacts on higher tropic levels aboveground (Wurst 2013). González-Megías and Müller (2010) investigated the effects of a detritivorous beetle larvae (*Morica hybrida*, Tenebrionidae) in interaction with belowground herbivorous beetle larvae (*Cebrio gypsicola*, Cebrionidae) and aboveground herbivores on the plant *Moricandia moricandioides* (Brassicaceae) and associated aboveground arthropod communities in the field. They documented higher parasitoid attack rates and abundances in the presence of the detritivorous insect larvae, while their mainly negative effects on the aboveground herbivores (aphids, plant-hoppers, seed predator, and leaf herbivores) depended on the presence of root and floral herbivores. Shifts in plant secondary metabolites (glucosinolates) due to the herbivore treatments were changed in the presence of the detritivores and might have mediated part of the indirect plant-mediated effects. Other studies on the impact of detritivorous macrofauna, belonging to millipedes, woodlice, and gastropods, on aboveground arthropod communities have not been conducted so far. If their impact on plant biomass and quality is significant, their effects might also cascade up to higher trophic levels above the ground. This link still remains to be elucidated.

8.5.3 Root Herbivores

Insect root herbivores are known to affect the performance of their shoot feeding counterparts via changes in plant chemistry. Performance parameters that have been investigated comprise relative growth rate, development time, mass or size gain, fecundity, offspring mass, and abundance (Johnson et al. 2012). Changes in plant chemistry that are believed to be involved in insect root herbivore effects on shoot herbivores include (1) stress-induced enhancement of nitrogen and carbohydrate content in plant foliage, following reduced water and nutrient uptake due to removal of root tissue by insects root herbivores (Masters et al. 1993) and (2) systemically induced plant defense (Bezemer and van Dam 2005; van Dam 2009). While increased N contents are expected to facilitate shoot herbivores, systemically induced defenses may negatively affect them. According to a meta-analysis by Johnson et al. (2012), the direction of insect root herbivore effects on their above-ground counterparts depends on the identity of the insect herbivores involved. Root feeding Diptera were shown to generally have negative impacts on shoot feeding insects. However, the authors point out that this trend may be ascribed to the fact that most studies were done on *Delia* spp. root flies feeding on highly inducible Brassica plants. Root feeding Coleoptera were shown to generally have negative effects on Hymenoptera but to increase the performance of Homoptera. The authors assume that the phloem sap feeding Homoptera benefit from enhanced plant N concentrations while avoiding any induced defense chemicals, as these are only transported in low amounts in the phloem sap. Besides herbivore identity, the meta-analysis found insect root herbivore effects on shoot herbivores to depend on the sequence of arrival at the host plant, with significant (positive) impact only occurring at simultaneous arrival. In contrast, insect root herbivore effects on aboveground herbivores were independent of the plant type (annual vs. perennial and domestic crop plant vs. natural species) and did not differ for the above-mentioned performance parameters of shoot herbivores.

Insect root herbivore effects may even cascade up to influence parasitoids of shoot herbivores. Mostly, parasitoid recruitment or performance has been found to be reduced by root herbivory (reviewed by A'Bear et al. 2014), with effects often mirroring changes in parasitoid prey quantity or quality as well as in herbivore-induced plant volatiles (HIPVs), which serve to attract parasitoids.

In contrast to effects on parasitoids, insect root herbivore effects on the frequency and duration of pollinator visitation appear to be predominantly positive (A'Bear et al. 2014). This trend is surprising as root herbivory is expected to reduce the amount of resources a plant can invest into floral display and production of pollinator-attracting HIPVs (A'Bear et al. 2014).

8.5.4 Predators

As far as we are aware, indirect, plant-mediated effects of predacious soil macrofauna on aboveground arthropod communities have not been documented so far. Changes in the abundances of other functional groups such as ecosystem engineers, detritivores, or herbivores through predation may affect plant performance with cascading effects on aboveground higher trophic levels. However, this bottom-up link remains to be experimentally tested. Predators not restricted to the soil may switch between below- and aboveground prey (Scheu 2001) and may affect aboveground arthropod communities by both top-down and bottom-up forces.

8.6 Future Perspectives

Thus far, studies on the impact of soil macrofauna on higher trophic levels above the ground mainly focus on earthworms and insect root herbivores. There are very few studies documenting the impacts of other soil macrofauna on aboveground arthropod communities. The effects range from positive, likely mediated by increased resource availability, to negative, often mediated by changes in plant secondary metabolites and/or defense gene expression. Since some of the effects are probably linked to changes in chemical, physical, and biological soil characteristics (Fig. 8.1), the effects may persist even when the active soil macrofauna is not present anymore. We propose that these legacy effects may play a role in shaping above–belowground interactions (Wurst and Ohgushi 2015). Recently, it has been shown that insect root herbivores can indirectly influence the aboveground food web via soil legacy effects (Kostenko et al. 2012). The authors reported that root herbivory by larvae of the click beetle (*Agriotes lineatus*) on ragwort (*Jacobaea vulgaris*) plants led to distinct changes in the composition of the soil fungal community that affected the composition of defense compounds (pyrrolizidine alkaloids) in newly grown ragwort plants, with negative consequences on the relative growth rate of shoot feeding *Mamestra brassicae* larvae. For other soil macrofauna, legacy effects on aboveground herbivores have been reported too. Earthworm-worked soil, the so-called vermicompost, has generally negative effects on aboveground herbivore performance (reviewed by Joshi et al. 2015). It is assumed that the organic matter in vermicomposts provides plants with a balanced source of nutrients that can influence the composition and physiology of plants (Arancon et al. 2005), leading to increased resistance. However, in general, it is largely unknown how long the impact of soil macrofauna on soil physics, chemistry, and other soil biota lasts and affects future biotic interactions of plants.

Although the modification of the soil habitat by soil macrofauna is important, their effects on soil microorganisms likely mediate their impact on plant growth and aboveground arthropod communities to a great extent. There is mounting evidence that root associations with symbiotic and free-living microorganisms affect plant growth and plant–insect interactions by various mechanisms involving hormonal pathways (Pangesti et al. 2013). It would be very interesting to include soil macrofauna in the mechanistic elucidation of microbe–plant–insect interactions, because soil macrofauna may modify the impact of single species of microbes and microbial communities on plant performance and associated insect communities. Knowledge on macrofauna–microbe interactions with impact on plant–herbivore interactions may also be crucial for the development of microbe-based plant protection strategies in agricultural fields.

Another important aspect to consider is the impact of global change factors on the soil macrofauna and how this affects their functional roles in ecosystems. For instance, increased atmospheric CO_2 concentrations have been shown to increase earthworm surface casting activity by 30%, mainly because of higher soil moisture under reduced evapotranspiration of vegetation under elevated CO_2 (Zaller and Arnone 1997). These nutrient-rich earthworm casts might have positive effects on some plant species (Zaller and Arnone 1999), with potential cascading effects on the diversity of plant communities and their associated aboveground arthropod communities. Several studies report that plant-mediated effects of soil macrofauna on shoot herbivores are changed under altered water availability. Johnson et al. (2011) showed that earthworms can mitigate the negative effects of drought on plant biomass and cause further declines in aphid populations under drought conditions. The performance of aphids and leaf miners has been documented to be reduced under the combined impact of drought stress and root-feeding insect larvae (Staley et al. 2007; Tariq et al. 2013a) and the effects might even cascade up to affect the performance of parasitoids (Tariq et al. 2013b). Enhanced precipitation was shown to influence the indirect interactions of belowground detritivorous beetles with generalist aboveground herbivores, with negative effects of the detritivores under enhanced rainfall (González-Megías and Menendez 2012). Another important global change factor is land use intensification. Recently, Sonnemann et al. (2016) documented that land use intensity can change the impact of insect root herbivores on aboveground herbivore damage. Under low mowing frequency, click beetle larvae (*Agriotes* spp.) enhanced leaf damage of plantain by aboveground herbivores, while under high mowing frequency the effect disappeared. To predict the consequences of global change on multitrophic interactions and indirect functions in terrestrial ecosystems, more studies in this respect are needed.

Another research challenge is the upscaling of the results from the predominantly small-scale studies to the landscape level. For example, controlled lab or small plot-scale studies have demonstrated that soil macrofauna such as earthworms can mobilize or transfer substantial quantities of nutrients to plants. However, the simple scaling up of such results to explain conditions on a large field scale is very much constrained by a lack of information on the spatio-temporal distribution of soil macrofauna in temperate ecosystems. Further, the numbers, diversity, and activity of soil macrofauna in agroecosystems are affected by agricultural management practices such as tillage. Still, our knowledge of the key organisms or groups of soil biota that contribute to nutrient cycling and crop production under different sets of management practices is limited (Whalen and Hamel 2004). However, this knowledge is crucial to estimate the role of soil macrofauna for crop production. Since soil macrofauna may affect not only plant productivity, but also resistance against pests, future research should focus on their impacts in a wider context, e.g., on the agricultural field level (Meyer-Wolfarth et al. 2017). In regard of the increasing demand for sustainable agriculture practices, more knowledge about the functional role of soil macrofauna in agroecosystems is urgently needed. Besides their role in impacting soil characteristics (in terms of physics, chemistry, and other soil biota), their present and past activity may have strong effects on plant performance, including interactions with aboveground pest organisms.

References

A'Bear AD, Johnson SN, Jones TH (2014) Putting the 'upstairs–downstairs' into ecosystem service: what can aboveground–belowground ecology tell us? Biol Control 75:97–107. https://doi.org/10.1016/j.biocontrol.2013.10.004

Aira M, Monroy F, Domínguez J (2005) Ageing effects on nitrogen dynamics and enzyme activities in casts of *Aporrectodea caliginosa* (Lumbricidae). Pedobiologia 49:467–473

Alphei J, Bonkowski M, Scheu S (1996) Protozoa, Nematoda and Lumbricidae in the rhizosphere of *Hordelymus europaeus* (Poaceae): faunal interactions, response of microorganisms and effects on plant growth. Oecologia 106:111–126

Amador JA, Gorres JH (2007) Microbiological characterization of the structures built by earthworms and ants in an agricultural field. Soil Biol Biochem 39:2070–2077

Antunes LFD, Scoriza RN, da Silva DG et al (2016) Production and efficiency of organic compost generated by millipede activity. Cienc Rural 46:815–819

Arancon NQ, Galvis PA, Edwards CA (2005) Suppression of insect pest populations and damage to plants by vermicomposts. Bioresour Technol 96:1137–1142

Arnone JA, Zaller JG (2014) Earthworm effects on native grassland root system dynamics under natural and increased rainfall. Front Plant Sci 5:152. https://doi.org/10.3389/fpls.2014.00152

Azcarate FM, Peco B (2007) Harvester ants (*Messor barbarus*) as disturbance agents in Mediterranean grasslands. J Veg Sci 18:103–110

Badorreck A, Gerke HH, Hüttl RF (2012) Effects of ground-dwelling beetle burrows on infiltration patterns and pore structure of initial soil surfaces. Vadose Zone J 11. https://doi.org/10.2136/vzj2011.0109

Bardgett R (2005) The biodiversity of soil. A community and ecosystem approach. Oxford University Press, Oxford

Baur ME, Kaya HK, Strong DR (1998) Foraging ants as scavengers on entomopathogenic nematode-killed insects. Biol Control 12:231–236

Baxter PF, Hole FD (1967) Ant (*Formica cinerea*) pedoturbation in a prairie soil. Soil Sci Soc Am Proc 31:425–428

Baylis JP, Cherrett JM, Ford JB (1986) A survey of the invertebrates feeding on living clover roots (*Trifolium repens* L.) using ^{32}P as a radiotracer. Pedobiologia 29:201–208

Bennett AE, Macrae AM, Moore BD et al (2013) Early root herbivory impairs mycorrhizal fungi colonisation and induces defence allocation in mature *Plantago lanceolata*. PLoS One 8:e66053

Bezemer TM, van Dam NM (2005) Linking aboveground and belowground interactions via induced plant defenses. Trends Ecol Evol 20:617–624

Blanchart E, Albrecht A, Brown G et al (2004) Effects of tropical endogeic earthworms on soil erosion. Agric Ecosyst Environ 104:303–315

Blossey B, Hunt-Joshi TR (2003) Belowground herbivory by insects: influence on plants and aboveground herbivores. Annu Rev Entomol 48:521–547

Blouin M, Zuily-Fodil Y, Pham-Thi AT et al (2005) Belowground organism activities affect plant aboveground phenotype, inducing plant tolerance to parasites. Ecol Lett 8:202–208
Bonkowski M, Griffiths BS, Ritz K (2000) Food preferences of earthworms for soil fungi. Pedobiologia 44:666–676
Boots B, Clipson N (2013) Linking ecosystem modification by the yellow meadow ant (*Lasius flavus*) to microbial assemblages in different soil environments. Eur J Soil Biol 55:100–106
Boots B, Keith AM, Niechoj R et al (2012) Unique soil microbial assemblages associated with grassland ant species with different nesting and foraging strategies. Pedobiologia 55:33–40
Bottinelli N, Jouquet P, Capowiez Y et al (2015) Why is the influence of soil macrofauna on soil structure only considered by soil ecologists? Soil Tillage Res 146:118–124
Bouché MB, Kretzschmar A (1974) Fonctions des lumbriciens. II. Recherches méthodologiques pour l'analyse du sol ingéré. Rev Écol Biol Sol 11:127–139
Boulton AM, Amberman KD (2006) How ant nests increase soil biota richness and abundance: a field experiment. Biodivers Conserv 15:69–82
Boulton AM, Jaffee BA, Scow KM (2003) Effects of a common harvester ant (*Messor andrei*) on richness and abundance of soil biota. Appl Soil Ecol 23:257–265
Brown VK, Gange AC (1990) Insect herbivory below ground. Adv Ecol Res 20:1–58
Brown J, Scholtz CH, Janeau JL et al (2010) Dung beetles (Coleoptera: Scarabaeidae) can improve soil hydrological properties. Appl Soil Ecol 46:9–16
Butenschoen O, Marhan S, Langel R et al (2009) Carbon and nitrogen mobilisation by earthworms of different functional groups as affected by soil sand content. Pedobiologia 52:263–272
Cammeraat ELH, Risch AC (2008) The impact of ants on mineral soil properties and processes at different spatial scales. J Appl Entomol 132:285–294
Campbell KU, Klompen H, Crist TO (2013) The diversity and host specificity of mites associated with ants: the roles of ecological and life history traits of ant hosts. Insect Soc 60:31–41
Cerda A, Jurgensen MF (2011) Ant mounds as a source of sediment on citrus orchard plantations in eastern Spain. A three-scale rainfall simulation approach. Catena 85:231–236
Chamberlain SA, Holland JN (2009) Quantitative synthesis of context dependency in ant-plant protection mutualisms. Ecology 90:2384–2392
Currie AF, Murray PJ, Gange AC (2006) Root herbivory by *Tipula paludosa* larvae increases colonization of *Agrostis capillaris* by arbuscular mycorrhizal fungi. Soil Biol Biochem 38: 1994–1997
Curry JP, Byrne D, Boyle KE (1995) The earthworm population of a winter cereal field and its effects on soil and nitrogen turnover. Biol Fertil Soils 19:166–172
D'Arcy-Burt S, Blackshaw RP (1991) Bibionids (Diptera: Bibionidae) in agricultural land: a review of damage, benefits, natural enemies and control. Ann Appl Biol 118:695–708
Dangerfield JM, Milner AE (1996) Millipede fecal pellet production in selected natural and managed habitats of Southern Africa: implications for litter dynamics. Biotropica 28:113–120
Darwin C (1881) The formation of vegetable mould through the action of worms. John Murray, London
Dauber J, Niechoj R, Baltruschat H et al (2008) Soil engineering ants increase grass root arbuscular mycorrhizal colonization. Biol Fertil Soils 44:791–796
Dawson LA, Grayston SJ, Murray PJ et al (2004) Impact of *Tipula paludosa* larvae on plant growth and the soil microbial community. Appl Soil Ecol 25:51–61
Dematheis F, Kurtz B, Vidal S et al (2013) Multitrophic interactions among Western Corn Rootworm, *Glomus intraradices* and microbial communities in the rhizosphere and endorhiza of maize. Front Microbiol 4:357. https://doi.org/10.3389/fmicb.2013.00357
Doube BM, Stephens PM, Davoren CW et al (1994) Interactions between earthworms, beneficial soil microorganisms and root pathogens. Appl Soil Ecol 1:3–10
Drager KI, Hirmas DR, Hasiotis ST (2016) Effects of ant (*Formica subsericea*) nests on physical and hydrological properties of a fine-textured soil. Soil Sci Soc Am J 80:364–375
Drouin M, Bradley R, Lapointe L et al (2014) Non-native anecic earthworms (*Lumbricus terrestris* L.) reduce seed germination and seedling survival of temperate and boreal trees species. Appl Soil Ecol 75:145–149
Dunger W (1964) Tiere im Boden. Ziemsen, Wittenberg

Eisenhauer N, König S, Sabais ACW et al (2009) Impacts of earthworms and arbuscular mycorrhizal fungi (*Glomus intraradices*) on plant performance are not interrelated. Soil Biol Biochem 41:561–567

Eisenhauer N, Butenschoen O, Radsick S et al (2010) Earthworms as seedling predators: importance of seeds and seedlings for earthworm nutrition. Soil Biol Biochem 42:1245–1252

Ernst G, Felten D, Vohland M et al (2009) Impact of ecologically different earthworm species on soil water characteristics. Eur J Soil Biol 45:207–213

Farmer EE, Dubugnon L (2009) Detritivorous crustaceans become herbivores on jasmonate-deficient plants. Proc Natl Acad Sci USA 106:935–940

Fernandez A, Farji-Brener AG, Satti P (2014) Moisture enhances the positive effect of leaf-cutting ant refuse dumps on soil biota activity. Austral Ecol 39:198–203

Fujimaki R, Sato Y, Okai N et al (2010) The train millipede (*Parafontaria laminata*) mediates soil aggregation and N dynamics in a Japanese larch forest. Geoderma 159:216–220

Gange AC (1993) Translocation of mycorrhizal fungi by earthworms during early succession. Soil Biol Biochem 25:1021–1026

Gilot-Villenave C, Lavelle P, Ganry F (1996) Effects of tropical geophageous earthworm, *Millsonia anomala*, on some soil characteristics, on maize-residue decomposition and on maize production in Ivory Coast. Appl Soil Ecol 4:201–212

Goncharov AA, Tsurikov SM, Potapov AM et al (2016) Short-term incorporation of freshly fixed plant carbon into the soil animal food web: field study in a spruce forest. Ecol Res 31:923–933

González-Megías A, Müller C (2010) Root herbivores and detritivores shape above-ground multitrophic assemblage through plant-mediated effects. J Anim Ecol 79:923–931

González-Megías A, Menendez R (2012) Climate change effects on above- and below-ground interactions in a dryland ecosystem. Philos Trans R Soc B 367:3115–3124

Gormsen D, Olsson PA, Hedlund K (2004) The influence of collembolans and earthworms on AM fungal mycelium. Appl Soil Ecol 27:211–220

Grayston SJ, Dawson LA, Treonis AM et al (2001) Impact of root herbivory by insect larvae on soil microbial communities. Eur J Soil Biol 37:277–280

Groombridge B (1992) Soil macrofauna. In: Groombridge B (ed) Global biodiversity: status of the earth's living resources. Springer, Dordrecht, pp 103–115

Gunn A, Cherrett JM (1993) The exploitation of food resources by soil meso- and macroinvertebrates. Pedobiologia 37:303–320

Guo XH, Petermann JS, Schittko C et al (2016) Root-knot nematodes (*Meloidogyne hapla*) can modify the positive plant intraspecific diversity-productivity effect on red clover in clover-grass communities. Appl Soil Ecol 102:26–35

Hawlena D, Strickland MS, Bradford MA et al (2012) Fear of predation slows plant-litter decomposition. Science 336:1434–1438

Hensen V (1877) Die Tätigkeit des Regenwurms für die Fruchtbarkeit des Erdbodens. Z wiss Zool 28:361

Holec M, Frouz J (2006) The effect of two ant species *Lasius niger* and *Lasius flavus* on soil properties in two contrasting habitats. Eur J Soil Biol 42:213–217

Honek A, Martinkova Z, Saska P et al (2009) Role of post-dispersal seed and seedling predation in establishment of dandelion (*Taraxacum* agg.) plants. Agric Ecosyst Environ 134:126–135

Hopkin SP, Read HJ (1992) The biology of millipedes. Oxford University Press, Oxford

Hunter M (2001) Out of sight, out of mind: the impact of root-feeding insects in natural and managed systems. Agric For Entomol 3:3–9

James AI, Eldridge DJ, Koen TB et al (2008) Landscape position moderates how ant nests affect hydrology and soil chemistry across a Chihuahuan Desert watershed. Landsc Ecol 23:961–975

Jana U, Barot S, Blouin M et al (2010) Earthworms influence the production of above- and belowground biomass and the expression of genes involved in cell proliferation and stress responses in *Arabidopsis thaliana*. Soil Biol Biochem 42:244–252

Johnson SN, Rasmann S (2015) Root-feeding insects and their interactions with organisms in the rhizosphere. Annu Rev Entomol 60:517–535

Johnson SN, Staley JT, McLeod FAL et al (2011) Plant-mediated effects of soil invertebrates and summer drought on above-ground multitrophic interactions. J Ecol 99:57–65
Johnson SN, Clark KE, Hartley SE et al (2012) Aboveground- belowground herbivore interactions: a meta-analysis. Ecology 93:2208–2215
Jones CG, Lawton JH, Shachak M (1997) Positive and negative effects of organisms as physical ecosystem engineers. Ecology 78:1946–1957
Joshi R, Singh J, Vig AP (2015) Vermicompost as an effective organic fertilizer and biocontrol agent: effect on growth, yield and quality of plants. Rev Environ Sci Biotechnol 14:137–159
Jouquet P, Dauber J, Lagerlöf J et al (2006) Soil invertebrates as ecosystem engineers: intended and accidental effects on soil and feedback loops. Appl Soil Ecol 32:153–164
Kevan DKM (1968) Soil animals. Witherby, London
Kim YN, Robinson B, Lee KA et al (2017) Interactions between earthworm burrowing, growth of a leguminous shrub and nitrogen cycling in a former agricultural soil. Appl Soil Ecol 110:79–87
Koprdova S, Saska P, Honek A et al (2010) Seed consumption by millipedes. Pedobiologia 54: 31–36
Körner K, Pfestorf H, May F et al (2014) Modelling the effect of belowground herbivory on grassland diversity. Ecol Model 273:79–85
Kostenko O, van de Voorde TFJ, Mulder PPJ et al (2012) Legacy effects of aboveground-belowground interactions. Ecol Lett 15:813–821
Kretzschmar A (1982) Description des gaieties de vers de terre et variations saisonnières des réseaux (observations en conditions naturelles). Rev Ecol Biol Sol 19:579–591
Laossi KR, Ginot A, Noguera DC et al (2010) Earthworm effects on plant growth do not necessarily decrease with soil fertility. Plant Soil 328:109–118
Lavelle P (1988) Earthworm activities and the soil system. Biol Fertil Soils 6:237–251
Lavelle P (1996) Diversity of soil fauna and ecosystem function. Biol Int 33:1–14
Lavelle P (2000) Ecological challenges for soil science. Soil Sci 165:73–86
Lavelle P, Spain AV (2001) Soil ecology. Kluwer Academic, Dordrecht
Lee KE (1985) Earthworms. Their ecology and relationships with soils and land use. Academic, Sydney
Lee KE, Foster RC (1991) Soil fauna and soil structure. Aust J Soil Res 29:745–775
Lee KK, Reddy MV, Wani SP et al (1996) Vesicular-arbuscular mycorrhizal fungi in earthworm casts and surrounding soil in relation to soil management of a semi-arid tropical Alfisol. Appl Soil Ecol 3:177–181
Lengyel S, Gove AD, Latimer AM et al (2009) Ants sow the seeds of global diversification in flowering plants. PLoS One 4(5):e5480. https://doi.org/10.1371/journal.pone.0005480
Li X, Feng Y, Boersma L (1994) Partition of photosynthates between shoot and root in spring wheat (*Triticum aestivum* L.) as a function of soil water potential and root temperature. Plant Soil 164: 43–50
Lobry De Bruyn LA, Conacher AJ (1990) The role of termites and ants in soil modification: a review. Aust J Soil Res 28:55–93
Lohmann M, Scheu S, Mueller C (2009) Decomposers and root feeders interactively affect plant defence in *Sinapis alba*. Oecologia 160:289–298
Ma Y, Dickinson NM, Wong MH (2006) Beneficial effects of earthworms and arbuscular mycorrhizal fungi on establishment of leguminous trees on Pb/Zn mine tailings. Soil Biol Biochem 38: 1403–1412
MacMahon JA, Mull JF, Crist TO (2000) Harvester ants (*Pogonomyrmex* spp.): their community and ecosystem influences. Annu Rev Ecol Syst 31:265–291
Maraun M, Scheu S (1996) Changes in microbial biomass, respiration and nutrient status of beech (*Fagus sylvatica*) leaf litter processed by millipedes (*Glomeris marginata*). Oecologia 107: 131–140
Marinissen JCY, Hillenaar SI (1997) Earthworm-induced distribution of organic matter in macro-aggregates from differently managed arable fields. Soil Biol Biochem 29:391–395
Martin A (1991) Short- and long-term effects of the endogeic earthworm *Millsonia anomala* (Omodeo) (Megascolecidae, Oligochaeta) of tropical savannas, on soil organic matter. Biol Fertil Soils 11:234–238

Masters GJ, Brown VK, Gange AC (1993) Plant mediated interactions between above- and below-ground insect herbivores. Oikos 66:148–151
Meyer-Wolfarth F, Schrader S, Oldenburg E et al (2017) Biocontrol of the toxigenic plant pathogen *Fusarium culmorum* by soil fauna in an agroecosystem. Mycotoxin Res 33:237–244
Milcu A, Schumacher J, Scheu S (2006) Earthworms (*Lumbricus terrestris*) affect plant seedling recruitment and microhabitat heterogeneity. Funct Ecol 20:261–268
Mortimer SR, van der Putten WH, Brown VK (1999) Insect and nematode herbivory below ground: interactions and role in vegetation succession. In: Olff H, Brown VK, Drent RH (eds) Herbivores: between plants and predators. Blackwell, Oxford, pp 205–238
Murray PJ, Clements RO (1998) Transfer of nitrogen between clover and wheat: effect of root herbivory. Eur J Soil Biol 34(I):25–30
Nichols E, Spector S, Louzada J et al (2008) Ecological functions and ecosystem services provided by Scarabaeinae dung beetles. Biol Conserv 141:1461–1474
Ortiz-Ceballos AI, Peña-Cabriales JJ, Fragoso C et al (2007) Mycorrhizal colonization and nitrogen uptake by maize: combined effect of tropical earthworms and velvet bean mulch. Biol Fertil Soils 44:181–186
Pangesti N, Pineda A, Pieterse CMJ et al (2013) Two-way plant mediated interactions between root-associated microbes and insects: from ecology to mechanisms. Front Plant Sci 4:414. https://doi.org/10.3389/fpls.2013.00414
Pfestorf H, Körner K, Sonnemann I et al (2016) Coupling experimental data with individual-based modelling reveals differential effects of root herbivory on grassland plant co-existence along a resource gradient. J Veg Sci 27:269–282
Pollierer MM, Langel R, Körner C et al (2007) The underestimated importance of belowground carbon input for forest soil animal food webs. Ecol Lett 10:729–736
Ponomareva SI (1950) The role of earthworms in the creation of a stable structure in ley rotations. Soviet Soil Sci 1:476–486
Poveda K, Steffan-Dewenter I, Scheu S et al (2005) Effects of decomposers and herbivores on plant performance and aboveground plant-insect interactions. Oikos 108:503–510
Puga-Freitas R, Barot S, Taconnat L et al (2012) Signal molecules mediate the impact of the earthworm *Aporrectodea caliginosa* on growth, development and defence of the plant *Arabidopsis thaliana*. PLoS One 7(12):e49504. https://doi.org/10.1371/journal.pone.0049504
Rasmann S, Köllner TG, Degenhardt J et al (2005) Recruitment of entomopathogenic nematodes by insect-damaged maize roots. Nature 434:732–737
Räty M, Huhta V (2003) Nematode communities of anthropogenous birch stands in central Finland. Nematology 5:629–639
Reddell P, Spain AV (1991) Earthworms as vectors of viable propagules of mycorrhizal fungi. Soil Biol Biochem 23:767–774
Rendon D, Whitehouse MEA, Taylor PW (2016) Consumptive and non-consumptive effects of wolf spiders on cotton bollworms. Entomol Exp Appl 158:170–183
Robert CAM, Erb M, Hibbard BE et al (2012) A specialist root herbivore reduces plant resistance and uses an induced plant volatile to aggregate in a density-dependent manner. Funct Ecol 26: 1429–1440
Rogers LE (1972) The ecological effects of the western harvester ant (*Pogonomyrmex occidentalis*) in the shortgrass plain ecosystem. USA/BP Grassland Biome, Technical Report 206
Rosumek FB, Silveira FAO, Neves FD et al (2009) Ants on plants: a meta-analysis of the role of ants as plant biotic defenses. Oecologia 160:537–549
Rypstra AL, Marshall SD (2005) Augmentation of soil detritus affects the spider community and herbivory in a soybean agroecosystem. Entomol Exp Appl 116:149–157
Sanders D, van Veen FJF (2011) Ecosystem engineering and predation: the multi-trophic impact of two ant species. J Anim Ecol 80:569–576
Saska P (2008) Granivory in terrestrial isopods. Ecol Entomol 33:742–747
Scheu S (1993) There is an earthworm mobilizable nitrogen pool in soil. Pedobiologia 37:243–249
Scheu S (1994) The influence of earthworms (Lumbricidae) on nitrogen mineralization in different ecosystems on limestone. Mitt hamburg zool Mus Inst 89:129–137

Scheu S (2001) Plants and generalist predators as links between the below-ground and above-ground system. Basic Appl Ecol 2:3–13

Scheu S (2003) Effects of earthworms on plant growth: patterns and perspectives. Pedobiologia 47: 846–856

Singh A, Braun J, Decker E et al (2014) Plant genetic variation mediates an indirect ecological effect between belowground earthworms and aboveground aphids. BMC Ecol 14:25. https://doi.org/10.1186/s12898-014-0025-5

Snyder SR, Crist TO, Friese CF (2002) Variability in soil chemistry and arbuscular mycorrhizal fungi in harvester ant nests: the influence of topography, grazing and region. Biol Fertil Soils 35: 406–413

Snyder BA, Boots B, Hendrix PF (2009) Competition between invasive earthworms (*Amynthas corticis*, Megascolecidae) and native North American millipedes (*Pseudopolydesmus erasus*, Polydesmidae): effects on carbon cycling and soil structure. Soil Biol Biochem 41:1442–1449

Sonnemann I, Hempel S, Beutel M et al (2013) The root herbivore history of the soil affects the productivity of a grassland plant community and determines plant response to new root herbivore attack. PLoS One 8(2):e56524. https://doi.org/10.1371/journal.pone.0056524

Sonnemann I, Buchholz Y, Rautenberg D et al (2016) Land use intensity modulates the impact of root herbivores on plant interactions with above- and below-ground organisms. Agric Ecosyst Environ 219:26–31

Spurgeon DJ, Keith AM, Schmidt O et al (2013) Land-use and land-management change: relationships with earthworm and fungi communities and soil structural properties. BMC Ecol 13:46. https://doi.org/10.1186/1472-6785-13-46

Staley JT, Mortimer SR, Morecroft MD et al (2007) Summer drought alters plant-mediated competition between foliar- and root-feeding insects. Glob Chang Biol 13:866–877

Stephens PM, Davoren CW, Ryder MH et al (1994) Influence of the earthworm *Aporrectodea trapezoides* (Lumbricidae) on the colonization of alfalfa (*Medicago sativa* L.) roots by *Rhizobium meliloti* L5-30R and the survival of *Rhizobium meliloti* L5-30R in soil. Biol Fertil Soils 18: 63–70

Stork NE, Eggleton P (1992) Invertebrates as determinants and indicators of soil quality. Am J Altern Agric 7:38–47

Swift MJ, Heal OW, Anderson JM (1979) Decomposition in terrestrial ecosystems. Blackwell, Oxford

Tariq M, Rossiter JT, Wright DJ et al (2013a) Drought alters interactions between root and foliar herbivores. Oecologia 172:1095–1104

Tariq M, Wright DJ, Bruce TJA et al (2013b) Drought and root herbivory interact to alter the response of above-ground parasitoids to aphid infested plants and associated plant volatile signals. PLoS One 8(7):e69013. https://doi.org/10.1371/journal.pone.0069013

Thorpe IS, Prosser JI, Glover LA et al (1996) The role of the earthworm *Lumbricus terrestris* in the transport of bacterial inocula through soil. Biol Fertil Soils 23:132–139

Tiunov AV, Scheu S (2004) Carbon availability controls the growth of detritivores (Lumbricidae) and their effect on nitrogen mineralization. Oecologia 138:83–90

Tiunov AV, Bonkowski M, Alphei J (2001) Microflora, protozoa and nematoda in *Lumbricus terrestris* burrow walls: a laboratory experiment. Pedobiologia 45:46–60

Toyota A, Kaneko N, Ito MT (2006) Soil ecosystem engineering by the train millipede *Parafontaria laminata* in a Japanese larch forest. Soil Biol Biochem 38:1840–1850

Treonis AM, Grayston SJ, Murray PJ et al (2005) Effects of root feeding, cranefly larvae on soil microorganisms and the composition of rhizosphere solutions collected from grassland plants. Appl Soil Ecol 28:203–215

Trouvé R, Drapela T, Frank T et al (2014) Herbivory of an invasive slug in a model grassland community can be affected by earthworms and mycorrhizal fungi. Biol Fertil Soils 50:13–23

van Dam NM (2009) Belowground herbivory and plant defenses. Annu Rev Ecol Evol Syst 40: 373–391

Van De Westeringh W (1972) Deterioration of soil structure in worm free orchards. Pedobiologia 12:6–15

van Groenigen JW, Lubbers IM, Vos HMJ et al (2014) Earthworms increase plant production: a meta-analysis. Sci Rep 4:6365. https://doi.org/10.1038/srep06365
Vos VCA, van Ruijven J, Berg MP et al (2011) Macro-detritivore identity drives leaf litter diversity effects. Oikos 120:1092–1098
Wagner D, Brown MJF, Gordon DM (1997) Harvester ant nests, soil biota and soil chemistry. Oecologia 112:232–236
Wallwork JA (1976) The distribution and diversity of soil fauna. Academic, London
Wardle DA, Bardgett RD, Klironomos JN et al (2004) Ecological linkages between aboveground and belowground biota. Science 304:1629–1633
Whalen JK, Hamel C (2004) Effects of key soil organisms on nutrient dynamics in temperate agroecosystems. J Crop Improv 11:175–207
Whalen JK, Parmelee RW, McCartney DA et al (1999) Movement of N from decomposing earthworm tissue to soil microbial and plant N pools. Soil Biol Biochem 31:487–492
Wiken EB, Boersma LM, Lavkulich LM et al (1976) Biosynthetic alteration in a British Columbia soil by ants (*Formica fusca* L). Soil Sci Soc Am J 40:422–426
Wolters V (2000) Invertebrate control of soil organic matter stability. Biol Fertil Soils 31:1–19
Wurst S (2010) Effects of earthworms on above- and belowground herbivores. Appl Soil Ecol 45: 123–130
Wurst S (2013) Plant-mediated links between detritivores and aboveground herbivores. Front Plant Sci 4:380. https://doi.org/10.3389/fpls.2013.00380
Wurst S, Jones TH (2003) Indirect effects of earthworms (*Aporrectodea caliginosa*) on an aboveground tritrophic interaction. Pedobiologia 47:91–97
Wurst S, Ohgushi T (2015) Do plant- and soil-mediated legacy effects impact future biotic interactions? Funct Ecol 29:1373–1382
Wurst S, Rillig MC (2011) Additive effects of functionally dissimilar above- and belowground organisms on a grassland plant community. J Plant Ecol 4:221–227
Wurst S, Dugassa-Gobena D, Langel R et al (2004a) Combined effects of earthworms and vesicular-arbuscular mycorrhizas on plant and aphid performance. New Phytol 163:169–176
Wurst S, Dugassa-Gobena D, Scheu S (2004b) Earthworms and litter distribution affect plant-defensive chemistry. J Chem Ecol 30:691–701
Wurst S, Langel R, Rodger S et al (2006) Effects of belowground biota on primary and secondary metabolites in *Brassica oleracea*. Chemoecology 16:69–73
Yeates GW (1981) Soil nematode populations depressed in the presence of earthworms. Pedobiologia 22:191–195
Yu X, Cheng J, Wong MH (2005) Earthworm-mycorrhiza interaction on Cd uptake and growth of ryegrass. Soil Biol Biochem 37:195–201
Zaller JG, Arnone JA (1997) Activity of surface-casting earthworms in a calcareous grassland under elevated atmospheric CO_2. Oecologia 111:249–254
Zaller JG, Arnone JA (1999) Interactions between plant species and earthworm casts in a calcareous grassland under elevated CO_2. Ecology 80:873–881
Zaller JG, Saxler N (2007) Selective vertical seed transport by earthworms: implications for the diversity of grassland ecosystems. Eur J Soil Biol 43:S86–S91
Zaller JG, Heigl F, Grabmaier A et al (2011a) Earthworm-mycorrhiza interactions can affect the diversity, structure and functioning of establishing model grassland communities. PLoS One 6: e29293. https://doi.org/10.1371/journal.pone.0029293
Zaller JG, Saccani F, Frank T (2011b) Effects of earthworms and mycorrhizal fungi on the growth of the medicinal herb *Calendula officinalis* (Asteraceae). Plant Soil Environ 57:499–450
Zaller JG, Parth M, Szunyogh I et al (2013) Herbivory of an invasive slug is affected by earthworms and the composition of plant communities. BMC Ecol 13:20
Zaller JG, Heigl F, Ruess L et al (2014) Glyphosate herbicide affects belowground interactions between earthworms and symbiotic mycorrhizal fungi in a model ecosystem. Sci Rep 4:5634. https://doi.org/10.1038/srep05634

Zaller JG, König N, Tiefenbacher A et al (2016) Pesticide seed dressings can affect the activity of various soil organisms and reduce decomposition of plant material. BMC Ecol 16:37. https://doi.org/10.1186/s12898-12016-10092-x

Zhao C, Griffin JN, Wu X, Sun S (2013) Predatory beetles facilitate plant growth by driving earthworms to lower soil layers. J Anim Ecol 82:749–758

Zhao C, Wu X, Griffin JN et al (2014) Territorial ants depress plant growth through cascading non-trophic effects in an alpine meadow. Oikos 123:481–487

Zhou XS, Kaya HK, Heungens K et al (2002) Response of ants to a deterrent factor(s) produced by the symbiotic bacteria of entomopathogenic nematodes. Appl Environ Microbiol 68:6202–6209

Chapter 9
The Feedback Loop Between Aboveground Herbivores and Soil Microbes via Deposition Processes

Cari A. Ritzenthaler, Caitlin E. Maloney, Audrey M. Maran, Eric A. Moore, Amanda Winters, and Shannon L. Pelini

9.1 Introduction

While aboveground vertebrates are significant contributors to the detrital pool (Wardle et al. 2002), this chapter will focus on invertebrate deposition of matter and subsequent effects on soil microbes, nutrient cycling, and feedbacks to plant growth (Fig. 9.1). The components discussed in this chapter are invertebrate herbivores, microbial communities, and vegetation, thus combining the above- and belowground food webs into a larger functioning ecosystem. Each component has a direct, or indirect, effect on the other, and in some cases, there is a cascading effect throughout the entire system. Invertebrate herbivores add frass, cadavers, honeydew, and alter litterfall (leaf fragments) and throughfall (water) inputs to the soil, thus altering the microbial community. Changes in the microbial community via invertebrate herbivore inputs can alter nutrient cycle, which can alter vegetation growth and architecture. Of course, growth and architecture of vegetation can also be altered by the consumption of the invertebrate herbivores themselves. Though some of the described effects will be positive and others will be negative, they all feed into each other and ultimately drive the larger above- and belowground feedback loop. We will discuss these effects, and more, throughout this chapter. Additionally, we will discuss the changes that could occur as the climate changes (e.g., rising temperatures and precipitation patterns). Finally, we'll end the chapter with a discussion of studies that have begun to examine the effect of a changing climate on above- and belowground processes.

C. A. Ritzenthaler · C. E. Maloney · A. M. Maran · E. A. Moore · A. Winters · S. L. Pelini (✉)
Department of Biological Sciences, Bowling Green State University, Bowling Green, OH, USA
e-mail: carir@bgsu.edu; cemalon@bgsu.edu; audreym@bgsu.edu; eamoore@bgsu.edu; winteam@bgsu.edu; spelini@bgsu.edu

T. Ohgushi et al. (eds.), *Aboveground–Belowground Community Ecology*, Ecological Studies 234, https://doi.org/10.1007/978-3-319-91614-9_9

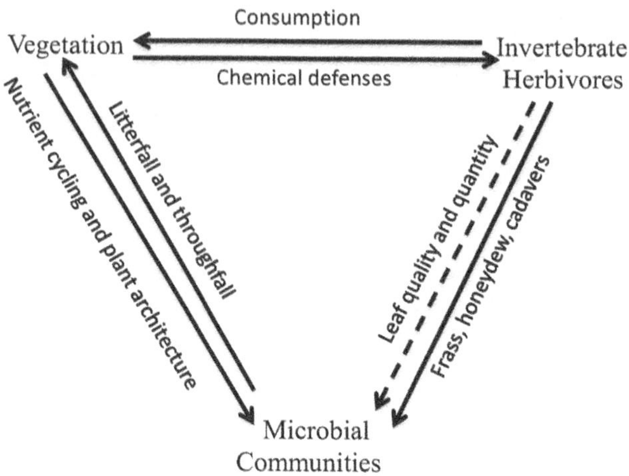

Fig. 9.1 Above- and belowground feedback loop. Arrows depict interactions (solid lines = direct, dashed lines = indirect) that are discussed in this chapter; however, there are likely many more interactions that are not included here

9.2 Herbivore Inputs and Microbial Community Response

Invertebrate herbivore feeding activity alters the timing, quantity, and quality of nutrients reaching the soil microbial communities by depositing feces (frass), cadavers, and plant matter. Though these inputs vary considerably in their nutrient qualities, they are relatively simple organic compounds and easily decomposed by soil microbes and are thus called "fast cycle" compounds. Nutrient dynamics are impacted by "fast cycle" compounds more quickly compared to herbivore-induced changes in litter quality, which result in slower responses ("slow cycle") (McNaughton et al. 1988). It means that when nutrients, such as through frass and green litterfall, are added to the soil, microbes respond to and decompose organic matter more quickly. On the other hand, when herbivores induce chemical changes in the material moving to the forest floor, such as through leaf chemical defenses and throughfall, the microbial community will decelerate the rate at which they are decomposing the organic matter. The nutrient acceleration hypothesis postulates that through selective feeding on easily decomposed vegetation, deposition of excrements, and increase in litter quality, herbivores speed up microbial activity, decomposition, and cycling of nutrients (Chapman et al. 2003; Ritchie et al. 1998). Whether herbivore-mediated alterations in timing, quantity, and quality of forest floor nutrients enhance (Chapman et al. 2003; Reynolds et al. 2000; Swank et al. 1981) or inhibit (Ritchie et al. 1998) nutrient cycling is under debate and may be dependent upon a number of biotic and abiotic factors.

9.2.1 Frass

Insect herbivores contribute nutrients to microbial communities by transforming recalcitrant plant matter into easily decomposed feces (Grace 1986). These relatively labile frass inputs represent a consistent source of carbon (C), phosphorus (P), and nitrogen (N) for soil organisms in spring and summer months located in temperate regions and throughout the year in tropical regions (Hunter 2001; Lovett and Ruesink 1995; Schowalter et al. 2011). Under severe herbivory, N input to soil from frass can exceed that of leaf litter (Fogal and Slansky 1985; Grace 1986).

Studies suggest that frass additions to soil have varied effects on soil microbes. In temperate systems, frass deposition to soil had no effect on soil respiration (Frost and Hunter 2004; Reynolds and Hunter 2001), which is a measure of microbial activity. However, the authors speculate that soil microbial respiration may have been masked by root respiration or the method of detection may have missed fine scale temporal alterations in respiration. However, frass inputs did result in greater surface soil microbial biomass, leading Frost and Hunter (2004) to hypothesize that soil microbial respiration is only affected at a certain threshold of frass quantity. Additionally, it has been noted that C from frass appears to stay in upper horizons of the soil, unless precipitation events occur immediately following deposition (Frost and Hunter 2004), an effect that has also been seen with N. For example, if rain events directly follow herbivore outbreaks, N from frass could leach out of soil before microbes are able to incorporate it into soil organic matter (Frost and Hunter 2004; Reynolds et al. 2000; Swank et al. 1981).

Several studies demonstrate that frass has positive effects on soil microbe activity; however, specific consequences for nutrient cycling vary and may be largely context-dependent. The diet quality of invertebrate herbivores has significant influences on frass quality (Kagata and Ohgushi 2012), which then can cause changes to the microbial community. For example, Fonte and Schowalter (2005) manipulated the degree of insect herbivory and found positive relationships between frass-related inputs and both nitrate (NO_3^-) concentration and leaf decomposition beneath trees. At endemic herbivore densities, the C:N of frass was positively correlated with soil respiration and soil NO_3^- was increased (Hunter et al. 2003). Another study found that while the C:N ratio of gypsy moth frass did not significantly differ from that of the soil in the study area, C in the frass was highly labile. This lability led to stimulation of microbial growth in laboratory incubation experiments, which resulted in immobilization of more than 90% of N within the first 10 days of incubation (Lovett and Ruesink 1995).

Additionally, the soil microbial response to labile C in frass inputs can conserve N within the soil, preventing it from leaching into waterways and potentially allowing plants to recover N lost to herbivory. However, the highly labile C in frass can also allow microbes to outcompete trees and rapidly immobilize available N, potentially reducing N mineralization (Lovett et al. 2002). Indeed, an isotopic study shows that only 1% of ^{15}N added to soil as frass was incorporated into oak seedlings, whereas 40% was incorporated into soils (Christenson et al. 2002).

Future studies seeking to understand effects of frass inputs on soil microbes and subsequent ecosystem processes should focus on threshold quantities and qualities of frass. These studies should include fine scale temporal monitoring of respiration while keeping in mind the importance of precipitation events for leaching of nutrients contained within frass.

Insect herbivory acts to redistribute N from tree leaves to frass, greenfall, and insect biomass. The fate of N in frass, and the degree of loss of frass N from the plant-soil system, is dependent upon a number of physical and biotic factors. While laboratory studies suggest that frass input to the detrital pool ultimately acts to conserve N within the plant-soil system through microbial immobilization, watershed studies indicate that even small quantities of frass N can lead to stream and lake acidification. This is because nitrate ions can transport calcium and magnesium cations from soil to bodies of water and also bring aluminum and hydrogen ions to the water surface (Lovett et al. 2002). Tree mortality, precipitation/run-off, and the N retention capacity of soils are all variables that may impact whether a forest ecosystem loses N following an herbivore outbreak. If defoliation events are severe enough to result in tree mortality, this loss of an uptake mechanism for N mobilized from frass (loss of a N sink), combined with increases in N availability from decaying roots, could overwhelm mechanisms for N retention, resulting in high rates of N leaching (Lovett et al. 2002). Large precipitation events following defoliation events and sandy or porous soils could result in run-off of N into bodies of water as well. Finally, a low N retention capacity of soil will result in leaching of N from the plant–soil system. Thin soils, soils with low organic matter content, and soil organisms that are not N limited could all result in low N retention capacity. Nitrogen limitation can be alleviated by chronic N deposition, fertilizers, and the presence of N fixers (Lovett et al. 2002). To understand factors that influence frass N loss from the plant–soil system, future watershed studies should focus on quantifying the above mentioned biotic and abiotic factors in conjunction with stream or lake chemistry, before and after defoliation events.

Frass inputs that are not leached out of the ecosystem can be taken back up into the nearby vegetation. Frost and Hunter (2007) applied ^{15}N labeled insect frass to soils surrounding experimentally damaged red oaks, *Quercus rubra*, and tracked its movement throughout two growing seasons. They found that much of ^{15}N was mineralized, taken up by the oak, returned to the foliage, and assimilated back into late-season herbivores within a single growing season. However, damaged oaks had lower levels of foliar N and ^{15}N in the following season, as well as reduced insect herbivore host preference. This suggests that herbivore damage may affect allocation over durations longer than a single growing season, but it should be noted that herbivore-induced plant defenses, as described later in the Sect. 9.3.2, may have acted to lower foliar N in damaged leaves in Frost and Hunter's study.

9.2.2 Honeydew

Excreta of insect herbivores that feed on plant fluids, such as aphids, take the form of honeydew. In contrast to typical insect frass, honeydew is a C-rich, sugary liquid with very low N content. Though the sugar and amino acid composition, as well as the amount of honeydew, differs depending on the temperature, aphid population, and the plant-resource's age (Stadler et al. 1998), it typically contains at least 100 times more C than N (Katayama et al. 2013). Aphids' release of honeydew has interesting effects on belowground functioning. Honeydew deposition has been shown to stimulate microbial growth and decrease available N belowground (Katayama et al. 2014; Milcu et al. 2015; Schmidt et al. 1997). This result is not unprecedented and supports the hypothesis that when C is not limiting, microbes are able to outcompete plants for N (Kaye and Hart 1997). Applying that hypothesis to aphid excretion, addition of C-rich honeydew releases microbes from C limitation and allows them to readily uptake and immobilize N. This effect has been shown to cascade up to the tree canopy, reducing N content of foliage. According to Katayama et al. (2014), the feedback effect on leaf N content may benefit aphids by harming other herbivores that cannot meet their N demands by feeding on the low quality leaves.

In addition to stimulating microbial growth, honeydew has been shown to increase microbial activity and, subsequently, respiration (Choudhury 1984; Dighton 1978; Milcu et al. 2015). Greater microbial biomass may extend to the invertebrate community, causing an increase in earthworms and decrease in mesofauna, such as springtails and mites (Milcu et al. 2015). Interestingly, the ecosystem type seems to affect the magnitude of microbial response. For example, in a study comparing microbial response to honeydew in woodland versus grassland soils, the woodland soil responded with 30% increase in the fungal community and a threefold increase of bacterial biomass, but the grassland soil had no significant increase in either group (Dighton 1978).

Recent studies have found that honeydew represents a significant input into soils that can cascade up trophic levels. The simple sugars inherent in honeydew have been shown to enhance aboveground production by indirectly bolstering the metabolism of microbes, which positively affects tree growth (Stadler et al. 1998). Milcu et al. (2015) manipulated honeydew deposition at the base of willows and found that applications increased soil microbial biomass, respiration, earthworm biomass, and litter mineralization. Effects of honeydew were also apparent in the willow flowering phenology and crown architecture. A greater ratio of tertiary to primary branches as well as greater shoot-to-root biomass ratios were observed in the honeydew treatments. Path analyses suggested that these observations might have been the result of increased endogeic earthworm effects. Enhanced microbial biomass at honeydew treated sites positively affected endogeic earthworms, which suggests that honeydew deposition may affect plant growth and architecture via indirect effects on soil biota. However, flowering was premature in the honeydew treatments, which is a common stress response in plants. The postulated mechanism for this observation is that honeydew may induce microbial nutrient immobilization, as discussed above, which results in nutrient limitation and altered allocation within the willows. However, it is worth noting that most studies discussed in this section directly applied honeydew, or a substitute, to soils; however, when honeydew is allowed to naturally deposit these effects may be weak or nonexistent (Stadler and Müller 1996). Most honeydew is deposited on leaves, and bacteria growing on the leaves may use the resources before rainfall brings it to the soil. Likely the effect of honeydew will depend on rainfall and the ability of the leaf structure to hold honeydew.

The mutualism between honeydew producing aphids and ants (Fischer and Shingleton 2001) has effects that cascade into soil processes as well. Variation in the chemical composition of honeydew is influenced by both interactions with ants and food preferences. Fischer and Shingleton (2001) also found that honeydew composition varies between and within aphid species, which suggests that interactions between aphids, their food sources, and mutualists directly influence the quality of resources entering belowground systems. Interactions between honeydew production and abiotic variables have also been documented. Nitrogen applications in cotton plants resulted in increased abundances of whiteflies, which resulted in increased honeydew production (Bi et al. 2001). Considering the sensitivity of honeydew composition, documenting how anthropogenic applications and disturbances influence the quantity and quality of invertebrate herbivore inputs to soils is a potentially impactful research direction.

9.2.3 Carcasses

Because of their small size, insect bodies have often been overlooked or are considered to play at most a minor role in nutrient transfer (Schowalter and Crossley 1983). However, there may be legacy effects of carcass deposition (Hawlena et al. 2012), and during outbreaks of herbivorous insects (e.g., locusts, cicadas) carcasses serve as a large influx of nutrients over a relatively short time (Brown and Chippendale 1973; Song et al. 2015). In fact, periodical cicadas reach a cumulative biomass similar to that of the largest of terrestrial animals (Whiles et al. 2001). Insect carcasses are also a high-quality resource, rich in N, potassium (K), magnesium (Mg), sodium (Na), calcium (Ca), lipids, and protein (Brown and Chippendale 1973; Schowalter and Crossley 1983), so large outbreaks can deposit a substantial amount of nutrients to the soil.

The belowground effects of herbivore carcasses are varied: they have impacts on soil invertebrates (see Chap. 8), microbes, and nutrients. Though carcasses are deposited in large pieces, they are more easily decomposed than other large detritus, such as leaf litter, especially in the presence of soil invertebrates (Seastedt et al. 1981; Yang 2006). Because of their high quality (Brown and Chippendale 1973; Fielding et al. 2013; Hawlena et al. 2012), soil microbes respond strongly to carcass deposition. Yang (2004) found that when cicadas were added to forest plots, bacterial and fungal biomass both increased significantly. Microbial respiration increased as well, a response also observed in other studies where bodies of cicadas, locusts, or grasshoppers were decomposed (Fielding et al. 2013; Hawlena et al. 2012; Song et al. 2015).

Research has reported that carcass deposition on nutrient pools has primarily focused on N availability and understandably so since under heavy defoliation 8% of foliar N may be transferred to the forest floor by herbivore carcasses (Hunter 2001). Unlike C-rich sources such as leaf litter or honeydew, insect bodies increase the plant-available ammonium and nitrate (Hunter 2001; Schimel and Bennett 2004; Song et al. 2015; Yang 2004). Yang (2004) found that the increase in N is surprisingly long-lived when cicada carcasses are decomposed; while ammonium availability increased 412% only for the first 30 days, nitrate (199% increase) remained elevated for at least 100 days after carcass addition. In a similar study where locust carcasses were added to plots in the temperate steppe, Song et al. (2015) found that soil N limitation was relieved, which ultimately resulted in a bottom-up effect on primary productivity that partly counteracted the negative impact of herbivory by locusts.

Though the literature is dominated by studies investigating outbreak conditions, there have been some studies looking at smaller-scale alterations to carcass depositions with potentially far-reaching consequences. Grasshoppers usually maintain a fairly stable N concentration in their tissues (10.49–10.71%) regardless of their food's N content (Fielding et al. 2013). However, they do exhibit changes to percent N in response to fear of predation. When Hawlena et al. (2012) exposed some grasshoppers to predation risk and kept others in safety, they found C:N in grasshoppers was 4% higher in grasshoppers that were exposed to predator risk than those that were not. Despite this small change, after decomposition of grasshopper carcasses, there were detectable impacts on ecosystem processes. In soil where the grasshoppers with higher C:N had decomposed, decomposition of plant material resulted in 62% greater CO_2 release than on soils where the lower C:N grasshoppers had decomposed. The impact of a 4% change in the C:N of grasshoppers led to an eventual 62% difference in CO_2 release which demonstrates the potential for small differences in carcass quality to have magnified effects on ecosystem functioning (Hawlena et al. 2012). Whether carcasses are being added to the soil in large quantities, as with outbreaks, or simply in normal quantities but with altered nutrient content, the consequences for belowground nutrient cycling are significant.

Insect carcasses often provide a pool of nutrients that can promote or inhibit production, depending on the context. Periodical cicadas have been shown to increase microbial biomass and N availability in North American forests. Yang (2004) described this massive influx of nutrients as a "bottom-up cascade" in which the emergence and subsequent senescence of cicadas positively affected plants and higher trophic levels. Insect emergence events can also be important vectors of nutrient and biomass transport across the aquatic–terrestrial interface. Insects that spend their larval stages in aquatic environments and their adult lives in terrestrial environments can subsidize terrestrial food webs. A long-term study from Lake Myvatn in Iceland has tracked how midge emergence and senescence acts as a fertilizer that affects the local heathland food web. Hoekman et al. (2011)

manipulated midge deposition in plots surrounding the lake and found a 108% increase in arthropod biomass over a 2-year period. This bottom-up amendment extends to plant communities as well. Bultman et al. (2014) compared two lakes with high and low levels of midge deposition. Tealeaf willow surrounding the lake that received high-density midge inputs had 8–11% higher foliar N content than was observed in willow surrounding the lake receiving low-density inputs (Bultman et al. 2014). Furthermore, July highflyer caterpillars that feed on tealeaf willow were observed at 4–6 times greater densities and at 72% greater biomass surrounding the lake with high-density midge deposition (Bultman et al. 2014).

Conversely, top-down trophic cascades can reduce the quality of herbivore cadavers. Hawlena et al. (2012) described how predator behavior in grassland indirectly decreased microbial decomposition. By inciting fear and physiological stress, certain behavioral traits of spiders goaded grasshoppers towards a more carbohydrate-rich diet, which elevated C:N in grasshopper bodies.

9.3 Herbivore Modifications to Leaf and Precipitation Inputs to Microbial Communities

Herbivore activity alters the timing, quantity, and quality of leaf litter and precipitation that falls from canopies. Leaf quality is altered through changes in types (green vs. brown) of leaf litter reaching the forest floor and through herbivore-induced chemical defense compounds. Herbivory also modifies water inputs to microbial communities by chemically enhancing precipitation that falls through the canopy (throughfall; Hunter 2001).

9.3.1 Leaf Quantity

The timing of litterfall can be dependent on the insect herbivore community. Grace (1986) compared defoliated and undefoliated tree stands and noted that overall quantity of litter was not altered by gypsy moth defoliation, but timing and quality of litter was influenced by herbivore feeding activity. Gypsy moth herbivory altered the seasonality of litter input to the forest floor. In defoliated plots, an initial deposition of leaf fragments (greenfall) occurred early in summer, followed by a natural deposition from senescence in the fall. In contrast, plots not experiencing gypsy moth herbivory followed a more natural pattern of senescence in the fall. Thus, herbivore activity can be an important factor in the timing of leaf inputs so that greenfall is a significant source of nutrients for soil organisms early in the summer, which could alter temporal patterns of nutrient cycling as well as productivity.

9.3.2 Leaf Quality

Insect herbivores can significantly increase nutrient availability in the soil by increasing inputs of greenfall (green leaves or leaf fragments) through their feeding activities. Greenfall decomposes faster and can have higher N and P concentrations and lower lignin to N ratios relative to senescent leaves (Fonte and Schowalter 2004; Grace 1986; Lodge et al. 1991; Risley and Crossley 1993). Reynolds and Hunter (2001) found significant decreases in soil respiration due to greenfall exclusion along an elevational gradient. Inputs of greenfall, therefore, may have important implications for ecosystem nutrient dynamics and productivity, at least in the short term (Fonte and Schowalter 2004).

Litterfall from plots defoliated by gypsy moth larvae had higher concentrations of N, P, and K and significantly lower concentrations of Ca than plots without caterpillars. Although this litterfall included frass inputs, the main contributors of these nutrient changes, accounting for 62.6%, were plant tissues (with greenfall fragments contributing the most). Gypsy moth caterpillar feeding on these trees induced premature senescence of leaves, resulting in trees being unable to translocate nutrients from leaves back into perennial tissues, leading to a nutrient-rich litterfall (Grace 1986).

Invertebrate herbivores alter quality of leaves by inducing synthesis of plant chemical defenses (e.g., tannins and polyphenols) within leaves and other plant parts. These chemical defenses can have significant effects on soil organisms and subsequent ecosystem processes (Fig. 9.2). Upon reaching the soil, tannins directly affect decomposers, change N availability by binding with proteins, and interact with other nutrients, all of which are processes that can influence decomposition and nutrient cycling (Chomel et al. 2016). Production of condensed tannins (CT) by *Populus* sp. is induced by herbivory, yet CT have little effect on plant herbivores, except those that utilize woody tissues (Schweitzer et al. 2008). Condensed tannins slow leaf litter decomposition rates, alter heterotrophic soil community compositions, and decrease nutrient availability in soil (Schweitzer et al. 2008). For example, Madritch and Lindroth (2015) show that condensed tannin concentration within leaves correlates positively with plant N recovery after severe herbivory.

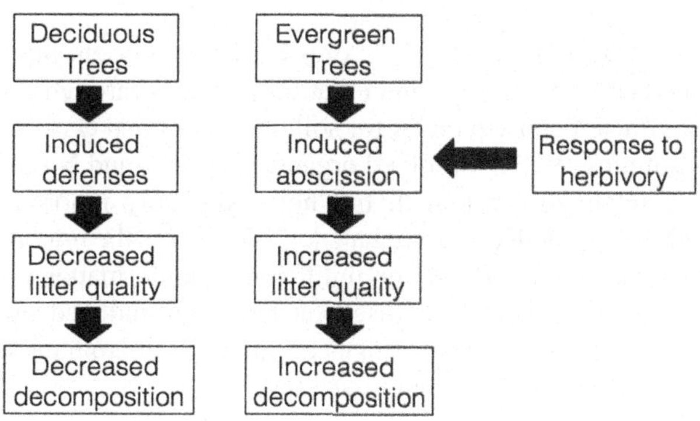

Fig. 9.2 Quality of litterfall deposited by herbivory is dependent on whether the tree is deciduous or evergreen (adapted from Chapman et al. 2006)

Specifically, high tannin concentrations in *Populus tremuloides* leaves correlated with recovery of frass containing ^{15}N within the same and following growing seasons. Additionally, this correlation with high leaf tannin concentrations and ^{15}N recovery held true for litter of trees that experienced herbivory in the season following defoliation (Madritch and Lindroth 2015); Schweitzer et al. (2008) proposes that rather than acting as a resistance mechanism by deterring herbivores, CT are produced by *Populus* as a mechanism of conserving N within soil, thereby tolerating herbivory. Additional evidence that plant defense compounds can slow soil processes (slow-cycle) comes from Findlay et al. (1996) who found that spider mite herbivory caused an increase in polyphenols and subsequently a 50% lower decomposition rate of litter. Future studies should consider the possibility that plant defense compounds act as a mechanism of tolerating defoliation through altering soil processes so as to retain nutrients in soil, rather than a defensive mechanism aimed at herbivore resistance.

Scale insects also play a role in leaf litter quality. Chapman et al. (2003) found that herbivory by scale insects and stem-boring moths altered litter quality by increasing N content and decreasing C:N ratios, resulting in more rapid decomposition (fast-cycle). Additionally, scale herbivory doubled P concentration in litter and increased the quantity of needle litterfall reaching the forest floor. This increase in litter quality due to herbivory extended the original "nutrient acceleration effect" (coined by Ritchie et al. 1998) that describes how herbivores can enhance nutrient cycling through selection of high quality plants and production of labile and energy-rich herbivore waste. Chapman et al. (2006) proposed that deciduous broadleaf and evergreen coniferous trees react differently to herbivory (Fig. 9.2). Deciduous trees generally respond to herbivory by producing defense compounds, which decrease litter quality and, therefore, decomposition rates. In contrast, evergreen trees induce premature leaf abscission, which increases litter quality because of reduced resorption of nutrients from leaves. Therefore, evergreens respond to herbivory in a manner, which leads to increases in decomposition rates (Chapman et al. 2006).

9.3.3 Throughfall

Defoliation by insect herbivores increases nutrient inputs to the soil via chemically modified throughfall, precipitation that falls through the canopy, which could have significant consequences for soil organisms and ecosystem processes (Hunter 2001; Kimmins 1972). Dissolved organic C (DOC) and N (DON) dynamics of throughfall are relatively unstudied, but not completely ignored (Le Mellec and Michalzik 2008). Le Mellec and Michalzik (2008) stress the importance of including particulate organic matter (POM) or unfiltered organic matter fraction, as well as dissolved nutrients, in studies focused on herbivore-induced nutrient fluxes in throughfall. Similar to frass inputs, enhanced throughfall contains labile forms of C and N, but effects on soil microbial processes vary.

A number of studies have found no effects of chemically enhanced throughfall on litter decomposition or soil microbial activity, but, interestingly, microbes living on aboveground portions of plants (phyllosphere) have been implicated in altering nutrient quality of throughfall reaching the forest floor. In a temperate system, sap-sucking herbivore abundance positively affected throughfall N, K, and Ca content during the early growing season, and this pattern held true for Ca throughout the entire growing season (Schowalter et al. 1991). However, effects of nutrient turnover did not result in differences in litter decomposition between herbivore abundance treatments. Furthermore, defoliation by the phytophagous pine lappet caterpillar resulted in the doubling of throughfall inputs of total DOC and DON compared to uninfested trees over a 6-month period, but no significant effects on soil microbes were found except for a marginally insignificant increase in microbial activity under infested trees (Le Mellec and Michalzik 2008). In a similar study, severe defoliation by two phytophagous herbivores increased DOC and DON in throughfall by a factor of 3 and 2.5, respectively, and total N bound and dissolved N bound by 1.4 and 1.3 times, respectively (Le Mellec et al. 2011). Leachate of forest floor total organic C, DOC, and NO_3-N were significantly lower at the infested site (Le Mellec et al. 2011). This decrease in NO_3-N in leachate under infested trees suggests that enhanced throughfall may increase biomass of phyllosphere microorganisms (filamentous fungi, bacteria, yeasts), which in turn immobilize inorganic N (Le Mellec et al. 2011; Stadler and Müller 1996; Stadler et al. 2001). These phyllosphere microbes potentially take advantage of labile C compounds leaching from frass-damaged leaves, acting as a temporal sink and subsequently altering the form of available nutrients that reach soil organisms. Reynolds and Hunter (2001) added throughfall containing NH_4Cl and KH_2PO_4 ions to soil in a temperate system and found reduced soil respiration; they hypothesized that soil microbes were outcompeted by ectomycorrhizal fungi, which may produce less CO_2 when assimilating mineral N (Table 9.1).

9.4 Plant and Herbivore Diversity, and Landscape Productivity as Predictors of Aboveground–Belowground Linkage Mechanisms

The effects of herbivory on vegetation vary by ecosystem and land use history but can significantly alter plant growth and community composition (Olff and Ritchie 1998). Intensive herbivory observed in low-productivity ecosystems will typically accelerate succession of well-defended plants with nutrient-poor material (Pastor et al. 1988). High-quality plant species are targeted, which allows less-preferred, lower quality plants to establish themselves (Bakker et al. 2006; Pastor et al. 1988). On the other hand, high-productivity, early successional systems with more soil resources generally exhibit plants with high growth rates, fast tissue turnover, high shoot leaf areas, fewer defensive compounds, and nutrient-rich foliage (Wardle and

Table 9.1 Effects of, based on lability of different inputs, invertebrate herbivores on belowground food webs

Input	Composition	Lability	Effects belowground
Frass	Frass quality is dependent on foliage eaten. Frass C:N is significantly lower than the foliage C:N, but they are correlated. Other foliage chemicals (e.g., tannins) can be present in frass	Highly labile	Increase in microbial biomass and soil respiration (Couture and Lindroth 2014; Frost and Hunter 2004; Hunter et al. 2003)
			Increase in litter mineralization and microbial immobilization (Fonte and Schowalter 2005; Lovett and Ruesink 1995)
			Increase or decrease in soil N depending on context (Fogal and Slansky 1985; Fonte and Schowalter 2005; Frost and Hunter 2004; Grace 1986; Hunter et al. 2003)
Honeydew	C-rich, low N sugar. Specific composition varies between insect species, plant food source, and environmental conditions	Highly labile	Stimulate microbial growth, immobilization, and reduction of soil N (Katayama et al. 2014; Milcu et al. 2015; Schmidt et al. 1997)
			Increase in soil respiration and litter mineralization (Choudhury 1984; Dighton 1978; Milcu et al. 2015)
			Increase in earthworm biomass (Milcu et al. 2015)
Insect Carcasses	Chitin, sclerotin, soluble protein, glycogen, and lipids. Also high in N, K, Mg, Na, and Ca	Moderately labile, particularly with invertebrate decomposer activity	Increase in microbial biomass and respiration (Fielding et al. 2013; Hawlena et al. 2012; Song et al. 2015; Yang 2004)
			Increase in plant-available N (Schimel and Bennett 2004; Song et al. 2015; Yang 2004)
Greenfall (herbivore-induced foliar litterfall)	Higher N and P concentrations and lower lignin-to-nitrogen ratios relative to senescent leaves	More labile than naturally senesced leaves	Decomposes faster than senescent leaves (Grace 1986; Lodge et al. 1991; Risley and Crossley 1993)

(continued)

Table 9.1 (continued)

Input	Composition	Lability	Effects belowground
Defended Leaves (herbivore-induced plant defense compounds)	High molecular weight phenolic compounds (large complex polymers, such as tannin and lignin)	Recalcitrant	Slow decomposition and C and N cycling (Hättenschwiler and Vitousek 2000)
			Can be toxic to microorganisms, inhibit enzyme activities, and affect N and C transformation in soils (Bradley et al. 2000; Fierer et al. 2001; Hättenschwiler and Vitousek 2000; Kraus et al. 2003; Schimel et al. 1996; Schofield et al. 2001)
			Delay microbial colonization of litter (Chomel et al. 2014; Ormeno et al. 2006)

Bardgett 2004). Therefore, habitats with nutrient-rich soil can support a greater overall herbivore biomass (McNaughton et al. 1989). Herbivory in these systems inhibits succession of nutrient poor, slow growing, well-defended plants with lignin-rich, woody tissues (Augustine and McNaughton 1998).

Herbivores induce trait responses in plants, in regards to nutrient allocation and growth, with contrasting effects. Mikola et al. (2009) noted that herbivore effects on plant growth and allocation were primarily explained by defoliation, and these effects can be positive or negative. Plants in high-productivity landscapes like grasslands exhibit compensatory growth responses, meaning plants will actively allocate resources towards remedying tissue damage. Direct defoliation may cause plants to reallocate available nutrients to remaining foliage, in an attempt to account for loss of photosynthetic potential (Hamilton and Frank 2001; Holland and Detling 1990). Aboveground grazing may also stimulate rhizosphere activity by promoting root exudates. That being said, intermediate levels of herbivory aboveground in high-productivity systems may be optimal for belowground communities, as microbial biomass can be maximized (Denton et al. 1998). Conversely, overgrazing may cause a reduction in mycorrhizal mutualisms as well as root biomass (Gehring and Whitham 2002; Guitian and Bardgett 2000; Mikola et al. 2001). This suggests that the optimization of decomposer organisms at intermediate intensities of herbivory does usually apply to mutualists at plant roots, and it is common for plants to compete with microbes for a limited pool of nutrients (Bardgett et al. 2003).

Herbivore-induced damage to specific plant tissues may also alter plant architecture. Rapid cellular division occurs in plant meristems; thus, nutrients are directed towards these areas. Consequently, meristems are susceptible to herbivory because of their high nutrient content. When the apical meristem, the region responsible for directing vertical growth, is damaged or consumed, the plant may be released from apical dominance (Aarssen 1995). This often causes branching of the stem with implications for canopy cover, photosynthetic potential, and reproductive success (Aarssen 1995). Selective consumption of nutrient-rich species or specific plant tissues also varies with herbivore diversity.

Classen et al. (2005) investigated the effects of two invertebrate herbivore species that differed in feeding preferences and subsequent effects on pinyon tree architecture. The pinyon needle scale insect selectively feeds on the needles of juvenile trees whereas the stem-boring moth focuses attention on mature pinyons. The differences between these two insects in terms of tissue preference and developmental stage had dramatic effects on plant architecture, with the needle scale insect generally causing crown opening and branching and the stem-boring moth generally inducing trees to form a more dense, compact canopy. These alterations to canopy structure ultimately had differing effects on local soil microclimate variables, where the effects of the needle scale insect were more pronounced in terms of increases in soil temperature and moisture. Scale insects also decreased canopy interception of precipitation more so than stem-boring moths, which implies that herbivore effects on throughfall as well as understory and soil microclimate are species-specific. However, it is difficult to model the functional diversity effects of herbivores on plants because, as mentioned before, these effects are environmentally context dependent (Wardle and Bardgett 2004).

Palatability of foliage and decomposability of litter are examples of physiological plant traits that are functionally diverse, meaning that they can vary widely across species (Grime et al. 1990). Relevant herbivore functional traits may include plant feeding preferences (which can vary in time and space), mobility, and reproductive rate. The ability to quantify and track these and other traits at the community level may help to model landscape changes in vegetation and herbivory over time.

Net primary production (NPP) and related plant-mediated processes typically peak when plant functional diversity is maximized, but herbivore functional diversity also plays a major role in governing diversity-functioning among plant communities. It is generally assumed that resource partitioning is maximized when the herbivore community is composed primarily of specialists representing functionally dissimilar guilds that utilize a variety of available resources efficiently, as opposed to a community of generalists competing for resources (Díaz and Cabido 2001).

The data to support this claim is thus far inconclusive; however, functional diversity does seem to play a substantial role in dictating inputs to the decomposer systems. Risch et al. (2015) used a series of different sized enclosure fences to assess the impact of herbivores on soil processes. Net N mineralization was highest in the treatment that excluded large fauna but retained the invertebrate community whereas mammals decreased total N mineralization. These studies collectively suggest that, in the short term, changes to the invertebrate herbivore community can have far-reaching impacts on nutrient cycling by affecting soil quality. Interactions between aboveground and belowground herbivores also present a challenge for determining effects on plant growth and diversity. A study by Ruijven et al. (2005)

revealed that the neutral effects of an aboveground herbivore and the positive effects of a belowground herbivore were eliminated when both species were present in a community. Overall plant diversity was reduced as a result of the interactive effect of selective aboveground defoliation and belowground root consumption.

However, herbivory does not only encompass foliar and root herbivores. A recent meta-analysis by Zvereva et al. (2010) covered herbivore functional diversity by consolidating studies on sap-feeding herbivore strategies. It was found that mesophyll and phloem feeders had greater negative effects on plants than xylem feeders, and that generalists exhibit stronger pressures on plant fitness than do specialists. Sap-feeding herbivores may also inadvertently contribute to belowground inputs by altering stemflow composition. Stemflow is analogous to throughfall in that it refers to the process by which precipitation reaches the ground by running down the length of a plant stem or trunk (Eaton et al. 1973). In this way, stemflow can accumulate defensive compounds that are leached from trees due to damage from sap-feeding invertebrates. Indeed, Michalzik et al. (2016) noted positive effects on stemflow nutrient and micronutrient concentrations in the presence of low-to-moderate pest aphid activity and further concluded that stemflow can be a useful metric to assess the impacts of leaf and sap-feeding invertebrates in forest ecosystems.

These examples have showcased that plant diversity and herbivore diversity have contrasting effects that are often difficult to separate in natural systems. However, our resolution of these effects may be improved upon further experimentation using multiple-species communities that directly manipulate functional traits and possible species interactions, as well as studies that address interactions between multiple trophic levels.

9.5 Global Change

As humans continue to modify the climate through increases in greenhouse gas emissions, understanding the complex linkages between aboveground herbivores and the belowground microbial community will become more important. With increases in greenhouse gas emissions, there is a subsequent predicted increase in temperature and soil respiration. Increased microbial respiration will lead to a feedback effect, exacerbating climate change. In fact, in their review, Bond-Lamberty and Thomson (2010) estimated that CO_2 release by the microbial community is about 10 times the amount released by humans and that it is increasing by 0.01% per year. This change can be attributed to the rise in global temperature (Li et al. 2013; Liu et al. 2016). As the climate continues to warm, this positive feedback loop will only continue to exacerbate global climate change.

Like microbes, invertebrate herbivore metabolism increases under warming conditions. Herbivores tend to consume more as temperatures increase and in turn may produce more frass (Bauerfeind and Fischer 2013; Cannon 1998; Gherlenda et al. 2016; Kozlov 2008). This leads to an increase in the amount of frass deposition and could affect the microbial community; however, this feedback loop becomes complicated because higher CO_2 levels in the atmosphere can lead to lower ratios of C to N within plants (Cannon 1998). Thus, the nutritional value of herbivore frass may change under higher CO_2 conditions. Additionally, this increase in consumption can be highly variable depending on the host-pest species and only holds true when temperatures are between 20 and 30 °C (Lemoine et al. 2014). After this point, temperature stops being a predictor of consumption, possibly due to heat stress.

To further complicate these interactions, temperature can change trophic interactions. For example, warming has been shown to decrease top-down controls on old field communities (Barton 2010). In this community, warming had an indirect, but positive, effect on grasshopper consumption rate because of behavioral changes of the spiders. As spiders retreated further down in the grassland canopy to escape heat, there was less overlap between the grasshopper and spider habitats, leading to less predation (Barton 2010).

Warming can also increase the maturation rate of insects. Many invertebrate herbivores experience greater growth rates when reared in higher temperature environments (Cannon 1998). Faster maturation times could lead to increases in the amount of herbivores within an area because multiple generations could occur within a season (Patterson et al. 1999), leading to greater inputs of frass and greenfall to the soil (Bale et al. 2002). Additionally, in a warmer world, pest species may flourish. For example, mountain pine beetle epidemics are becoming increasingly prevalent and are spreading into previously unaffected territories due to increases in temperature (Bentz et al. 2010; Creeden et al. 2014; Cudmore et al. 2010). The epidemic of beetles may lead to an influx of C as woody debris falls to the soil from the dead trees. Furthermore, Lepidoptera pests are expected to see increases in population and distribution (Vanhanen et al. 2007). These increases in Lepidoptera populations may lead to increases in invertebrate herbivore mediated litterfall, including an increase in carcasses and frass.

Because of these predicted changes to herbivore population with climate change, it will become increasingly important to study community level interactions to fully understand how insect herbivore deposition impacts the microbial community in an ever-changing world.

9.6 Conclusion

Throughout this chapter, we've discussed the components in the above- and below-ground feedback loop (Fig. 9.1). Additionally, we've discussed how interconnected these processes are on each other. The invertebrate herbivore exhibits top-down control on the microbes by depositing frass, cadavers, and honeydew, but also by changing the quality and quantity of litter and precipitation reaching the forest floor. However, the microbes are also exhibiting bottom-up control on the vegetation by altering the quality and quantity of nutrients reaching the plants, inevitably altering the invertebrate herbivores that consume the affected plants. The vegetation itself also has a bottom-up control on the invertebrate herbivores because of the production of chemical defenses.

Whether an effect is top-down or bottom-up depends entirely on how the system is being examined. We argue that the system should be examined as a whole because of all these complicated relationships. Without considering the entire above- and belowground feedback loop, studies could miss vital ecosystem functions and a more complete knowledge of how they will withstand environmental change.

References

Aarssen LW (1995) Hypotheses for the evolution of apical dominance in plants: implications for the interpretation of overcompensation. Oikos 74:149–156. https://doi.org/10.2307/3545684

Augustine DJ, McNaughton SJ (1998) Ungulate effects on the functional species composition of plant communities: herbivore selectivity and plant tolerance. J Wildl Manage 62:1165–1183

Bakker ES, Ritchie ME, Olff H et al (2006) Herbivore impact on grassland plant diversity depends on habitat productivity and herbivore size. Ecol Lett 9:780–788

Bale JS, Masters GJ, Hodkinson ID et al (2002) Herbivory in global climate change research: direct effects of rising temperature on insect herbivores. Glob Chang Biol 8:1–16. https://doi.org/10.1046/j.1365-2486.2002.00451.x

Bardgett RD, Streeter TC, Bol R (2003) Soil microbes compete effectively with plants for organic-nitrogen inputs to temperate grasslands. Ecology 84:1277–1287

Barton BT (2010) Climate warming and predation risk during herbivore ontogeny. Ecology 91:2811–2818

Bauerfeind SS, Fischer K (2013) Increased temperature reduces herbivore host-plant quality. Glob Chang Biol 19:3272–3282

Bentz BJ, Régnière J, Fettig CJ et al (2010) Climate change and bark beetles of the western United States and Canada: direct and indirect effects. Bioscience 60:602–613

Bi J, Ballmer G, Hendrix D et al (2001) Effect of cotton nitrogen fertilization on Bemisia argentifolii populations and honeydew production. Entomol Exp Appl 99:25–36

Bond-Lamberty B, Thomson A (2010) Temperature-associated increases in the global soil respiration record. Nature 464:579–U132. https://doi.org/10.1038/nature08930

Bradley R, Titus B, Preston C (2000) Changes to mineral N cycling and microbial communities in black spruce humus after additions of (NH 4) 2 SO 4 and condensed tannins extracted from *Kalmia angustifolia* and balsam fir. Soil Biol Biochem 32:1227–1240

Brown J, Chippendale G (1973) Nature and fate of the nutrient reserves of the periodical (17 year) cicada. J Insect Physiol 19:607–614

Bultman H, Hoekman D, Dreyer J et al (2014) Terrestrial deposition of aquatic insects increases plant quality for insect herbivores and herbivore density. Ecol Entomol 39:419–426

Cannon RJ (1998) The implications of predicted climate change for insect pests in the UK, with emphasis on non-indigenous species. Glob Chang Biol 4:785–796

Chapman SK, Hart SC, Cobb NS et al (2003) Insect herbivory increases litter quality and decomposition: an extension of the acceleration hypothesis. Ecology 84:2867–2876

Chapman SK, Whitham TG, Powell M (2006) Herbivory differentially alters plant litter dynamics of evergreen and deciduous trees. Oikos 114:566–574

Chomel M, Fernandez C, Bousquet-Mélou A et al (2014) Secondary metabolites of Pinus halepensis alter decomposer organisms and litter decomposition during afforestation of abandoned agricultural zones. J Ecol 102:411–424

Chomel M, Guittonny-Larchevêque M, Fernandez C et al (2016) Plant secondary metabolites: a key driver of litter decomposition and soil nutrient cycling. J Ecol 104:1527–1541

Choudhury D (1984) Aphids and plant fitness: a test of Owen and Wiegert's hypothesis. Oikos 43:401–402
Christenson LM, Lovett GM, Mitchell MJ et al (2002) The fate of nitrogen in gypsy moth frass deposited to an oak forest floor. Oecologia 131:444–452
Classen AT, Hart SC, Whitman T et al (2005) Insect infestations linked to shifts in microclimate. Soil Sci Soc Am J 69:2049–2057
Couture JJ, Lindroth RL (2014) Atmospheric change alters frass quality of forest canopy herbivores. Arthropod-Plant Interact 8:33–47
Creeden EP, Hicke JA, Buotte PC (2014) Climate, weather, and recent mountain pine beetle outbreaks in the western United States. For Ecol Manage 312:239–251
Cudmore TJ, Björklund N, Carroll AL et al (2010) Climate change and range expansion of an aggressive bark beetle: evidence of higher beetle reproduction in naïve host tree populations. J Appl Ecol 47:1036–1043
Denton CS, Bardgett RD, Cook R et al (1998) Low amounts of root herbivory positively influence the rhizosphere microbial community in a temperate grassland soil. Soil Biol Biochem 31:155–165
Díaz S, Cabido M (2001) Vive la difference: plant functional diversity matters to ecosystem processes. Trends Ecol Evol 16:646–655
Dighton J (1978) Effects of synthetic lime aphid honeydew on populations of soil organisms. Soil Biol Biochem 10:369–376
Eaton JS, Likens GE, Bormann FH (1973) Throughfall and stemflow chemistry in a northern hardwood forest. J Ecol 61:495–508
Fielding DJ, Trainor E, Zhang M (2013) Diet influences rates of carbon and nitrogen mineralization from decomposing grasshopper frass and cadavers. Biol Fertil Soils 49:537–544
Fierer N, Schimel JP, Cates RG et al (2001) Influence of balsam poplar tannin fractions on carbon and nitrogen dynamics in Alaskan taiga floodplain soils. Soil Biol Biochem 33:1827–1839
Findlay S, Carreiro M, Krischik V et al (1996) Effects of damage to living plants on leaf litter quality. Ecol Appl 6:269–275
Fischer M, Shingleton A (2001) Host plant and ants influence the honeydew sugar composition of aphids. Funct Ecol 15:544–550
Fogal W, Slansky F Jr (1985) Contribution of feeding by European pine sawfly larvae to litter production and element flux in scots pine plantations. Can J For Res 15:484–487
Fonte SJ, Schowalter TD (2004) Decomposition of greenfall vs. senescent foliage in a tropical forest ecosystem in Puerto Rico. Biotropica 36:474–482
Fonte S, Schowalter T (2005) The influence of a neotropical herbivore (*Lamponius portoricensis*) on nutrient cycling and soil processes. Oecologia 146:423–431
Frost CJ, Hunter MD (2004) Insect canopy herbivory and frass deposition affect soil nutrient dynamics and export in oak mesocosms. Ecology 85:3335–3347
Frost CJ, Hunter MD (2007) Recycling of nitrogen in herbivore feces: plant recovery, herbivore assimilation, soil retention, and leaching losses. Oecologia 151:42–53
Gehring CA, Whitham TG (2002) Mycorrhizae-herbivore interactions: population and community consequences. Mycorrhizal ecology. Springer, New York, pp 295–320
Gherlenda AN, Crous KY, Moore BD et al (2016) Precipitation, not CO_2 enrichment, drives insect herbivore frass deposition and subsequent nutrient dynamics in a mature Eucalyptus woodland. Plant Soil 399:29–39
Grace J (1986) The influence of gypsy moth on the composition and nutrient content of litter fall in a Pennsylvania oak forest. For Sci 32:855–870
Grime JP, Hodgson JG, Hunt R (1990) Abridged comparative plant ecology. Springer, New York
Guitian R, Bardgett RD (2000) Plant and soil microbial responses to defoliation in temperate semi-natural grassland. Plant Soil 220:271–277
Hamilton EW, Frank DA (2001) Can plants stimulate soil microbes and their own nutrient supply? Evidence from a grazing tolerant grass. Ecology 82:2397–2402

Hättenschwiler S, Vitousek PM (2000) The role of polyphenols in terrestrial ecosystem nutrient cycling. Trends Ecol Evol 15:238–243
Hawlena D, Strickland MS, Bradford MA et al (2012) Fear of predation slows plant-litter decomposition. Science 336:1434–1438
Hoekman D, Dreyer J, Jackson RD et al (2011) Lake to land subsidies: experimental addition of aquatic insects increases terrestrial arthropod densities. Ecology 92:2063–2072
Holland EA, Detling JK (1990) Plant response to herbivory and belowground nitrogen cycling. Ecology 71:1040–1049
Hunter MD (2001) Insect population dynamics meets ecosystem ecology: effects of herbivory on soil nutrient dynamics. Agric For Entomol 3:77–84
Hunter MD, Linnen CR, Reynolds BC (2003) Effects of endemic densities of canopy herbivores on nutrient dynamics along a gradient in elevation in the southern Appalachians. Pedobiologia 47:231–244
Kagata H, Ohgushi T (2012) Non-additive effects of leaf litter and insect frass mixture on decomposition processes. Ecol Res 27:69–75
Katayama N, Tsuchida T, Hojo M et al (2013) Aphid genotype determines intensity of ant attendance: do endosymbionts and honeydew composition matter? Ann Entomol Soc Am 106:761–770
Katayama N, Silva AO, Kishida O et al (2014) Herbivorous insect decreases plant nutrient uptake: the role of soil nutrient availability and association of below-ground symbionts. Ecol Entomol 39:511–518
Kaye JP, Hart SC (1997) Competition for nitrogen between plants and soil microorganisms. Trends Ecol Evol 12:139–143
Kimmins J (1972) Relative contributions of leaching, litter-fall and defoliation by Neodiprion sertifer (Hymenoptera) to the removal of cesium-134 from red pine. Oikos 23:226–234
Kozlov MV (2008) Losses of birch foliage due to insect herbivory along geographical gradients in Europe: a climate-driven pattern? Clim Change 87:107–117
Kraus TE, Dahlgren RA, Zasoski RJ (2003) Tannins in nutrient dynamics of forest ecosystems-a review. Plant Soil 256:41–66
Le Mellec A, Michalzik B (2008) Impact of a pine lappet (*Dendrolimus pini*) mass outbreak on C and N fluxes to the forest floor and soil microbial properties in a Scots pine forest in Germany. Can J For Res 38:1829–1841
Le Mellec A, Gerold G, Michalzik B (2011) Insect herbivory, organic matter deposition and effects on belowground organic matter fluxes in a central European oak forest. Plant Soil 342:393–403
Lemoine NP, Burkepile DE, Parker JD (2014) Variable effects of temperature on insect herbivory. PeerJ 2:e376
Li Q, Zhou BZ, Wang XM et al (2013) Effects of throughfall exclusion on soil respiration in a Moso Bamboo Forest Soil in Southeast China. In: Advanced materials research. Trans Tech, Zürich, pp 3762–3766
Liu Y, Liu S, Wan S et al (2016) Differential responses of soil respiration to soil warming and experimental throughfall reduction in a transitional oak forest in central China. Agric For Meteorol 226:186–198
Lodge DJ, Scatena F, Asbury C et al (1991) Fine litterfall and related nutrient inputs resulting from Hurricane Hugo in subtropical wet and lower montane rain forests of Puerto Rico. Biotropica 23:336–342
Lovett GM, Ruesink AE (1995) Carbon and nitrogen mineralization from decomposing gypsy moth frass. Oecologia 104:133–138
Lovett GM, Christenson LM, Groffman PM et al (2002) Insect defoliation and nitrogen cycling in forests: laboratory, plot, and watershed studies indicate that most of the nitrogen released from forest foliage as a result of defoliation by insects is redistributed within the ecosystem, whereas only a small fraction of nitrogen is lost by leaching. Bioscience 52:335–341
Madritch MD, Lindroth RL (2015) Condensed tannins increase nitrogen recovery by trees following insect defoliation. New Phytol 208:410–420

McNaughton S, Ruess R, Seagle S (1988) Large mammals and process dynamics in African ecosystems. Bioscience 38:794–800
McNaughton SJ, Oesterheld M, Frank DA et al (1989) Ecosystem-level patterns of primary productivity and herbivory in terrestrial habitats. Nature 341:142–144
Michalzik B, Levia DF, Bischoff S et al (2016) Effects of aphid infestation on the biogeochemistry of the water routed through European beech (Fagus sylvatica L.) saplings. Biogeochemistry 129:197–214
Mikola J, Yeates GW, Barker GM et al (2001) Effects of defoliation intensity on soil food-web properties in an experimental grassland community. Oikos 92:333–343
Mikola J, Setälä H, Virkajärvi P et al (2009) Defoliation and patchy nutrient return drive grazing effects on plant and soil properties in a dairy cow pasture. Ecol Monogr 79:221–244
Milcu A, Bonkowski M, Collins CM et al (2015) Aphid honeydew-induced changes in soil biota can cascade up to tree crown architecture. Pedobiologia 58:119–127
Olff H, Ritchie ME (1998) Effects of herbivores on grassland plant diversity. Trends Ecol Evol 13:261–265
Ormeno E, Baldy V, Ballini C et al (2006) Effects of environmental factors and leaf chemistry on leaf litter colonization by fungi in a Mediterranean shrubland. Pedobiologia 50:1–10
Pastor J, Naiman RJ, Dewey B et al (1988) Moose, microbes, and the boreal forest. Bioscience 38:770–777
Patterson D, Westbrook J, Joyce R et al (1999) Weeds, insects, and diseases. Clim Change 43:711–727
Reynolds B, Hunter M (2001) Responses of soil respiration, soil nutrients, and litter decomposition to inputs from canopy herbivores. Soil Biol Biochem 33:1641–1652
Reynolds BC, Hunter MD, Crossley D Jr (2000) Effects of canopy herbivory on nutrient cycling in a northern hardwood forest in western North Carolina. Selbyana 21:74–78
Risch AC, Schütz M, Vandegehuchte ML et al (2015) Aboveground vertebrate and invertebrate herbivore impact on net N mineralization in subalpine grasslands. Ecology 96:3312–3322
Risley LS, Crossley D Jr (1993) Contribution of herbivore-caused greenfall to litterfall nitrogen flux in several southern Appalachian forested watersheds. Am Midl Nat 129:67–74
Ritchie ME, Tilman D, Knops JM (1998) Herbivore effects on plant and nitrogen dynamics in oak savanna. Ecology 79:165–177
Ruijven J, De Deyn GB, Raaijmakers CE et al (2005) Interactions between spatially separated herbivores indirectly alter plant diversity. Ecol Lett 8:30–37
Schimel JP, Bennett J (2004) Nitrogen mineralization: challenges of a changing paradigm. Ecology 85:591–602
Schimel JP, Van Cleve K, Cates RG et al (1996) Effects of balsam poplar (*Populus balsamifera*) tannins and low molecular weight phenolics on microbial activity in taiga floodplain soil: implications for changes in N cycling during succession. Can J Bot 74:84–90
Schmidt IK, Michelsen A, Jonasson S (1997) Effects of labile soil carbon on nutrient partitioning between an arctic graminoid and microbes. Oecologia 112:557–565
Schofield P, Mbugua D, Pell A (2001) Analysis of condensed tannins: a review. Anim Feed Sci Technol 91:21–40
Schowalter T, Crossley D (1983) Forest canopy arthropods as sodium, potassium, magnesium and calcium pools in forests. For Ecol Manage 7:143–148
Schowalter T, Sabin T, Stafford S et al (1991) Phytophage effects on primary production, nutrient turnover, and litter decomposition of young Douglas-fir in western Oregon. For Ecol Manage 42:229–243
Schowalter T, Fonte SJ, Geaghan J et al (2011) Effects of manipulated herbivore inputs on nutrient flux and decomposition in a tropical rainforest in Puerto Rico. Oecologia 167:1141–1149
Schweitzer JA, Madritch MD, Bailey JK et al (2008) From genes to ecosystems: the genetic basis of condensed tannins and their role in nutrient regulation in a Populus model system. Ecosystems 11:1005–1020

Seastedt T, Mameli L, Gridley K (1981) Arthropod use of invertebrate carrion. Am Midl Nat 105:124–129

Song J, Wu D, Shao P et al (2015) Ecosystem carbon exchange in response to locust outbreaks in a temperate steppe. Oecologia 178:579–590

Stadler B, Müller T (1996) Aphid honeydew and its effect on the phyllosphere microflora of Picea abies (L.) Karst. Oecologia 108:771–776

Stadler B, Michalzik B, Müller T (1998) Linking aphid ecology with nutrient fluxes in a coniferous forest. Ecology 79:1514–1525

Stadler B, Müller T, Sheppard L et al (2001) Effects of *Elatobium abietinum* on nutrient fluxes in Sitka spruce canopies receiving elevated nitrogen and sulphur deposition. Agric For Entomol 3:253–261

Swank W, Waide J, Crossley D et al (1981) Insect defoliation enhances nitrate export from forest ecosystems. Oecologia 51:297–299

Vanhanen H, Veteli TO, Paivinen S et al (2007) Climate change and range shifts in two insect defoliators: gypsy moth and nun moth-a model study. Silva Fenn 41:621

Wardle DA, Bardgett RD (2004) Indirect effects of invertebrate herbivory on the decomposer subsystem. In: Weisser W, Siemann E (eds) Insects and ecosystem function. Springer, Heidelberg, pp 53–69

Wardle D, Bonner K, Barker G (2002) Linkages between plant litter decomposition, litter quality, and vegetation responses to herbivores. Funct Ecol 16:585–595

Whiles MR, Callaham MA Jr, Meyer CK et al (2001) Emergence of periodical cicadas (*Magicicada cassini*) from a Kansas riparian forest: densities, biomass and nitrogen flux. Am Midl Nat 145:176–187

Yang LH (2004) Periodical cicadas as resource pulses in North American forests. Science 306:1565–1567

Yang LH (2006) Interactions between a detrital resource pulse and a detritivore community. Oecologia 147:522–532

Zvereva EL, Lanta V, Kozlov MV (2010) Effects of sap-feeding insect herbivores on growth and reproduction of woody plants: a meta-analysis of experimental studies. Oecologia 163:949–960

Chapter 10
Eco-evolutionary Factors Driving Plant-Mediated Above–Belowground Invertebrate Interactions Along Elevation Gradients

Alan Kergunteuil, Moe Bakhtiari, and Sergio Rasmann

10.1 Introduction

The fossil record suggests that invertebrate-centred aboveground (AG) and belowground (BG) interactions may have evolved since more than 350 Ma ago (Labandeira 2013). In this context, research conducted for more than three decades, has indeed acknowledged that plants act as biological conduits, allowing herbivores damaging one compartment of soil-atmosphere to affect invertebrate animals on the other compartment (Bardgett and Wardle 2003; van Dam et al. 2003; Wardle et al. 2004). AG and BG communities can affect each other via modifications at the leaves and root levels. Such AG–BG modifications at the community level are widespread in nature and can trickle up to affect ecosystem functioning (Wardle et al. 2004). While several landmark papers have offered a theoretical framework of how AG or BG insect communities influence each other (Masters et al. 1993; Hooper et al. 2000; Wardle et al. 2004; de Deyn and van der Putten 2005; Johnson et al. 2012), the context-dependency of how plants respond to AG and BG attacks has often offered deflections from general theory (Hooper et al. 2000; van der Putten et al. 2009; van Geem et al. 2013). We, thus, here advocate that incorporating ecological gradients in AG–BG studies can stimulate further research and provide a more generalizable working framework in this regard.

In nature, a single plant often shelters several herbivores simultaneously, resulting in important competition dynamics for resource acquisition (Denno et al. 1995). However, direct interference competition does not apply to herbivorous insects occupying spatially separated organs, such as between roots and shoots

---

A. Kergunteuil · M. Bakhtiari · S. Rasmann (✉)
Laboratory of Functional Ecology, Institute of Biology, University of Neuchâtel, Neuchâtel, Switzerland
e-mail: alan.kergunteuil@unine.ch; sergio.rasmann@unine.ch

T. Ohgushi et al. (eds.), *Aboveground–Belowground Community Ecology*, Ecological Studies 234, https://doi.org/10.1007/978-3-319-91614-9_10

specialists. Resource allocation and variations in plant primary metabolism have first been hailed to be the central mechanism underpinning AG and BG herbivore competition and thereby driving community assemblage across the soil boundary (Masters et al. 1993). More recently, the need to move toward more integrative analyses of physiological changes in plants has also been advocated, for instance by measuring plant secondary metabolism related to plant defences against herbivores in both roots and shoots (Bezemer and van Dam 2005). These two mechanisms involved in plant-mediated AG–BG interactions have been, respectively, associated with the "stress response hypothesis" (i.e. predominantly driven by plant primary metabolism) and the "defence induction hypothesis" (i.e. predominantly driven by plant secondary metabolism) (Bezemer et al. 2002).

Here, we aim at providing new perspectives for expanding the conceptual framework of AG–BG interactions of invertebrate communities colonizing the plants. Overall, we propose to move beyond AG–BG interactions based on static and bi-trophic viewpoints by embedding ecological/habitat variations and multi-trophic interactions. Based on the "stress response hypothesis" and the "defence induction hypothesis", we first explore, through an extensive literature review, the main outcomes in AG–BG interactions, including herbivores and their natural enemies. Then, we present how variations in environmental conditions along ecological gradients shape selective forces driving plant defences and how it consequently affects AG–BG interactions. As final conceptual model, we address variations in AG–BG interactions along elevation gradients based on the combination of fluctuations in ecological factors (biotic and abiotic) and evolutionary mechanisms driving plant defence syndromes (local adaptation and plasticity). Elevation gradients are particularly well adapted to disentangle the resulting eco-evolutionary factors affecting AG–BG outcomes due to steady changes in abiotic factors along which biotic interaction variations can unfold.

10.2 Interactions Between AG and BG Herbivores Are Regulated by Nutrients' Allocation and Plant Defences

10.2.1 The Stress Response Hypothesis

The initial theoretical framework for predicting AG–BG interactions was initiated by Masters et al. (1993). Their predictions for the potential outcomes in the cross-effects between AG and BG herbivores were mainly based on plant primary metabolism. While BG herbivory should benefit the AG plant feeders, AG herbivores, on the contrary, should generally decrease performance of BG root feeders. These effects were mainly predicted based on resource allocation in response to herbivore attack and, ultimately, nutrient availability for insects. Specifically, it has been generally postulated that root feeder attack on BG tissues leads to drought stress. Consequently, water content in leaves decreases, and nutrients, such as nitrogen, amino

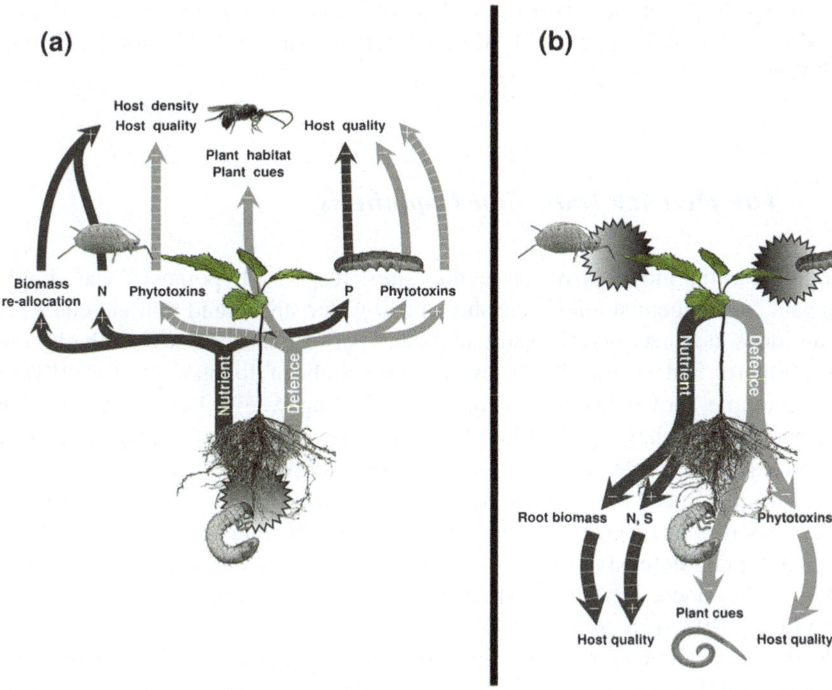

Fig. 10.1 Plant-mediated above–belowground interactions including insects and nematodes. (**a**) Effects of root herbivores on foliar herbivores and their natural enemies. So far, studied root herbivores include phytophagous nematodes, coleopteran, and dipteran larvae. Shoot herbivores include piercer-suckers and chewers from different insect orders (Hymenoptera, Lepidoptera, Diptera, Coleoptera, Hemiptera). Natural enemies belong to the Hymenoptera insect order. (**b**) Effects of foliar herbivores on root herbivores and their natural enemies. Studied shoot herbivores include piercer-suckers (Hemiptera) and chewers (Lepidoptera). Root herbivores include hemipteran aphids, coleopteran, and dipteran larvae. Natural enemies include Hymenoptera and entomopathogenic nematodes. Arrows indicate plant-mediated effects via the "stress response hypothesis" (dark grey) or the "defence induction hypothesis" (light grey). Solid arrows refer to the examples discussed in the text. Broken arrows are putative and have not been tested. The sign in the arrowhead reflects whether the plant-mediated effects on herbivores and their natural enemies are positive (+) or negative (−). The text associated to each arrow indicates the mechanisms altering invertebrate populations and/or individual performances. *N* nitrogen, *P* phosphorus, *S* sulphur

acids, and carbohydrates, are mobilized in shoots, therefore improving the performance of foliar herbivores, specifically phloem-feeders (Fig. 10.1a). Johnson et al. (2013) for example showed that root feeders (*Otiorhynchus sulcatus* beetle larvae) favoured the growth of aphid populations in the field, which was mediated by changes in carbon, nitrogen, and phosphorus. In the opposite direction, by reducing the production of photosynthates, AG herbivores have been postulated to mediate a decrease in root biomass, in turn having negative effects on the root herbivores (Fig. 10.1b). These patterns were mainly restricted to primary successional plant communities and with a particular interest in annual plants such as *Sonchus oleraceus*, *Capsella bursa-pastoris*, or *Chenopodium album* (Gange and Brown 1989; Moran and Whithman 1990; Masters and Brown 1992; Blossey and Hunt-Joshi 2003).

10.2.2 The Defence Induction Hypothesis

Although the initial model provided by the "stress response hypothesis" was simple and elegant, subsequent studies have shown that water stress (and nutrient changes) in plants does not necessarily result in better performance of several herbivore groups (Huberty and Denno 2004). In the same study of Johnson et al. (2013), it was for example shown that *O. sulcatus* root-feeding weevil larvae decrease the performance of leaf-chewing sawflies (*Nematus olfaciens*). Also, as a wider array of plant species were implemented in AG–BG interactions, predictions were more unstable and new mechanisms have been proposed to fully understand these interactions. Several authors have thus suggested that AG–BG interactions could only be unravelled if plant defensive traits—particularly secondary metabolites—were also included in the model predictions (Bezemer et al. 2002; Kaplan et al. 2008b; van Dam and Heil 2011; Erb et al. 2015).

Indeed, plant defences play a crucial role in the balance between plants and herbivores' populations (Schoonhoven et al. 2005). The general consensus posits that a relatively few number of conserved hormonal signal pathways, which govern the expression of the incredible phytochemical diversity in the plant kingdom, have allowed the evolution of a plethora of plant defence syndromes, in which both the localized and the systemic induction of defences are of first importance (Pieterse et al. 2012). Five different hormones related to plant defences have been reported as mobile signals between AG and BG compartments and are potentially responsible for AG–BG outcomes: jasmonic acid, auxins, abscisic acid, ethylene, and cytokinine (Soler et al. 2013). Particularly, since the systemic induction of chemical defences relies on phytohormones conveyed through the vascular architecture of the entire plants, and sometimes involved in cross talks, we can expect important cross-effects between AG and BG defences induced by herbivores sharing a single plant (Johnson et al. 2016; Papadopoulou and van Dam 2017).

10.3 The Integrative Framework of AG–BG Bi-trophic Interactions

While, both the "stress response hypothesis" and the "defence induction hypothesis" are valid, it is clear that the most relevant framework is the one combining all explanatory factors based on both nutrient allocation and induced plant defences, and additionally, distinguishing herbivore guilds, such as phloem-feeders or chewers

(Bezemer et al. 2002; Bezemer and van Dam 2005; Soler et al. 2013). Hereafter, we present recent evidence supporting this integrative model, particularly outlined by Soler et al. (2013), first from roots to shoots (Fig. 10.1a) and second from shoots to roots (Fig. 10.1b). Given the asymmetry in research concerned by AG and BG insect herbivores, we discriminate plant-mediated effects affecting AG herbivore guilds although we do not distinguish BG herbivore guilds since the overwhelming majority of studies were interested in insect root chewers. Nonetheless, we also extend the dialogue to other soil-dwelling invertebrates such as the nematodes. Most of the studies interested in AG–BG interactions put emphasis on insects and microorganisms such as pathogens, but generally ignoring soil-dwelling nematodes despite their crucial contributions to soil ecosystem functioning and the linkage between AG and BG compartments. Indeed, field surveys have clearly shown that dynamics in root-feeding nematodes and leaf-feeding insects are interconnected through plant-mediated effects (Kaplan et al. 2009). The numerous roles played by nematodes in BG energy flows rely on a wide range of feeding groups such as herbivore, bacterivore, fungivore, or substrate decomposer (Yeates et al. 1993). In this section, we implement herbivorous nematode in AG–BG interactions, while the effects of additional trophic guilds, such as entomopathogenic nematodes, are discussed in Sects. 10.4 and 10.5.

10.3.1 From BG to AG

Generally, BG chewing herbivores have been reviewed to negatively affect leaf-chewers, while they generally facilitate phloem-feeders located on AG compartments (Fig. 10.1a). In the former case, for instance, the larvae of the western corn rootworm (*Diabrotica virgifera virgifera*) attacking corn plants (*Zea mays*) are responsible for the activation of genes related to abscisic acid in leaves, thereby increasing both plant defence and plant priming and limiting the development of AG chewers such as *Spodoptera littoralis* caterpillars (Erb et al. 2009, 2011). However, depending on the organs where phytotoxins are produced, we can expect different patterns. Indeed, since nicotine production in tobacco roots is strongly disrupted by root-feeding nematodes (*Meloidogyne incognita*), nicotine cannot be mobilized in shoots under nematode attack, thus resulting in facilitation of leaf-chewers such as *Manduca sexta* (Kaplan et al. 2008a) (Fig. 10.1a). In parallel, independently from plant defences, assimilate translocation and nitrogen mobilization from root to shoot were shown to enhance the abundance of leaf aphid (e.g. *Brevicoryne brassicae*) compared to control plants when *Sinapis arvensis* is attacked by wireworms (*Agriotes* sp.) (Poveda et al. 2005) (Fig. 10.1a). Interestingly, while both the "stress response hypothesis" and the "defence induction hypothesis" are generally considered as completely separated processes, Kaplan et al. (2011) highlighted that they are not necessarily mutually exclusive and can be even inter-dependent. In their study conducted on tobacco plants, they showed that root-knot nematodes (*M. incognita*) drastically diminished the performance of foliar-feeding aphids (*Myzus persicae*).

Since these two different herbivores locally promote nutrient sinks for their own benefits, strong leaf–root sink competitions were expected. However, this hypothesis was not conclusive. Kaplan et al. (2011) consequently advocated for defence-mediated mechanisms for explaining these patterns, while they also stressed the interdependency of nutrient sinks and defensive metabolism. Indeed, local accumulation of photosynthates can also provide carbohydrates for the production of defensive metabolites.

10.3.2 From AG to BG

A proportionally smaller number of studies have addressed the effects of shoot feeders on root herbivores. AG herbivores, including chewers and phloem-feeders, are generally reported to decrease the performance of BG herbivores belonging to their respective trophic guilds (Fig. 10.1b). The mechanisms underpinning the negative effects of AG herbivores on root feeders rely on nutrient allocation and/or chemical defences depending on the biological models. In *Z. mays,* perhaps the most studied system in this context, it has been shown that the negative effects of AG herbivores on BG pests result from different processes related to plant defence. Indeed, foliar chewers such as *S. littoralis* alter profiles of both soluble phenolic compounds and root volatiles, thereby modifying antibiosis and antixenosis of maize roots towards the western corn rootworm *D. v. virgifera* (Robert et al. 2012; Erb et al. 2015) (Fig. 10.1b). For brassicaceous plants, similarly, leaf-chewers such as *Pieris brassicae* decrease the performance of cabbage root maggot (*Delia radicum*), without affecting root biomass. This outcome is likely supported by the effect of AG herbivores on root variations in defensive compounds such as specific glucosinolates (Soler et al. 2007). Along the same lines, foliar aphids (*Hayhurstia atriplicis*) feeding on *Chenopodium album* also reduce root aphids' populations (*Pemphigus betae*), although the proposed underlying mechanisms relied on nutrient availability, since host plant attacked by AG herbivores reduced root biomass by 21% (Moran and Whithman 1990) (Fig. 10.1b). On the contrary, it was shown that AG chewers (e.g. *M. sexta*) or aphids (*Rhopalosiphum padi*) can also increase root biomass or root mineral content (e.g. C, S), consequently benefitting root-knot nematodes (*M. incognita*) (Kaplan et al. 2008a), or root chewer (*Agriotes* spp.) performance (Johnson et al. 2009) (Fig. 10.1b).

In sum, although many examples support the AG–BG integrative framework from root to shoot and in the opposite direction, few studies also reported difficulties in seeking to apply basic assumptions of the "stress response hypothesis" (see, e.g. Fig. 10.1b showing positive effect of nutrient dynamics from shoots to roots) and the "defence induction hypothesis" (see, e.g. Fig. 10.1a showing positive effect of relaxation in defence induction in shoots following root attack).

10.4 Integrating the Third Trophic Level on AG–BG Interactions

As initially recognized by Price et al. (1980), most species interactions involve at least three trophic levels: plant, herbivore, and their natural enemies. Population densities of herbivores and their natural enemies are intimately correlated, and ultimately influence plant biomass available for both AG and BG herbivores. In addition, plants under herbivore attacks can produce information-rich cues enabling top predators and parasitoids to locate preys, and consequently setting the stage for tri-trophic interactions to unfold (Dicke and Baldwin 2010; Kessler and Heil 2011). Indeed, research performed over almost 30 years has shown that indirect plant defences, where plant traits regulate the recruitment of natural enemies, are widespread in nature in both AG (Mumm and Dicke 2010) and BG (Johnson and Rasmann 2015) compartments. Therefore, understanding the ecological significance of AG–BG interactions for ecosystem functioning requires a multi-trophic perspective, whereby herbivores and natural enemies could directly and indirectly interact through density-dependant and trait-dependent mechanisms (Soler et al. 2012).

10.4.1 From BG to AG

To date, the majority of studies that included the third trophic level in AG–BG interactions have mainly focused on the effect of root feeders on AG tri-trophic interactions. BG herbivores have been shown to drive either the abundance or the performance of AG natural enemies in several systems. Concerning the abundance, two different field studies have reported an increase in AG natural enemies' population following root herbivore attacks (Masters et al. 2001; Johnson et al. 2013), while one study showed a strong negative effect of root herbivores on parasitoid abundance (White and Andow 2006) (Fig. 10.1a). The authors suggested different reasons for these contrasted patterns. In the two former cases, density-dependent mechanisms were retained, in which the increase in AG herbivores abundance (*Terellia ruficauda* or *N. olfaciens*) facilitated by the root feeders (*Phyllopertha horticola* or *O. sulcatus,* respectively) merely caused a subsequent increase in natural enemies' population (*Torymus chloromerus*, *Aphidius* spp., respectively). In the second case, the negative correlation between root herbivory and abundance of AG parasitoids (*Macrocentrus grandii*) regulating the European corn borer (*Ostrinia nubilalis*) relied on plant-habitat modification by BG herbivores (*D. v. virgifera*). Specifically, in corn fields, *D. v. virgifera* rootworm feeding reduced plant height and plant density, resulting in open habitats, which are detrimental for natural enemies of the European corn borer (White and Andow 2006).

Concerning plant secondary metabolites, one experiment conducted on brassicaceous plants has underlined a negative effect of BG herbivores (*D. radicum*) on the development time of AG parasitoids (*Cotesia glomerata*) of

P. brassicae due to cascading effects of phytotoxins along tri-trophic food chain (Fig. 10.1a; Soler et al. 2005). It was also shown that BG herbivores can also alter the foraging behaviour of AG natural enemies through the modification of plant traits mediating their recruitment (Fig. 10.1a). For instance, *Triadica sebifera* plants produced lower amount of extra-floral nectar when they are co-attacked by both adults (AG) and larvae (BG) of *Bikasha collaris* leaf beetles compared to plants only facing AG adult herbivory (Huang et al. 2015). In addition, BG herbivores have also been shown to affect information-providing plant traits, such as the foliar emissions of herbivore-induced plant volatiles (HIPVs), thereby disrupting the attraction of AG natural enemies. For instance, changes in volatile profiles released by *Brassica nigra* exposed to dual infestation in both AG (*P. brassicae*) and BG (*D. radicum*) compartments have been suggested to be used by *C. glomerata* to optimize parasitism behaviour (Soler et al. 2012). Finally, volatiles induced by *Spodoptera littoralis* feeding, and subsequent *Cotesia marginiventris* parasitoid wasps recruitment were lowered by the presence of root feeders (*D. v. virgifera*) (Rasmann and Turlings 2007). Together, this suggests that root herbivores can affect aboveground food-web dynamics, ultimately inflecting plant fitness, but the net outcomes remain to date largely unpredictable.

10.4.2 From AG to BG

To our knowledge only two studies addressed the effect of AG herbivores on BG natural enemies and tri-trophic interactions. One example supports the hypothesis of the cascading effects of host quality from the second to the third trophic level (Fig. 10.1b). Indeed, systemic changes and accumulation of toxic glucosinolates in the roots of *Brassica nigra* in response to shoot-feeding herbivores (*P. brassicae*) lead to a reduction in the performance of both root feeders (*D. radicum*) and their natural enemies such as *Trybliographa rapae* parasitoid wasps (Soler et al. 2007). Secondly, it was shown that aboveground feeding by *S. littoralis* caterpillars resulted in the inhibition of the sesquiterpene (*E*)-β-caryophyllene production in the roots after corn rootworm (*D. v. virgifera*) attack. Such a reduction of the HIPVs production was correlated with a reduction in the recruitment of *Heterorhabditis megidis* entomopathogenic nematodes to the damaged roots (Rasmann and Turlings 2007).

Because of the paucity of work on such interactions, we have recently performed similar AG–BG studies on additional two naturally-occurring systems: the plant species *Cardamine impatiens* and *C. pratensis* (Brassicaceae), the cabbage root fly *Delia radicum* (Diptera: Anthomyiidae), and the entomopathogenic nematodes *Heterorhabditis megidis* (Rhabdita: Herorhabditidae). *Cardamine impatiens* and *C. pratensis* generally occupy low- to mid-elevation forested habitat and grasslands of Central Europe, respectively. They have been shown to vary in chemical profiles of leaf glucosinolates and volatile organic compounds production within the *Cardamine* complex of 16 species commonly growing in Switzerland (Pellissier et al. 2016). We used a custom-made four-arms BG olfactometer, modified after the

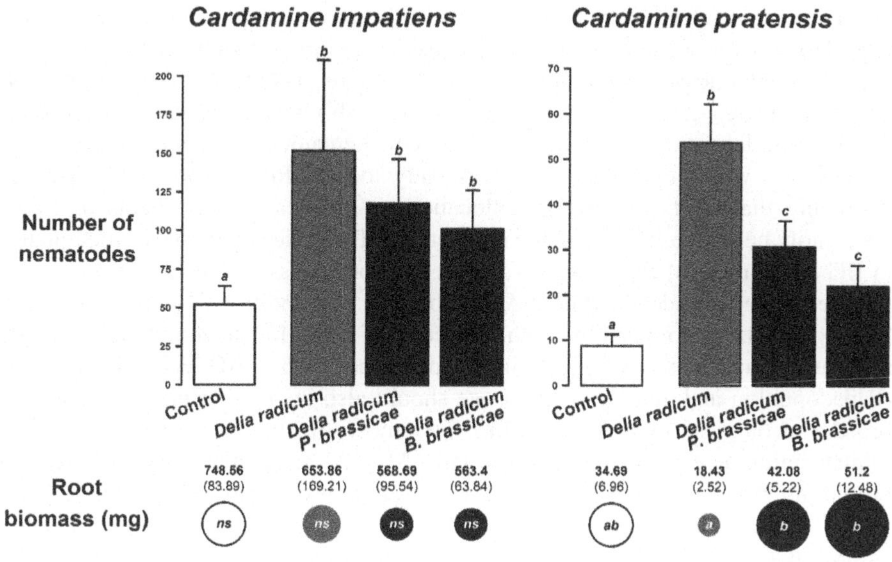

Fig. 10.2 Aboveground herbivore effect on belowground tritrophic interactions. Shown are the means (± SE) of the number of entomopathogenic nematodes (*Heterorhabditis megidis*) recruited in each of the four arms of the olfactometers [(1) control plants without infestation, (2) plants attacked with *Delia radicum* larvae only, (3) plants attacked by both *D. radicum* and the caterpillars *Pieris brassicae*, and (4) plants attacked by both *D. radicum* and the aphid *Brevicoryne brassicae*] across two *Cardamine* species. For each species, differences among the numbers of nematodes recruited are represented with different letters above bars (GLM, "quasipoisson" family, $P \leq 0.05$). Circles below the bars represent root biomass for each treatment (mean ± SE). For each species, circle areas are proportional to root biomass of control plants. Differences in root biomass between treatments are represented with different letters inside the corresponding circles (LM after square root transformation, $P \leq 0.05$)

original six-arm belowground olfactometer (Rasmann et al. 2005), to test the attraction of entomopathogenic nematodes toward four different odour sources: (1) control plants without herbivory, (2) plants infested with the specialist root chewers *Delia radicum*, (3) plants co-infested with *D. radicum* and AG leaf-chewers (*Pieris brassicae*; Lepidoptera: Pieridae), and (4) plant co-infested with *D. radicum* and AG phloem-feeders (*Brevicoryne brassicae*; Hemiptera: Aphididae). As shown in Fig. 10.2, we tested the two different *Cardamine* species separately. Overall, we observed consistent patterns in the BG attraction of natural enemies toward each treatment (Fig. 10.2). Although the presence of root herbivores strongly enhanced the attraction of natural enemies, this pattern was negatively affected by AG herbivores, indicating that AG herbivores, independent of their feeding guilds, reduce the ability of roots to signal distress when damaged by root herbivores. The exact mechanisms of such reduction in predator recruitment are yet to be elucidated, but hypotheses could be advanced. On one hand, it is important to note that far less nematodes were attracted by *C. pratensis* whose root biomass represents only 4.5% of *C. impatiens*. Hence, this lower attraction of BG natural enemies could be due to lower, biomass-related root emissions of informative compounds (e.g. exudates, CO_2, HIPVs). On the other hand, when considering independently each *Cardamine* species, nematode attraction and root biomass were no longer correlated. Indeed, despite an increase in root biomass due to AG herbivores, *C. pratensis* with foliar feeders were less attractive to entomopathogenic nematodes in comparison to plants infested only with root feeders. Thus, this study does not fully support the hypothesis predicting that shifts in resource allocation, in this case from shoots to roots, necessarily benefit to BG tri-trophic interactions. To elucidate the role of root quality in BG recruitment of EPN, further analyses of root emissions of HIPVs are surely required to better understand how AG herbivores alter indirect BG defences.

In general, in conjunction with the current work for shedding light on the molecular mechanisms governing plant-mediated AG–BG interactions (Papadopoulou and van Dam 2017), we should also increase efforts to expand on broader ecological approaches. Thus far, most of the research interested in implementing tri-trophic interactions into AG–BG interactions has focused on bottom-up effects of herbivory on the third trophic levels. However, because parasitoids and predators modulate the magnitude of herbivory, we can reasonably assume that top-down forces exerted by herbivore's natural enemies might also fashion AG–BG interactions. Along this line, further research should take into account the influence of natural enemies according to their life history traits (e.g. koinobiont or idiobiont) and the host stage attacked (e.g. egg or larva). Indeed, those ecological parameters may affect herbivore pressures and plant stoichiometry differentially across soil boundary. More generally, studying AG–BG interactions along ecological gradients could greatly contribute to unravelling the selective forces exerted on such interactions, thereby providing a better understanding of the outcomes of such interactions in plant-associated invertebrate communities.

10.5 Ecological and Evolutionary Drivers of Plant Defences

To fully grasp the role of different ecological and evolutionary factors responsible for shaping the strength of plant-mediated AG–BG interactions, we propose to expand the current thinking into a spatially and temporally explicit model. Indeed, different physical characteristics of the AG and BG compartments can differently influence the temporal and spatial processes that shape interactions between plants and organisms. Taken together, these differences result in infrequent attacks in roots compared to shoots, even if they are more intense and more persistent when they occur (Rasmann and Agrawal 2008). As a consequence, plants may have evolved alternative strategies to respond to herbivores in each compartment (van Geem et al. 2013; Johnson et al. 2016). Ecological gradients are optimal systems for studying shifts in species interactions in varying environmental conditions (Körner 2007; Beier et al. 2012; Pellissier et al. 2012; Rasmann et al. 2014a), in which variations in biotic and abiotic factors could be harnessed for better evaluation of the ecological and evolutionary forces driving the strengths of AG–BG interactions. Elevation gradients, particularly, because they impose a sharp variation in abiotic factors over relatively short distances, act as natural experiments under which biotic interactions can be evaluated (Darrow and Bowers 1997; Salmore and Hunter 2001; Zehnder et al. 2009; Rasmann et al. 2014a). Below, we first discuss how elevation gradients, specifically, affect plant defences in light of classic plant defence theories. Subsequently, we aim at dissecting the selective forces exerted on plant defences along steep elevation slopes and, ultimately, governing AG–BG interactions in mountainous ecosystems.

10.5.1 Plant Defence Variation Along Ecological Gradients

The type of defence and the relative amount of energy allocation by plants should represent an optimal strategy considering the local abiotic conditions (e.g. soil resources, climate, photosynthesis activity) and the abundance and identity of attackers (Herms and Mattson 1992; Fine et al. 2004). Specifically, along elevation gradients, explanations for the variability in plant defence deployment rely on two contrasting different predictions. First, the "resource availability hypothesis" postulates that low-resource habitats should favour plants with slow growth rates, but with high level of defences, due to the high cost of tissue loss (Coley et al. 1985). Therefore, it is expected that high elevation plants, living in poorer soils and in harsher and colder environments, should exhibit high levels of defences. Such postulate was corroborated by several studies. Using the 16 species of *Cardamine* (Brassicaceae) growing in Switzerland, Pellissier et al. (2016) showed that high elevation species invest in higher levels of constitutive production of glucosinolates, and are less inducible, than their low-elevation counterparts. Along the same lines, another study showed that both herbivory and plant phenolics in an oak species (*Quercus* sp.) increased at higher altitudes and temperature was the most important factor associated with the altitudinal increase in herbivory, indicating that altitudinal gradients in herbivory are associated with abiotic factors independently of altitudinal clines in plant defences (Abdala-Roberts et al. 2016). However, given the relatively low upper limit of the altitudinal range sampled (maxima at 900 m), high-altitude sites likely enjoyed a more moderate climate relative to the hotter and drier low-elevation sites, likely imposing greater abiotic stress on insect herbivores and plants (Abdala-Roberts et al. 2016). Finally, Salgado et al. (2016) also found that high-elevation host plant populations of *Lotus corniculatus* produced higher levels of cyanogenic glycosides compared with low-elevation populations.

Second, the "optimal plant defence hypothesis" suggests that, since the production of chemical defences is supposed to be costly (Gershenzon 1994; Cipollini et al. 2003), defences should decrease at high elevation where herbivore abundance and attack rates are lower. Therefore, where herbivore pressure is low, such as at high elevation, plant defences should decline. In accordance, two studies along elevation gradients showed that entire plant communities at high elevation are composed of species more palatable to herbivores, and this shift of plant palatability along elevation gradients may be the outcome of a relaxation of herbivore pressure and changes in abiotic condition (Descombes et al. 2016; Callis-Duehl et al. 2017). Moreover, another study showed a clear correlation between the decrease of herbivory pressure with increasing elevation and the production of toxic iridoid glycosides in *Plantago lanceolata* plants (Pellissier et al. 2014). Furthermore, *S. littoralis* caterpillars experienced lower survival and grew significantly less rapidly on low-elevation *P. lanceolata* ecotypes when compared with high-elevation ecotypes growing in common garden, likely suggesting that plants were locally adapted to variation in herbivore pressure, rather than purely the product of the available resources during the growing season (Rasmann et al. 2014b). Finally, a phylogenetically controlled experiment across 16 pairs of high and low elevation species showed that high-elevation plants are overall less resistant than their congeneric low-elevation relatives (Pellissier et al. 2012).

Plant indirect defences also vary along elevation gradients due to fluctuations in the presence and abundance of the third trophic level which are expected to influence the expression and adaptive value of plant defence strategies (Núñez-Farfán et al. 2007). A recent study of tri-trophic interactions and indirect shoot defences along elevation gradient indicated that plants at high elevations produced lower constitutive volatile organic compounds, and subsequently were less attractive for natural enemies such as ants, whereas inducible defences increased with elevation. These trade-offs between constitutive and inducible indirect defences were correlated with the variation in abundance of ants along the elevation gradient (Rasmann et al. 2014b). When considering root indirect defences, a decline in predatory nematode communities (Kergunteuil et al. 2016) is correlated with high elevation *Festuca rubra* plants producing different blends of HIPVs and a general reduction in nematode recruitment, when compared to low-elevation ecotypes (Kerguteuil et al., unpublished).

In sum, more than 40 studies that have dealt with plant defences along elevation gradients (Moreira et al. 2017) suggest contrasted patterns of defence investment along elevation gradients across species, indicating that different species may show dissimilar sensitivities to herbivore abundance and abiotic conditions, leading to a lack of a clear trend when considering all species individually (Rasmann et al. 2014a). Therefore, research models focusing on biotic interactions should include abiotic conditions, and, undoubtedly, more empirical data on leaf and root damage, herbivore pressure, and insect communities that have colonized the different environments are needed to test whether herbivores and/or abiotic conditions are the main drivers of plant defence evolution (Pellissier et al. 2012).

10.5.2 Evolutionary Forces Driving Plant Defences

Several compelling studies have shown that insect herbivory can exert strong selection on the evolution of anti-herbivore defences (Ehrlich and Raven 1964; Mauricio and Rausher 1997; Fine et al. 2004; Agrawal et al. 2012; Zust et al.

2012). However, studies measuring the selection on plant defence traits in response to AG–BG herbivory and estimating evolutionary response of plants to such interactions are currently lacking, perhaps due to the complexity of the experimental manipulation needed. Nonetheless, few attempts have been made at investigating the fitness impact of AG–BG herbivory. A study showed that while both AG and BG herbivory affects fitness in cucumber plants, root herbivory affected the fitness, such as fruit size and production as well as seed production, more consistently (Barber et al. 2011). In a 3-year study on *Lupinus arboreus*, Maron (1998) showed that the suppression of AG herbivory increased mean seed production by 45%, suppression of BG herbivory by 28%, while the combined suppression of AG–BG herbivory increased the mean cumulative fecundity by 94%.

Spatial variations in both biotic and abiotic conditions AG and BG generate strong variation in plant phenotypes, and these factors in turn are able to cause strong changes in plant biomass, nutrient allocation, water content, or concentration of defence compounds, while genetic adaptations and/or phenotypic plasticity enables plants to cope with heterogeneous environments. In turn, the resulting changes in plant quality may strongly interact with behaviour and population dynamics of associated arthropod community. Taken together, plants' defence syndrome toward herbivores is the outcome of positive and negative integrations at the plant physiological level and their emerging ecological properties (Agrawal et al. 2009). In this context, the deployment of defensive strategies by plants growing in shared ecological niches, and experiencing similar biotic and/or abiotic stresses, has been supposed to converge, thus providing an alternative mechanism to phylogenetically conserved patterns of plant defences. In general, evolutionary response to environmental changes can be transmitted through both genetic and nongenetic inheritance (Bonduriansky et al. 2012), thus providing complex mechanisms for adaptive radiation of plant defence syndromes along ecological gradients. For instance, epigenetics has been reported as an interesting mechanism involved in plant transgenerational resistance towards herbivores. Such evolutionary process is likely indicative of the ability of plants to cope with rapid changes in the environment with non-permanent strategy (Rasmann et al. 2012a, b). However, it is largely unknown whether the effect of BG communities on AG plant–insect interactions, and vice versa, also depends on epi-genotypic variation (Kabouw et al. 2011; Vandegehuchte et al. 2011).

Parallel to ecological factors and genetic variability affecting plant material carrying hereditary information, variation in defence traits in plants may be also forged by energy demand-driven trade-offs at the phenotypic level (van der Putten et al. 2001). The notion of trade-off has been used to explain why so few plants—if any—are free from herbivory, as the co-expression of defences and other vital physiological traits are limited by shared resources (Futuyma and Moreno 1988). Trade-offs occur either through opposing selection placed on a single trait by different selective agents, referred to as one-trait trade-offs, or as a multiple-trait trade-off where multiple traits share same limiting resources and that are under directional selection to increase. Negative correlations between traits can be interpreted as either trade-off or an adaptation. However, even if it is often difficult to distinguish the outcome of negative correlation, adaptation is a more likely scenario when negative correlation is detected between traits that are both costly and redundant (Agrawal et al. 2010). For example, the negative correlation between constitutive and induced defence maybe an adaptive correlation where expression of both may be redundant, as it may provide little benefits, and is costly. On the other hand, constitutive and induced defence may trade-off as they may compete for same limiting resources that prevent the simultaneous expression of both. Among populations, different plant traits can be selected depending on the local biotic and abiotic conditions. Consequently, the resulting local adaptation means that individual plants have higher fitness at their local sites compared to other sites. In other words, local adaptation could result from trade-offs at key loci where native alleles show a fitness advantage relative to foreign alleles (antagonistic pleiotropy). Alternatively, multiple independent loci could interact to produce local adaptation at the organismal level, if alleles at some loci are beneficial in only one environment, but neutral in the contrasting environment, and alleles at other loci show the opposite pattern (conditional neutrality) (Anderson et al. 2011). Trade-off in local adaptation can be caused by allocation costs, limited resources, or ecological or genetic constraints (Kalske et al. 2012). Due to the constraining effect of limited resources that can be allocated to different traits by plants, different trade-offs have been reported among growth and defence (Herms and Mattson 1992), tolerance and defence (Strauss and Agrawal 1999), and among defensive traits such as between direct and indirect defence (Balhorn et al. 2008), or as mentioned above, between constitutive and inducible defence (Heil et al. 2004).

Overall, while variations in ecological niches along ecological gradients may influence differentially AG and BG compartments, shoots and roots are expected to respond differently to environmental fluctuations (van Dam 2009; Gargallo-Garriga et al. 2015). In the last section, we aim at developing a conceptual model to predict AG–BG outcomes along ecological gradients based on variations in AG and BG plant defences, and associated trade-offs, under changing environments.

10.6 A Conceptual Model of AG–BG Interactions Along Ecological Gradients

The different types of plant defences presented above (e.g. constitutive/inducible, direct/indirect) are not mutually exclusives and act in concert to shape the full defence syndromes against herbivores. We here propose to integrate the main defence theories to design a conceptual model for predicting plant defences in shoots and roots along ecological gradients, i.e. mountain slopes in our case, and ultimately estimating the consequences on AG–BG interactions. This could represent a promising roadmap for disentangling eco-evolutionary factors driving plant defences and AG–BG outcomes.

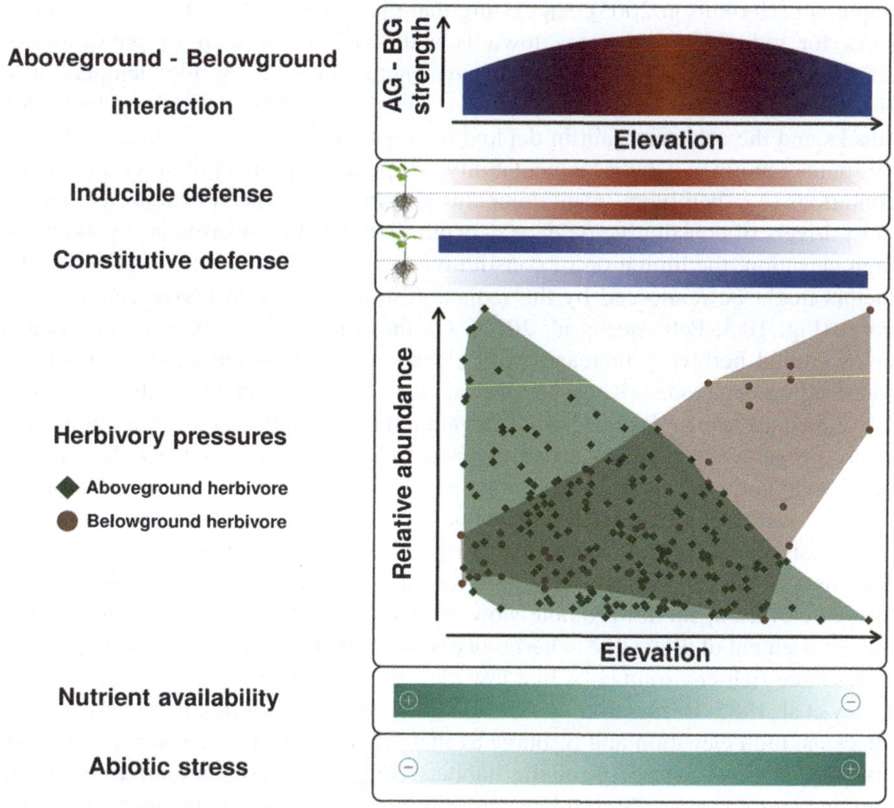

Fig. 10.3 Conceptual model of aboveground–belowground (AG–BG) interactions along elevation gradients. Moving from low to high elevation, plants should experience an increase in abiotic stressors, a decline in nutrient availably, a decline in aboveground herbivore pressure, but an increase in root colonization by herbivorous nematodes. For herbivore pressures, the relative abundances of aboveground and belowground herbivores have been computed based on field surveys as described in Pellissier et al. (2012) which surveyed butterflies and Kergunteuil et al. (2016) which surveyed soil-dwelling herbivorous nematodes. See the text for details on predicting the evolution of both constitutive and inducible defences and finally the strength of AG–BG interactions along the gradient

10.6.1 Constitutive Production of Defences

Basic considerations of optimal defence theory predict that plants should be less defended against herbivores at high elevation (cf. details in Sect. 10.5). Costly defences such as constitutive defences should fit particularly well with this pattern. On one hand, as indicated in Fig. 10.3, nutrient availability decreases with elevation (e.g. Thébault et al. 2014), thereby reducing the pool of resources that can be mobilized by plants independently from their allocation (e.g. defence, growth, reproduction). On the other hand, at high elevation, abiotic stresses become more important (Hodkinson 2005), suggesting that plants might be under selective pressures for redirecting resources towards additional needs such as physiological response to partial pressures, short wavelength radiation, or low temperatures. However, optimal defence theory also requires to consider the risk of herbivore attacks and the selective gain in deploying constitutive defences (Stamp 2003). At this point, abiotic (e.g. nutrient availability, temperature, partial pressures) and biotic conditions (e.g. herbivory, plant competition) can act in concert or represent antagonist drivers of constitutive defence depending on plant compartment. In shoots, at high elevation, the limitation of constitutive anti-herbivore defences due to abiotic factors might be reinforced by the reduction in abundance of aboveground herbivores (Fig. 10.3; Pellissier et al. 2012). On the contrary, Fig. 10.3 highlights that belowground herbivory increases with elevation. This was particularly shown for root-feeding nematodes (Kergunteuil et al. 2016). In this context, predicting constitutive defence levels of roots at high elevation remains challenging since abiotic and biotic properties of habitats exert opposite selective pressures from the optimal defence hypothesis perspective. Nonetheless, shifting the conceptual framework towards additional plant defence hypothesis, such as the resource availability hypothesis, could represent a wealthy perspective to avoid a dead-end issue. Indeed, according to this latter hypothesis, poor resource environments, such as those found at high elevation, should promote slow-growth plants that are well-defended since the replacement of tissue due to herbivory is more costly compared to plants growing in resource-rich environments, like low elevation habitats (cf. details in Sect. 10.5; Coley et al. 1985; Rasmann et al. 2014a). Thus, when considering higher risk of root attacks at high elevation and resource availability hypothesis, we can expect higher constitutive level of root in alpine habitats (Fig. 10.3). To summarize, based on current natural history observations, constitutive defence should be higher in shoots of low elevation plants, while they should be higher in roots of high elevation plants.

10.6.2 Inducibility of Defences

Two different plant defence hypotheses concomitantly support higher expression of inducible defences at intermediate elevation for both plant compartments (Fig. 10.3). First, at the genetic level, the optimal defence theory states that "low probability of attack are expected to exhibit greater inducibility than those with a high probability of attack because the latter would benefit more by high levels of constitutive defences" (Stamp 2003). In this context, shoots and roots that are less exposed to their respective herbivores from intermediate elevation are presumed to redirect defence investments from constitutive to inducible defences at mid-elevation. For shoots, above mid-elevation, inducible defences should consequently prevail relatively to constitutive defences. On the contrary, for roots, plants located below mid-elevation should express more inducible than constitutive defences. Second, at the phenoytypic level, the "carbon-nutrient balance hypothesis" assumes that defences rely on a combination of baselines and flexible allocation of resources, whose proportions vary along environments (Bryant et al. 1983; Stamp 2003). Per this hypothesis, more resources could be supplied to inducible defences against insects at mid-elevation. Indeed, at high and low elevation, growth requires most carbohydrates produced by plants due to harsh conditions or plant competition, respectively, resulting in lower amount of resources available for inducible defences in roots and shoots, respectively.

10.6.3 Plant Defence and Plant Tolerance Mediate the Strength of AG–BG Interactions Along Elevation Gradients

Here, we propose that AG–BG interactions should be enhanced at intermediate elevation based on both "defence induction hypothesis" and "stress response hypothesis". In the former hypothesis, the elevation patterns in constitutive and inducible defences discussed above support the idea that phytochemical connectivity between plant compartments should be strengthened at mid-elevation (Fig. 10.3). At extreme elevations, constitutive expression of plant defences in shoots and roots in low and high habitats, respectively, could locally monopolize phytotoxins and reduce bi-directional exchanges between plant compartments. Reversely, at intermediate elevation, higher levels of inducible defences in both roots and shoots should favour translocation of secondary metabolites between plant compartments and promote mechanisms underlying the "defence induction hypothesis". Moreover, the overall production of secondary metabolites could also be enhanced at intermediate elevation according to the "growth-differentiation balance hypothesis" (Herms and Mattson 1992). Indeed, secondary metabolites should tend to accumulate in plants located in intermediate resource conditions like mid-elevation habitats. In such environmental conditions, while abiotic factors do not drastically restrict photosynthesis yet, resource availability already slows down growth, thereby enabling the accumulation of secondary metabolites (Herms and Mattson 1992; Stamp 2003).

Stronger AG–BG interactions at intermediate elevation could also rely on mechanisms related to primary metabolism and the "stress response hypothesis". Considering trade-offs between plant tolerance and plant defences, two different selective drivers could enhance nutrient exchanges across plant compartments. First, when tolerance prevails, plants require diverting resources away from attacked area for later re-growth. In this context, several studies have demonstrated increased flow of resources from shoots to roots, and vice-versa, following AG and BG herbivory (Schultz et al. 2013; Robert et al. 2014). At mid-elevation, given that plant's accumulation of resources occurs in habitats with already reduced nutrient availability but subsisting herbivory pressures, we can expect that this "bunkering" of resources remains under important selective pressures optimizing re-growth efficiency. Second, when plant anti-herbivore syndromes are mainly composed by defensive strategies, we believe that mid-elevation versatile risks in herbivory in both root and shoots (Fig. 10.3) could also promote nutrient exchanges between plant compartments. In such habitats, characterized by unpredictability in herbivory, efficient mobility of nutrients (but also signals or secondary metabolites) across AG and BG compartment could be well adapted for fast induction of defences and local de-novo synthesis of phytotoxins in attacked plant compartment (Schultz et al. 2013).

10.7 Conclusion

Research in plant-mediated AG–BG interactions is generating a wealth of novel information in community ecology. While most of studies have focused on AG and BG herbivore–plant interaction, some authors have already paved the way for future research integrating the third trophic levels. However, there is still a major gap in knowledge about the top-down forces exerted by herbivore's natural enemies on AG–BG interactions. This book chapter also stresses the interest of elevation gradients to study eco-evolutionary drivers of plant-mediated AG–BG interactions. Based on variations in abiotic and biotic conditions along mountain slopes, we assume that the defence induction and the stress response hypotheses may concomitantly enhance AG–BG interactions at mid-elevation. Parallel to mountain slopes, various ecological gradients, such as latitudinal gradients, host similar variations in abiotic and/or biotic factors. When those variations occur in similar patterns as compared to elevation gradients, our predictive model could be used to infer the resulting outcomes of AG–BG interactions for a wider range of large-scale ecological gradients.

Acknowledgements We thank Takayuki Ohgushi, Susanne Wurst, and Scott Johnson for inviting us to contribute to this book chapter. Data for Fig. 10.2 were collected with the help of Lila Siegfried and Megane Rohrer. We are grateful to Anne Marie Cortesero (UMR IGEPP, Rennes, France) for providing eggs of *Delia radicum*. Work on elevation gradients of plant defences against herbivores has been supported by several grants from the Swiss National Science Foundation to SR (31003A_159869 and PZ00P3_131956).

References

Abdala-Roberts L, Moreira X, Rasmann S et al (2016) Test of biotic and abiotic correlates of latitudinal variation in defences in the perennial herb *Ruellia nudiflora*. J Ecol 104(2):580–590

Agrawal AA, Fishbein M, Jetter R et al (2009) Phylogenetic ecology of leaf surface traits in the milkweeds (*Asclepias* spp.): chemistry, ecophysiology, and insect behavior. New Phytol 183 (3):848–867

Agrawal AA, Conner JK, Rasmann S (2010) Tradeoffs and negative correlations in evolutionary ecology. In: Bell MA, Eanes WF, Futuyma DJ, Levinton JS (eds) Evolution after Darwin: the first 150 years. Sinauer Associates, Sunderland, pp 243–268

Agrawal AA, Hastings AP, Johnson MTJ et al (2012) Insect herbivores drive real-time ecological and evolutionary change in plant populations. Science 338(6103):113–116

Anderson P, Sadek M, Wäckers FL (2011) Root herbivory affects oviposition and feeding behavior of a foliar herbivore. Behav Ecol 22(6):1272–1277

Balhorn D, Kautz S, Lion U et al (2008) Trade-offs between direct and indirect defences of lima bean (*Phaseolus lunatus*). J Ecol 96(5):971–980

Barber NA, Adler LS, Bernardo HL (2011) Effects of above- and belowground herbivory on growth, pollination, and reproduction in cucumber. Oecologia 165(2):377–386

Bardgett RD, Wardle DA (2003) Herbivore-mediated linkages between aboveground and belowground communities. Ecology 84(9):2258–2268

Beier C, Beierkuhnlein C, Wohlgemuth T et al (2012) Precipitation manipulation experiments – challenges and recommendations for the future. Ecol Lett 15(8):899–911

Bezemer TM, van Dam NM (2005) Linking aboveground and belowground interactions via induced plant defenses. Trends Ecol Evol 20(11):617–624

Bezemer TM, Wagenaar R, van Dam NM et al (2002) Interactions between root and shoot feeding insects are mediated by primary and secondary plant compounds. Proc Exper Appl Entomol 13:117–121

Blossey B, Hunt-Joshi T (2003) Belowground herbivory by insects: influence on plants and aboveground herbivores. Annu Rev Entomol 48:521–547

Bonduriansky R, Crean A, Day T (2012) The implications of nongenetic inheritance for evolution in changing environments. Evol Appl 5(2):192–201

Bryant JP, Chapin FS, Klein D (1983) Carbon/nutrient balance of boreal plants in relation to vertebrate herbivory. Oikos 40:357–368

Callis-Duehl K, Vittoz P, Defossez E et al (2017) Community-level relaxation of plant defenses against herbivores at high elevation. Plant Ecol 218(3):291–304

Cipollini D, Purrington C, Bergelson J (2003) Costs of induced response in plants. Basic Appl Ecol 4(1):79–89

Coley PD, Bryant JP, Chapin FS (1985) Resource availability and plant antiherbivore defense. Science 230(4728):895–899

Darrow K, Bowers M (1997) Phenological and population variation in iridoid glycosides of *Plantago lanceolata* (Plantaginaceae). Biochem Syst Ecol 25(1):1–11

de Deyn GB, van der Putten WH (2005) Linking aboveground and belowground diversity. Trends Ecol Evol 20(11):625–633

Denno RF, McClure MS, Ott JR (1995) Interspecific interactions in phytophagous insects: competition reexamined and resurrected. Annu Rev Entomol 40:297–331

Descombes P, Marchon J, Pradervand J-N et al (2016) Community-level plant palatability increases with elevation as insect herbivore abundance declines. J Ecol 105(1):142–151

Dicke M, Baldwin IT (2010) The evolutionary context for herbivore-induced plant volatiles: beyond the “cry for help”. Trends Plant Sci 15(3):167–175

Ehrlich PR, Raven PH (1964) Butterflies and plants: a study in coevolution. Evolution 18 (4):586–608

Erb M, Flors V, Karlen D et al (2009) Signal signature of aboveground-induced resistance upon belowground herbivory in maize. Plant J 59(2):292–302

Erb M, Köllner TG, Degenhardt J et al (2011) The role of abscisic acid and water stress in root herbivore-induced leaf resistance. New Phytol 189(1):308–320

Erb M, Robert CAM, Marti G et al (2015) A physiological and behavioral mechanism for leaf herbivore-induced systemic root resistance. Plant Physiol 169(4):2884–2894

Fine P, Mesones I, Coley PD (2004) Herbivores promote habitat specialization by trees in amazonian forests. Science 305(5684):663–665

Futuyma DJ, Moreno G (1988) The evolution of ecological specialization. Annu Rev Ecol Syst 19:207–233

Gange A, Brown V (1989) Effects of root herbivory by an insect on a foliar-feeding species, mediated through changes in the host plant. Oecologia 81(1):38–42

Gargallo-Garriga A, Sardans J, Pérez-Trujillo M et al (2015) Warming differentially influences the effects of drought on stoichiometry and metabolomics in shoots and roots. New Phytol 207 (3):591–603

Gershenzon J (1994) Metabolic costs of terpenoid accumulation in higher plants. J Chem Ecol 20 (6):1281–1328

Heil M, Greiner S, Meimberg H et al (2004) Evolutionary change from induced to constitutive expression of an indirect plant resistance. Nature 430:205–208

Herms DA, Mattson WJ (1992) The dilema of plants: to grow or defend. Q Rev Biol 67(3):283–335

Hodkinson ID (2005) Terrestrial insects along elevation gradients: species and community responses to altitude. Biol Rev Camb Philos Soc 80(3):489–513

Hooper DU, Bignell DE, Brown VK et al (2000) Interactions between aboveground and below-ground biodiversity in terrestrial ecosystems: patterns, mechanisms, and feedbacks. BioScience 50(12):1049–1061

Huang W, Siemann E, Carillo J et al (2015) Below-ground herbivory limits induction of extrafloral nectar by above-ground herbivores. Ann Bot 115(5):841–846

Huberty A, Denno R (2004) Plant water stress and its consequences for herbivorous insects: a new synthesis. Ecology 85(5):1383–1398

Johnson SN, Rasmann S (2015) Root-feeding insects and their interactions with organisms in the rhizosphere. Annu Rev Entomol 60:517–535

Johnson SN, Hawes C, Karley A (2009) Reappraising the role of plant nutrients as mediators of interactions between root- and foliar-feeding insects. Funct Ecol 23(4):699–706

Johnson SN, Clark KE, Hartley SE et al (2012) Aboveground–belowground herbivore interactions: a meta-analysis. Ecology 93(10):2208–2215

Johnson SN, Mitchell C, McNicol JW et al (2013) Downstairs drivers – root herbivores shape communities of above-ground herbivores and natural enemies via changes in plant nutrients. J Anim Ecol 82(5):1021–1030

Johnson SN, Erb M, Hartley SE (2016) Roots under attack: contrasting plant responses to below- and aboveground insect herbivory. New Phytol 210(2):413–418

Kabouw P, Kos M, Kleine S et al (2011) Effects of soil organisms on aboveground multitrophic interactions are consistent between plant genotypes mediating the interaction. Entomol Exp Appl 139(3):197–206

Kalske A, Muola A, Laukkanen L et al (2012) Variation and constraints of local adaptation of a long-lived plant, its pollinators and specialist herbivores. J Ecol 100(6):1359–1372

Kaplan I, Halitschke R, Kessler A et al (2008a) Physiological integration of roots and shoots in plant defense strategies links above- and belowground herbivory. Ecol Lett 11(8):841–851

Kaplan I, Halitschke R, Kessler A et al (2008b) Constitutive and induced defenses to herbivory in above- and belowground plant tissues. Ecology 89(2):392–406

Kaplan I, Sardanelli S, Denno RF (2009) Field evidence for indirect interactions between foliar-feeding insect and root-feeding nematode communities on *Nicotiana tabacum*. Ecol Entomol 34 (2):262–270

Kaplan I, Sardanelli S, Rehill BJ et al (2011) Toward a mechanistic understanding of competition in vascular-feeding herbivores: an empirical test of the sink competition hypothesis. Oecologia 166 (3):627–636

Kergunteuil A, Campos-Herrera R, Sanchez-Moreno S et al (2016) The abundance, diversity, and metabolic footprint of soil nematodes is highest in high elevation alpine grasslands. Front Ecol Evol. https://doi.org/10.3389/fevo.2016.00084

Kessler A, Heil M (2011) The multiple faces of indirect defences and their agents of natural selection. Funct Ecol 25(2):348–357

Körner C (2007) The use of "altitude" in ecological research. Trends Ecol Evol 22(11):569–574
Labandeira CC (2013) A paleobiologic perspective on plant–insect interactions. Curr Opin Plant Biol 16(4):414–421
Maron JL (1998) Insect herbivory above- and belowground: individual and joint effects on plant fitness. Ecology 79(4):1281–1293
Masters GJ, Brown VK (1992) Plant-mediated interactions between two spatially separated insects. Funct Ecol 6(2):175–179
Masters GJ, Brown VK, Gange AC (1993) Plant mediated interactions between above- and below-ground insect herbivores. Oikos 66(1):148–151
Masters GJ, Jones TH, Rogers M (2001) Host-plant mediated effects of root herbivory on insect seed predators and their parasitoids. Oecologia 127(2):246–250
Mauricio R, Rausher M (1997) Experimental manipulation of putative selective agents provides evidence for the role of natural enemies in the evolution of plant defense. Evolution 51 (5):1435–1444
Moran NA, Whithman TG (1990) Interspecific competition between root-feeding and leaf-galling aphids mediated by host-plant resistance. Ecology 71(3):1050–1058
Moreira X, Petry WK, Mooney KA et al (2017) Elevational gradients in plant defences and insect herbivory. Ecography. https://doi.org/10.1111/ecog.03184
Mumm R, Dicke M (2010) Variation in natural plant products and the attraction of bodyguards involved in indirect plant defense. Can J Zool 88(7):628–667
Núñez-Farfán J, Fornoni J, Valverde PL (2007) The evolution of resistance and tolerance to herbivores. Annu Rev Ecol Evol Syst 38:541–566
Papadopoulou GV, van Dam NM (2017) Mechanisms and ecological implications of plant-mediated interactions between belowground and aboveground insect herbivores. Ecol Res 32 (1):13–26
Pellissier L, Fiedler K, Ndribe C et al (2012) Shifts in species richness, herbivore specialization, and plant resistance along elevation gradients. Ecol Evol 2(8):1818–1825
Pellissier L, Roger A, Bilat J et al (2014) High elevation *Plantago lanceolata* plants are less resistant to herbivory than their low elevation conspecifics: is it just temperature? Ecography 37(10):950–959
Pellissier L, Moreira X, Danner H et al (2016) The simultaneous inducibility of phytochemicals related to plant direct and indirect defences against herbivores is stronger at low elevation. J Ecol 104(4):1116–1125
Pieterse CM, Van der Does D, Zamioudis C et al (2012) Hormonal modulation of plant immunity. Annu Rev Cell Dev Biol 28:489–521
Poveda K, Steffan-Dewenter I, Scheu S et al (2005) Effects of decomposers and herbivores on plant performance and aboveground plant-insect interactions. Oikos 108(3):503–510
Price PW, Bouton CE, Gross P et al (1980) Interactions among three trophic levels: influence of plants on interactions between insect herbivores and natural enemies. Annu Rev Ecol Syst 11:41–65
Rasmann S, Agrawal AA (2008) In defense of roots: a research agenda for studying plant resistance to belowground herbivory. Plant Physiol 146(3):875–880
Rasmann S, Turlings TCJ (2007) Simultaneous feeding by aboveground and belowground herbivores attenuates plant-mediated attraction of their respective natural enemies. Ecol Lett 10 (10):926–936
Rasmann S, Köllner TG, Degenhardt J et al (2005) Recruitment of entomopathogenic nematodes by insect-damaged maize roots. Nature 434:732–737
Rasmann S, De Vos M, Casteel C et al (2012a) Herbivory in the previous generation primes plants for enhanced insect resistance. Plant Physiol 158(2):854–563
Rasmann S, De Vos M, Jander G (2012b) Ecological role of transgenerational resistance against biotic threats. Plant Signal Behav 7(4):447–449

Rasmann S, Alvarez N, Pellissier L (2014a) The altitudinal niche-breadth hypothesis in insect plant interactions. Annu Plant Rev 47:339–360. https://doi.org/10.1002/9781118829783
Rasmann S, Buri A, Gallot-Lavallée M et al (2014b) Differential allocation and deployment of direct and indirect defences by *Vicia sepium* along elevation gradients. J Ecol 102(4):930–938
Robert CAM, Erb M, Duployer M et al (2012) Herbivore-induced plant volatiles mediate host selection by a root herbivore. New Phytol 194(4):1061–1069
Robert CAM, Ferrieri R, Schirmer S et al (2014) Induced carbon reallocation and compensatory growth as root herbivore tolerance mechanisms. Plant Cell Environ 37(11):2613–2622
Salgado AL, Suchan T, Pellissier L et al (2016) Differential phenotypic and genetic expression of defence compounds in a plant–herbivore interaction along elevation. R Soc Open Sci 3:160226
Salmore AK, Hunter MD (2001) Elevational trends in defense chemistry, vegetation, and reproduction in *Sanguinaria canadensis*. J Chem Ecol 27(9):1713–1727
Schoonhoven LM, van Loon JJA, Dicke M (2005) Insect-plant biology, 2nd edn. Oxford University Press, New York
Schultz JC, Appel HM, Ferrieri AP et al (2013) Flexible resource allocation during plant defense responses. Front Plant Sci. https://doi.org/10.3389/fpls.2013.00324
Soler R, Bezemer TM, van der Putten WH et al (2005) Root herbivore effects on above-ground herbivore, parasitoid and hyperparasitoid performance via changes in plant quality. J Anim Ecol 74:1121–1130
Soler R, Bezemer TM, Cortesero AM et al (2007) Impact of foliar herbivory on the development of a root-feeding insect and its parasitoid. Oecologia 152(2):257–264
Soler R, Van der Putten WH, Harvey JA et al (2012) Root herbivore effects on aboveground multitrophic interactions: patterns, processes and mechanisms. J Chem Ecol 38(6):755–767
Soler R, Erb M, Kaplan I (2013) Long distance root–shoot signalling in plant–insect community interactions. Trends Plant Sci 18(3):149–156
Stamp N (2003) Out of the quagmire of plant defense hypotheses. Q Rev Biol 78(1):23–55
Strauss SY, Agrawal AA (1999) The ecology and evolution of plant tolerance to herbivory. Trends Ecol Evol 14(5):179–185
Thébault A, Clément J-C, Ibanez S et al (2014) Nitrogen limitation and microbial diversity at the treeline. Oikos 123(6):729–740
van Dam NM (2009) Belowground herbivory and plant defenses. Annu Rev Ecol Evol Syst 40:373–391
van Dam NM, Heil M (2011) Multitrophic interactions below and above ground: *en route* to the next level. J Ecol 99(1):77–88
van Dam NM, Harvey JA, Wäckers FL et al (2003) Interactions between aboveground and belowground induced responses against phytophages. Basic Appl Ecol 4(1):63–77
van der Putten WH, Vet LE, Harvey JA et al (2001) Linking above- and belowground multitrophic interactions of plants, herbivores, pathogens, and their antagonists. Trends Ecol Evol 16 (10):547–554
van der Putten WH, Bardgett RD, de Ruiter PC et al (2009) Empirical and theoretical challenges in aboveground–belowground ecology. Oecologia 161(1):1–14
van Geem M, Gols R, van Dam NM et al (2013) The importance of aboveground–belowground interactions on the evolution and maintenance of variation in plant defense traits. Front Plant Sci 4:431. https://doi.org/10.3389/fpls.2013.00431
Vandegehuchte M, de la Pena E, Bonte D (2011) Contrasting covariation of above- and below-ground invertebrate species across plant genotypes. J Anim Ecol 80(1):148–158
Wardle DA, Bardgett RD, Klironomos JN et al (2004) Ecological linkages between aboveground and belowground biota. Science 304(5677):1629–1633
White JA, Andow DA (2006) Habitat modification contributes to associational resistance between herbivores. Oecologia 148(3):482–490

Yeates GW, Bongers T, de Goede RG et al (1993) Feeding habits in soil nematode families and genera – an outline for soil ecologists. J Nematol 25(3):315–331
Zehnder CB, Stodola KW, Joyce BL et al (2009) Elevational and seasonal variation in the foliar quality and arthropod community of *Acer pensylvanicum*. Environ Entomol 38(4):1161–1167
Zust T, Heichinger C, Grossniklaus U et al (2012) Natural enemies drive geographic variation in plant defenses. Science 338(6103):116–119

Chapter 11
Cross-Compartment Herbivory Effects on Antagonists and Mutualists and Their Consequences for Plant Fitness

Nicholas A. Barber

11.1 Introduction

The rapid expansion of belowground ecology research has led to an appreciation of the ubiquity of interactions between organisms in the rhizosphere and those aboveground (Bardgett and Wardle 2010; Soler et al. 2013; A'Bear et al. 2014). Plants frequently play important roles in these interactions as links between the two compartments of terrestrial ecosystems. Plant stems, leaves, and flowers aboveground are intimately linked to roots below the soil surface, making individual plants mediators of cross-compartment indirect effects (Soler et al. 2008, 2013). Increased knowledge of these effects, and their incorporation into ecological theory (Masters et al. 1993; van der Putten et al. 2009; Vannette and Hunter 2011), has paralleled the broader recognition among ecologists of indirect effects in general (Strauss 1991; Wooton 1994; Ohgushi 2005).

Indirect effects mediated by a shared host plant can involve diverse and taxonomically unrelated organisms that have either positive or negative effects on plant growth and fitness (Karban and Baldwin 1997; Ohgushi 2005). Herbivore–herbivore interactions have been studied most often, but indirect interactions can also involve mutualists. Mutualisms are demonstrably important at organismal, population, community, and ecosystem levels (Holland et al. 2002; Bronstein 2009; Wilson et al. 2009; Prior et al. 2015), so their inclusion in aboveground–belowground research is certainly warranted. Twenty years ago, Sharon Strauss and colleagues demonstrated that leaf herbivores, through their direct effects on *Raphanus raphanistrum* (wild mustard, Brassicaceae) floral characteristics, could indirectly alter the behavior of pollinators (Strauss et al. 1996; Lehtila and Strauss 1997). Such shifts in pollinator

N. A. Barber (✉)
Department of Biological Sciences and Institute for the Study of the Environment, Sustainability, and Energy, Northern Illinois University, DeKalb, IL, USA
e-mail: nbarber@niu.edu

T. Ohgushi et al. (eds.), *Aboveground–Belowground Community Ecology*, Ecological Studies 234, https://doi.org/10.1007/978-3-319-91614-9_11

visitation patterns showed that herbivores had the potential to impact plant fitness through both male and female functions if they altered pollen export or receipt (Strauss 1997). This indirect effect of an antagonist (herbivore) on a mutualist (pollinator) has since been documented in other plant–herbivore–pollinator study systems (e.g., Mothershead and Marquis 2000; Barber et al. 2012).

Not surprisingly, research combining these two topics—indirect effects of herbivores and the aboveground–belowground paradigm—has proliferated. The increasing number of systems investigated, and the increasing complexity of these studies, may allow us to draw general conclusions about how herbivory influences both antagonists (other herbivores, pathogens) and mutualists (pollinators, microbial root symbionts) in the opposite compartment and how these interactions vary with different plant species or environmental context. Further, indirect effects on other community members may be expected to have indirect feedbacks on plant growth and fitness, although this is not studied as often. When considering indirect effects of herbivory on both mutualists and antagonists, there is the possibility of responses with conflicting consequences to the plant. For example, if a belowground herbivore reduces aboveground herbivory but also pollinator visitation, these would be expected to have positive and negative effects on the plant, respectively. On the other hand, effects on mutualists and antagonists do not necessarily result in changes in plant fitness, so it is important for ecologists to determine when cross-system indirect effects cause fitness changes that can ultimately drive changes in population growth or selection on plant responses (van Geem et al. 2013).

Here, I review studies of above- and belowground herbivory that examine indirect effects, through a shared host plant, on community members in the opposite compartment (i.e., on the opposite side of the soil surface boundary) (Fig. 11.1). I draw attention to general conclusions that have emerged, highlighting recent studies. I also discuss what is known about the consequences of these indirect effects for plant growth and fitness, emphasizing how limited the number of studies that have measured plant fitness effects is. Finally, I highlight pressing questions and useful approaches identified throughout the chapter that may be incorporated into future aboveground–belowground experiments to build on the existing literature.

11.2 Aboveground Herbivory Effects

The majority of research on plant–herbivore interactions has examined aboveground herbivores feeding on aboveground plant tissues. Although soil communities that interact with plant roots are highly diverse, most studies of belowground impacts following aboveground herbivory have measured the responses of root herbivores and root-associated microbes, particularly mycorrhizal fungi, and I discuss these below (Fig. 11.2). Understanding of how leaf or shoot damage influences root pathogens, as well as symbiotic and nonpathogenic bacterial communities and other fungi, is considerably more limited (Biere and Goverse 2016), although aboveground herbivores are able to increase resistance to root pathogens in some cases (e.g., Lee et al. 2012). However, newer sequence-based studies of microbial

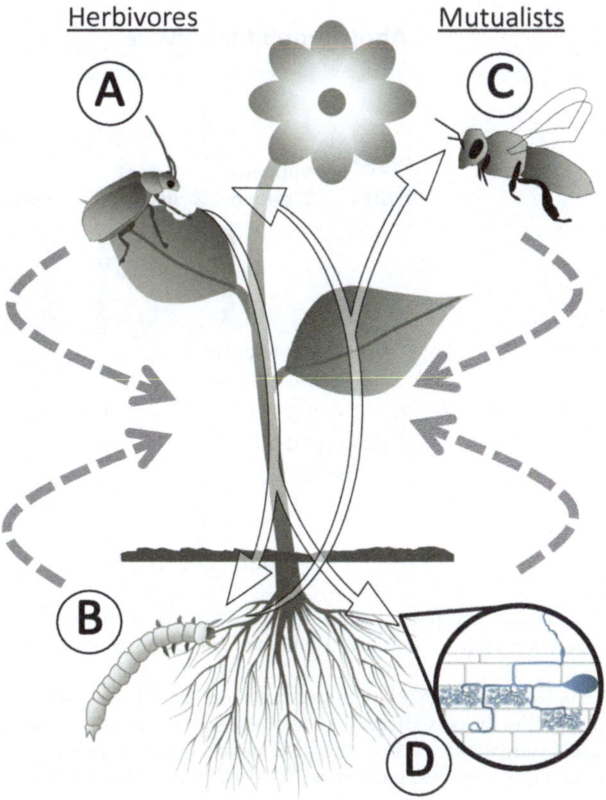

Fig. 11.1 Schematic of interactions included in this chapter. Solid arrows indicate the indirect effects of aboveground herbivores (**a**) on belowground herbivores (**b**) and arbuscular mycorrhizal fungi (**d**), and effects of belowground herbivores on aboveground herbivores and pollinators (**c**). These indirect effects are transmitted through a shared host plant. When indirect effects alter the abundance or impacts of the recipient on the plant, there is potential for feedbacks on plant growth and fitness, indicated by dashed arrows. For example, if belowground herbivores hypothetically reduce the abundance of aboveground herbivores, it could reduce the extent or impacts of leaf herbivory, indicated by the dashed arrow from the aboveground herbivore back to the host plant

communities are shedding light on these effects in the soil (Yang et al. 2013) and represent a promising avenue for extending knowledge of community indirect impacts on these hyperdiverse taxa.

11.2.1 Effects of Aboveground Herbivory on Root Herbivores

Experiments manipulating aboveground herbivory to measure impacts on below-ground herbivores have produced a wide range of results that vary in direction and magnitude, including demonstrations of facilitation by aboveground herbivores that benefit root-feeder growth (Erwin et al. 2014), as well as reductions in preference (Lu et al. 2016) and performance variables such as oviposition (McCarville et al. 2014; Wei et al. 2016) and growth (Clark et al. 2011). Herbivore–herbivore indirect interactions can be mediated by both changes in host plant nutrients and by induced defense responses controlled by plant hormone pathways (Karban and Baldwin 1997). In this section, I review the different ways these indirect effects have been

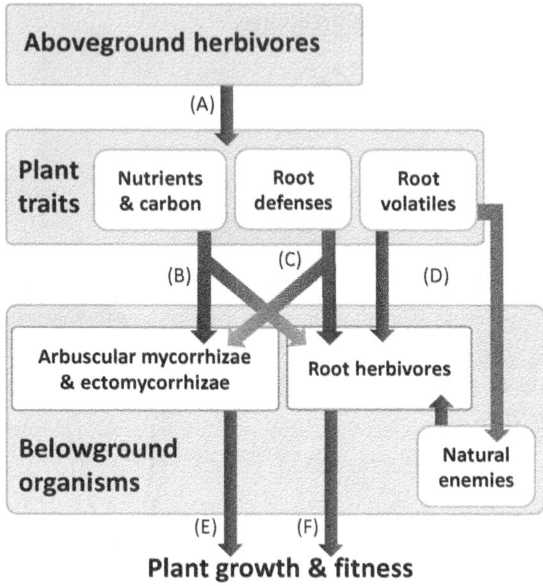

Fig. 11.2 Conceptual framework of aboveground herbivory indirect effects on belowground herbivores and mycorrhizae. Herbivore damage to aboveground tissues alters root nutrients and carbon, defenses, and volatile emissions (**a**); this is most likely to lead to negative impacts on root herbivores when the aboveground herbivore attacks the plant first. Changes in carbon and nutrient availability in roots affects both herbivores and mycorrhizal fungi (**b**), with environmental context being particularly important in determining fungal responses. Induced root defenses are likely to reduce root herbivore preference, growth, or survival, but induction pathways can also interact with mycorrhizal colonization, shifting plant–fungal interactions as well (**c**). Herbivore-driven variation in root volatiles may affect attraction of both herbivores and their natural enemies in the soil, such as entomopathogenic nematodes (**d**). These indirect effects of aboveground herbivory on belowground organisms can feedback to affect plant growth and fitness (**e, f**), leading to potential changes in population dynamics and plant community structure

measured, the importance of order of damage, and plant systems where root herbivores benefit from shoot damage.

Effects on root herbivores have been measured in many ways, perhaps due to the challenges of studying belowground processes. Studies have measured root herbivore abundance (Kaplan et al. 2009; Barber et al. 2012), host preference (Milano et al. 2015; Erb et al. 2015), survival (Soler et al. 2007a), growth (Masters and Brown 1992; Bezemer et al. 2003), or some combination of these responses (Clark et al. 2011; Erb et al. 2011b). For example, Clark et al. (2011) recorded both survival and mass of weevil larvae on *Rubus idaeus* (European raspberry, Rosaceae) with or without leaf-feeding adult weevils. Leaf damage resulted in lower larval mass, but no significant change in survival, although there was a trend for reduced survival in these smaller larvae. Preference assays have not been used widely, perhaps because single-host plant pot experiments are most amenable to no-choice assays. Choice experiments with olfactometers (e.g., Johnson et al. 2004) may better mimic how belowground herbivores select among multiple potential hosts in nature.

This variety of responses examined has resulted in a wide range of results. One important factor that has emerged to explain these variable results is the timing or order of damage. Johnson et al. (2012) performed a meta-analysis of aboveground–belowground herbivore interactions, focusing on whether one group of herbivores attacked plants before or after the other (or simultaneously). They found that aboveground herbivory significantly reduced belowground herbivore performance (survival and growth in particular) when the aboveground herbivory took place first. For example, leaf chewing damage by either a generalist caterpillar or a specialist beetle on *Cucumis sativus* (cucumber, Cucurbitaceae) reduced subsequent preference for these plants when root-feeding larvae of the same beetle species were later allowed to choose between feeding on damaged or undamaged control plants (Milano et al. 2015).

Order-dependent effects also have been documented in *Zea mays* (corn, Poaceae), with the same generalist leaf herbivore (*Spodoptera frugiperda*) as in the cucumber experiment. *Spodoptera* leaf damage reduced the preference and growth of western corn rootworm larvae (*Diabrotica virgifera virgifera*) feeding belowground (Erb et al. 2011b). Additional work to identify the mechanisms behind these effects demonstrated that there are biochemical changes in roots which drive rootworm preference behaviors (Erb et al. 2015). Measurements of root volatiles revealed that *Spodoptera* damage to leaves reduces ethylene emissions, which acts as an attractant to *Diabrotica* larvae (Robert et al. 2012). Characterization of root tissues following leaf herbivory also showed significant changes in a variety of free and hydrolysable phenolic acids (Erb et al. 2015). When phenolic acid biosynthesis was prevented using biochemical and genetic manipulations, the reduced larval preference for leaf-damaged plants disappeared. These larvae also responded most strongly to root extract fractions that contained high concentrations of hydrolysable phenolic acids. Together these studies provide strong evidence for the chemicals that mediate root herbivore host plant choice, both from a distance when volatile chemicals play an attractant role, and when corn rootworm larvae initiate feeding on root tissues where phenolic acids would be consumed (Johnson and Gregory 2006). They also demonstrate how powerful metabolomic approaches can be used to elucidate host plant biochemical mechanisms underlying ecological interactions between herbivores and other community members.

The meta-analysis by Johnson et al. (2012) also verified the previous observation (Blossey and Hunt-Joshi 2003) that negative impacts of aboveground herbivores on root feeding herbivores are primarily documented in annual plants and in relatively short-term lab or greenhouse experiments, while no consistent negative or positive effects were observed in studies that focused on perennial plant species or studies that took place in a field setting. These patterns were not observed in studies that examined belowground herbivory impacts on aboveground herbivores; indirect effects of belowground herbivores did not have a consistent direction and did not differ with plant life history or experimental setting (see Sect. 11.3).

Positive indirect effects of shoot herbivores on root feeders have been documented recently in two perennial plant systems. In an experiment using *Triadica sebifera* (tallow tree, Euphorbiaceae) from populations in both the native Chinese range and the introduced US range, feeding by adult leaf beetles increased survival of conspecific root-feeding larvae (Huang et al. 2012). A similar beneficial effect of adult leaf feeding on conspecific larvae belowground has also been documented in *Asclepias syriaca* (common milkweed, Apocynaceae) (Erwin et al. 2014). Investigation in *T. sebifera* roots revealed two ways that larvae benefited from adult leaf feeding (Huang et al. 2013). First, leaf damage caused increased nitrogen content in roots, likely making these tissues more nutritious because nitrogen is frequently a limiting resource for insect herbivores. Second, although root feeding induced higher levels of tannins in root tissues, which would be expected to reduce larval growth, this induction was suppressed on plants that also experienced leaf damage. Thus, adult herbivory can shield offspring and other conspecific larvae from chemical resistance, further increasing their survival. This facilitative effect did not occur with other species of aboveground herbivores: a weevil and two Lepidoptera caterpillars all increased root tannins and lowered larval survival, while feeding by an aphid had no effects (Huang et al. 2014).

Although it may be predicted that selection will favor facilitation mechanisms when adults feed on the same host plants prior to conspecific larvae, as in *T. sebifera* and *A. syriaca*, this has not often been tested. Larvae have reduced preference for *C. sativus* plants in the field and greenhouse after adult feeding (Barber et al. 2012; Milano et al. 2015), but studies measuring performance variables like growth, survival, emergence, or later fitness might provide a more accurate picture of leaf herbivory effects on conspecific root feeders. Related to this, aboveground herbivory has the potential to influence root herbivore performance through belowground indirect defenses and the recruitment of herbivores' natural enemies, such as entomopathogenic nematodes, but research on this topic is just beginning (see A'Bear et al. 2014 for a review).

11.2.2 Effects of Aboveground Herbivory on Soil Mutualists

Relatively few studies have considered how herbivory may alter plant–mutualist interactions in the soil, compared to the body of literature examining the complex ways that soil microorganisms influence aboveground interactions. Mycorrhization affects plant susceptibility and induced response to aboveground antagonists, including both insect herbivores and pathogenic microbes (Pozo and Azcón-Aguilar 2007; Jung et al. 2012; Pieterse et al. 2014; Balmer et al. 2015). But given the potential benefits of nutrient and water uptake by symbiotic fungi, the opposite interaction, herbivore-driven changes in mycorrhizal effects, could have important consequences for plant growth and fitness (Smith and Read 2008). If aboveground herbivory removes photosynthetic leaf tissue, it can reduce a host plant's ability to assimilate carbon into carbohydrates and alter carbon allocation patterns. For this reason, herbivory has traditionally been predicted to reduce root colonization rates by mycorrhizae, which rely on their plant hosts for carbohydrates, when that herbivory leads to carbon limitations (Wardle et al. 2004). Additionally, induction of defenses by leaf pathogens can temporarily reduce arbuscular mycorrhizal fungi (AMF) colonization (de Román et al. 2011), indicating that aboveground elicitation of defenses also may reduce plant–fungal interactions. However, experimental studies of herbivory effects on both AMF and ectomycorrhizal colonization have produced a range of results, including increased colonization following herbivory.

This motivated a meta-analysis to identify potential factors explaining variation in fungal responses to herbivory, including both AMF and ectomycorrhizae (ECM) (Barto and Rillig 2010). Overall, colonization reductions were more common, but the average magnitude of this effect was quite small, which the authors concluded to be "not biologically meaningful" for both AMF and ECM. The nature of the herbivory treatment applied in each experiment was a significant variable, with artificial shoot herbivory that removed the apical meristem tending to increase colonization or have no effect, in contrast to real shoot feeding, which consistently reduced colonization. This finding is a reminder that simulated herbivory often fails to accurately mimic the feeding of live herbivores, so the ability to observe more realistic responses of plants and AMF may be worth the logistical challenges of imposing real herbivory treatments (Barber et al. 2012; Vannette and Hunter 2014). Interestingly, colonization significantly increased when the focal plants were a mix of grasses and forbs, although this finding was based on just five studies from two publications. It is very likely that more complex community dynamics may occur in natural or more realistic communities where fungi are able to colonize or select from multiple individual hosts; these effects would be overlooked in greenhouse or laboratory studies where replicates are individual, isolated plants.

Other research has identified environmental conditions contributing to the context-dependence of herbivore effects on mycorrhizae. In northern Finland, *Solidago virgaurea* (European goldenrod, Asteraceae) were transplanted to two tundra habitats: nonacidic, fertile sites with plant communities dominated by species that form AMF symbioses and acidic, low-fertility sites where AMF-associating plants are scarce (Ruotsalainen and Eskelinen 2011). Mammalian herbivory increased arbuscular colonization in the fertile site and decreased it in the acidic site, but this was apparently not driven by soil fertility differences, because fertilization treatments did not affect the results. Rather, the authors attribute the difference to the availability of mycorrhizal spores and hyphae in the surrounding soil environment.

Because carbon availability and allocation is considered an important factor in plant–AMF interactions, it is not surprising that CO_2 may also influence herbivore effects on fungal colonization. In *A. syriaca*, herbivory effects on colonization varied significantly among plant genetic families, and these differences interacted with CO_2 treatments (Vannette and Hunter 2014). Caterpillar herbivory, which may have generated a demand for mycorrhiza-derived nutrients, led to increased colonization that was even stronger when plants were exposed to elevated CO_2 levels. Aphids, which as phloem-feeders could compete with AMF by acting as a carbon sink, drove changes in colonization that varied in direction between families, and this variance was increased by CO_2.

These two studies, as well as others showing variation in AMF responses to herbivory among different plants (e.g., Gange et al. 2002), underscore the potential variation in effects and response within and between different plants, herbivores, and fungi. Colonization, even when broken down into different structures for AMF (hyphae, arbuscules, vesicles), is a coarse measurement that often pools multiple fungi species and overlooks potential variation within species. Defoliation of *Pinus sylvestris* (Scots pine, Pinaceae) tended to decrease high-biomass morphotypes of ECM and increase low-biomass morphotypes, presumably because these latter species had lower carbon demands (Saravesi et al. 2008). Total ECM colonization and measures of fungal biomass that lump morphotypes together could not detect these patterns, demonstrating the importance of investigating species-specific fungal responses. Similarly, in a Tibetan grazing system, sheep grazing did not affect AMF root colonization levels, spore density, or extra-radical hyphal density (Yang et al. 2013). However, sequence-based analysis of the fungal community in roots showed significant shifts in OTU composition following grazing even while OTU richness was unaffected. Molecular techniques like those used in this study can characterize fungal community changes with high precision. When combined with traditional measures of colonization and nutrient transfer, they will likely advance ecological understanding by building on past research of both above- (Ba et al. 2012; Petipas and Brody 2014) and belowground (Rodríguez-Echeverría et al. 2009) herbivory effects. In particular, these techniques may reveal how plant–herbivore–mycorrhizae interactions vary along environmental gradients that shift community composition.

11.3 Belowground Herbivory Effects

The indirect effects of aboveground herbivores on organisms in the opposite compartment are mirrored by indirect effects of root feeders on shoot or leaf herbivores and on pollinators (Fig. 11.3). Indeed more experiments have manipulated belowground herbivores and assessed aboveground herbivore responses than vice versa (Johnson et al. 2012), perhaps due to a greater ease of manipulating belowground organisms than measuring them. In addition to induced effects of belowground herbivores on direct plant defenses, ecologists have demonstrated that root damage can affect indirect defenses by influencing parasitoid behavior. Although few experimenters have examined how root herbivory impacts plant–pollinator interactions, root damage affects a variety of floral traits that mediate pollinator visitation.

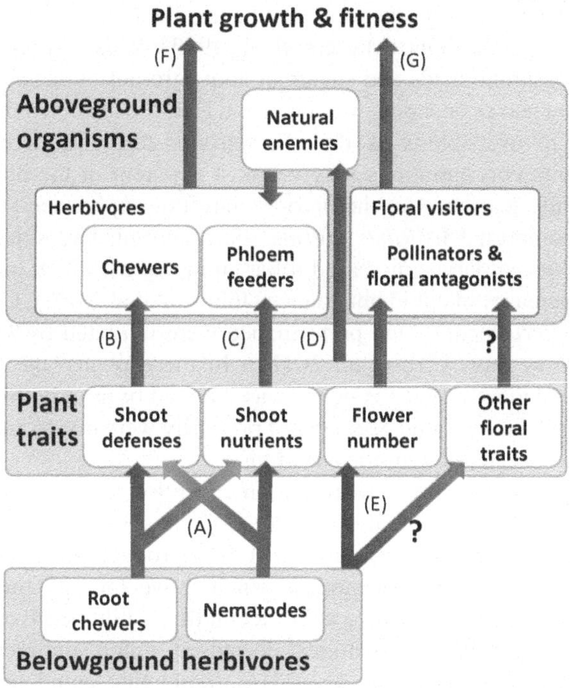

Fig. 11.3 Conceptual framework of belowground herbivory indirect effects on aboveground herbivores and floral visitors. While root-chewing herbivores and root-feeding nematodes both influence shoot defenses and shoot nutrients (**a**), indirect effects on aboveground chewing herbivores may be more affected by defenses (**b**), while phloem-feeding herbivores like aphids are driven by nutrient effects (**c**). Root herbivory tends to decrease recruitment and effects of herbivore natural enemies like parasitoids (**d**), and this may occur via changes in plant traits or changes in herbivore abundance and development. Root herbivory can also reduce flower number (**e**), with consequences for flower visitation by both pollinators and floral antagonists like nectar larcenists. Herbivory effects on other floral traits such as nectar quality and floral volatiles may also occur, but these have not been studied much. Indirect effects on both aboveground herbivores and floral visitors may affect plant growth and fitness (**f, g**), potentially driving population- and community-level changes

11.3.1 Effects of Belowground Herbivory on Shoot Herbivores

In experiments, the effects of root herbivores on aboveground herbivores have included both positive and negative outcomes, with researchers often attributing the former to greater resource content in the plant, which would benefit herbivores feeding on these more nutritious plant tissues, and the latter to increased plant resistance traits (reviewed by Blossey and Hunt-Joshi 2003; van Dam and Heil 2011). Early studies (Gange and Brown 1989; Moran and Whitham 1990; Masters and Brown 1992) found positive effects on shoot and leaf feeders, leading to a conceptual model that proposed root feeding resulted in a plant stress response similar to drought conditions (Masters et al. 1993). This response increases the concentrations of amino acids and sugars in aboveground tissues, benefiting herbivores that feed on leaves or shoots. van Dam and Heil (2011) noted that these earliest studies tended to use aphids as the aboveground-response herbivore and that increased resource concentrations may be most apparent in the phloem on which aphids feed, while leaf-chewing herbivores may not experience such facilitation. This has been supported in *Ribes nigrum* (blackcurrant, Grossulariaceae), where root-feeding by weevil larvae increased foliar amino acids, which was accompanied by significant increases in aphid abundance (Johnson et al. 2013). Conversely, leaf-chewing *Spodoptera* feeding and performance were inhibited by *D. virgifera* herbivory on *Z. mays* roots (Erb et al. 2011a). In these plants, the transcription of abscisic acid-dependent resistance genes was induced by root herbivory, but negative effects on the aboveground *Spodoptera* primarily were due to water deficiencies in the plants that interfered with their feeding.

Root-feeding nematodes may represent an exception to this general supposition that root herbivory is beneficial for phloem-feeders (Soler et al. 2012). Their presence frequently reduces aphid preference or reproduction, and a vote-counting review noted that migratory nematodes, which move through root tissues while feeding, were more likely to reduce aphid fecundity, while herbivory by sedentary species such as root-knot nematodes decreased aphid preference for plants (Wondafrash et al. 2013). Hol et al. (2016) found that aphid populations were lower on *Brassica nigra* (black mustard, Brassicaceae) infected with one nematode species. By tracking life history characteristics, they showed that reduced population growth was due to lower number of offspring per female. Nematode feeding may create nutrient sinks in the host plant that redirect resources away from shoot tissues on which aphids feed, preventing them from establishing their own aboveground nutrient sinks (Kaplan et al. 2011; Soler et al. 2012). A root-knot nematode also reduced whiteflies on *Solanum lycopersicum* (tomato, Solanaceae), which the authors attribute to significantly increased expression of several resistance genes and enhancement of both salicylic acid-dependent and jasmonic acid-dependent defenses, as well as reduced leaf nitrogen (Guo and Ge 2017).

Belowground herbivory frequently induces defense chemicals in aboveground tissues, including root herbivory by both nematodes (van Dam et al. 2005; Kaplan et al. 2008; Mundim et al. 2016; Guo and Ge 2017) and insects (Bezemer et al. 2003; van Dam et al. 2005; Huang et al. 2013, 2014; Kumar et al. 2016). Negative impacts of these induced defenses on performance are particularly common when the aboveground response herbivore is a leaf-chewer (van Dam and Heil 2011). For example, in a recent study, a tuber-feeding Lepidoptera larva in *S. tuberosum* (potato, Solanaceae) induced defenses in leaves and reduced the growth of two *Spodoptera* species (Kumar et al. 2016). Tuber herbivory led to higher concentrations of leaf chlorogenic acid and two glycoalkaloids, and the expression of genes in the synthesis pathways of these compounds was also elevated. When these defense chemicals were incorporated into artificial diet, *Spodoptera* growth was similarly reduced, providing further evidence that induction of these chemicals is the mechanism of this cross-system indirect effect.

Similar effects of root herbivory have been documented in *T. sebifera* (tallow tree), in which a specialist chrysomelid beetle feeds on roots as larvae and leaves as adults. Larval feeding reduced both survival rates of adults and the amount of leaf damage those adults imposed on host plants (Huang et al. 2012). Adult beetle feeding elevated the concentration of leaf tannins, which have defense capabilities, but this induction was even stronger when larvae were present on roots (Huang et al. 2013). Thus, while adult feeding benefited larvae (see above), larval feeding had the opposite effect on adults. Further, this negative impact on aboveground herbivores is not restricted to conspecifics, as larval feeding reduced the abundance and leaf damage by the local caterpillar community that was composed of both tallow-specialists and generalist species from several families (Huang et al. 2014). This negative impact of larvae on all aboveground herbivores, through enhanced aboveground induced defenses, contrasts with the positive effects of conspecific adults, which increased nutrients and suppressed induction in roots.

Reduced performance of chewing herbivores as a result of root herbivory could also be due to leaf nutrition changes. Sawfly larval growth was significantly lower on *R. nigrum* fed on by root-feeding weevils, and this coincided with lower phosphorus content of leaves (Johnson et al. 2013). Reduced leaf phosphorus is a novel mechanism for these effects because insect herbivory research has often focused on nitrogen as a mediating nutrient. Phosphorus is often the most important nutrient transferred to plants by AMF, so these results raise the intriguing possibility that indirect root herbivore effects on aboveground herbivores could interact with root fungal root symbionts when phosphorus is an important limiting resource.

Although they most often act as plant mutualists, I address natural enemies such as parasitoids here because of their pertinence to aboveground herbivore performance and survival. In contrast to the limited research on how aboveground herbivory affects belowground indirect defense, the impacts of root herbivory on aboveground herbivore–parasitoid interactions are better studied. Root herbivory generally leads to reduced preference and performance of parasitoids on aboveground herbivores (Soler et al. 2012; A'Bear et al. 2014). This might result from changes in host herbivore abundance, with reduced parasitoid recruitment to plants with fewer hosts present, although increased recruitment by high aphid densities also occurs (Poveda et al. 2005; Johnson et al. 2013). Root herbivory can also interfere with parasitoids' ability to locate hosts by altering leaf herbivore-induced plant volatile blends (Rasmann and Turlings 2007; Soler et al. 2007b) or plant architecture (White and Andow 2006). Finally, negative impacts on parasitoid development may also occur, particularly when root damage induces chemical changes to the tissues on which parasitoids' hosts are feeding (Soler et al. 2005). If host herbivores encounter well-defended plants, their own stress or delayed development is likely to reduce developing parasitoids' growth and likelihood of survival or emergence. Thus, root herbivory may lead to a trade-off faced by aboveground herbivores: host plants uninfested by root feeders may be more nutritious or may lack induced chemical defenses, but these plants also represent greater risk of attack by parasitoid enemies (Soler et al. 2008, 2009, 2010).

11.3.2 Effects of Belowground Herbivory on Pollinators

Belowground herbivores may also affect aboveground interactions between plants and their mutualists, particularly pollinators. Although few studies have directly examined the impacts of root herbivory on plant–pollinator interactions, the general root herbivory literature demonstrates effects on floral traits that can mediate pollinator behaviors. When pollinator responses to belowground herbivory have been measured, the results have been complex and include both positive and negative responses. A recent review emphasized how limited this past work has been but identified potential traits (e.g., floral number and size, nectar composition, and floral volatiles) that might influence flower visitation (Barber and Soper Gorden 2014). For example, a soil insecticide treatment to remove root herbivores increased inflorescence size in a thistle species, which might attract more pollinators (Masters et al. 2001). The review identified only four studies that measured pollinator response to root herbivory, including three that reported positive responses by honeybees (*Apis mellifera*) (Poveda et al. 2003, 2005; Barber et al. 2011) and one that found no effects (Hladun and Adler 2009). Here, I review two additional new studies that contribute to this still-developing field.

In an earlier study, honeybees spent significantly more time probing individual flowers in *C. sativus* that had experienced enhanced root herbivory (Barber et al. 2011). In a follow-up experiment with more root herbivore treatment levels (no, low, medium, and high infestation) (Barber et al. 2015), increases in root damage led to fewer flowers, particularly female flowers, but there were no changes in the two other attraction traits measured, flower size and floral volatile composition. In contrast to the earlier study, pollinator responses were consistently negative or neutral. Pollinators in general and the two most common floral visitors, honeybees and a butterfly, visited fewer plants with high levels of root herbivory. The time spent probing each flower and the proportion of flowers on each plant probed were unaffected. The fact that pollinator visitation frequency reflected flower production suggests this may be the primary plant trait to which insect visitors responded and that, in this study, other floral traits were of minor or no importance.

In a similar experiment, Ghyselen et al. (2016) allowed weevils to oviposit on potted *Cynoglossum officinale* (Boraginaceae) so the weevil larvae would feed on the plants' roots. These and control plants were arranged in a field setting to allow pollinators to visit. As in *C. sativus*, root herbivores reduced both flower size and the number of pollinator visits to *C. officinale*. However, this study did not examine any other potential floral attraction traits.

Not all floral visitors are mutualists, and floral antagonists like florivores and nectar larcenists, which feed on nectar without providing pollination services, can reduce plant fitness (Irwin et al. 2001; McCall and Irwin 2006). Floral antagonists are attracted to the same floral traits as "honest" pollinators (Soper Gorden and Adler 2013), so if root herbivory alters flower characteristics, both mutualistic and antagonistic visitors may be affected. For example, fertilization of *Impatiens capensis* (common jewelweed, Balsaminaceae) increased flower and nectar production, leading to more visits by both pollinators and nectar thieves, with thieves responding more strongly than pollinators (Soper Gorden and Adler 2013). However, root herbivore impacts on floral antagonists remained unstudied.

Clearly, there is a need for more studies examining root herbivory impacts on plant–pollinator interactions, and drawing general conclusions at this point is difficult, although changes in floral display due to increased or decreased flower number seem like an important mediator. Studies manipulating AMF presence have found parallel effects on floral traits and pollinator behavior, with AMF increasing flower number and visitation in several plant species (Wolfe et al. 2005; Gange and Smith 2005; Varga and Kytöviita 2010). Barber and Soper Gorden (2014) propose several important features of potential future studies that would maximize the impact of future root herbivore–pollination research. These include comprehensive measurements of floral traits, careful quantification of pollinator behaviors of multiple species in addition to total visitation, and assessments of pollinator efficacy and resultant plant fitness. Given contemporary concerns about declines of both wild and domestic pollinators (Biesmeijer et al. 2006; National Research Council of the National Academies 2007), increasing our knowledge of how aboveground–belowground linkages influence pollination takes on particular urgency.

11.4 Plant Feedbacks

Indirect interactions crossing the soil surface, discussed above, have the potential to affect plant growth and fitness. Whether the effects occur aboveground or belowground, increased interaction frequency or strength with mutualists (pollinators or soil symbionts, respectively) might be predicted to increase plant performance. Conversely, increased attraction of, abundance of, or damage by herbivores represents likely negative impacts on host plants. However, ecological studies examining indirect effects of herbivory on the opposite soil compartment often do not include assessments of plant performance, limiting our understanding of how above- and belowground interactions ultimately affect individual host plants, plant populations, and plant communities.

Cross-compartment indirect effects of herbivory may not necessarily drive plant growth and fitness responses, especially if the direct negative effects of the initial herbivore damage are weak. In a meta-analysis of studies where multiple herbivores (not specifically in opposite compartments) attacked plants, herbivory effects on plant performance were additive in the majority of cases (Stephens et al. 2013). That is, despite the potential for indirect effects of one herbivore on another, impacts on plant growth or fitness by both herbivores in combination were not different from the sum of their independent effects. Synergistic effects, where herbivores in combination would have enhanced negative effects on plants, were uncommon and not predicted under any of the scenarios investigated. Antagonistic effects, in which herbivory effects on plants are relatively reduced by both herbivores in combination, were most likely only when herbivores fed on the same plant tissues in the same above- or belowground compartment and so were in direct competition. The authors suggest that the rarity of nonadditive effects could be due to competitive or facilitative effects between two herbivores being stronger than their effects on plants, or to plants' abilities to compensate for damage, which may override nonadditive effects. Thus in cases of herbivory effects crossing the soil surface to indirectly impact other herbivores, the focus of this chapter, overall increased or reduced herbivory due to antagonistic (non-additive) effects would not be expected. Testing this hypothesis will require experiments that factorially manipulate root and shoot herbivores and that carefully measure herbivory in both compartments.

In a pair of studies on *C. sativus* (Barber et al. 2012, 2015), cross-compartment effects had little impact on plant fitness not because herbivory impacts were weak, but apparently because initial herbivory effects were very strong. In these experiments, both leaf and root herbivory reduced prevalence of antagonists on opposite-compartment tissues. Increased leaf damage reduced the likelihood of root herbivore presence (Barber et al. 2012) and increased root herbivory reduced symptoms of a leaf pathogen (Barber et al. 2015). In both cases, the indirect effect of a reduced antagonist would be considered a "benefit" to the host plant, but in both studies overall effects on plant growth, fruit production, and seed production were strongly negative. That is, the direct effects of the initial herbivore appeared much stronger than the secondary impacts through another antagonist. Similarly, A'Bear et al. (2014) note that even in similar cases where root herbivores reduce leaf herbivore performance, the effect on plants is still expected to be negative and possibly strengthened by lower effectiveness of aboveground parasitoids.

Few studies have evaluated plant feedbacks resulting from indirect effects of herbivores on plant mutualists. For soil mutualists like AMF, this is probably due to the difficulty of separating the specific effects of mycorrhizae on nutrient uptake and plant performance. As a result, feedbacks are presumed based on the extent of total or arbuscular colonization (Ruotsalainen and Eskelinen 2011; Barber et al. 2012, 2015), with greater colonization hypothetically representing increased plant benefit. However, the knowledge that different AMF species or strains can differ greatly in their effects on plant nutrient status or growth (Johnson et al. 1997), and on plant defense signaling (Fernández et al. 2014), again underscores the challenges of identifying mechanisms in diverse soil communities. In a manipulation of AMF abundance, intermediate colonization rates maximized plant growth and defense benefits for one AMF species, but in another species higher colonization reduced growth (Vannette and Hunter 2011). In either case, if leaf herbivory were to reduce high AMF colonization to more moderate levels, it would be predicted to have a beneficial feedback on plant growth and a potentially negative feedback on herbivores by increasing resistance traits.

Identifying the relative contributions of pollinator feedbacks following herbivory are more feasible because of the ability to manipulate pollinator access or to supplement pollen receipt of host plants. When combined with herbivory manipulations, ecologists can determine whether the indirect effects of herbivores on pollinators lead to pollen limitation of plants that would reduce reproduction. In *C. officinale*, root damage by weevil larvae reduced the number of pollinator visits at both the per-plant and per-flower levels, with reductions in seed set as well (Ghyselen et al. 2016). However, supplemental hand-pollination did not eliminate the reduced seed set, and pollen loads and pollen tube development did not differ between plants with and without root herbivory. This indicates that plants were not pollen-limited for seed set and that, even on plants with root herbivory, the limited numbers of visiting pollinators were still delivering sufficient pollen. Rather, root-damaged plants apparently lacked the resources necessary for seed development, a finding that was bolstered by the observation that seed maturation rates and presumed preferential seed abortion were higher in root-infested plants (Ghyselen et al. 2015). These findings from *C. officinale* mirror the effects of both above- and belowground herbivory in *C. sativus* (Barber et al. 2012, 2015). Enhancing both leaf damage and the abundance of root herbivores led to fewer pollinator visits and fewer fruits produced, but supplemental hand-pollination did not affect fruit production. Damage either above- or belowground, which occurred early in the growing season, apparently restricted plant growth such that they were unable to catch up with undamaged plants later in the season when herbivory treatments had ceased.

If cross-compartment effects influence plant fitness and population dynamics, they may also lead to changes in plant community composition, but this has not been studied, although herbivory on both sides of the soil surface can directly affect plant community structure. For example, an early experiment comparing the direct effects of above- and belowground herbivores on early successional communities demonstrated that feeding by both groups influenced plant community succession trajectories, but neither indirect effects on the opposite-side herbivores nor interactions between them were investigated (Brown and Gange 1989). These same authors showed that AMF also plays a role, slowing succession while root herbivory advanced it (Gange and Brown 2002). Given that cross-compartment indirect herbivory effects do not necessarily lead to fitness changes in plants (see above), systems in which these indirect effects are particularly strong may be the most likely to also see these interactions translate to plant community differences. These studies suggest that, in scenarios where leaf herbivory facilitates root feeders and/or decreases AMF colonization, it could lead to more rapid plant species turnover.

11.5 Future Directions

The sections above review what is known about cross-compartment herbivory effects, but throughout I have tried to emphasize outstanding questions, potential hypotheses, and promising approaches to better understand these indirect effects. Here, I highlight these possible next steps to outline how ecologists may develop a more comprehensive knowledge of herbivore impacts and further contribute to aboveground–belowground ecology. Specifically, ecologists should study feedbacks of indirect effects on plants, incorporate a wider range of organisms in studies, and "scale-up" research by examining different environmental and community contexts.

First, although cross-compartment effects of herbivory have been well-documented, the implications of these for plant growth and fitness are often not included in studies, as noted above. Careful measurement of impacts on plants, including levels of herbivore damage to shoots and roots, growth effects, and plant fitness, will help clarify the importance of indirect herbivore effects. For example, the finding that spatially separated herbivores feeding on different plant tissues tend to have additive effects (Stephens et al. 2013) warrants further investigation. Factorially manipulating both above- and belowground herbivores, and carefully measuring both herbivore and plant responses, could help determine if indirect effects on other herbivores are indeed stronger than direct effects on plants, as the authors predict. Determining plant fitness effects will also strengthen predictions of how above- and belowground herbivory may influence plant population dynamics.

Second, including a wider range of organisms, representing different taxonomic groups or other life-history strategies, will be necessary to more fully understand indirect herbivory effects. I have noted how few researchers have investigated root herbivory impacts on pollinators; by including pollen manipulations to detect pollen limitation and measurements of floral traits to identify potential mediating mechanisms, further studies like these may clarify the factors that affect this mutualism that is important in both natural and agro-ecosystems. Similarly, there is a dearth of studies on how aboveground herbivores affect soil organisms other than herbivores and mycorrhizae. Impacts on root pathogens have been studied in agricultural plants (reviewed by Biere and Goverse 2016), but the consequences for plant fitness or population dynamics of wild species are unknown. Although it is not a focus of this chapter, belowground interactions between root herbivores and mutualists are understudied compared to aboveground indirect interactions, so their inclusion in future research may strengthen studies as well. Incorporating higher-order consumers may also be important. Two intriguing studies have shown that aboveground herbivores reduce the recruitment of root herbivore natural enemies [parasitoids (Pierre et al. 2011) and entomopathogenic nematodes (Rasmann and Turlings 2007), reviewed by A'Bear et al. (2014)], demonstrating that these soil organisms deserve further attention.

Studying species that vary in life history characteristics or other traits is also necessary to understand how generalizable results may be. For example, few aboveground-belowground herbivory experiments have used perennial plants. Positive effects of shoot-feeders on root feeders may be more common on long-lived host plants than annuals (Huang et al. 2012; Johnson et al. 2012; Erwin et al. 2014), but studies from more perennial species would help determine if this is due to increased nutrients or reduced defenses (Huang et al. 2013). Papadopoulou and van Dam (2017) recently drew attention to the importance of plant vascular morphology in cross-compartment induction by root herbivores. The monocot root vasculature of *Z. mays* connects roots more directly to shoot tissues than in dicots, resulting in drought-dependent induction of aboveground defenses by root herbivores that are absent in dicotyledonous *Brassica* study species. Some herbivore characteristics, discussed above, are also known to be important determinants of indirect effects, like chewing vs. sucking feeding strategies. Herbivore effects may vary with their host plant specialization, although the limited number of cross-compartment herbivore studies to examine specialist vs. generalist effects have not found differences (Huang et al. 2014; Milano et al. 2015).

Third, our understanding of above- and belowground interactions can be expanded by examining herbivory effects in wider community and environmental contexts. The herbivore–AMF studies discussed above exemplify this, demonstrating how mycorrhizal responses to herbivory can depend on habitat (Ruotsalainen and Eskelinen 2011) or CO_2 conditions (Vannette and Hunter 2014). Because mycorrhizae species differ in the benefits they provide to plants, combining better measures of fungal effects on host plants with molecular tools to detect changes in fungal community composition due to herbivory may be a particularly promising approach to understanding and linking changes at the individual plant fitness level and the community level (Yang et al. 2013). Pairing field studies with more traditional, carefully controlled lab or greenhouse studies may be useful to determine both the context-dependence of indirect interactions and the mechanisms driving them. Understanding pollinator responses to belowground herbivory will also require a community view because most pollinators forage on multiple plant species in a community. Cahill et al. (2008) documented shifts in visitation rates among plants in a community when AMF was altered. Root herbivory would likely cause similar shifts in pollination rates when it alters floral displays, but this has not been investigated.

11.6 Conclusions

This chapter illustrates that indirect effects of herbivory, transmitted through shared host plants to both mutualists and antagonists in the opposite soil compartment, are widespread in natural and agro-ecosystems. The diversity of above- and below-ground communities, especially invertebrate and microbial communities, provides myriad pathways along which these effects can occur, resulting in a variety of potential outcomes for all the organisms involved. Despite this variation, both antagonists and mutualists most commonly respond negatively to herbivory in the opposite compartment. These negative impacts are particularly evident when the opposite-compartment herbivore feeds first, when the response herbivore is a chewer, and when the mediating host plant is an annual species. Phloem-feeding aphids are an exception to this, although they respond negatively to root-feeding nematodes. The two focal mutualist groups here, mycorrhizae and pollinators, are usually reduced by herbivory, at least in experiments utilizing natural herbivores (mycorrhizae) or in plant species where root herbivory reduces flower production (pollinators). Current research is investigating not just whether leaf herbivores affect root herbivores and vice versa, but how nutritional and defensive chemical changes in plants mediate these interactions and how they vary among different environmental contexts. Research on the effects of herbivores on mutualists is also poised to advance in coming years, taking advantage of sequence-based descriptions of soil communities and more comprehensive evaluations of floral trait changes. The approaches described in recent studies above, including factorial manipulations of herbivores, root symbionts, and pollinators or pollination, in combination with measurement of plant performance, will help elucidate the study systems and contexts in which indirect effects are most important.

References

A'Bear AD, Johnson SN, Jones TH (2014) Putting the "upstairs–downstairs" into ecosystem service: what can aboveground–belowground ecology tell us? Biol Control 75:97–107. https://doi.org/10.1016/j.biocontrol.2013.10.004

Ba L, Ning J, Wang D et al (2012) The relationship between the diversity of arbuscular mycorrhizal fungi and grazing in a meadow steppe. Plant Soil 352: 143–156. https://doi.org/10.1007/s11104-011-0985-6

Balmer A, Pastor V, Gamir J et al (2015) The "prime-ome": towards a holistic approach to priming. Trends Plant Sci 20:443–452. https://doi.org/10.1016/j.tplants.2015.04.002

Barber NA, Soper Gorden NL (2014) How do belowground organisms influence plant–pollinator interactions? J Plant Ecol 8:1–11. https://doi.org/10.1093/jpe/rtu012

Barber NA, Adler LS, Bernardo HL (2011) Effects of above- and belowground herbivory on growth, pollination, and reproduction in cucumber. Oecologia 165:377–386. https://doi.org/10.1007/s00442-010-1779-x

Barber NA, Adler LS, Theis NA et al (2012) Herbivory reduces plant interactions with above- and belowground antagonists and mutualists. Ecology 93:1560–1570. https://doi.org/10.1890/11-1691.1

Barber NA, Milano NJ, Kiers ET et al (2015) Root herbivory indirectly affects above-and belowground community members and directly reduces plant performance. J Ecol 103:1509–1518. https://doi.org/10.1111/1365-2745.12464

Bardgett RD, Wardle DA (2010) Aboveground-belowground linkages: biotic interactions, ecosystem processes, and global change. Oxford University Press, Oxford

Barto EK, Rillig MC (2010) Does herbivory really suppress mycorrhiza? A meta-analysis. J Ecol 98:745–753. https://doi.org/10.1111/j.1365-2745.2010.01658.x

Bezemer TM, Wagenaar R, Van Dam NM et al (2003) Interactions between above- and belowground insect herbivores as mediated by the plant defense system. Oikos 101:555–562. https://doi.org/10.1034/j.1600-0706.2003.12424.x

Biere A, Goverse A (2016) Plant-mediated systemic interactions between pathogens, parasitic nematodes, and herbivores above- and belowground. Annu Rev Phytopathol 54:499–527. https://doi.org/10.1146/annurev-phyto-080615-100245

Biesmeijer JC, Roberts SPM, Reemer M et al (2006) Parallel declines in pollinators and insect-pollinated plants in Britain and the Netherlands. Science 313:351–354. https://doi.org/10.1126/science.1127863

Blossey B, Hunt-Joshi TR (2003) Belowground herbivory by insects: influence on plants and aboveground herbivores. Annu Rev Entomol 48:521–547. https://doi.org/10.1146/annurev.ento.48.091801.112700

Bronstein JL (2009) The evolution of facilitation and mutualism. J Ecol 97:1160–1170. https://doi.org/10.1111/j.1365-2745.200901566.x

Brown VK, Gange AC (1989) Differential effects of below- and aboveground herbivory during early plant succession. Oikos 54:67–76. https://doi.org/10.2307/3565898

Cahill JF, Elle E, Smith GR et al (2008) Disruption of a belowground mutualism alters interactions between plants and their floral visitors. Ecology 89:1791–1801. https://doi.org/10.1007/s11284-016-1410-7

Clark KE, Hartley SE, Johnson SN (2011) Does mother know best? The preference–performance hypothesis and parent–offspring conflict in aboveground–belowground herbivore life cycles. Ecol Entomol 36:117–124. https://doi.org/10.1111/j.1365-2311.2010.01248.x

de Román M, Fernández I, Wyatt T et al (2011) Elicitation of foliar resistance mechanisms transiently impairs root association with arbuscular mycorrhizal fungi. J Ecol 99:36–45. https://doi.org/10.1111/j.1365-2745.2010.01752.x

Erb M, Köllner TG, Degenhardt J et al (2011a) The role of abscisic acid and water stress in root herbivore-induced leaf resistance. New Phytol 189: 308–320. https://doi.org/10.1111/j.1469-8137.2010.03450.x

Erb M, Robert CAM, Hibbard BE et al (2011b) Sequence of arrival determines plant-mediated interactions between herbivores. J Ecol 99:7–15. https://doi.org/10.1111/j.1365-2745.2010.01757.x

Erb M, Robert CAM, Marti G et al (2015) A physiological and behavioral mechanism for leaf herbivore-induced systemic root resistance. Plant Physiol 169:2884–2894. https://doi.org/10.1104/pp.15.00759

Erwin AC, Züst T, Ali JG et al (2014) Above-ground herbivory by red milkweed beetles facilitates above-and below-ground conspecific insects and reduces fruit production in common milkweed. J Ecol 102:1038–1047. https://doi.org/10.1111/1365-2745.12248

Fernández I, Merlos M, López-Ráez JA et al (2014) Defense related phytohormones regulation in arbuscular mycorrhizal symbioses depends on the partner genotypes. J Chem Ecol 40:791–803. https://doi.org/10.1007/s10886-014-0473-6

Gange AC, Brown VK (1989) Effects of root herbivory by an insect on a foliar-feeding species, mediated through changes in the host plant. Oecologia 81:38–42. https://doi.org/10.1007/BF00377007

Gange AC, Brown VK (2002) Soil food web components affect plant community structure during early succession. Ecol Res 17:217–227. https://doi.org/10.1046/j.1440-1703.2002.00481.x

Gange A, Bower E, Brown V (2002) Differential effects of insect herbivory on arbuscular mycorrhizal colonization. Oecologia 131:103–112. https://doi.org/10.1007/s00442-001-0863-7

Gange AC, Smith AK (2005) Arbuscular mycorrhizal fungi influence visitation rates of pollinating insects. Ecol Entomol 30:600–606. https://doi.org/10.1111/j.0307-6946.2005.00732.x

Ghyselen C, Bonte D, Brys R (2015) The influence of herbivory on pre-and postzygotic stages of reproduction following open, self, and outcross pollination. Am J Bot 102:2013–2019. https://doi.org/10.3732/ajb.1500291

Ghyselen C, Bonte D, Brys R (2016) Is there a missing link? Effects of root herbivory on plant–pollinator interactions and reproductive output in a monocarpic species. Plant Biol 18:156–163. https://doi.org/10.1111/plb.12325

Guo H, Ge F (2017) Root nematode infection enhances leaf defense against whitefly in tomato. Arthropod-Plant Interact 11:23–33. https://doi.org/10.1007/s11829-016-9462-8

Hladun KR, Adler LS (2009) Influence of leaf herbivory, root herbivory, and pollination on plant performance in *Cucurbita moschata*. Ecol Entomol 34:144–152. https://doi.org/10.1111/j.1365-2311.2008.01060.x

Hol WHG, Raaijmakers CE, Mons I et al (2016) Root-lesion nematodes suppress cabbage aphid population development by reducing aphid daily reproduction. Front Plant Sci 7:111. https://doi.org/10.3389/fpls2016.00111

Holland JN, DeAngelis DL, Bronstein JL (2002) Population dynamics and mutualism: functional responses of benefits and costs. Am Nat 159:231–244. https://doi.org/10.1086/338510

Huang W, Carrillo J, Ding J et al (2012) Invader partitions ecological and evolutionary responses to above-and belowground herbivory. Ecology 93:2343–2352. https://doi.org/10.1890/11-1964.1
Huang W, Siemann E, Yang X et al (2013) Facilitation and inhibition: changes in plant nitrogen and secondary metabolites mediate interactions between above-ground and below-ground herbivores. Proc R Soc Lond B Biol Sci 280:20131318. https://doi.org/10.1098/rspb.2013.1318
Huang W, Siemann E, Xiao L et al (2014) Species-specific defence responses facilitate conspecifics and inhibit heterospecifics in above–belowground herbivore interactions. Nat Commun 5:4851. https://doi.org/10.1038/ncomms5851
Irwin RE, Brody AK, Waser NM (2001) The impact of floral larceny on individuals, populations, and communities. Oecologia 129:161–168. https://doi.org/10.1007/s004420100739
Johnson SN, Gregory PJ (2006) Chemically-mediated host-plant location and selection by root-feeding insects. Physiol Entomol 31:1–13. https://doi.org/10.1111/j.1365-3032.2005.00487.x
Johnson NC, Graham J, Smith FA (1997) Functioning of mycorrhizal associations along the mutualism–parasitism continuum. New Phytol 135:575–585. https://doi.org/10.1046/j.1469-8137.1997.00729.x
Johnson SN, Gregory PJ, Murray PJ et al (2004) Host plant recognition by the root-feeding clover weevil, *Sitona lepidus* (Coleoptera: Curculionidae). Bull Entomol Res 94: 433–439. https://doi.org/10.1079/BER2004317
Johnson SN, Clark KE, Hartley SE et al (2012) Aboveground–belowground herbivore interactions: a meta-analysis. Ecology 93:2208–2215. https://doi.org/10.1890/11-2272.1
Johnson SN, Mitchell C, McNicol JW et al (2013) Downstairs drivers—root herbivores shape communities of above-ground herbivores and natural enemies via changes in plant nutrients. J Anim Ecol 82:1021–1030. https://doi.org/10.1111/1365-2656.12070
Jung SC, Martinez-Medina A, Lopez-Raez JA et al (2012) Mycorrhiza-induced resistance and priming of plant defenses. J Chem Ecol 38:651–664. https://doi.org/10.1007/s10886-012-0134-6
Kaplan I, Halitschke R, Kessler A et al (2008) Constitutive and induced defenses to herbivory in above- and belowground plant tissues. Ecology 89:392–406. https://doi.org/10.1890/07-0471.1
Kaplan I, Sardanelli S, Denno RF (2009) Field evidence for indirect interactions between foliar-feeding insect and root-feeding nematode communities on *Nicotiana tabacum*. Ecol Entomol 34:262–270. https://doi.org/10.1111/j.1365-2311.2008.01062.x
Kaplan I, Sardanelli S, Rehill BJ et al (2011) Toward a mechanistic understanding of competition in vascular-feeding herbivores: an empirical test of the sink competition hypothesis. Oecologia 166:627–636. https://doi.org/10.1007/s00442-010-1885-9
Karban R, Baldwin IT (1997) Induced responses to herbivory. University of Chicago Press, Chicago
Kumar P, Ortiz EV, Garrido E et al (2016) Potato tuber herbivory increases resistance to aboveground lepidopteran herbivores. Oecologia 182:177–187. https://doi.org/10.1007/s00442-016-3633-2
Lee B, Lee S, Ryu C-M (2012) Foliar aphid feeding recruits rhizosphere bacteria and primes plant immunity against pathogenic and non-pathogenic bacteria in pepper. Ann Bot 110:281–290. https://doi.org/10.1093/aob/mcs055
Lehtila K, Strauss SY (1997) Leaf damage by herbivores affects attractiveness to pollinators in wild radish, *Raphanus raphanistrum*. Oecologia 111:396–403. https://doi.org/10.1007/s004420050251
Lu J, Robert CA, Lou Y et al (2016) A conserved pattern in plant-mediated interactions between herbivores. Ecol Evol 6:1032–1040. https://doi.org/10.1002/ece3.1922
Masters GJ, Brown VK (1992) Plant-mediated interactions between two spatially separated insects. Funct Ecol 6:175–179. https://doi.org/10.2307/2389752
Masters GJ, Brown VK, Gange AC (1993) Plant mediated interactions between above- and belowground insect herbivores. Oikos 66:148–151. https://doi.org/10.2307/3545209
Masters GJ, Jones TH, Rogers M (2001) Host-plant mediated effects of root herbivory on insect seed predators and their parasitoids. Oecologia 127:246–250. https://doi.org/10.1007/s004420000569

McCall AC, Irwin RE (2006) Florivory: the intersection of pollination and herbivory. Ecol Lett 9: 1351–1365. https://doi.org/10.1111/j.1461-0248.2006.00975.x

McCarville MT, Soh DH, Tylka GL et al (2014) Aboveground feeding by soybean aphid, Aphis glycines, affects soybean cyst nematode, Heterodera glycines, reproduction belowground. PLoS One 9:e86415. https://doi.org/10.1371/journal.pone.0086415

Milano NJ, Barber NA, Adler LS (2015) Conspecific and heterospecific aboveground herbivory both reduce preference by a belowground herbivore. Environ Entomol 44:317–324. https://doi.org/10.1093/ee/nvv003

Moran NA, Whitham TG (1990) Interspecific competition between root-feeding and leaf-galling aphids mediated by host-plant resistance. Ecology 71:1050–1058. https://doi.org/10.2307/1937373

Mothershead K, Marquis RJ (2000) Fitness impacts of herbivory through indirect effects on plant-pollinator interactions in *Oenothera macrocarpa*. Ecology 81:30–40. https://doi.org/10.1890/0012-9658(2000)081[0030:FIOHTI]2.0.CO;2

Mundim FM, Alborn HT, Vieira-Neto EH et al (2016) A whole-plant perspective reveals unexpected impacts of above-and belowground herbivores on plant growth and defense. Ecology 98:70–78. https://doi.org/10.1002/ecy.1619

National Research Council of the National Academies (2007) Status of pollinators in North America. National Academies, Washington, DC. https://doi.org/10.17226/11761

Ohgushi T (2005) Indirect interaction webs: herbivore-induced effects through trait change in plants. Annu Rev Ecol Evol Syst 36:81–105. https://doi.org/10.1146/annurev.ecolsys.36.091704.175523

Papadopoulou GV, van Dam NM (2017) Mechanisms and ecological implications of plant-mediated interactions between belowground and aboveground insect herbivores. Ecol Res 32:13–26

Petipas RH, Brody AK (2014) Termites and ungulates affect arbuscular mycorrhizal richness and infectivity in a semiarid savanna. Botany 92:233240. https://doi.org/10.1139/cjb-2013-0223

Pierre PS, Dugravot S, Ferry A et al (2011) Aboveground herbivory affects indirect defences of brassicaceous plants against the root feeder *Delia radicum*: laboratory and field evidence. Ecol Entomol 36:326–334. https://doi.org/10.1111/j.1365-2311.2011.01276.x

Pieterse CMJ, Zamioudis C, Berendsen RL et al (2014) Induced systemic resistance by beneficial microbes. Annu Rev Phytopathol 52:347–375. https://doi.org/10.1146/annurev-phyto-082712-102340

Poveda K, Steffan-Dewenter I, Scheu S et al (2003) Effects of below- and aboveground herbivores on plant growth, flower visitation, and seed set. Oecologia 135:601–605. https://doi.org/10.1007/s00442-003-1228-1

Poveda K, Steffan-Dewenter I, Scheu S et al (2005) Effects of decomposers and herbivores on plant performance and aboveground plant-insect interactions. Oikos 108:503–510. https://doi.org/10.1111/j.0030-1299.2005.13664.x

Pozo MJ, Azcón-Aguilar C (2007) Unraveling mycorrhiza-induced resistance. Curr Opin Plant Biol 10:393–398. https://doi.org/10.1016/j.pbi.2007.05.004

Prior KM, Robinson JM, Meadley Dunphy SA et al (2015) Mutualism between co-introduced species facilitates invasion and alters plant community structure. Proc R Soc Lond B Biol Sci 282:20142846. https://doi.org/10.1098/rspb.2014.2846

Rasmann S, Turlings TC (2007) Simultaneous feeding by aboveground and belowground herbivores attenuates plant-mediated attraction of their respective natural enemies. Ecol Lett 10: 926–936. https://doi.org/10.1111/j.1461-0248.2007.01084.x

Robert CA, Erb M, Duployer M et al (2012) Herbivore-induced plant volatiles mediate host selection by a root herbivore. New Phytol 194:1061–1069. https://doi.org/10.1111/j.1469-8137.2012.04127.x

Rodríguez-Echeverría S, de la Peña E, Moens M et al (2009) Can root-feeders alter the composition of AMF communities? Experimental evidence from the dune grass *Ammophila arenaria*. Basic Appl Ecol 10:131–140. https://doi.org/10.1016/j.baae.2008.01.004

Ruotsalainen AL, Eskelinen A (2011) Root fungal symbionts interact with mammalian herbivory, soil nutrient availability and specific habitat conditions. Oecologia 166:807–817. https://doi.org/10.1007/s00442-011-1928-x

Saravesi K, Markkola A, Rautio P et al (2008) Defoliation causes parallel temporal responses in a host tree and its fungal symbionts. Oecologia 156:117–123. https://doi.org/10.1007/s00442-008-0967-4

Smith SE, Read DJ (2008) Mycorrhizal symbiosis, 3rd edn. Academic, London

Soler R, Bezemer T, Van Der Putten WH et al (2005) Root herbivore effects on above-ground herbivore, parasitoid and hyperparasitoid performance via changes in plant quality. J Anim Ecol 74:1121–1130. https://doi.org/10.1111/j.1365-2656.2005.01006.x

Soler R, Bezemer TM, Cortesero AM et al (2007a) Impact of foliar herbivory on the development of a root-feeding insect and its parasitoid. Oecologia 152: 257–264. https://doi.org/10.1007/s00442-006-0649-z

Soler R, Harvey JA, Kamp AF et al (2007b) Root herbivores influence the behaviour of an aboveground parasitoid through changes in plant-volatile signals. Oikos 116:367–376. https://doi.org/10.1111/j.0030-1299.2007.15501.x

Soler R, Harvey JA, Bezemer TM et al (2008) Plants as green phones. Plant Signal Behav 3: 519–520

Soler R, Schaper SV, Bezemer T et al (2009) Influence of presence and spatial arrangement of belowground insects on host-plant selection of aboveground insects: a field study. Ecol Entomol 34:339–345. https://doi.org/10.1111/j.1365-2311.2008.01082.x

Soler R, Harvey JA, Rouchet R et al (2010) Impacts of belowground herbivory on oviposition decisions in two congeneric butterfly species. Entomol Exp Appl 136:191–198. https://doi.org/10.1111/j.1570-7458.2010.01015.x

Soler R, van der Putten WH, Harvey JA et al (2012) Root herbivore effects on aboveground multitrophic interactions: patterns, processes and mechanisms. J Chem Ecol 38:755–767. https://doi.org/10.1007/s10886-012-0104-z

Soler R, Erb M, Kaplan I (2013) Long distance root–shoot signalling in plant–insect community interactions. Trends Plant Sci 18:149–156. https://doi.org/10.1016/j.tplants.2012.08.010

Soper Gorden NL, Adler LS (2013) Abiotic conditions affect floral antagonists and mutualists of *Impatiens capensis* (Balsaminaceae). Am J Bot 100:679–689. https://doi.org/10.3732/ajb.1200460

Stephens AEA, Srivastava DS, Myers JH (2013) Strength in numbers? Effects of multiple natural enemy species on plant performance. Proc R Soc Lond B Biol Sci 280:20122756. https://doi.org/10.1098/rspb.2012.2756

Strauss SY (1991) Indirect effects in community ecology: their definition, study and importance. Trends Ecol Evol 6:206–210. https://doi.org/10.1016/0169-5347(91)90023-Q

Strauss SY (1997) Floral characters link herbivores, pollinators, and plant fitness. Ecology 78: 1640–1645. https://doi.org/10.1890/0012-9658(1997)078[1640:FCLHPA]2.0.CO;2

Strauss SY, Conner JK, Rush SL (1996) Foliar herbivory affects floral characters and plant attractiveness to pollinators: implications for male and female plant fitness. Am Nat 147: 1098–1107. https://doi.org/10.1086/285896

van Dam NM, Heil M (2011) Multitrophic interactions below and above ground: en route to the next level. J Ecol 99:77–88. https://doi.org/10.1111/j.1365-2745.2010.01761.x

van Dam NM, Raaijmakers CE, Van Der Putten WH (2005) Root herbivory reduces growth and survival of the shoot feeding specialist *Pieris rapae* on Brassica nigra. Entomol Exp Appl 115: 161–170. https://doi.org/10.1111/j.1570-7458.2005.00241.x

van der Putten WH, Bardgett RD, Ruiter PC et al (2009) Empirical and theoretical challenges in aboveground–belowground ecology. Oecologia 161:1–14. https://doi.org/10.1007/s00442-009-1351-8

van Geem M, Gols R, van Dam NM et al (2013) The importance of aboveground–belowground interactions on the evolution and maintenance of variation in plant defense traits. Front Plant Sci 4:431. https://doi.org/10.3389/fpls.2013.00431

Vannette RL, Hunter MD (2011) Plant defence theory re-examined: nonlinear expectations based on the costs and benefits of resource mutualisms. J Ecol 99:66–76. https://doi.org/10.1111/j.1365-2745.2010.01755.x

Vannette RL, Hunter MD (2014) Genetic variation in plant below-ground response to elevated CO_2 and two herbivore species. Plant Soil 384:303–314. https://doi.org/10.1007/s11104-014-2203-9

Varga S, Kytöviita M-M (2010) Gender dimorphism and mycorrhizal symbiosis affect floral visitors and reproductive output in *Geranium sylvaticum*. Funct Ecol 24:750–758. https://doi.org/10.1111/j.1365-2435.2010.01708.x

Wardle DA, Bardgett RD, Klironomos JN et al (2004) Ecological linkages between aboveground and belowground biota. Science 204:1629–1633. https://doi.org/10.1126/science.1094875

Wei H, He M, Lu X et al (2016) Differences in interactions of aboveground and belowground herbivores on the invasive plant *Alternanthera philoxeroides* and native host *A. sessilis*. Biol Invasions 18:3437–3447. https://doi.org/10.1007/s10530-016-1234-y

White JA, Andow DA (2006) Habitat modification contributes to associational resistance between herbivores. Oecologia 148:482–490. https://doi.org/10.1007/s00442-006-0388-1

Wilson GWT, Rice CW, Rillig MC et al (2009) Soil aggregation and carbon sequestration are tightly correlated with the abundance of arbuscular mycorrhizal fungi: results from long-term field experiments. Ecol Lett 12:452–461. https://doi.org/10.1111/j.1461-0248.2009.01303.x

Wolfe BE, Husband BC, Klironomos JN (2005) Effects of a belowground mutualism on an aboveground mutualism. Ecol Lett 8:218–223. https://doi.org/10.1111/j.1461-0248.2004.00716.x

Wondafrash M, Van Dam NM, Tytgat TO (2013) Plant systemic induced responses mediate interactions between root parasitic nematodes and aboveground herbivorous insects. Front Plant Sci 4:87. https://doi.org/10.3389/fpls.2013.00087

Wooton JT (1994) The nature and consequences of indirect effects in ecological communities. Annu Rev Ecol Syst 25:443–466. https://doi.org/10.1146/annurev.es.25.110194.002303

Yang W, Zheng Y, Gao C et al (2013) The arbuscular mycorrhizal fungal community response to warming and grazing differs between soil and roots on the Qinghai-Tibetan Plateau. PLoS One 8 (9):e76447. https://doi.org/10.1371/journal.pone.0076447

Chapter 12
Eco-evolutionary Dynamics of Above- and Belowground Herbivores and Invasive Plants

Wei Huang, Evan Siemann, and Jianqing Ding

12.1 Introduction

Herbivory is a major determinant of plant growth, reproduction, and defense. Ecological changes in abundance and composition of herbivores may alter plant phenotypic traits but also serve as an important selective agent triggering adaptive evolution in these traits, which in turn may alter interactions with surrounding organisms, in particular herbivores that exerted the selective pressure (Utsumi 2011; Ohgushi 2016). These interdependent ecological and evolutionary processes are often viewed as "eco-evolutionary dynamics" and have been documented in many different systems aboveground, but little attention has been paid to belowground (Fussmann et al. 2007; Pelletier et al. 2009; Hendry 2016). Moreover, above- and belowground herbivores can indirectly interact with each other via the shared host plant, resulting in a wide range of impacts on plant phenotypic traits depending on whether their combined impacts are independent, synergistic, or offsetting (Kaplan et al. 2008b; Erwin et al. 2014; Mundim et al. 2017). Thus, it is imperative to include above- and belowground herbivores in eco-evolutionary dynamics of plant and herbivore interactions in order to extrapolate how ecological changes in herbivores drive plant trait evolution and how evolution of plant traits influences

W. Huang
Key Laboratory of Aquatic Plant and Watershed Ecology, Wuhan Botanical Garden, Chinese Academy of Sciences, Wuhan, Hubei, China

Institute of Plant Sciences, University of Bern, Bern, Switzerland
e-mail: huangwei0519@wbgcas.cn

E. Siemann
Department of Biosciences, Rice University, Houston, TX, USA
e-mail: siemann@rice.edu

J. Ding (✉)
School of Life Sciences, Henan University, Kaifeng, Henan, China

T. Ohgushi et al. (eds.), *Aboveground–Belowground Community Ecology*, Ecological Studies 234, https://doi.org/10.1007/978-3-319-91614-9_12

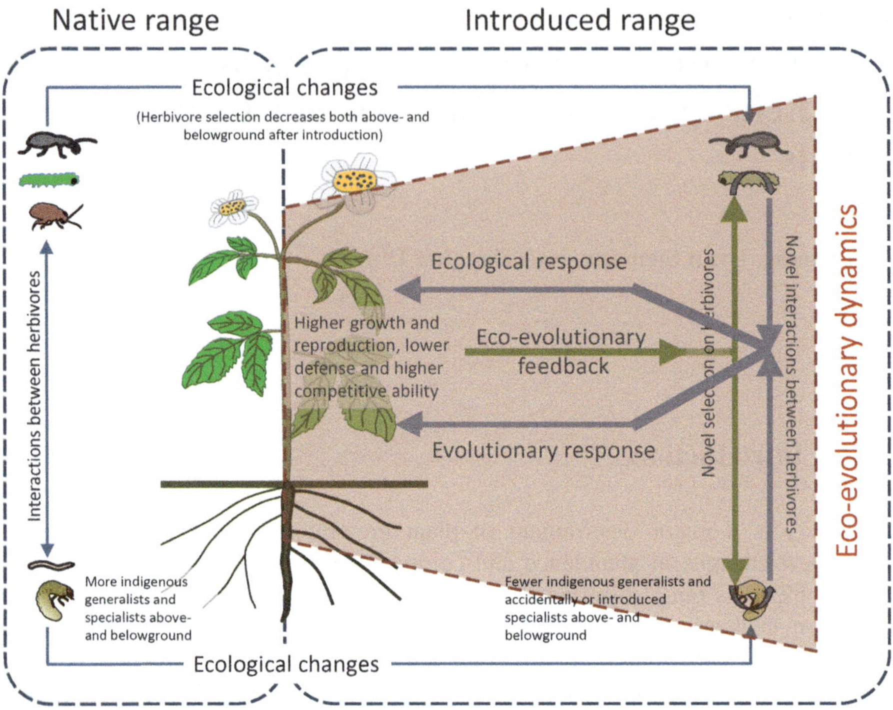

Fig. 12.1 The eco-evolutionary dynamics of above- and belowground interactions in invasive plants. Invasive plants often experience lower above- and belowground herbivore loads in the introduced range than in the native range. This ecological variation in herbivore pressure may drive evolutionary changes in plant traits, for example, increasing growth and reproduction and decreasing chemical defenses. Simultaneously, evolutionary changes in invasive plants may influence ecological interactions between above- and belowground herbivores

ecological processes in plant–herbivore interactions (van der Putten et al. 2009; van Geem et al. 2013).

Invasive plants are excellent candidates for examining the interplay between ecological and evolutionary processes in the context of above- and belowground interactions (Fig. 12.1) (van der Putten et al. 2009; Harvey et al. 2010; Vestergård et al. 2015). Firstly, invasive plants are often released from their coevolved above- and belowground specialist herbivores and may experience attack from a different group of generalists in the introduced range (Elton 1958; Maron and Vilà 2001; Keane and Crawley 2002; but see Chun et al. 2010). The variations in abundance and composition of herbivores in the introduced range may change selective pressure of herbivory on invasive plants. Secondly, these ecological variations in herbivore pressure may drive evolutionary changes in plant traits. Specifically, lower herbivore loads in the introduced range may select for increased competitive ability of invasive plants by evolutionary shifts in allocation from defense against herbivores to growth and/or reproduction (Blumenthal and Hufbauer 2007; Feng et al. 2009; Huang et al.

2010; but see Felker-Quinn et al. 2013). Lastly, evolutionary changes in invasive plants may influence ecological interactions, when the interactions with above- and belowground herbivores in the introduced range are modified by these genetic changes in plant defense and growth traits (Pearse and Altermatt 2013; Tanner et al. 2013; Bezemer et al. 2014). Thus, studies on the response of invasive and native populations of invasive plant species to above- and belowground herbivores may provide new insight into the interplay of ecological and evolutionary processes in altering the interactions among aboveground herbivores, belowground herbivores, and plants.

In this chapter, we review existing knowledge about eco-evolutionary dynamics of aboveground herbivores, belowground herbivores, and invasive plants. We aim to (1) provide an overview of the variation in herbivore communities associated with invasive plants in both above- and belowground compartments, (2) identify how above- and belowground herbivores drive selection on growth and defense traits of invasive plants, and (3) discuss whether genetic differences in growth and defense between native and invasive populations affect above- and belowground communities differently. In addition, we present a case study to illustrate interactions and feedbacks in eco-evolutionary dynamics.

12.2 Variations in Above- and Belowground Herbivores Between Introduced and Native Ranges

Herbivores can drastically influence plant growth, reproduction, abundance, and distribution. The Enemy Release Hypothesis (ERH) postulates that exotic plants will gain a competitive advantage over native plants through a plastic phenotypic (ecological) response to escaping suppression by coevolved natural enemies (Elton 1958; Maron and Vilà 2001; Keane and Crawley 2002). Although invasive plants commonly escape their co-evolved specialists, they may also be attacked by generalist natural enemies in the introduced range (Maron and Vilà 2001; Bezemer et al. 2014). The Biotic Resistance Hypothesis (BRH) emphasizes the importance of generalists in limiting invasions and posits that some exotic plants will be constrained by generalists because they can recognize, exploit, and suppress exotic plants in the introduced range (Parker and Hay 2005).

In addition to varying by diet breadth, herbivores attacking invasive plants vary in feeding guild, such as chewing vs. sucking feeders, foliar vs. seed feeders as well as gall formers and miners. Many studies have found herbivore community composition differs between invasive plants and related resident plants. For example, Ando et al. (2010) showed that herbivore species richness on invasive *Solidago altissima* and the native congener *S. virgaurea* were similar in the introduced range, but *S. altissima* plants were predominantly attacked by sucking feeders, while *S. virgaurea* plants were mainly attacked by foliar chewers and miners. In a manipulative common garden experiment, Burghardt and Tallamy (2013) found that the impact of plant origin (native vs. invasive plants) on abundance of herbivores differed among herbivore feeding guilds. Plant origin had a stronger effect on abundance of chewing feeders than sucking feeders, while xylem feeder abundance was unaffected by plant origin.

To date, studies on the variations in herbivores between introduced and native ranges of invasive plants have mainly focused on aboveground enemies. Belowground enemies have received little attention, despite the fact that belowground enemies are pervasive in most terrestrial ecosystems and play critical roles in mediating the abundance and spread of plants and plant-associated organisms (van Dam 2009; van der Putten et al. 2009; Johnson and Rasmann 2015). In a biogeographical field survey, Cripps et al. (2006) showed that invasive plant *Lepidium draba* is attacked by root chewers and galls in the native range, but no root herbivores feed on *L. draba* in the introduced range, indicating *L. draba* completely escaped from suppression by belowground herbivores. Although direct evidence of escaping belowground herbivores is scarce, classical biological control provides clear information that belowground herbivores may be a major driver of plant invasions (Blossey 1993; Gerber et al. 2007; Huang et al. 2011). A review by Blossey and Hunt-Joshi (2003) showed that a total of 49 belowground herbivores have been released to control 19 invasive plants and more than half of them suppress their host plants.

Taken together, previous studies clearly demonstrate that above- and belowground herbivore communities differ between native and introduced ranges in their composition, abundance, and species richness (Maron and Vilà 2001; Blossey and Hunt-Joshi 2003; Cripps et al. 2006). Therefore, plant invasions offer an excellent opportunity to investigate eco-evolutionary dynamics in both above- and belowground compartments and future studies examining the role of natural enemies in plant invasions should benefit from combined above- and belowground perspectives.

12.3 Evolution of Plant Defense and Changes in Plant–Herbivore Interactions During Plant Invasions

12.3.1 Impact of Aboveground Herbivores

12.3.1.1 Trade-Off Between Plant Growth and Defense

Herbivores feed on almost all parts of plants, including leaves, stems, roots, flowers, fruits, and seeds. Thus, herbivory is considered an important selective agent in the evolution of many plant traits, such as growth, reproduction, and defense (Agrawal et al. 2012; Züst et al. 2012; Huber et al. 2016a). As a result, release from co-evolved natural enemies may not only lead to ecological benefits for invasive plants, but also drive evolution in a suite of traits of invasive plants (Bossdorf et al. 2005; Lin et al. 2015b; Uesugi and Kessler 2016). As an extension of the ERH, the Evolution of

Increased Competitive Ability hypothesis (EICA) posits that invasive plants that escape from specialist herbivores may increase their competitive ability through an evolutionary shift in resource allocation away from defense against herbivores toward traits conferring increased competitive ability, such as growth and reproduction (Blossey and Nötzold 1995).

Many studies have tested the predictions of EICA hypothesis, but found mixed results (Bossdorf et al. 2005; Chun et al. 2010; Felker-Quinn et al. 2013). Some studies supported the EICA hypothesis and found a trade-off between plant growth and defense (Joshi and Tielbörger 2012; Huang and Ding 2016). However, other studies did not support the EICA hypothesis and showed that invasive plants had either greater performance or lower defense (Meyer et al. 2005; Caño et al. 2009). The mixed results may be due to overlooking the abundance and composition of herbivores in the introduced range (Müller-Schärer et al. 2004; Orians and Ward 2010; Prior et al. 2015). Invasive plants often escape specialists, but may encounter generalists in the introduced range. Furthermore, invasive plants may reestablish associations with coevolved specialists or generalists due to accidental or intended introductions by human activities. Thus, herbivores in the introduced range, regardless of origins, may also have the potential to affect the evolutionary direction and magnitude of plant defense and growth.

12.3.1.2 Plant Resistance

Resistance is a defensive trait that protects a plant from herbivores by reducing the performance and/or preference of the herbivores. Specialist and generalist herbivores can exert opposite selection pressures on plant resistance (van der Meijden 1996; Lankau 2007; Ali and Agrawal 2012). The Shifting Defense Hypothesis (SDH) argued that invasive plants should maintain or increase their less-costly, toxic defense compounds (qualitative defenses) to defend against generalists and decrease their more-costly, digestibility-reducing compounds (quantitative defenses) which are more important in defense against specialists (Müller-Schärer et al. 2004). A meta-analysis and some empirical studies supported SDH (Joshi and Vrieling 2005; Doorduin and Vrieling 2011). However, invasive plants may reacquire their resistance against herbivores, including specialists and generalists, when they are introduced from native ranges, or when herbivores from introduced range could adapt to the invasive plants (Siemann et al. 2006; Fukano and Yahara 2012; Sakata et al. 2014).

Plant resistance to herbivory is not only expressed constitutively, but can also be induced upon herbivore attack (Karban and Myers 1989; Agrawal 2005; Kant et al. 2015). This induced resistance may be a cost-saving defense strategy, because plants can increase resistance when herbivores are present, while shifting resources from defense to growth and reproduction when herbivores are absent (Agrawal and Karban 1999; Cipollini and Heil 2010; Karban 2011). Many studies have demonstrated trade-offs between constitutive and induced resistance and trade-offs between defense and growth (Kempel et al. 2011). It is, therefore, reasonable to expect that invasive populations that are rarely attacked by herbivores in the introduced range should have higher induced resistance and lower constitutive resistance than their native conspecifics. This shift in defense strategies may favor invasive plants in competition with native plants. Although previous studies have demonstrated changes in induced resistance of invasive plants (Cipollini et al. 2005; Eigenbrode et al. 2008; Wang et al. 2012), theory that predicts evolutionary directions and consequences is still in its infancy. Thus, we need more detailed comparisons between native and invasive populations to fully evaluate how induced resistance changes during plant invasion.

Furthermore, some plants utilize indirect defenses [e.g., extrafloral nectar (EFNs) and volatile organic compounds (VOCs)] to attract predators or parasitoids of herbivores for reducing damage levels (Arimura et al. 2005; Heil 2008; Kessler and Heil 2011). To date, these indirect defenses have been demonstrated in many plant species under both laboratory and field conditions (Poelman et al. 2011; Mathur et al. 2013; Huang et al. 2015). In contrast to the evolution of direct defenses which are affected mainly by herbivores alone, the evolution of indirect defenses may be determined by herbivores and their natural enemies simultaneously (Poelman and Kessler 2016). Novel herbivore communities or differences in the predator and parasitoid communities in the introduced range may each influence selection on indirect defense, resulting in indirect defense being more sensitive to the changes in interaction network structure than is direct defense (Carrillo et al. 2012a; Wang et al. 2013). Based on limited available information and mixed results in plant invasion, it is still unclear how selection by herbivores affects indirect defense of invasive plants. Furthermore, changes in indirect defense compounds in the introduced range may also be the result of other selection pressures since VOCs and EFNs are also affected by many other biotic factors, such as plant neighbor identity and pollinators (Heil and Karban 2010; Heil 2011; Karban et al. 2014).

12.3.1.3 Plant Tolerance

In addition to resisting herbivore attack, plants also tolerate damage by herbivores. Tolerance is the ability to prevent or attenuate the negative impacts of herbivores through compensatory growth (Strauss and Agrawal 1999; Agrawal 2011; Fornoni 2011). The high growth rate of plants from invasive populations may lead to higher tolerance since plant growth rate is often positively correlated with tolerance to herbivory (Agrawal 2011). Also, negative correlations between herbivore resistance and tolerance have been detected in many agricultural and wild plant species (Núñez-Farfán et al. 2007); as a result invasive plants with lower resistance may have higher tolerance (Wang et al. 2011). Furthermore, invasive plants are still attacked by some herbivores in the introduced range that could favor a strategy of increased tolerance (Fornoni 2011). A growing body of research has indeed found that invasive populations maintained or increased tolerance compared to conspecific native populations after artificial damage, specialist or generalist herbivory, or in field conditions (Bossdorf et al. 2004; Huang et al. 2010; Gard et al. 2013; Huang and Ding 2016). However, a few studies have found lower tolerance in invasive populations (Oduor et al. 2011; Lin et al. 2015b). Testing the traits underlying such differences in tolerance will reveal a better understanding of the role of herbivore tolerance in plant invasions.

12.3.2 Impact of Belowground Herbivores

Up to now, investigation of the impact of herbivores on the evolutionary trajectories of invasive plants has mostly focused on aboveground interactions and plant traits. There is comparatively little known about whether and how belowground herbivores affect root traits such as growth and belowground defense strategies of invasive plants. It is likely belowground herbivores would affect plant traits because they also have the potential to affect plant growth and defense (Pierre et al. 2012; Erwin et al. 2013; Huber et al. 2016a), and many invasive plants are released from suppression by belowground herbivores (Blossey and Hunt-Joshi 2003; Cripps et al. 2006; Knochel et al. 2010).

12.3.2.1 Plant Growth

Among plant root traits, root branching and specific root length (root length to mass ratio) are two important indicators of environmental changes, such as temperature, precipitation, and fertilization (Ostonen et al. 2007; Arredondo and Johnson 2011; Postma et al. 2014). Greater branching and higher specific root length may lead plants to absorb soil water and nutrients more efficiently, but may also render plants more vulnerable to belowground herbivores. Recently, Dawson and Schrama (2016) predicted that invasive plants should evolve to have greater branching and higher specific root length when released from their belowground enemies because such variations in root traits could increase their competitive ability through more resource uptake. So far, however, no empirical study has tested this hypothesis. In contrast, root biomass has been extensively studied and many studies have demonstrated that plants from the introduced range invest relatively fewer resources to belowground than to aboveground, leading to invasive populations that have lower root-to-shoot ratio than native populations (Huang et al. 2012b; Liao et al. 2013; Lin et al. 2015a).

12.3.2.2 Plant Defense

Plants are known to defend against belowground herbivores through increasing root toxins after attack, releasing volatile chemicals to attract the enemies of belowground herbivores, and/or compensatory growth (Rasmann and Agrawal 2008; van Dam 2009; Huber et al. 2016b). For plant root defense, in the study that put forward the EICA hypothesis, Blossey and Nötzold (1995) tested the performance of root feeding larvae of the weevil *Hylobius transversovittatus*, on potted plants of *Lythrum salicaria* from introduced and native ranges. They found that larval weight and survival were significantly higher on invasive plants than on native conspecifics, indicating that *L. salicaria* may have evolved lower resistance to belowground herbivores in the introduced range. For plant root tolerance, Huang et al. (2012b) demonstrated that there was no significant difference in root tolerance between native and invasive populations of Chinese tallow tree after root herbivory. However, to date, research on invasive plant root growth and defense is so limited that it is unlikely to predict how root growth and defense of invasive plants evolve under new selections. Thus, it is imperative to include different root traits and defensive strategies into studies of invasive plants in order to extrapolate the evolutionary trajectories of root growth and defensive strategies during the process of invasion.

12.3.3 Impact of Above- and Belowground Herbivore Interactions

Above- and belowground herbivores are linked through induced responses of the shared host plant. First, above- and belowground herbivores can interact through plant direct resistance which can influence herbivore growth and/or foraging behavior (Erb et al. 2009; Robert et al. 2012). Second, interactions between above- and belowground herbivores can be mediated by plant indirect resistance [e.g., herbivore induced plant volatiles (HIPVs)] which can attract the natural enemies of herbivores (Rasmann and Turlings 2007; Soler et al. 2007). Finally, plant tolerance also has potential to affect above- and belowground herbivores interactions via shifting allocation of primary metabolites between above- and belowground structures (Kaplan et al. 2008a; Johnson et al. 2009). Thus, it is reasonable to predict that variation in selection on plant defense strategies may not only depend on the abundance and identity of herbivores but also on the interactions among herbivores. Genetic variation in plant defense may lead to different plant genotypes showing different physiological responses to above- and belowground herbivores that in turn alter the outcome of their interactions (Hol et al. 2004; Wurst et al. 2008; Kafle et al. 2014). Furthermore, the outcome of above- and belowground interactions with different plant genotypes likely depends on the feeding guild, modes of feeding, and diet breadth of herbivores with interactions potentially varying among specific combinations of herbivores (Johnson et al. 2012; Singh et al. 2014).

Under these conditions, invasive plants are likely to confront new combinations of both above- and belowground herbivores in terms of the taxa present as well as the feeding guilds, especially when some guilds are lacking in the introduced range. Thus, changes in above- and belowground herbivore interactions may also play a critical role in driving adaptive evolution of defense strategies for invasive plants. However, to date, most studies examining the role of herbivores in the evolution of defense during plant invasion focused on herbivore release and/or gain and treated above- and belowground herbivores separately (if they included belowground herbivores). Furthermore, our current understanding of how genetic variation in invasive plant defenses affects above- and belowground herbivores is quite limited. As a consequence, we know little about feedbacks resulting from eco-evolutionary dynamics. Thus, investigating the difference in above- and belowground herbivore interactions between native and introduced ranges and feedback of genetic variation in defense to above- and belowground herbivores would be two important steps to understanding evolutionary trajectories of invasive plant defenses and corresponding ecological consequences.

12.4 Case Studies: Above- and Belowground Herbivore Interactions in *Triadica sebifera*

Triadica sebifera (synonyms include *Sapium sebiferum*) is a rapidly growing Euphorb tree (Zhang and Lin 1994). It is native to China and has become a severe invader in the southeastern United States (Siemann and Rogers 2003a; Pattison and Mack 2008). In China, *T. sebifera* is attacked by a diversity of specialist and generalist herbivores from both above- and belowground compartments (Zheng et al. 2005; Huang et al. 2014). However, only a few foliar chewing generalists (no sucking feeders or seed predators) and no root herbivores are detected in the USA (Siemann and Rogers 2003b, c), indicating *T. sebifera* experiences low above- and belowground herbivore loads after invasion. A recent apparently accidental introduction of a specialist leaf miner and roller from Asia has expanded the feeding modes of herbivores attacking *T. sebifera* (Davis et al. 2013). Recent studies on *Triadica sebifera* showed that *T. sebifera* generally had lower resistance to both above- and belowground herbivores, higher tolerance to aboveground herbivores, and comparable tolerance to belowground herbivores after invasion (Table 12.1). Furthermore, con- and heterospecific above- and belowground herbivore interactions were more intense on invasive populations than on native ones (Table 12.1). These results suggest that invasive plants evolve different growth and defense strategies to above- and belowground herbivores after invasion and feedback of these changes to herbivores interactions is stronger after invasion (see below for details).

12.4.1 *Aboveground Herbivores*

In a 14-year common garden experiment in North America, Siemann and Rogers (2001) found that invasive populations of *T. sebifera* had greater basal area and produced more seeds, but had lower foliar tannins than native populations. These

Table 12.1 Summary of reports on above–belowground herbivores interactions and plant responses to herbivory in invasive plant *Triadica sebifera*

	No-herbivory	Aboveground herbivory		Belowground herbivory	Both herbivory	
		Generalist	Specialist	Specialist	Conspecifics (above facilitate below, below inhibit above)	Heterospecifics (reciprocal inhibition)
Insect performance		– or ↑	↑	↑	Above –, below ↑	Above ↓, Below ↓
Plant growth	↑	↑	↑	–	–	
Direct resistance	↓	– or ↓	↓	↓	Leaf, –, root ↓	Root ↑, leaf ↑
Indirect resistance	↓, – or ↑	–	↑			
Tolerance		↑	↑	–	–	
References	Siemann and Rogers (2001), Carrillo et al. (2012a, b), Wang et al. (2013), Siemann et al. (2017)	Siemann and Rogers (2003b, c), Rogers and Siemann (2005), Huang et al. (2010, 2012a), Carrillo et al. (2012b, 2014) Wang et al. (2013)	Zou et al. (2008b), Huang et al. (2010, 2012b), Wang et al. (2011, 2013)	Huang et al. (2012b, 2014), Li et al. (2016)	Huang et al. (2012b, 2014)	Li et al. (2016)

For insects, ↓, – and ↑ indicate insect performance decreased, no changed or increased on invasive populations, relative to native populations, respectively. For plants, ↓, – and ↑ indicate the growth, resistance, or tolerance of invasive populations was lower; had no significant change; or was higher than native populations, respectively results were consistent with the EICA hypothesis and provided clear evidence that release from herbivores facilitates evolutionary changes in resource allocation between growth, reproduction, and defense. In another introduced site in Hawaii, Siemann and Rogers (2003b) found that a generalist beetle from the native range caused greater damage on plants from invasive populations. Similarly, caged North American generalist grasshoppers (Siemann and Rogers 2003c) and Asian specialist beetles (Zou et al. 2008b) caused more damage to plants from invasive populations when given a choice between plants from invasive or native populations. Overall, these studies indicate that *T. sebifera* decreases resistance to aboveground herbivores after invasion. However, the greater performance of plants from invasive populations than native populations in common gardens in North America, Hawaii, and Asia in which aboveground herbivores were suppressed suggests that the link from herbivore damage to plant performance may not be simple (Siemann et al. 2017).

In contrast to resistance, invasive *T. sebifera* exhibits higher tolerance to aboveground herbivory than native conspecifics when plants are exposed to simulated defoliation (Rogers and Siemann 2005), generalist herbivores (Rogers and Siemann 2005; Huang et al. 2012a; Carrillo et al. 2014), specialist herbivores (Zou et al. 2008b; Wang et al. 2011; Huang et al. 2012a), and natural herbivore communities (Zou et al. 2008a, b). Huang et al. (2010) examined the resistance and tolerance of *T. sebifera* from introduced and native ranges to specialist and generalist caterpillars. Bioassays and chemical analyses demonstrated that invasive populations had lower resistance to specialist caterpillars than native populations, but similar resistance to the generalist caterpillar. Furthermore, a common garden experiment showed that invasive populations had higher herbivore tolerance than native ones, especially for generalists (Huang et al. 2010). Taken together, changes in composition (specialist vs. generalist) and abundance (lower generalist loads) of aboveground herbivores have the potential to drive *T. sebifera* to evolve lower resistance and higher tolerance to herbivory.

In addition, *T. sebifera* produces EFN in glands at the base and underside margins of leaves that potentially act as an indirect defense through attracting arthropod predators and parasitoids of herbivores. Several studies have investigated EFN production of *T. sebifera* populations from introduced and native ranges, but results were mixed. For example, invasive populations had less (Carrillo et al. 2012a), similar, (Carrillo et al. 2012b) or more constitutive EFN production (Wang et al. 2013) than native populations in different studies. These contrasting results may result from different methodology used because EFNs are affected by environmental conditions and plant physiological status (Heil 2008; Izaguirre et al. 2013; Jones and Koptur 2015). Although EFN production can be induced by aboveground herbivory, studies showed that *T. sebifera* EFN production did not differ between native and invasive populations after simulated leaf herbivory (Rogers et al. 2003; Carrillo et al. 2012a) or generalist caterpillar damage (Carrillo et al. 2012b). Wang et al. (2013) investigated the impact of generalist and specialist herbivory on EFNs and found similar responses to generalist herbivory, while specialist caterpillars elicited more EFNs on plants from native populations than from invasive populations. These studies indicated that changes in aboveground herbivores between introduced and native ranges may be also able to exert selection pressure on indirect resistance. Plants may retain constitutive and induced EFN in the introduced range to efficiently defend against generalists through attracting organisms in the higher trophic levels, while induced indirect resistance to specialist herbivores is attenuated because of lack of specialists.

12.4.2 Belowground Herbivores

Despite the fact that belowground herbivores strongly affect *T. sebifera* in the native range (Zheng et al. 2005), the role of belowground herbivores in driving the evolution of *T. sebifera* traits has received less attention. To date, such studies mainly focused on the response of *T. sebifera* to potential biological control agents or simulated root herbivory. For example, Huang et al. (2012b) and Li et al. (2016) found that larvae of a specialist flea beetle developed better on roots of plants from invasive populations than native populations. Chemical analyses showed that the invasive populations had lower root tannins than native populations, which may underlie the observed changes in larval performance between invasive and native populations (Huang et al. 2014). These results indicate that invasive *T. sebifera* decreases the investment of resources in belowground resistance, displaying the same evolutionary pattern as aboveground resistance. However, in contrast to increasing tolerance to aboveground herbivores, invasive populations had comparable compensatory growth to native populations after feeding by larvae of specialist flea beetle or simulated root damage (Huang et al. 2012b; Carrillo and Siemann 2016). These studies suggest that invasive plants, such as *T. sebifera*, have evolved lower belowground resistance and maintained their tolerance to belowground herbivores, thus supporting the EICA hypothesis predictions that invasive plants invest less resource into defense.

12.4.3 Above- and Belowground Herbivore Interactions

Aboveground herbivores may influence the induced response elicited by belowground herbivores, and vice versa, resulting in plant responses to single above- or belowground herbivores differing from their responses to multiple herbivores (Erb et al. 2008; Kaplan et al. 2008b; Huang et al. 2013, 2017; Soler et al. 2013). The specialist flea beetle, *Bikasha collaris*, is a common herbivore attacking *T. sebifera* in the native range (Huang et al. 2011). The flea beetle has aboveground adult and belowground larval life stages that cause serious damage to leaves and roots, respectively. In a recent study, Huang et al. (2014) found that both larvae and adults performed better on plants from invasive populations than from native populations, suggesting invasive *T. sebifera* decreased resistance to herbivores in both above- and belowground compartments. However, adult feeding significantly decreased root tannins and increased larval survival, and these effects were stronger on invasive populations than on native populations. In contrast, larval feeding significantly increased leaf tannins and decreased adult survival, but plant origin and larvae feeding had no interactive effect. Apart from conspecific species, *T. sebifera* is also attacked by heterospecific above- and belowground herbivores in the native range. Li et al. (2016) examined the interaction between aboveground specialist leaf-rolling weevil *Heterapoderopsis bicallosicollis* and/or belowground *B. collaris* larvae on *T. sebifera* from introduced and native ranges. In contrast to conspecific species, the weevil and beetle inhibited each other. In addition, such reciprocal negative feedback between weevil and beetle species was stronger in invasive populations than in native populations. Overall, these studies show that the contrasting patterns of asymmetric feedback (facilitation and inhibition) in conspecific species and reciprocal negative feedback in heterospecific species are stronger in invasive populations. However, how changed selective pressure drives observed resistance strategies of invasive *T. sebifera* is still unknown.

Above- and belowground herbivore interactions also affect invasive plant tolerance. Huang et al. (2012b) examined plant tolerance to *B. collaris* adult and larval herbivory and found that invasive populations had higher tolerance to adult herbivory than native populations, while tolerance to larval herbivory was comparable. But when both adults and larvae were present, tolerance was still not different between invasive and native populations as there was no significant difference in biomass between invasive and native populations. In a recent study using simulated above- and belowground herbivory, Carrillo and Siemann (2016) also found there was no difference in tolerance to combined above- and belowground damage between invasive and native populations. These studies indicate that the presence of belowground herbivores strongly affects plant tolerance to aboveground herbivores, but this effect only occurs in invasive populations.

Changes in resistance and tolerance may in turn influence invasive plant resource investment into growth. In a study using *B. collaris* adults and larvae, Huang et al. (2012b) found adults and larvae each significantly decreased plant biomass. But adults more strongly affected aboveground biomass, while larvae more strongly affected belowground biomass. Furthermore, when plants were exposed to both herbivore stages, plants had lower biomass than predicted by the independent effects of each herbivore, suggesting simultaneous above- and belowground herbivory had a non-additive effect on plant growth.

Taken together, by examining the combined effects of above- and belowground herbivores on growth and defense of invasive plant and evaluating the feedbacks of invasive plant to above- and belowground herbivores simultaneously, these studies on *T. sebifera* and its herbivores exhibited eco-evolutionary dynamics of above- and belowground plant–herbivore interactions in biological invasion. These results suggest that selection pressure imposed by both above- and belowground herbivores is different from selection pressure imposed by either above- or belowground herbivores alone, especially for invasive plant resistance and tolerance. Compared with plants from native populations, plants from invasive populations had lower resistance to above- and belowground herbivory by generalists or specialists, but higher tolerance to aboveground herbivory only. This in turn leads to invasive populations that have greater total and aboveground biomass, but comparable belowground biomass. These results indicate invasive plants may adopt an "above-ground first" strategy, allocating more resources aboveground in response to selection for increased competitive ability, which increases aboveground tolerance to herbivory (Huang et al. 2012a, b). Furthermore, evolution of invasive plant growth and defense affects aboveground, belowground herbivores, and their interactions. Invasive plants intensify the herbivores interactions, regardless of asymmetric feedback in conspecific species or reciprocal negative feedback in heterospecific species. These intensified feedbacks may considerably change the population dynamics and community compositions of herbivores in the introduced range.

12.5 Conclusions

The effect of combined above- and belowground herbivores on eco-evolutionary dynamics of invasive plants is largely different from the effect of each single herbivore (Fig. 12.1). Therefore, without integration of herbivores in both above- and belowground compartments, it is hard to make accurate predictions of how variation in herbivores contributes to the success of invasive plants. Furthermore, changes in growth and defense of invasive plants have profound impacts on above- and belowground herbivores, not only affecting herbivores in each compartment but also their interactions (Fig. 12.1). Thus, without evaluation of the impacts of invasive plants on herbivores in both above- and belowground compartments, it is impossible to have full understanding of how invasive plants affect population dynamics and community composition of herbivores in the introduced range. Together, future studies should focus on the impacts of and feedbacks to herbivores during plant invasions from both above- and belowground perspectives.

Our chapter also emphasizes that invasive plants may be excellent models to explore fundamental ecological and evolutionary questions regarding multispecies plant–herbivore interactions. This reflects, in part, that invasive plants experience different herbivore pressure in the introduced range compared to their native range and such changes in herbivore pressure may drive evolution of invasive plants in the new range. Invasive plants may change defense and growth strategies in both the above- and belowground compartments. The novel defense and growth strategies may be adaptive for invasive plants when they suffer lower above- and belowground herbivory in the introduced range compared with the native range. The resources saved from lower defense may be used to increase plant growth and reproduction and facilitate further invasion. Furthermore, novel defense and growth strategies may alter the outcome of above- and belowground herbivore interactions in the introduced range, for instance, strengthening or weakening the facilitation or inhibition between herbivores. As a result, changed interactions between herbivores may directly influence organisms that are closely associated with above- or belowground herbivores in higher trophic levels. Alternatively, it may also indirectly affect competition between invasive plants and resident plants through host shifts of herbivores between invasive and resident plants. Thus, the interactions and feedbacks of above- and belowground herbivores may play an important role in plant invasions and determine the magnitude of negative impacts on resident communities.

Furthermore, studies on interactions of above- and belowground herbivores on invasive plants also have practical implications for management of invasive species (Huang et al. 2012b; Vestergård et al. 2015; Li et al. 2016). Biological control by releasing host-specific herbivores of invasive plants has long been recognized as an efficient and sustainable method of managing invasive plants, but the success rate is not high (van Driesche et al. 2010). Simultaneously releasing both above- and belowground host-specific herbivores or a single herbivore with above- and belowground life stages may make control more efficient, because herbivores attacking in one plant compartment could modify plant defense (e.g., tolerance) in another compartment (Huang et al. 2012b; Carrillo and Siemann 2016).

The current state of research also has important implications for the impacts of herbivores on plant evolution of both above- and belowground traits. The selection pressure of herbivores may not only affect plant parts where herbivores feed, but also in distant parts that are not sites of herbivore feeding through resource allocation trade-offs or plant systemic induced responses (Erb et al. 2008; Kaplan et al. 2008a; Huang et al. 2012b; Biere and Goverse 2016). Therefore, aboveground herbivores may not only be able to shape plant evolutionary trajectories of aboveground traits but also traits of roots, and vice versa. Furthermore, selection pressure of herbivory in above- and belowground compartments may be not constant, varying temporally and spatially (Siemann and Rogers 2003b; Agrawal et al. 2006; Huber et al. 2016b). Thus, temporal and spatial variability of above- and belowground herbivores may yield different patterns of eco-evolutionary dynamics of herbivores and plants. For example, native resident generalists may accumulate over time on invasive plants and co-evolved specialists may be introduced for biological control of some invasive plants (Siemann et al. 2006; Bezemer et al. 2014; Gruntman et al. 2017). Thus, experiments at multiple temporal and spatial scales may help to better understand the ecological and evolutionary processes of invasive plants both above- and belowground.

Acknowledgements We thank Takayuki Ohgushi, Susanne Wurst, and Scott Johnson for the invitation to contribute to the book "Aboveground and Belowground Community Ecology." We are grateful for comments by Takayuki Ohgushi and two anonymous referees that improved the early version of this manuscript. This work was supported by National Natural Science Foundation of China (31470447 to WH, 31370404 to JD), the Youth Innovation Promotion Association of the Chinese Academy of Sciences (Y329351H03 to WH), and the Foreign Visiting Professorship of Chinese Academy of Sciences (2015VBA025 to ES).

References

Agrawal AA (2005) Future directions in the study of induced plant responses to herbivory. Entomol Exp Appl 115:97–105
Agrawal AA (2011) Current trends in the evolutionary ecology of plant defence. Funct Ecol 25:420–432
Agrawal AA, Karban R (1999) Why induced defenses may be favored over constitutive strategies in plants. In: Tollrian R, Harvell CD (eds) The ecology and evolution of inducible defenses. Princeton University Press, Princeton, pp 45–61
Agrawal AA, Lau JA, Hambäck PA (2006) Community heterogeneity and the evolution of interactions between plants and insect herbivores. Q Rev Biol 81:349–376
Agrawal AA, Hastings AP, Johnson MTJ et al (2012) Insect herbivores drive real-time ecological and evolutionary change in plant populations. Science 338:113–116
Ali JG, Agrawal AA (2012) Specialist versus generalist insect herbivores and plant defense. Trends Plant Sci 17:293–302
Ando Y, Utsumi S, Ohgushi T (2010) Community structure of insect herbivores on introduced and native *Solidago* plants in Japan. Entomol Exp Appl 136:174–183
Arimura G, Kost C, Boland W (2005) Herbivore-induced, indirect plant defences. BBA-Mol Cell Biol L 1734:91–111
Arredondo JT, Johnson DA (2011) Allometry of root branching and its relationship to root morphological and functional traits in three range grasses. J Exp Bot 62:5581–5594
Bezemer TM, Harvey JA, Cronin JT (2014) Response of native insect communities to invasive plants. Annu Rev Entomol 59:119–141
Biere A, Goverse A (2016) Plant-mediated systemic interactions between pathogens, parasitic nematodes, and herbivores above- and belowground. Annu Rev Phytopathol 54:499–527
Blossey B (1993) Herbivory below ground and biological weed control: life history of a root-boring weevil on *Purple loosestrife*. Oecologia 94:380–387
Blossey B, Hunt-Joshi TR (2003) Belowground herbivory by insects: influence on plants and aboveground herbivores. Annu Rev Entomol 48:521–547
Blossey B, Nötzold R (1995) Evolution of increased competitive ability in invasive nonindigenous plants: a hypothesis. J Ecol 83:887–889
Blumenthal DM, Hufbauer RA (2007) Increased plant size in exotic populations: a common-garden test with 14 invasive species. Ecology 88:2758–2765
Bossdorf O, Schroder S, Prati D et al (2004) Palatability and tolerance to simulated herbivory in native and introduced populations of *Alliaria petiolata* (Brassicaceae). Am J Bot 91:856–862
Bossdorf O, Auge H, Lafuma L et al (2005) Phenotypic and genetic differentiation between native and introduced plant populations. Oecologia 144:1–11
Burghardt KT, Tallamy DW (2013) Plant origin asymmetrically impacts feeding guilds and life stages driving community structure of herbivorous arthropods. Divers Distrib 19:1553–1565
Caño L, Escarré J, Vrieling K et al (2009) Palatability to a generalist herbivore, defence and growth of invasive and native *Senecio* species: testing the evolution of increased competitive ability hypothesis. Oecologia 159:95–106
Carrillo J, Siemann E (2016) A native plant competitor mediates the impact of above- and belowground damage on an invasive tree. Ecol Appl 26:2060–2071
Carrillo J, Wang Y, Ding J et al (2012a) Decreased indirect defense in the invasive tree, *Triadica sebifera*. Plant Ecol 213:945–954
Carrillo J, Wang Y, Ding J et al (2012b) Induction of extrafloral nectar depends on herbivore type in invasive and native Chinese tallow seedlings. Basic Appl Ecol 13:449–457
Carrillo J, McDermott D, Siemann E (2014) Loss of specificity: native but not invasive populations of *Triadica sebifera* vary in tolerance to different herbivores. Oecologia 174:863–871
Chun YJ, van Kleunen M, Dawson W (2010) The role of enemy release, tolerance and resistance in plant invasions: linking damage to performance. Ecol Lett 13:937–946

Cipollini D, Heil M (2010) Costs and benefits of induced resistance to herbivores and pathogens in plants. CAB Rev Perspect Agric Vet Sci Nutr Nat Resour 5:1–25

Cipollini D, Mbagwu J, Barto K et al (2005) Expression of constitutive and inducible chemical defenses in native and invasive populations of *Alliaria petiolata*. J Chem Ecol 31:1255–1267

Cripps MG, Schwarzländer M, McKenney JL et al (2006) Biogeographical comparison of the arthropod herbivore communities associated with *Lepidium draba* in its native, expanded and introduced ranges. J Biogeogr 33:2107–2119

Davis DR, Fox MS, Hazen RF (2013) Systematics and biology of *Caloptilia triadicae* (Lepidoptera: Gracillariidae), a new species of leaf-mining moth of the invasive Chinese tallow tree (*Triadica sebifera* (L.) Euphorbiaceae). J Lepid Soc 67:281–290

Dawson W, Schrama M (2016) Identifying the role of soil microbes in plant invasions. J Ecol 104:1211–1218

Doorduin L, Vrieling K (2011) A review of the phytochemical support for the shifting defence hypothesis. Phytochemistry 10:99–106

Eigenbrode S, Andreas J, Cripps M et al (2008) Induced chemical defenses in invasive plants: a case study with *Cynoglossum officinale* L. Biol Invasions 10:1373–1379

Elton CS (1958) The ecology of invasions by animals and plants. Methuen, London

Erb M, Ton J, Degenhardt J et al (2008) Interactions between arthropod-induced aboveground and belowground defenses in plants. Plant Physiol 146:867–874

Erb M, Flors V, Karlen D et al (2009) Signal signature of aboveground-induced resistance upon belowground herbivory in maize. Plant J 59:292–302

Erwin AC, Geber MA, Agrawal AA (2013) Specific impacts of two root herbivores and soil nutrients on plant performance and insect-insect interactions. Oikos 122:1746–1756

Erwin AC, Züst T, Ali JG et al (2014) Above-ground herbivory by red milkweed beetles facilitates above- and below-ground conspecific insects and reduces fruit production in common milkweed. J Ecol 102:1038–1047

Felker-Quinn E, Schweitzer JA, Bailey JK (2013) Meta-analysis reveals evolution in invasive plant species but little support for evolution of increased competitive ability (EICA). Ecol Evol 3:739–751

Feng YL, Lei YB, Wang RF et al (2009) Evolutionary tradeoffs for nitrogen allocation to photosynthesis versus cell walls in an invasive plant. P Natl Acad Sci USA 106:1853–1856

Fornoni J (2011) Ecological and evolutionary implications of plant tolerance to herbivory. Funct Ecol 25:399–407

Fukano Y, Yahara T (2012) Changes in defense of an alien plant *Ambrosia artemisiifolia* before and after the invasion of a native specialist enemy *Ophraella communa*. PLoS One 7:e49114

Fussmann GF, Loreau M, Abrams PA (2007) Eco-evolutionary dynamics of communities and ecosystems. Funct Ecol 21:465–477

Gard B, Bretagnolle F, Dessaint F et al (2013) Invasive and native populations of common ragweed exhibit strong tolerance to foliar damage. Basic Appl Ecol 14:28–35

Gerber E, Hinz HL, Blossey B (2007) Impact of the belowground herbivore and potential biological control agent, *Ceutorhynchus scrobicollis*, on *Alliaria petiolata* performance. Biol Control 42:355–364

Gruntman M, Segev U, Glauser G et al (2017) Evolution of plant defences along an invasion chronosequence: defence is lost due to enemy release – but not forever. J Ecol 105:255–264

Harvey JA, Bukovinszky T, van der Putten WH (2010) Interactions between invasive plants and insect herbivores: a plea for a multitrophic perspective. Biol Conserv 143:2251–2259

Heil M (2008) Indirect defence via tritrophic interactions. New Phytol 178:41–61

Heil M (2011) Nectar: generation, regulation and ecological functions. Trends Plant Sci 16:191–200

Heil M, Karban R (2010) Explaining evolution of plant communication by airborne signals. Trends Ecol Evol 25:137–144

Hendry AP (2016) Eco-evolutionary dynamics. Princeton University Press, Princeton

Hol WHG, Macel M, van Veen JA et al (2004) Root damage and aboveground herbivory change concentration and composition of pyrrolizidine alkaloids of *Senecio jacobaea*. Basic Appl Ecol 5:253–260
Huang W, Ding J (2016) Effects of generalist herbivory on resistance and resource allocation by the invasive plant, *Phytolacca americana*. Insect Sci 23:191–199
Huang W, Siemann E, Wheeler GS et al (2010) Resource allocation to defence and growth are driven by different responses to generalist and specialist herbivory in an invasive plant. J Ecol 98:1157–1167
Huang W, Wheeler GS, Purcell MF et al (2011) The host range and impact of *Bikasha collaris* (Coleoptera: Chrysomelidae), a promising candidate agent for biological control of Chinese tallow, *Triadica sebifera* (Euphorbiaceae) in the United States. Biol Control 56:230–238
Huang W, Carrillo J, Ding J et al (2012a) Interactive effects of herbivory and competition intensity determine invasive plant performance. Oecologia 170:373–382
Huang W, Carrillo J, Ding J et al (2012b) Invader partitions ecological and evolutionary responses to above- and belowground herbivory. Ecology 93:2343–2352
Huang W, Siemann E, Yang X et al (2013) Facilitation and inhibition: changes in plant nitrogen and secondary metabolites mediate interactions between above-ground and below-ground herbivores. Proc R Soc Lond B Biol Sci 280:20131318
Huang W, Siemann E, Xiao L et al (2014) Species-specific defence responses facilitate conspecifics and inhibit heterospecifics in above-belowground herbivore interactions. Nat Commun 5:4851
Huang W, Siemann E, Carrillo J et al (2015) Below-ground herbivory limits induction of extrafloral nectar by above-ground herbivores. Ann Bot 115:841–846
Huang W, Robert CAM, Hervé MR et al (2017) A mechanism for sequence specificity in plant-mediated interactions between herbivores. New Phytol 214:169–179
Huber M, Bont Z, Fricke J et al (2016a) A below-ground herbivore shapes root defensive chemistry in natural plant populations. Proc R Soc Lond B Biol Sci 283:20160285
Huber M, Epping J, Schulze Gronover C et al (2016b) A latex metabolite benefits plant fitness under root herbivore attack. PLoS Biol 14:e1002332
Izaguirre MM, Mazza CA, Astigueta MS et al (2013) No time for candy: passionfruit (*Passiflora edulis*) plants down-regulate damage-induced extra floral nectar production in response to light signals of competition. Oecologia 173:213–221
Johnson SN, Rasmann S (2015) Root-feeding insects and their interactions with organisms in the rhizosphere. Annu Rev Entomol 60:517–535
Johnson SN, Hawes C, Karley AJ (2009) Reappraising the role of plant nutrients as mediators of interactions between root- and foliar-feeding insects. Funct Ecol 23:699–706
Johnson SN, Clark KE, Hartley SE et al (2012) Aboveground-belowground herbivore interactions. A meta-analysis. Ecology 93:2208–2215
Jones IM, Koptur S (2015) Quantity over quality: light intensity, but not red/far-red ratio, affects extrafloral nectar production in *Senna mexicana* var. *chapmanii*. Ecol Evol 5:4108–4114
Joshi S, Tielbörger K (2012) Response to enemies in the invasive plant *Lythrum salicaria* is genetically determined. Ann Bot 110:1403–1410
Joshi J, Vrieling K (2005) The enemy release and EICA hypothesis revisited: Incorporating the fundamental difference between specialist and generalist herbivores. Ecol Lett 8:704–714
Kafle D, Krähmer A, Naumann A et al (2014) Genetic variation of the host plant species matters for interactions with above- and belowground herbivores. Insects 5:651–667
Kant MR, Jonckheere W, Knegt B et al (2015) Mechanisms and ecological consequences of plant defence induction and suppression in herbivore communities. Ann Bot 115:1015–1051
Kaplan I, Halitschke R, Kessler A et al (2008a) Physiological integration of roots and shoots in plant defense strategies links above- and belowground herbivory. Ecol Lett 11:841–851
Kaplan I, Halitschke R, Kessler A et al (2008b) Constitutive and induced defenses to herbivory in above- and belowground plant tissues. Ecology 89:392–406
Karban R (2011) The ecology and evolution of induced resistance against herbivores. Funct Ecol 25:339–347

Karban R, Myers JH (1989) Induced plant-responses to herbivory. Annu Rev Ecol Syst 20:331–348
Karban R, Yang LH, Edwards KF (2014) Volatile communication between plants that affects herbivory: a meta-analysis. Ecol Lett 17:44–52
Keane RM, Crawley MJ (2002) Exotic plant invasions and the enemy release hypothesis. Trends Ecol Evol 17:164–170
Kempel A, Schädler M, Chrobock T et al (2011) Tradeoffs associated with constitutive and induced plant resistance against herbivory. P Natl Acad Sci USA 108:5685–5689
Kessler A, Heil M (2011) The multiple faces of indirect defences and their agents of natural selection. Funct Ecol 25:348–357
Knochel D, Monson N, Seastedt T (2010) Additive effects of aboveground and belowground herbivores on the dominance of spotted knapweed (*Centaurea stoebe*). Oecologia 164:701–712
Lankau RA (2007) Specialist and generalist herbivores exert opposing selection on a chemical defense. New Phytol 175:176–184
Li X, Guo W, Siemann E et al (2016) Plant genotypes affect aboveground and belowground herbivore interactions by changing chemical defense. Oecologia 182:1107–1115
Liao Z-Y, Zhang R, Barclay GF et al (2013) Differences in competitive ability between plants from nonnative and native populations of a tropical invader relates to adaptive responses in abiotic and biotic environments. PLoS One 8:e71767
Lin T, Doorduin L, Temme A et al (2015a) Enemies lost: parallel evolution in structural defense and tolerance to herbivory of invasive *Jacobaea vulgaris*. Biol Invasions 17:2339–2355
Lin T, Klinkhamer PGL, Vrieling K (2015b) Parallel evolution in an invasive plant: effect of herbivores on competitive ability and regrowth of *Jacobaea vulgaris*. Ecol Lett 18:668–676
Maron JL, Vilà M (2001) When do herbivores affect plant invasion? Evidence for the natural enemies and biotic resistance hypotheses. Oikos 95:361–373
Mathur V, Wagenaar R, Caissard J-C et al (2013) A novel indirect defence in Brassicaceae: structure and function of extrafloral nectaries in *Brassica juncea*. Plant Cell Environ 36:528–541
Meyer G, Clare R, Weber E (2005) An experimental test of the evolution of increased competitive ability hypothesis in goldenrod, *Solidago gigantea*. Oecologia 144:299–307
Müller-Schärer H, Schaffner U, Steinger T (2004) Evolution in invasive plants: implications for biological control. Trends Ecol Evol 19:417–422
Mundim FM, Alborn HT, Vieira-Neto EHM et al (2017) A whole-plant perspective reveals unexpected impacts of above- and belowground herbivores on plant growth and defense. Ecology 98:70–78
Núñez-Farfán J, Fornoni J, Valverde PL (2007) The evolution of resistance and tolerance to herbivores. Annu Rev Ecol Evol Syst 38:541–566
Oduor AMO, Lankau RA, Strauss SY et al (2011) Introduced *Brassica nigra* populations exhibit greater growth and herbivore resistance but less tolerance than native populations in the native range. New Phytol 191:536–544
Ohgushi T (2016) Eco-evolutionary dynamics of plant–herbivore communities: incorporating plant phenotypic plasticity. Curr Opin Insect Sci 14:40–45
Orians CM, Ward D (2010) Evolution of plant defenses in nonindigenous environments. Annu Rev Entomol 55:439–459
Ostonen I, Püttsepp Ü, Biel C et al (2007) Specific root length as an indicator of environmental change. Plant Biosyst 141:426–442
Parker JD, Hay ME (2005) Biotic resistance to plant invasions? Native herbivores prefer non-native plants. Ecol Lett 8:959–967
Pattison RR, Mack RN (2008) Potential distribution of the invasive tree *Triadica sebifera* (Euphorbiaceae) in the United States: evaluating CLIMEX predictions with field trials. Glob Chang Biol 14:813–826
Pearse IS, Altermatt F (2013) Predicting novel trophic interactions in a non-native world. Ecol Lett 16:1088–1094
Pelletier F, Garant D, Hendry AP (2009) Eco-evolutionary dynamics. Philos Trans R Soc B 364:1483–1489

Pierre PS, Dugravot S, Cortesero A-M et al (2012) Broccoli and turnip plants display contrasting responses to belowground induction by *Delia radicum* infestation and phytohormone applications. Phytochemistry 73:42–50

Poelman EH, Kessler A (2016) Keystone herbivores and the evolution of plant defenses. Trends Plant Sci 21:477–485

Poelman EH, Gols R, Snoeren TAL et al (2011) Indirect plant-mediated interactions among parasitoid larvae. Ecol Lett 14:670–676

Postma JA, Schurr U, Fiorani F (2014) Dynamic root growth and architecture responses to limiting nutrient availability: linking physiological models and experimentation. Biotechnol Adv 32:53–65

Prior KM, Powell THQ, Joseph AL et al (2015) Insights from community ecology into the role of enemy release in causing invasion success: the importance of native enemy effects. Biol Invasions 17:1283–1297

Rasmann S, Agrawal AA (2008) In defense of roots: a research agenda for studying plant resistance to belowground herbivory. Plant Physiol 146:875–880

Rasmann S, Turlings TCJ (2007) Simultaneous feeding by aboveground and belowground herbivores attenuates plant-mediated attraction of their respective natural enemies. Ecol Lett 10:926–936

Robert CAM, Erb M, Duployer M et al (2012) Herbivore-induced plant volatiles mediate host selection by a root herbivore. New Phytol 194:1061–1069

Rogers WE, Siemann E (2005) Herbivory tolerance and compensatory differences in native and invasive ecotypes of Chinese tallow tree (*Sapium sebiferum*). Plant Ecol 181:57–68

Rogers WE, Siemann E, Lankau RA (2003) Damage induced production of extrafloral nectaries in native and invasive seedlings of Chinese tallow tree (*Sapium sebiferum*). Am Midl Nat 149:413–417

Sakata Y, Yamasaki M, Isagi Y et al (2014) An exotic herbivorous insect drives the evolution of resistance in the exotic perennial herb *Solidago altissima*. Ecology 95:2569–2578

Siemann E, Rogers WE (2001) Genetic differences in growth of an invasive tree species. Ecol Lett 4:514–518

Siemann E, Rogers WE (2003a) Herbivory, disease, recruitment limitation, and success of alien and native tree species. Ecology 84:1489–1505

Siemann E, Rogers WE (2003b) Increased competitive ability of an invasive tree may be limited by an invasive beetle. Ecol Appl 13:1503–1507

Siemann E, Rogers WE (2003c) Reduced resistance of invasive varieties of the alien tree *Sapium sebiferum* to a generalist herbivore. Oecologia 135:451–457

Siemann E, Rogers WE, DeWalt SJ (2006) Rapid adaptation of insect herbivores to an invasive plant. Proc R Soc Lond B Biol Sci 273:2763–2769

Siemann E, DeWalt SJ, Zou J et al (2017) An experimental test of the EICA hypothesis in multiple ranges: invasive populations outperform those from the native range independent of insect herbivore suppression. Ann Bot Plants plw087

Singh A, Braun J, Decker E et al (2014) Plant genetic variation mediates an indirect ecological effect between belowground earthworms and aboveground aphids. BMC Ecol 14:25

Soler R, Harvey JA, Kamp AFD et al (2007) Root herbivores influence the behaviour of an aboveground parasitoid through changes in plant-volatile signals. Oikos 116:367–376

Soler R, Erb M, Kaplan I (2013) Long distance root-shoot signalling in plant-insect community interactions. Trends Plant Sci 18:149–156

Strauss SY, Agrawal AA (1999) The ecology and evolution of plant tolerance to herbivory. Trends Ecol Evol 14:179–185

Tanner RA, Varia S, Eschen R et al (2013) Impacts of an invasive non-native annual weed, *Impatiens glandulifera*, on above- and below-ground invertebrate communities in the United Kingdom. PLoS One 8:e67271

Uesugi A, Kessler A (2016) Herbivore release drives parallel patterns of evolutionary divergence in invasive plant phenotypes. J Ecol 104:876–886

Utsumi S (2011) Eco-evolutionary dynamics in herbivorous insect communities mediated by induced plant responses. Popul Ecol 53:23–34
van Dam NM (2009) Belowground herbivory and plant defenses. Annu Rev Ecol Evol Syst 40:373–391
van der Meijden E (1996) Plant defence, an evolutionary dilemma: contrasting effects of (specialist and generalist) herbivores and natural enemies. Entomol Exp Appl 80:307–310
van der Putten WH, Bardgett RD, de Ruiter PC et al (2009) Empirical and theoretical challenges in aboveground – belowground ecology. Oecologia 161:1–14
van Driesche RG, Carruthers RI, Center T et al (2010) Classical biological control for the protection of natural ecosystems. Biol Control 54(Suppl 1):S2–S33
van Geem M, Gols R, van Dam NM et al (2013) The importance of aboveground-belowground interactions on the evolution and maintenance of variation in plant defence traits. Front Plant Sci 4:431
Vestergård M, Rønn R, Ekelund F (2015) Above-belowground interactions govern the course and impact of biological invasions. AoB Plants 7:plv025
Wang Y, Huang W, Siemann E et al (2011) Lower resistance and higher tolerance of host plants: biocontrol agents reach high densities but exert weak control. Ecol Appl 21:729–738
Wang Y, Siemann E, Wheeler GS et al (2012) Genetic variation in anti-herbivore chemical defences in an invasive plant. J Ecol 100:894–904
Wang Y, Carrillo J, Siemann E et al (2013) Specificity of extrafloral nectar induction by herbivores differs among native and invasive populations of tallow tree. Ann Bot 112:751–756
Wurst S, Dam N, Monroy F et al (2008) Intraspecific variation in plant defense alters effects of root herbivores on leaf chemistry and aboveground herbivore damage. J Chem Ecol 34:1360–1367
Zhang KD, Lin YT (1994) Chinese tallow. China Forestry Press, Beijing (In Chinese)
Zheng H, Wu Y, Ding J et al (2005) Invasive plants established in the United States that are found in Asia and their associated natural enemies. Forest Health Technology Enterprise Team, Morgantown
Zou JW, Rogers WE, Siemann E (2008a) Increased competitive ability and herbivory tolerance in the invasive plant *Sapium sebiferum*. Biol Invasions 10:291–302
Zou JW, Siemann E, Rogers WE et al (2008b) Decreased resistance and increased tolerance to native herbivores of the invasive plant *Sapium sebiferum*. Ecography 31:663–671
Züst T, Heichinger C, Grossniklaus U et al (2012) Natural enemies drive geographic variation in plant defenses. Science 338:116–119

Chapter 13
Soil Biota as Drivers of Plant Community Assembly

Paul Kardol, Jonathan R. De Long, and Pierre Mariotte

13.1 Introduction

Soil organisms affect ecosystem processes, such as carbon (C) and nutrient cycling (Nielsen et al. 2011; De Graaff et al. 2015), plant productivity (van der Heijden et al. 1998; Wagg et al. 2011), and the response of ecosystems to climate change and increased atmospheric nitrogen (N) deposition (de Vries and Shade 2013), through their influences on plants (Wardle et al. 2004). The effects of soil biota on plant growth, performance, and productivity are relatively well studied (e.g., Bradford et al. 2002; Kulmatiski et al. 2014), but less is known about how soil biota drive plant community assembly (Bennett and Cahill 2016; Sikes et al. 2016), namely, how local assemblages are constructed from the regional species pool. This knowledge is crucial in predicting long-term and large-scale ecosystem processes and how plant communities reassemble in response to environmental changes (Kardol et al. 2013).

P. Kardol (✉)
Department of Forest Ecology and Management, Swedish University of Agricultural Sciences, Umeå, Sweden
e-mail: Paul.Kardol@slu.se

J. R. De Long
Department of Terrestrial Ecology, Netherlands Institute of Ecology, Wageningen, The Netherlands
e-mail: J.DeLong@nioo.knaw.nl

P. Mariotte
Ecole Polytechnique Fédérale de Lausanne (EPFL), School of Architecture, Civil and Environmental Engineering (ENAC), Laboratory of Ecological Systems (ECOS), Lausanne, Switzerland

Swiss Federal Institute for Forest, Snow and Landscape Research (WSL), Lausanne, Switzerland

T. Ohgushi et al. (eds.), *Aboveground–Belowground Community Ecology*, Ecological Studies 234, https://doi.org/10.1007/978-3-319-91614-9_13

Soil biota can affect plants both directly and indirectly (Wardle et al. 2004). Direct effects can result from the accumulation of pathogens and root herbivores (e.g., van der Putten et al. 1993) and from associations with mutualistic soil organisms such as mycorrhizal fungi or N-fixing bacteria (e.g., Teste et al. 2017). Indirect effects can result from the impacts of soil organisms on soil properties and nutrient dynamics (e.g., Zhang et al. 2016). In this chapter, we discuss how these direct and indirect effects of soil biota modify how plant communities assemble, from local-scale priority effects to broad-scale vegetation dynamics. We discuss soil biota-mediated *niche modification* and *soil legacy effects* and how these forces can drive plant community assembly by affecting (1) plant-competitive and facilitative interactions, (2) species replacements, and (3) patterns of coexistence and community diversity. Finally, we discuss primary and secondary succession, biodiversity conservation, plant invasion, and how current and projected global environmental changes may alter the impacts of soil biota on plant community assembly.

13.2 Niche Modification, Priority Effects, and Soil Legacies

Soil organisms can sway the selection of species able to coexist in the community through niche modification, such as changes in soil properties and microclimate and associations with plant roots that favor or suppress certain species in the plant community. Figure 13.1 depicts how different functional groups of soil organisms and soil biodiversity affect plant community composition through the selective assembly of species.

Soil bioturbators, such as earthworms and termites, can affect soil properties directly by altering the physical structure of the soil environment, such as soil porosity (Lavelle et al. 1997). These physical alterations influence the movement of water though the soil profile and often result in increased soil N availability (van Groenigen et al. 2014). As such, bioturbators can stimulate plant growth, but their effects are species-specific, favoring resource-acquisitive species that are the strongest competitors for N (Wurst et al. 2005) (Fig. 13.1a). Furthermore, saprotrophic soil organisms change soil nutrient availability by decomposing and mineralizing dead plant-, animal-, and microbe-derived organic matter, making nutrients available again for plant uptake (Wardle et al. 2004). In general, saprotrophs stimulate plant growth, allowing more individuals to coexist (Fig. 13.1b). However, an increase in the ratio between fungal and bacterial energy channels in the decomposer food web is typically associated with reduced nutrient cycling and low rates of nutrient supply to plants (Wardle et al. 2004), reducing the total number of individual plants and increasing the competitive ability of slow-growing, resource-conservative species over fast-growing, nutrient-acquisitive species within the plant community (Fig. 13.1c).

Plant species vary in their susceptibility to root pathogens and herbivores (Brown and Gange 1990). Fast-growing species are often more susceptible than slow-growing species because of trade-offs between investment in growth and

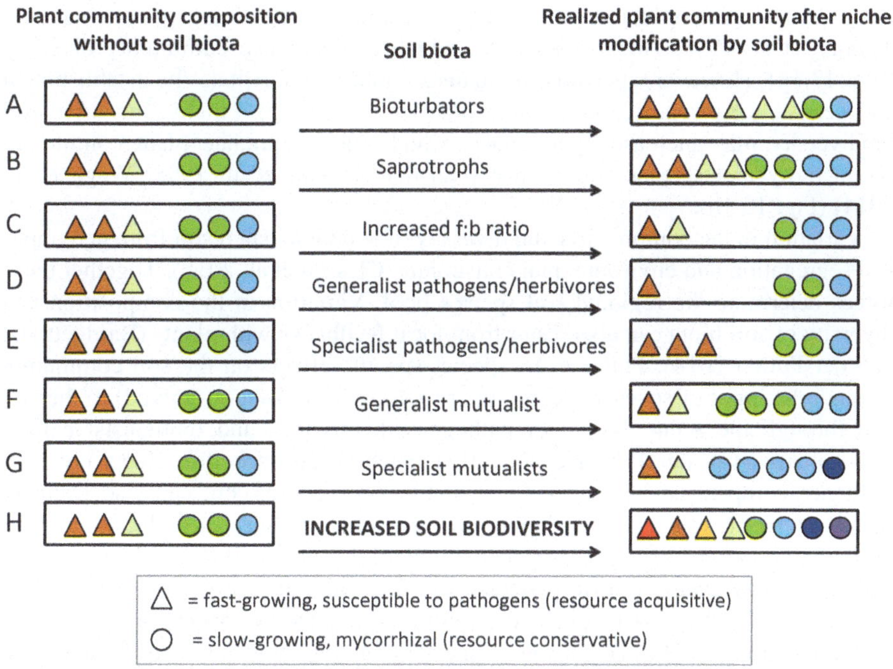

Fig. 13.1 Hypotheses on how soil organisms may modify niches and affect plant community assembly through niche modification. Shown are hypothetical plant communities where each symbol represents one individual plant. Triangles represent fast-growing, resource-acquisitive species which are susceptible to root herbivores and pathogens. Circles represent slow-growing, resource-conservative species which are less susceptible to root pathogens but may more strongly depend on root mutualists such as mycorrhizal fungi. Different colors denote different species. Soil biota can shift community density or productivity as indicated by the number of individuals present in the realized community and by shifting species composition. (**a**) Bioturbation increases soil N availability, which promotes fast-growing plant species at the expense of slow-growing species. (**b**) By decomposing litter and soil organic matter, saprotrophic organisms increase nutrient availability (i.e., increased niche size), which allow more individuals to coexist. (**c**) An increase in the soil fungal to bacteria ratio slows down nutrient cycling, which both constrains the number species able to coexist and reduces the performance of nutrient-demanding resource-acquisitive plant species. (**d–e**) Pathogens tend to reduce the abundance of fast-growing species while mutualists (**f–g**) often promote the abundance of slow-growing species. (**h**) Overall, the increase in soil biota diversity favors species coexistence and plant community diversity

reproduction versus investment in defense mechanisms and longevity, respectively (Reynolds et al. 2003). Hence, generalist pathogens and herbivores would reduce available niche space with stronger consequences for fast-growing than for slow-growing species (Fig. 13.1d). Specialist root pathogens and herbivores, such as certain plant-feeding nematodes, may cause local species extinction, which would release the remaining species from competition (Fig. 13.1e). Generalist mutualists enhance plant community resource utilization, thereby increasing the number of individuals and species able to coexist and favoring those species associating with the mutualists (Fig. 13.1f). In contrast, specialist mutualists selectively benefit their host species, gaining them competitive advantage over other species (van der Heijden et al. 1998; Hoeksema 2005), which could reduce community diversity (Fig. 13.1g). However, specialist mutualists could also facilitate the establishment and persistence of species that would otherwise be absent. Finally, increased soil biodiversity may increase niche space, which could stimulate plant community productivity and species coexistence (van der Heijden et al. 1998; Wagg et al. 2014) (Fig. 13.1h).

Variation in the composition and diversity of soil biota can result from geographical segregation and environmental constraints (Classen et al. 2015). Together these forces determine the regional soil species pool. Variation in the composition and diversity of soil biota can also result from local feedbacks with plants (Bardgett and van der Putten 2014). Plants exert species-specific effects on the soil community through the input of species-specific litter (both root- and shoot-derived) to the soil and through the accumulation of pathogens, herbivores, and mutualists in their rhizospheres (van der Putten et al. 2013). Such species-specific effects not only influence competitive interactions and dominance patterns, but can also result in persistent biotic soil legacies, thereby driving temporal species replacements and possibly longer-term community dynamics (Kardol et al. 2007) (Fig. 13.2). As such, soil biota may cause priority effects, where plant-competitive interactions and community dynamics depend on the timing and order of species arrival (Chase 2003; Fukami 2015). Early-arriving species can preempt and modify available niches for later-arriving species through their effects on soil biota, causing long-lasting historical contingency effects (Kardol et al. 2007; van de Voorde et al. 2011) (Fig. 13.2). For example, plant species-specific effects on the soil microbial community may carry over to affect the relative abundance of plant species at a later stage (Kardol et al. 2007). Along the same lines, mycorrhizal fungi can modify the priority effects of early-arriving species on plant community diversity and community composition (Burkle and Belote 2015). Moreover, root and shoot herbivores may directly and indirectly affect soil microbial communities, which in turn can create soil legacies for the next generation of plants (Kostenko et al. 2012; Sonnemann et al. 2013).

Can we scale species-specific effects up to the ecosystem level? It is not well understood how the above-described priority and historical contingency effects resulting from plant interactions with soil biota persist in longer-term plant community dynamics and whether historical contingency effects would be amplified over time (Fukami and Nakajima 2013; Fukami 2015). However, feedbacks with soil biota have been shown to explain relative species abundance in tropical forests (Mangan et al. 2010) and grasslands (Klironomos 2002) and maintain the high diversity in Mediterranean-climate shrublands (Teste et al. 2017) (Fig. 13.3). These feedback effects are often explained by soil-borne pathogens promoting local plant species coexistence and diversity via conspecific negative density dependence (Bever et al. 2015). Studies have shown how local soil pathogen abundance can explain the distribution of neighboring plant species (Gómez-Aparicio et al. 2012) and the assembly of plant communities across large spatial scales (Blomqvist et al. 2000). Other studies have shown that soil biota can affect plant communities

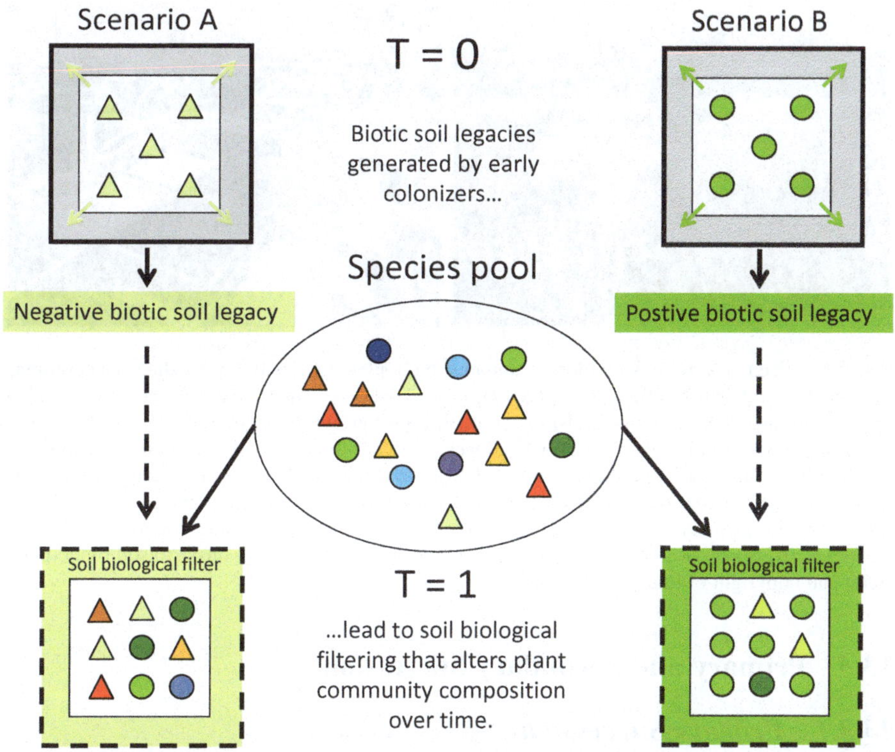

Fig. 13.2 Hypotheses on how soil biota may affect temporal plant community assembly through plant–soil feedback and the provisioning of biotic soil legacies. Shown are hypothetical plant communities where each symbol represents one individual plant. Triangles represent fast-growing, resource-acquisitive species which are susceptible to root herbivores and pathogens. Circles represent slow-growing, resource-conservative species which are less susceptible to root pathogens, but may more strongly depend on root mutualists such as mycorrhizal fungi. Scenario A indicates negative plant–soil feedback for the early-arriving species ($t = 0$), creating more available niche space for the establishment of other, heterospecific individuals at the later successional stage ($t = 1$). Scenario B, on the other hand, indicates positive plant–soil feedback for the early-arriving species ($t = 0$), creating more available niche space for conspecific individuals to establish at the later successional stage and, hence, less available niche space for heterospecific individuals ($t = 1$). Further, the species-specific effects of the early-arriving species on the soil biological communities determine which heterospecific species are able to establish at the later successional stage

through altering plant-facilitative interactions (Montesinos-Navarro et al. 2016). For example, common mycorrhizal networks can connect multiple plant species and promote the establishment of other plants near the original host plant (Moora and Zobel 2009). The effects of soil pathogens and mycorrhizal networks on neighboring plant species vary among plant species (van der Heijden and Horton 2009). In other words, the resident plant community creates a soil biological filter, which could drive larger-scale plant community dynamics.

Fig. 13.3 Feedback with soil biota maintains high plant diversity of Mediterranean-climate shrublands, as shown by Teste et al. (2017). In a glasshouse experiment, 16 co-occurring plant species with different nutrient-acquisition strategies were grown in the presence of soil biota from under their own or other plant species. Feedbacks with soil biota varied with nutrient-acquisition strategy, and computer simulations showed that these feedbacks were strong enough to promote species coexistence and maintain high plant community diversity. (**a**) Example of a hyperdiverse shrubland near Lesueur National Park, Western Australia (photo credit and copyright © Etienne Laliberté). (**b**) Impression of the glasshouse experiment testing plant responses to soil biota (photo credit and copyright © François P. Teste)

13.3 Primary and Secondary Succession

13.3.1 Primary Succession

Vascular plants are generally slow to establish during primary succession (Fig. 13.1) on newly exposed surfaces such as glacial moraine, cooled volcanic lava, mobile sand dunes, or mine tailings (e.g., Bardgett et al. 2007). Slow establishment cannot be explained solely by dispersal limitation. Harsh conditions and low soil N availability are among the primary constraints (Vitousek et al. 1993; Laliberté et al. 2012; Titus and del Moral 1998). Soil organisms can facilitate the initial stage of plant community assembly though niche modification. In primary succession, plant community assembly is generally preceded by a heterotrophic phase, i.e., aerial deposition of highly dispersive invertebrates and/or allochthonous detritus that provides the initial inputs of energy and nutrients (e.g., Hodkinson et al. 2002; Bardgett et al. 2007). Soil organisms, including microbes, testate amoebae, large invertebrates, are often windborne and may establish on newly exposed surfaces well before vascular plants (Hodkinson et al. 2002; Wanner et al. 2015), facilitated by allochthonous inputs of organic matter (Brankatschk et al. 2011).

Initial soil communities may conserve nutrients (particularly N) and as such facilitate the establishment of vascular plants. Further, the defecation and death of early-arriving soil animals initiates decomposition and thus gradual soil formation, further steering the trajectory of plant community assembly. For example, Bardgett et al. (2007) found that early-colonizing microbes were initially supported by "old" C and that only after more than 50 years of organic matter accumulation was the soil microbial community primarily supported by plant-derived C. Community assembly of saprotrophic microbes (Kazemi et al. 2016) and testate amoebae (Wanner et al. 2015) can be highly variable initially and converge later, likely as a response to plant colonization. Inputs of nutrients and organic matter during early ecosystem development are generally dominated by microbial C and N fixation (Schmidt et al. 2008). In addition, the biomass of the soil meso- and macrofauna can increase rapidly (Kaufmann 2001; Bokhorst et al. 2017) long before plant arrival and may further contribute to niche modification and diversification during primary succession.

In primary succession, soil nutrient concentrations are often low. Therefore, early-arriving plant species may depend on associations with soil organisms for capturing nutrients. For example, the earliest colonizing plants on infertile volcanic substrates are often legumes, in part because of their ability to fix atmospheric N_2 via symbiosis with rhizobia bacteria (Wang et al. 2012). The N fixed by these legumes is ultimately incorporated in the terrestrial N cycle. Since many legume species have poor long-distance dispersal mechanisms, (re-)colonization of volcanic substrates may proceed slowly (del Moral 1983). Once legumes have established, they may facilitate the establishment of other species through increased niche availability (Vitousek et al. 1987). During later stages of succession, rhizobia may further shape community assembly by influencing soil N cycling and, hence, plant resource competition. For example, it is known from mesocosms simulating early-successional plant communities that rhizobia specializing on the dominant plants can shift competitive interactions and community composition. Thereby, such specializations can negatively impact on community diversity through a combination of both inhibitory and facilitative effects (Keller 2014). Additionally, mycorrhizal fungi can also affect plant community assembly after volcanic eruptions. Initial plant colonization after volcanic eruption often includes plants without obligate dependency on mycorrhizal partnerships, but facultative species also colonize soon after (Obase et al. 2008). Mycorrhizal spores may be brought in by wind dispersal, and previous studies showed that facultative mycorrhizal species arrive and establish relatively early during primary succession (Allen et al. 1992; Nara 2006). Once established, short-distance dispersal occurs and the mycorrhizal network may expand, facilitating the establishment of other plant species that depend on mycorrhizae.

In foredune succession, soil-borne diseases and root herbivores modify plant-competitive interactions, feedbacks, and species replacements, driving initial plant community assembly (e.g., van der Putten et al. 1993; Brinkman et al. 2005). For example, in the Dutch foredunes, the first-arriving plant species, *Ammophila arenaria*, suffers from species-specific soil-borne diseases, giving room for succeeding species that are more tolerant of or immune to the soil-borne diseases of their predecessors (van der Putten et al. 1993). Root-feeding nematodes also affect plant community dynamics (Brinkman et al. 2015). For example, van der Stoel et al. (2002) used bioassays to demonstrate that the endoparasitic nematodes *Heterodera arenaria* and *Pratylenchus* spp. rapidly built up in the root zone of *A. arenaria* and coincided with a negative soil feedback. However, subsequent work by Brinkman et al. (2005) indicated that negative feedback generated by *A. arenaria* could not solely be explained by root-feeding nematodes and most likely were caused by more complex interactions among different groups of soil organisms.

13.3.2 Secondary Succession, Ecosystem Restoration, and Biodiversity Conservation

While primary succession starts from newly exposed surfaces, virtually lacking any soil life, secondary succession is the process of species colonization and replacement which begins with some biological legacy following an initial disturbance or the cessation of agricultural or forest management. The soil biotic legacy (Fig. 13.2) depends on the pre-disturbance condition or the former land use. For example, after abandonment of agricultural land, densities of soil pathogens and root herbivores may initially be high, and the saprotrophic microbial community is often dominated by bacteria rather than by fungi (i.e., a low F:B ratio, see Fig. 13.1). This is mostly because agricultural management negatively affects saprotrophic and mycorrhizal fungi (van der Wal et al. 2006) (see also Sect. 5.2). Similarly, other natural or human-induced disturbances such as mining, wildfire, wind throws, and clear-cuts affect soil communities. These disturbances affect some taxa or functional groups more than others, confronting early-arriving plant species with a distinct soil biotic legacy (e.g., Banning et al. 2011). For example, after coalmine reclamation, degraded mycorrhizal communities could delay or inhibit the establishment of later-successional species (Bauman et al. 2013).

After initial plant establishment, soil biota can further control succession through the influence of early-successional species on the soil community. In secondary succession on ex-arable land in the Netherlands, early-successional species create negative soil feedbacks through species-specific buildup of soil microbial pathogens. These feedbacks reduced the competitive strength of early-successional species and thus accelerated succession (Kardol et al. 2006). In addition, early-arriving species can have selective soil legacy effects on the growth of other plant species (Kardol et al. 2007; van de Voorde et al. 2011). Moreover, the legacy effects of early-successional grasses on mid-successional grasses through soil microbial community development were more pronounced than effects on forbs (Kardol et al. 2007). As such, soil biotic legacies can drive the direction in which a plant community develops. It is unclear how long plant species-specific soil legacies last (Wurst and Ohgushi 2015), but plant community dynamics at any stage of ecosystem development may in part reflect soil community influences from the past.

Soil biota affect the rate and direction of community assembly during secondary succession, but can also drive community diversity. In a greenhouse experiment, De Deyn et al. (2003) showed how invertebrate soil fauna from a series of secondary sward successional stages selectively suppressed fast-growing, early-successional dominant plant species, and thereby enhanced the relative abundances of subordinate species and species from later succession stages. These effects were most likely due to root-feeding nematodes and Elateridae larvae selectively feeding on the roots of otherwise dominant grasses. A follow-up experiment further suggested that root-feeding nematodes reduced the effects of nutrient supply on plant dominance, leading to an increase in plant community diversity (De Deyn et al. 2004). Additionally, plant diversity can also be enhanced through subordinate plants benefitting from associations with mycorrhizal fungi (Gange et al. 1990; de Leon et al. 2016). Kardol et al. (2006) showed that late successional plant species displayed positive feedbacks, particularly with soil organisms, most likely mycorrhizal fungi, from later successional stages. This suggests that the absence of appropriately compatible mycorrhizal spores after disturbances, such as an intensive agricultural land-use (see Sect. 13.5.2), could delay plant community succession.

Can the rate and direction of changes in plant community assembly be manipulated? Evidence for the importance of soil biota in successional plant communities (Kardol et al. 2007) and the notion that plant community development may be constrained by slow dispersal of soil mutualists (e.g., mycorrhizae) (Eschen et al. 2009; Zobel and Opik 2014) suggests that manipulation of soil biota could be used in restoration of degraded plant communities. Ecosystem restoration after human disturbances represents a special case of secondary succession where the initial assembly order of plant communities may determine whether or not communities develop towards the desired "target state." Here, plant community assembly can be assisted by artificially modifying species arrival order by sowing or planting species that would not naturally colonize the restoration site (Kardol et al. 2009; Burkle and Belote 2015). However, establishment of the desired species can depend on the absence or presence of soil biota (Hodkinson et al. 2002). Simultaneous introduction of plant propagules and soil organisms from a target site may enhance plant community succession (Middleton and Bever 2012) and, moreover, may be more effective than introduction of plant propagules or soil organisms alone.

Attempts to use soil inoculation to enhance ecosystem restoration have not always been successful (Hedlund and Gormsen 2002; Aprahamian et al. 2016), often because of a lack of compatibility in abiotic soil conditions between the "donor" and the "receptor" sites (Kardol et al. 2009). Some studies, however, have shown how soil inoculations can be used to assist the restoration of plant assemblages. In a greenhouse study, Carbajo et al. (2011) found, for example, that inoculation of soil from late-successional grassland into ex-arable land promoted the competitive ability of target plant species. Furthermore, there is evidence that these beneficial effects of late-successional soil may result from enhanced mycorrhizal associations. Indeed, in tallgrass prairies, Koziol and Bever (2015, 2016) found that late-successional plant species were more responsive to mycorrhizal inoculation than early-successional species and had a higher degree of specificity to different mycorrhizal species. In line with these findings, Helm and Carling (1993) demonstrated that soil transfer from intermediate and late-successional stages was effective in establishing plant-mycorrhizal associations in reforestation on abandoned mine lands. Similar results were found for the establishment of late-successional target species on ex-arable land (Eschen et al. 2009) and for seedling establishment in degraded drylands (Pineiro et al. 2013).

Fig. 13.4 Effects of inoculation with heathland soil on plant community assembly during secondary succession on ex-arable land in the Netherlands after top soil removal (Wubs et al. 2016). (**a**) Control plot without soil inoculation, and (**b**) Plot inoculated with heathland soil (photo credits and copyright © Jasper Wubs). (**c**) Aerial view of the experimental area: in the lighter areas, the top soil is removed. Within these areas, some darker parts are inoculated with heathland soil (Google Earth). All photos were taken approximately 7 years after set up of the experiment

However, detailed knowledge on which soil communities would promote a certain successional trajectory (Perring et al. 2015) is still lacking. Probably the best example so far of the manipulation of soil biota to control plant community development comes from a recent study by Wubs et al. (2016) (Fig. 13.4). They established a large-scale experiment on ex-arable land and tested for the effects of soil inocula from different origins (i.e., a grassland restored more than two decades ago or a dry heathland) on plant community development. Different origins of the soil inocula steered plant communities into the directions one would anticipate (i.e., towards grassland or heathland vegetation, respectively, depending on inocula origin) (Fig. 13.4).

Finally, what is the role of soil biota in the conservation of diverse plant communities? In Australian shrubland, Teste et al. (2017) found that plants of contrasting nutrient strategies [i.e., N-fixers, arbuscular mycorrhizal (AM), ectomycorrhizal (EM)] experienced different plant–soil feedbacks. These feedbacks were strong enough to at least partly explain the high levels of plant diversity observed in this system (Fig. 13.3). Additionally, Rodriguez-Echeverria et al. (2013) found that specific bacterial communities were associated with benefactor and beneficiary plants and that the positive effects of the soil biota on the beneficiary plants were independent of the benefactor. This shows that the legacy effect of a specific plant on soil biota can have far-reaching consequences. Essentially, specific groups of soil organisms can shape facilitation relationships between plants and might, therefore, promote community diversity. Furthermore, soil biota can play a role in modifying competitive interactions between plants in a way that can enhance diversity. For example, Pendergast et al. (2013) showed that in North American old fields, soil communities from contrasting origins determined whether or not intra- versus interspecific competition was stronger, all the while maintaining the dominance of the forb *Solidago canadensis*. This indicates that the diversity of subordinate species within a plant community may be maintained by complex biotic plant–soil feedback and competitive interactions. In line with this, Maron et al. (2016) used alpine grassland species to show that biotic plant–soil feedbacks were the likely cause of maintained diversity, while competition between species was less important in heterospecific soils versus conspecific soils. This could mean that in this alpine system, the soil community is a stronger player in maintaining plant community diversity than competition. In summary, there is a need to consider complex plant–soil biota interactions when developing programs to preserve plant biodiversity.

13.4 Plant Invasions and Native Range Expansions

13.4.1 Invasive Plants and Priority Effects via Soil Biota

Invasive plants cause ecological and economic damage (van Kleunen et al. 2015) and pose one of the largest threats to global biodiversity (McGeoch et al. 2010). One way that invasive plants might interact with the soil community to affect their own future dominance in a system is through soil-mediated priority effects (Fig. 13.2). Priority effects may involve competition between plants for resources (Grman and Suding 2010) or accumulation of allelochemicals in the soil (Bais et al. 2003). However, the advantage of enhanced competitive ability amongst invasive plants tends to diminish with increasing time since invasion (Gioria and Osborne 2014). For example, Lankau et al. (2009) found that in North America, the invader *Alliaria petiolata* produced chemicals that the local soil community is initially incapable of breaking down. However, with increasing time since invasion, the soil decomposer community adapted to cope with the chemical compounds produced by *A. petiolata*. This finding indicates that over time, the soil community might mitigate the allelopathic effects of an invader, allowing for the resurgence of native plants, subsequent compositional changes to the plant community, and further niche modification.

Soil legacy effects (Fig. 13.2) may in part explain why invasive plant dominance might persist. Such soil legacies can manifest in different ways. For example, Kuebbing et al. (2016) found that the nonnative woody shrubs *Lonicera maackii* and *Ligustrum sinense* generated significant (likely biotic) soil legacies. This led to decreased native plant community root biomass, possibly resulting in reduced presence of native plants in the community over time. In further support of this, Eppinga et al. (2006) found that the invasive grass *Ammophila arenaria* promoted the accumulation of native pathogens, which subsequently excluded native plant species. Moreover, Stinson et al. (2006) found that *A. petiolata* produced chemicals that disrupted the mutualistic relationship between tree seedlings and mycorrhizal fungi, essentially changing the types of plants that could establish. Exclusion of native tree seedlings due to changes in the soil mycorrhizal community could potentially alter the overstory plant community composition or even pave the way for the establishment of additional invaders, generating an "invasional meltdown" (Simberloff and Von Holle 1999; Simberloff 2006; O'Dowd et al. 2003; Green et al. 2011), although evidence for such invasional meltdowns involving plants and the soil community is rare (but see Paudel et al. 2016).

Alternatively, the relaxation of the soil biotic filter may contribute to plant invasion. The enemy release hypothesis suggests that invasive plants proliferate in new ranges because they have escaped above- and belowground pathogens from their home ranges (Maron and Vila 2001; Keane and Crawley 2002). For example, Gundale et al. (2014) found that the North American native *Pinus contorta* performed better when grown in soils from its nonnative range in Sweden than when grown in soils from its home range. These differences in growth were the result of the soil biotic community. Although *P. contorta* has yet to become invasive in Sweden, it is invasive in other nonnative ranges, such as New Zealand and Argentina (Dickie et al. 2014; Bravo-Monasterio et al. 2016). Therefore, there is concern that its highly competitive growth could eventually allow it to outcompete native species in Sweden, such as *Pinus sylvestris* and *Picea abies* (Despain 2001). This could result in an altered tree species community composition due to a cascade effect initiated by the soil biotic community, particularly if *P. contorta* invades areas above the tree line or grasslands that are naturally devoid of trees (Bravo-Monasterio et al. 2016). Essentially, invasion by *P. contorta* leads to the occupation of an unfulfilled niche, resultantly modifying the realized niche of other plant species (i.e., grasses, forbs) that used to dominate the plant community.

Maron et al.'s (2014) study supports the enemy release hypothesis at a regional scale. Six perennial European forb species exhibited no negative plant–soil feedbacks in their introduced North American range, while negative feedbacks tended to occur when they were grown in European soils. Soils were collected from across a wide regional scale, suggesting that these effects were generated by soil biota with a large spatial distribution. In another study, Diez et al. (2010) found in their survey of New Zealand invasive plants that with increasing time since invasion, the soil biotic community generated greater negative feedbacks. Although this study showed higher invasive plant abundance to be associated with increasingly negative soil feedbacks, such negative feedbacks may lead to suppression of the invaders.

Changes to the soil biotic filter wrought by invasive plants might not always be proportional or consistent; namely, benefits or detriments of the soil biotic filter might affect both native and invasive species, but the effects might not be equivalent within a group. In old fields in Tennessee, USA, Stuble and Souza (2016) found that the effects of early arrival were beneficial to both native and exotic plants, but the effects of later arrival were disproportionately negative for natives compared to exotics. Further, Sikes et al. (2016) found that the growth rates of the exotic grass *Melinis repens* and the native grass *Schizacharyium niveum* were negatively affected when each species was grown in soils with its "home" fungi. However, this negative priority effect was negated when other plants or fungi from different sites were introduced earlier, illustrating how the assembly of plant and soil communities interact. Conversely, Molina-Montenegro et al. (2015) showed that the Chilean alpine cushion plant *Laretia acaulis* promoted fungal endophytes that facilitated the survival of both native and exotic plant seedlings. These findings indicate that both negative and positive conditioning of soil biotic communities can occur. Even if initial facilitative relationships do not occur, over time invasive plant species might develop mutualistic relationships with belowground organisms such as mycorrhizae (Callaway et al. 2004) or decomposers (Liao et al. 2008). These examples demonstrate that invasive plants interact with the soil biota in ways that could change biotic filters, leading to their persistence (i.e., via filtering that favors invasive plants due to enhanced mutualisms) (Callaway et al. 2004) or decline (i.e., via filtering that disfavors invasive plants due to rapid evolution of soil pathogens) (Bardgett and van der Putten 2014).

What role do biotic communities play during restoration efforts that aim to eradicate invasive plants? Simply removing invasive plant species from an ecosystem may not be enough to affect restoration. The soil legacy effect of an invasive plant can alter the soil biota, which could make it impossible for native plants to reestablish. For example, in disturbed and pasture sites in Florida, USA, that were, respectively, invaded by the grasses *Melinis repens* and *Paspalum notatum*, Hamman and Hawkes (2013) found that restoration efforts were more successful when a soil inoculum from under native plants was added. However, the effect of soil biotic inoculum was site-specific, with sites that were more heavily invaded experiencing less successful restoration. Furthermore, removing *Pinus contorta* trees from a New Zealand forest left a legacy of EM fungi, which facilitated the establishment of another invasive tree, *Pseudotsuga menziesii*, and shifted the soil community to bacteria-based channels (Dickie et al. 2014). In line with these findings, Kulmatiski et al. (2006) discovered that invasive plants tend to dominate in abandoned agricultural fields in Washington, USA, due to beneficial fungal populations and faster nutrient cycling (likely due to changes in soil biota), with competition playing a relatively minor role in determining plant community composition.

13.4.2 Range Expansion

In addition to inter-continental plant invasion, global climate change and alterations to land use (for further details, see Sect. 13.5) have caused intra-continental plant range expansion (van der Putten et al. 2010). Range expansion of specific plant species can lead to alterations to the local plant community (van der Putten 2012) and plant–soil organism interactions (van der Putten et al. 2013). Plants that expand their ranges might benefit from lower accumulation of rhizosphere pathogens, resulting in reduced negative plant–soil feedback effects (Engelkes et al. 2008), similar to what invasive plants experience in line with the enemy release hypothesis. For example, Morriën and van der Putten (2013) found that range-expanding plants tended to harbor fewer *Fusarium* spp. (a genus of pathogenic fungi) in their rhizospheres compared to native congeners. Smaller negative soil effects could in part explain range expansion. However, certain plant species might expand into a range where they find a mismatch with the local soil community, given that soil organisms are presumably relatively poor dispersers (Classen et al. 2015). A literature survey by Berg et al. (2010) showed that most organisms across trophic groups appeared to have contrasting dispersal abilities, but little evidence has been obtained to determine if mismatches between plant and soil communities exist (van der Putten 2012).

The relationship between range-expanding plants and the soil biotic community is often not unidirectional. For example, top-down and bottom-up control of parasitic nematodes can vary at the interspecific level in range-expanding plants and their native congeners (Wilschut et al. 2016). Additionally, van Grunsven et al. (2007) found that range-expanding plants experienced less negative plant–soil feedbacks in their expanded range than in their native range. However, nutrient availability altered these effects. Therefore, broad conclusions may not be possible when predicting these interactions. Furthermore, range-expanding plants might fill a modified realized niche in their new range, leading to changes such as altered nutrient cycling or primary productivity (van der Putten 2012). Finally, although the soil community in a plant's new range can often control the success or failure of expansion, range-expanding native plants themselves can have both direct and indirect impacts on the soil microbial community. Collins et al. (2016) showed that the native sagebrush *Artemisia rothrockii* increased soil bacterial diversity both directly and indirectly via changes to soil organic N and C stocks. Changes to the microbial community as driven by a single range-expanding species could have far-reaching impacts on the assembly of the entire plant community over time that trump the effects of the original soil community. On the other hand, the loss of soil biota could result in plant range contraction, or in certain circumstances the presence of soil biota might even delay plant range contraction (Lankau et al. 2015). However, experimental evidence testing these hypotheses is rare.

13.5 Global Environmental Change

13.5.1 Climate Change

Climatic change is expected to modify the structure of plant communities. Numerous forms of climate perturbations, such as drought, warming, and elevated [CO_2] are occurring simultaneously, generating uncertainties in predicting the reassembly of plant communities. For example, increases in atmospheric CO_2 concentrations can accelerate succession by promoting the growth of late successional plant species and thus increasing their dominance in the community (Polley et al. 2003). By contrast, shifts in precipitation regimes can delay succession as disturbances, such as floods, drought, and fire often favor early successional communities and limit the arrival or the growth of late successional species (Laflower et al. 2016). Overall, the response of plant communities to climate change and the consequences for successional pathways has mostly been observed aboveground. Less emphasis has been put on soil microbial communities and their potential effects on plant communities under climate change (Fig. 13.5).

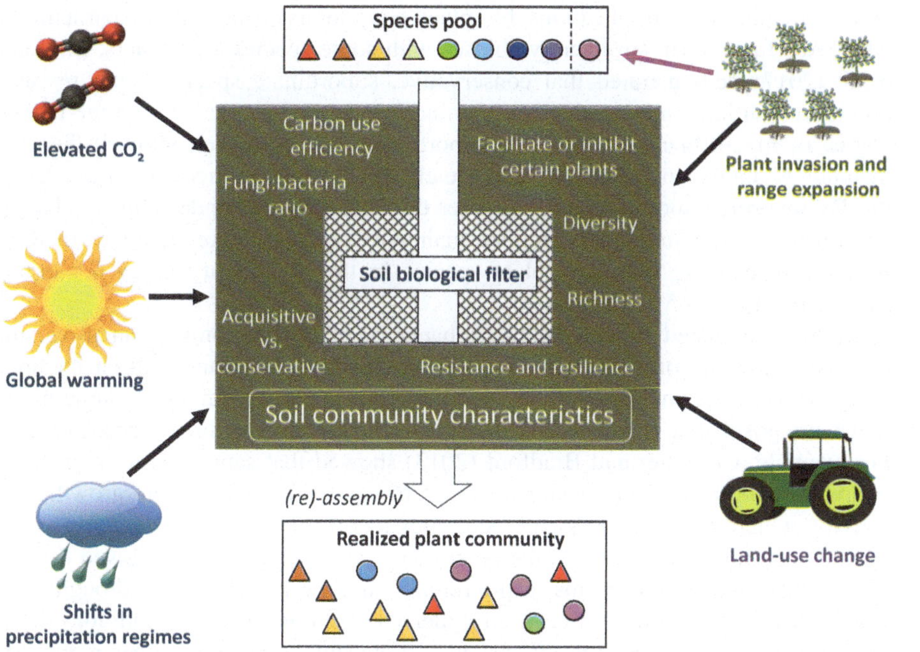

Fig. 13.5 Global and land-use change factors can determine how plant communities assemble and/or reassemble. However, soil biota can impact on *how* these factors affect plant community assembly through a variety of mechanisms. For example, shifts from conservative to acquisitive soil communities as the result of intensive agricultural practices might favor plants with similar (i.e., acquisitive) traits. Further, soil communities that are resistant to changes in precipitation regimes might be better able to buffer the effect of, e.g., drought on the plant community, thereby helping to maintain community diversity. These examples highlight the dynamic, interactive nature between global and land use change, the soil community, and the plant community, pulling focus on the necessity of considering how changes to one component could lead to alterations of the others. To note, plant invasion and range expansion can affect plant community assembly either directly by influencing the species pool (pink arrow) or indirectly through influencing the soil biological filter

What is the role of soil biota in plant community assembly under forecasted climate change perturbations? Soil microbial community composition and biodiversity are being altered by climate change, but evidence suggests that some soil organisms can respond and adapt faster than plants. This may have direct impacts on plant communities. For example, root colonization by mutualistic arbuscular mycorrhizal (AM) fungi often increases under drought, warming, and elevated [CO_2] (e.g., Staddon et al. 2004), thus creating increased resistance and resilience to disturbance for the host plant. A meta-analysis by Augé et al. (2015) suggests that root colonization by AM fungi increased stomatal conductance in plants during water stress, especially under moderate and severe drought. Further, in more than 90% of 32 studies, plant productivity and growth under drought was improved by inoculation with AM fungi (Gehring et al. 2017). Beneficial effects of AM fungi under drought have implications for plant community composition, particularly when AM fungi form stronger symbioses with some species than others. Mariotte et al. (2017) demonstrated that conservative, subordinate species better resisted drought conditions than acquisitive, dominant species. This was due to the higher degree of mycorrhizal symbiosis of subordinate species, which allowed them to maintain N uptake and increase water use efficiency under reduced water availability. By favoring subordinate species over dominant species under climate change perturbations, AM fungi can, therefore, maintain or increase plant diversity, thereby potentially reducing dominant plant species in the community or even altering succession (Fig. 13.5).

Microbe-mediated effects of climate change on plant communities can also result from changes in decomposer communities, which affects soil C and nutrient cycling. However, it is yet unclear if climate-induced changes in decomposer communities will cause positive or negative effects on plants. So far, studies show variable results. For example, Crowther and Bradford (2013) showed that saprotrophic fungi from temperate woodland soils acclimate to warming by reducing their growth and activity, which would consequently slow down soil organic matter (SOC) decomposition and mineralization. On the other hand, Yuste et al. (2011) showed that, in dry Mediterranean ecosystems, saprotrophic fungi were tolerant to drought and maintained SOC decomposition even under harsh environmental conditions. This suggests that climate change effects on saprotrophic fungi can either reduce or maintain levels of plant-available nutrients. Jassey et al. (2013) showed how climate change effects on saprotrophic microbes can affect plant community assembly. In a warming experiment in peatland, they showed that warming changed the regulatory role of *Sphagnum* polyphenols (i.e., niche modification), which inhibited top predators and promoted bacterial communities. This led to increases in nutrient cycling and increased the abundance of vascular plants at the expense of mosses. Vascular plants produce more easily degradable litter that promotes decomposers, thus potentially creating positive feedbacks that might further accelerate nutrient cycling and shift plant community structure.

In addition, soil nematodes and earthworms help make nutrients available to the plant community through their waste, a function that could be maintained despite forecasted changes in precipitation patterns. Indeed, a recent study showed that nematodes can withstand deep soil frost that occurred due to simulated reduced snow cover meant to mimic climate change-induced shifts in winter precipitation regimes (De Long et al. 2016). Furthermore, Cesarz et al. (2015) found that the negative effects of reduced precipitation on the nematode community were mitigated by increased [CO_2]. This finding demonstrates that interactions between global change factors can cancel out the effects of individual factors, highlighting the need to consider multiple factors in a single experiment. Finally, soil bioturbators have been shown to be particularly resistant and resilient to climatic changes, with their abundance remaining constant or increasing in response to climatic perturbations. For example, earthworms have been shown to increase in abundance under elevated [CO_2] (Milcu et al. 2011). Moreover, Mariotte et al. (2016) showed that the widely dominant earthworm *Lumbricus terrestris* increased in abundance after drought and as such contributed to plant community recovery.

In summary, the effects of the soil biota seem to depend on the type of climate change factor considered (i.e., warming, drought, elevated [CO_2]; see Fig. 13.5). However, interactive effects of multiple climatic factors on soil biota and consequent feedbacks to the plant community remain largely unknown (but see Cesarz et al. 2015). Too often studies evaluated the effects of climate change on plant and soil communities separately and only a few experiments have been aimed at testing how climate change-induced shifts in soil biota affected plant community composition, reassembly, and succession.

13.5.2 Land-Use Change

Land-use intensification and conversion from natural to agricultural land may reduce plant species diversity (Thébault et al. 2014). During the Green Revolution, human-based processes such as deep soil tillage, fertilizer application, and chemical pest control have replaced soil biological processes. Increased physical perturbations, in combination with the expansion of monoculture cropping, have disrupted AM fungal communities and networks (Hamel and Plenchette 2017). Similarly, the abundance and functional diversity of earthworms has strongly declined under agricultural intensification (Chan 2001). This suggests that the role of soil biota in improving plant nutrient uptake (Hamel and Plenchette 2017) and recycling plant inputs and protecting plants against diseases has rapidly decreased (Weller et al. 2002; Pieterse et al. 2014). At the same time, the homogeneous genetic and physical environment of agro-systems (e.g., low plant diversity) has promoted the build-up of new and host-specific soil pathogens (McDonald and Stukenbrock 2016). Furthermore, the reduction in abundance of soil mutualists and bioturbators and the increase in pathogen abundance could ultimately affect the reassembly of plant communities on ex-arable lands (Fig. 13.1), for example through niche modification or by altering species-specific soil–plant interactions.

Can we restore the soil biodiversity in current or ex-arable agricultural systems? The development of AM fungal inoculation technologies suggests that bringing back mycorrhizal-based biological processes is possible. For example, an experiment on chickpeas (*Cicer arietinum*) by Pellegrino and Bedini (2014) demonstrated that inoculation with AM fungi increased crop yield, as well as the nutritional value of the grain. The inoculation of soils with AM fungi can also improve soil fertility by reducing nutrient loss from the soil (Cavagnaro et al. 2015), especially phosphorus and N (van der Heijden 2010). In low-input grasslands or abandoned agro-systems, mycorrhizal-mediated increases in nutrient availability (i.e., increased niche size) could affect plant community reassembly by promoting fast-growing species, similar to how saprotrophs increase nutrient availability and thereby favor fast-growing species (Fig. 13.1b). Van der Heijden et al. (1998) also showed that soil inoculation with a diverse AM fungal community promoted plant diversity because contrasting plant species benefited from different AM fungal taxa. Inoculating ex-arable fields with high diversity mycorrhizal communities could, therefore, encourage the reassembly of a diverse plant community. Moreover, AM fungal inoculation might both improve the sustainability of crop systems and impact on community reassembly after land abandonment. Additionally, Pieterse et al. (2014) showed that induced systemic resistance might be a mechanism by which plant growth-promoting soil organisms improve defenses against pathogens or root herbivores. For example, they showed that mutualist species of *Bacillus*, *Pseudomonas*, and *Trichoderma* activate the plant immune system, which enhances defense without any cost for the host plant. Additionally, *Fusarium*, *Streptomyces*, and *Actinomyces* can suppress soil pathogens (Weller et al. 2002) through competition for space, hyperparasitism, or antagonism via production of secondary metabolites. Pathogens often build up in crop systems (McDonald and Stukenbrock 2016), and inoculation with disease-suppressive organisms could help restore a more "balanced" soil biotic community composition by limiting the influence of generalist and specialist pathogens (e.g., Fig. 13.1). Therefore, microbial inoculation, alongside increased multispecies intercropping and organic farming practices, could promote plant community diversity in agricultural systems.

Soil biological legacies can last for a long time after abandonment of agricultural practices. For example, Jangid et al. (2011) showed that 17 years after conversion of croplands to grasslands, microbial communities of early-successional grasslands were more similar to microbial communities of the original cropland compared to those of long-term mowed grasslands. This suggests a lasting impact of land-use history on the soil microbial community that continues to affect successive plant communities (de la Peña et al. 2016). Intensive land use generally favors bacterium-dominated communities while reducing the abundance of fungi (e.g., Bardgett et al. 2001; van der Wal et al. 2006). For example, Fraterrigo et al. (2006) showed that 50 years after farming abandonment, soil communities of the recovering forest were still characterized by higher abundance of bacteria and lower abundance of fungi compared to the control unperturbed forest site. Bacterial-dominated soil communities (i.e., low fungal to bacterial ratio) are associated with more rapid decomposition rates and faster nutrient cycling, which promotes N mineralization (Wardle et al. 2004; de Vries et al. 2013). A lasting reduction in the soil fungal to bacterial ratio after cessation of agricultural activities is, therefore, expected to favor resource-acquisitive plant species over conservative species (Fig. 13.1c; Grigulis et al. 2013), thus increasing plant dominance and productivity while reducing diversity. Agricultural soil legacies can also have strong impacts on the assembly of the recovering vegetation by differentially affecting the success of early-colonizing species (Kardol et al. 2007; Bauman et al. 2013) (see Sect. 3.2). However, it is important to note that management practices related to land use do not always have major impacts on the soil biota and subsequently plant community assembly. For example, it has been shown that regular clear cut harvests reduced the diversity of EM communities but did not affect the growth of regenerating tree seedlings and, thus, did not affect community succession toward mature forests (Kranabetter 2004).

Overall, intensification and land-use change are largely responsible for the loss of soil biota diversity in agro-systems, creating soil legacies that continue to affect plant communities long after land-use abandonment. Certain management practices can mediate negative effects of land-use change on soil biota. Recent research also suggests that soil biota-based technologies (e.g., AM fungi inoculants) can contribute to more sustainable agricultural practices (e.g., Cavagnaro et al. 2015). Moreover, soil inoculation with soil mutualists and disease-suppressive organisms could help counterbalance soil legacies after land-use abandonment, through niche modification, specific effects on pathogens, and improved plant resistance. Finally, adopting practices such as no/low-till agriculture, multispecies intercropping, and natural pest management will also help restore soil community function and diversity, with knock-on effects for plant community reassembly (Fig. 13.5).

13.6 Synthesis and the Way Forward

In this chapter, we explored how soil biota drive plant community assembly through niche modification and through soil legacies. Effects of soil biota on plant performance have been studied intensively (e.g., Wardle et al. 2004), but our understanding of how soil biota affect the processes by which plant species are filtered from the species pool into local communities is still limited. As we discussed in this chapter, this understanding can help us predict how plant communities reassemble under altered land use, climatic change, and plant invasion. This will further allow us to develop strategies for ecosystem restoration and conservation.

Where to go from here?—A major challenge in studying the effects of soil biota in plant community assembly is our limited knowledge of the ecology of soil biota. Novel developments in the analyses of soil communities promise to provide new insights into the enormous genetic and taxonomic diversity belowground and the divergent niches occupied by soil biota (e.g., Taylor et al. 2014). However, in addition to molecular sequencing, we advocate prioritizing empirical studies manipulating the order and timing of plant species arrival (or, in case of eradication of invasive plants, the order of species removal) in combination with varying the composition of soil communities. Such experiments would benefit from trait-based approaches which are now common in plant ecology (van Kleunen et al. 2010), but still in the early stages of development when it comes to soil biota (e.g., George and Lindo 2015; Widenfalk et al. 2016). Further work needs to be done, but such trait-based approaches should also help in teasing apart the relative contribution of different groups of soil organisms, e.g., mutualists versus pathogens.

A further challenge lies in actively using soil biota (e.g., microbial inocula) for sustainable agricultural practices, restoring and conserving endangered plant communities, and combating the effects of climatic change and plant invasions. Accounting for interactions between different global change factors that simultaneously affect plant and soil communities when designing experiments will enhance the real-life applicability of the results obtained (Kardol et al. 2012). So far, most experiments testing the effects of soil biota on plant community assembly have been run under controlled (i.e., greenhouse) conditions. Therefore, our knowledge on the relative importance of soil biota and how their effects vary in relation to environmental factors is still limited. Field experiments manipulating soil biota come with their own challenges, but are nevertheless valuable due to an added sense of realism. Alternative approaches would include controlled experiments using larger mesocosms and adding environmental manipulations (e.g., soil chemistry or climate) to the experimental design. Future studies should also focus on the importance of local and regional species pools (both for plant and soil organisms) and how community membership depends on dispersal.

Acknowledgements We thank Tadashi Fukami and Benjamin Sikes for their helpful comment on an earlier version of the manuscript. Financial support to PK was provided by the Swedish Research Council (Vetenskapsrådet).

References

Allen MF, Crisafulli C, Friese CF et al (1992) Re-formation of mycorrhizal symbiosis on Mount St Helens, 1980–1990 – interaction of rodents and mycorrhizal fungi. Mycol Res 96:447–453

Aprahamian AM, Lulow ME, Major MR et al (2016) Arbuscular mycorrhizal inoculation in coastal sage scrub restoration. Botany 94:493–499

Augé RM, Toler HD, Saxton AM (2015) Arbuscular mycorrhizal symbiosis alters stomatal conductance of host plants more under drought than under amply watered conditions: a meta-analysis. Mycorrhiza 25:13–24

Bais HP, Vepachedu R, Gilroy S et al (2003) Allelopathy and exotic plant invasion: from molecules and genes to species interactions. Science 301:1377–1380

Banning NC, Gleeson DB, Grigg AH et al (2011) Soil microbial community successional patterns during forest ecosystem restoration. Appl Environ Microbiol 77:6158–6164

Bardgett RD, van der Putten WH (2014) Belowground biodiversity and ecosystem functioning. Nature 515:505–511

Bardgett RD, Jones AC, Jones DL et al (2001) Soil microbial community patterns related to the history and intensity of grazing in sub-montane ecosystems. Soil Biol Biochem 33:1653–1664

Bardgett RD, Richter A, Bol R et al (2007) Heterotrophic microbial communities use ancient carbon following glacial retreat. Biol Lett 3:487–490

Bauman JM, Keiffer CH, Hiremath S et al (2013) Soil preparation methods promoting ectomycorrhizal colonization and American chestnut *Castanea dentata* establishment in coal mine restoration. J Appl Ecol 50:721–729

Bennett JA, Cahill JF (2016) Fungal effects on plant-plant interactions contribute to grassland plant abundances: evidence from the field. J Ecol 104:755–764

Berg MP, Kiers ET, Driessen G et al (2010) Adapt or disperse: understanding species persistence in a changing world. Glob Change Biol 16:587–598

Bever JD, Mangan SA, Alexander HM (2015) Maintenance of plant species diversity by pathogens. Annu Rev Ecol Evol Syst 46:305–325

Blomqvist MM, Olff H, Blaauw MB et al (2000) Interactions between above- and belowground biota: importance for small-scale vegetation mosaics in a grassland ecosystem. Oikos 90:582–598

Bokhorst S, Kardol P, Bellingham PJ et al (2017) Responses of communities of soil organisms and plants to soil aging at two contrasting long-term chronosequences. Soil Biol Biochem 106:69–79

Bradford MA, Jones TH, Bardgett RD, et al (2002) Impacts of soil faunal community composition on model grassland ecosystems. Science 298:615–618

Brankatschk R, Towe S, Kleineidam K et al (2011) Abundances and potential activities of nitrogen cycling microbial communities along a chronosequence of a glacier forefield. ISME J 5:1025–1037

Bravo-Monasterio P, Pauchard A, Fajardo A (2016) *Pinus contorta* invasion into treeless steppe reduces species richness and alters species traits of the local community. Biol Invasions 18:1883–1894

Brinkman EP, Troelstra SR, van der Putten WH (2005) Soil feedback effects to the foredune grass Ammophila arenaria by endoparasitic root-feeding nematodes and whole soil communities. Soil Biol Biochem 37:2077–2087

Brinkman EP, Duyts H, Karssen G et al (2015) Plant-feeding nematodes in coastal sand dunes: occurrence, host specificity and effects on plant growth. Plant Soil 397:17–30

Brown VK, Gange AC (1990) Insect herbivory below ground. Adv Ecol Res 20:1–58

Burkle LA, Belote RT (2015) Soil mutualists modify priority effects on plant productivity, diversity, and composition. Appl Veg Sci 18:332–342

Callaway RM, Thelen GC, Rodriguez HWE (2004) Soil biota and exotic plant invasion. Nature 427:731–733

Carbajo V, den Braber B, van der Putten WH et al (2011) Enhancement of late successional plants on ex-arable land by soil inoculations. PLoS One 6:e21943

Cavagnaro TR, Bender SF, Asghari HR et al (2015) The role of arbuscular mycorrhizas in reducing soil nutrient loss. Trends Plant Sci 20:283–290

Cesarz S, Reich PB, Scheu S et al (2015) Nematode functional guilds, not trophic groups, reflect shifts in soil food webs and processes in response to interacting global change factors. Pedobiologia 58:23–32

Chan KY (2001) An overview of some tillage impacts on earthworm population abundance and diversity – implications for functioning in soils. Soil Tillage Res 57:179–191

Chase JM (2003) Community assembly: when should history matter? Oecologia 136:489–498

Classen AT, Sundqvist MK, Henning JA et al (2015) Direct and indirect effects of climate change on soil microbial and soil microbial-plant interactions: what lies ahead? Ecosphere 6:1–21

Collins CG, Carey CJ, Aronson EL et al (2016) Direct and indirect effects of native range expansion on soil microbial community structure and function. J Ecol 104:1271–1283

Crowther TW, Bradford MA (2013) Thermal acclimation in widespread heterotrophic soil microbes. Ecol Lett 16:469–477

De Deyn GB, Raaijmakers CE, Zoomer HR et al (2003) Soil invertebrate fauna enhances grassland succession and diversity. Nature 422:711–713

De Deyn GB, Raaijmakers CE, van der Putten WH (2004) Plant community development is affected by nutrients and soil biota. J Ecol 92:824–834

de Graaff MA, Adkins J, Kardol P et al (2015) A meta-analysis of soil biodiversity impacts on the carbon cycle. SOIL 1:257–271

de la Peña E, Baeten L, Steel H et al (2016) Beyond plant–soil feedbacks: mechanisms driving plant community shifts due to land-use legacies in post-agricultural forests. Funct Ecol 30:1073–1085

de Leon DG, Moora M, Opik M et al (2016) Symbiont dynamics during ecosystem succession: co-occurring plant and arbuscular mycorrhizal fungal communities. FEMS Microbiol Ecol 92: fiw097

De Long JR, Laudon H, Blume-Werry G et al (2016) Nematode community resistant to deep soil frost in boreal forest soils. Pedobiologia 59:243–251

de Vries FT, Shade A (2013) Controls on soil microbial community stability under climate change. Front Microbiol 4:265

de Vries FT, Thébault E, Liiri M et al (2013) Soil food web properties explain ecosystem services across European land use systems. Proc Natl Acad Sci USA 110:14296–14301

del Moral R (1983) Initial recovery of subalpine vegetation on Mount St Helens, Washington. Am Midl Nat 109:72–80

Despain DG (2001) Dispersal ecology of lodgepole pine (*Pinus contorta* Dougl.) in its native environment as related to Swedish forestry. For Ecol Manage 141:59–68

Dickie IA, St John MG, Yeates GW et al (2014) Belowground legacies of *Pinus contorta* invasion and removal result in multiple mechanisms of invasional meltdown. AoB Plants 6:15
Diez JM, Dickie I, Edwards G et al (2010) Negative soil feedbacks accumulate over time for non-native plant species. Ecol Lett 13:803–809
Engelkes T, Morriën E, Verhoeven KJF et al (2008) Successful range-expanding plants experience less above-ground and below-ground enemy impact. Nature 456:946–948
Eppinga MB, Rietkerk M, Dekker SC et al (2006) Accumulation of local pathogens: a new hypothesis to explain exotic plant invasions. Oikos 114:168–176
Eschen R, Müller-Schärer H, Schaffner U (2009) Aboveground environment type, soil nutrient content and arbuscular mycorrhizal fungi explain establishment success of *Centaurea jacea* on ex-arable land and in late-successional grasslands. Plant Soil 322:115–123
Fraterrigo JM, Balser TC, Turner MG (2006) Microbial community variation and its relationship with nitrogen mineralization in historically altered forests. Ecology 87:570–579
Fukami T (2015) Historical contingency in community assembly: integrating niches, species pools, and priority effects. Annu Rev Ecol Evol Syst 46:1–23
Fukami T, Nakajima M (2013) Complex plant-soil interactions enhance plant species diversity by delaying community convergence. J Ecol 101:316–324
Gange AC, Brown VK, Farmer LM (1990) A test of mycorrhizal benefit in an early successional plant community. New Phytol 115:85–91
Gehring CA, Swaty R, Deckert E (2017) Mycorrhizas, drought, and host-plant mortality. In: Johnson NC, Gehring CA, Jansa J (eds) Mycorrhizal mediation of soil. Fertility, structure, and carbon storage. Elsevier, Amsterdam, pp 279–298
George PBL, Lindo Z (2015) Application of body size spectra to nematode trait-index analyses. Soil Biol Biochem 84:15–20
Gioria M, Osborne BA (2014) Resource competition in plant invasions: emerging patterns and research needs. Front Plant Sci 5:21
Gómez-Aparicio L, Ibáñez B, Serrano MS et al (2012) Spatial patterns of soil pathogens in declining Mediterranean forests: implications for tree species regeneration. New Phytol 194:1014–1024
Green PT, O'Dowd DJ, Abbott KL et al (2011) Invasional meltdown: invader-invader mutualism facilitates a secondary invasion. Ecology 92:1758–1768
Grigulis K, Lavorel S, Krainer U et al (2013) Relative contributions of plant traits and soil microbial properties to mountain grassland ecosystem services. J Ecol 101:47–57
Grman E, Suding KN (2010) Within-year soil legacies contribute to strong priority effects of exotics on native California grassland communities. Restor Ecol 18:664–670
Gundale MJ, Kardol P, Nilsson M-C et al (2014) Interactions with soil biota shift from negative to positive when a tree species is moved outside its native range. New Phytol 202:415–421
Hamel C, Plenchette C (2017) Implications of past, current, and future agricultural practices for mycorrhiza-mediated nutrient flux. In: Johnson NC, Gehring C, Jansa J (eds) Mycorrhizal mediation of soil. Fertility, structure, and carbon storage. Elsevier, Amsterdam, pp 175–186
Hamman ST, Hawkes CV (2013) Biogeochemical and microbial legacies of non-native grasses can affect restoration success. Restor Ecol 21:58–66
Hedlund K, Gormsen D (2002) Mycorrhizal colonization of plants in set-aside agricultural land. Appl Soil Ecol 19:71–78
Helm DJ, Carling DE (1993) Use of soil transfer for reforestation on abandoned mined lands in Alaska. 2. Effects of soil transfer from different successional stages on growth and mycorrhizal formation by *Populus balsamifera* and *Alnus crispa*. Mycorrhiza 3:107–114
Hodkinson ID, Webb NR, Coulson SJ (2002) Primary community assembly on land - the missing stages: why are the heterotrophic organisms always there first? J Ecol 90:569–577
Hoeksema JD (2005) Plant-plant interactions vary with different mycorrhizal fungus species. Biol Lett 1:439–442

Jangid K, Williams MA, Franzluebbers AJ et al (2011) Land-use history has a stronger impact on soil microbial community composition than aboveground vegetation and soil properties. Soil Biol Biochem 43:2184–2193

Jassey VE, Chiapusio G, Binet P et al (2013) Above- and belowground linkages in *Sphagnum* peatland: climate warming affects plant-microbial interactions. Glob Change Biol 19:811–823

Kardol P, Bezemer TM, van der Putten WH (2006) Temporal variation in plant-soil feedback controls succession. Ecol Lett 9:1080–1088

Kardol P, Cornips NJ, van Kempen MML et al (2007) Microbe-mediated plant-soil feedback causes historical contingency effects in plant community assembly. Ecol Monogr 77:147–162

Kardol P, Bezemer TM, van der Putten WH (2009) Soil organism and plant introductions in restoration of species-rich grassland communities. Restor Ecol 17:258–269

Kardol P, De Long JR, Sundqvist MK (2012) Crossing the threshold: the power of multi-level experiments in identifying global change responses. New Phytol 196:323–326

Kardol P, De Deyn GB, Laliberté E et al (2013) Biotic plant-soil feedbacks across temporal scales. J Ecol 101:309–315

Kaufmann R (2001) Invertebrate succession on an alpine glacier foreland. Ecology 82:2261–2278

Kazemi S, Hatam I, Lanoil B (2016) Bacterial community succession in a high-altitude subarctic glacier foreland is a three-stage process. Mol Ecol 25:5557–5567

Keane RM, Crawley MJ (2002) Exotic plant invasions and the enemy release hypothesis. Trends Ecol Evol 17:164–170

Keller KR (2014) Mutualistic rhizobia reduce plant diversity and alter community composition. Oecologia 176:1101–1109

Klironomos JN (2002) Feedback with soil biota contributes to plant rarity and invasiveness in communities. Nature 417:67–70

Kostenko O, van de Voorde TFJ, Mulder PPJ et al (2012) Legacy effects of aboveground-belowground interactions. Ecol Lett 15:813–821

Koziol L, Bever JD (2015) Mycorrhizal response trades off with plant growth rate and increases with plant successional status. Ecology 96:1768–1774

Koziol L, Bever JD (2016) AMF, phylogeny, and succession: specificity of response to mycorrhizal fungi increases for late-successional plants. Ecosphere 7(11):e01555

Kranabetter JM (2004) Ectomycorrhizal community effects on hybrid spruce seedling growth and nutrition in clearcuts. Can J Bot 82:983–991

Kuebbing SE, Patterson CM, Classen AT et al (2016) Co-occurring nonnative woody shrubs have additive and non-additive soil legacies. Ecol Appl 26:1896–1906

Kulmatiski A, Beard KH, Stark JM (2006) Soil history as a primary control on plant invasion in abandoned agricultural fields. J Appl Ecol 43:868–876

Kulmatiski A, Hines J, Eisenhauer N (2014) Soil organism effects on grassland production and diversity. In: Mariotte P, Kardol P (eds) Grassland biodiversity and conservation in a changing world. Nova Science Publishers Inc., New York, pp 27–49

Laflower DM, Hurteau MD, Koch GW et al (2016) Climate-driven changes in forest succession and the influence of management on forest carbon dynamics in the Puget Lowlands of Washington State, USA. For Ecol Manage 362:194–204

Laliberté E, Turner BL, Costes T et al (2012) Experimental assessment of nutrient limitation along a 2-million-year dune chronosequence in the south-western Australia biodiversity hotspot. J Ecol 100:631–642

Lankau RA, Nuzzo V, Spyreas G et al (2009) Evolutionary limits ameliorate the negative impact of an invasive plant. Proc Natl Acad Sci USA 106:15362–15367

Lankau RA, Zhu K, Ordonez A (2015) Mycorrhizal strategies of tree species correlate with trailing range edge responses to current and past climate change. Ecology 96:1451–1458

Lavelle P, Bignell A, Lepage M et al (1997) Soil function in a changing world: the role of invertebrate ecosystem engineers. Eur J Soil Biol 33:159–193

Liao CZ, Peng RM, Luo YQ et al (2008) Altered ecosystem carbon and nitrogen cycles by plant invasion: a meta-analysis. New Phytol 177:706–714

Mangan SA, Schnitzer SA, Herre EA et al (2010) Negative plant-soil feedback predicts tree-species relative abundance in a tropical forest. Nature 466:752–755

Mariotte P, Le Bayon R-C, Eisenhauer GC et al (2016) Subordinate plant species moderate drought effects on earthworm communities in grasslands. Soil Biol Biochem 96:119–127

Mariotte P, Canarini A, Dijkstra FA (2017) Stoichiometric N:P flexibility and mycorrhizal symbiosis favor plant resistance against drought. J Ecol 105:958–967

Maron JL, Vila M (2001) When do herbivores affect plant invasion? Evidence for the natural enemies and biotic resistance hypotheses. Oikos 95:361–373

Maron JL, Klironomos J, Waller L et al (2014) Invasive plants escape from suppressive soil biota at regional scales. J Ecol 102:19–27

Maron JL, Smith AL, Ortega YK et al (2016) Negative plant-soil feedbacks increase with plant abundance, and are unchanged by competition. Ecology 97:2055–2063

McDonald BA, Stukenbrock EH (2016) Rapid emergence of pathogens in agro-ecosystems: global threats to agricultural sustainability and food security. Philos Trans R Soc B 371:20160026

McGeoch MA, Butchart SHM, Spear D et al (2010) Global indicators of biological invasion: species numbers, biodiversity impact and policy responses. Divers Distrib 16:95–108

Middleton EL, Bever JD (2012) Inoculation with a native soil community advances succession in a grassland restoration. Restor Ecol 20:218–226

Milcu A, Paul S, Lukac M (2011) Belowground interactive effects of elevated CO_2, plant diversity and earthworms in grassland microcosms. Basic Appl Ecol 12:600–608

Molina-Montenegro MA, Oses R, Torres-Diaz C et al (2015) Fungal endophytes associated with roots of nurse cushion species have positive effects on native and invasive beneficiary plants in an alpine ecosystem. Perspect Plant Ecol Evol Syst 17:218–226

Montesinos-Navarro A, Segarra-Moragues JG, Valiente-Banuet A et al (2016) Fungal phylogenetic diversity drives plant facilitation. Oecologia 181:533–541

Moora M, Zobel M (2009) Arbuscular mycorrhiza and plant-plant interactions – impact of invisible world on visible patterns. In: Pugnaire FI (ed) Positive interactions and plant community dynamic. CRC Press, Boca Ration, pp 79–98

Morriën E, van der Putten WH (2013) Soil microbial community structure of range-expanding plant species differs from co-occurring natives. J Ecol 101:1093–1102

Nara K (2006) Pioneer dwarf willow may facilitate tree succession by providing late colonizers with compatible ectomycorrhizal fungi in a primary successional volcanic desert. New Phytol 171:187–198

Nielsen UN, Ayres E, Wall DH et al (2011) Soil biodiversity and carbon cycling: a review and synthesis of studies examining diversity-function relationships. Eur J Soil Sci 62:105–116

O'Dowd DJ, Green PT, Lake PS (2003) Invasional 'meltdown' on an oceanic island. Ecol Lett 6:812–817

Obase K, Tamai Y, Yajima T et al (2008) Mycorrhizal colonization status of plant species established in an exposed area following the 2000 eruption of Mt. Usu, Hokkaido, Japan. Landsc Ecol Eng 4:57–61

Paudel S, Wilson GWT, MacDonald B et al (2016) Predicting spatial extent of invasive earthworms on an oceanic island. Divers Distrib 22:1013–1023

Pellegrino E, Bedini A (2014) Enhancing ecosystem services in sustainable agriculture: biofertilization and biofortification of chickpea (*Cicer arietinum* L.) by arbuscular mycorrhizal fungi. Soil Biol Biochem 68:429–439

Pendergast TH, Burke DJ, Carson WP (2013) Belowground biotic complexity drives aboveground dynamics: a test of the soil community feedback model. New Phytol 197:1300–1310

Perring MP, Standish RJ, Price JN et al (2015) Advances in restoration ecology: rising to the challenges of the coming decades. Ecosphere 6:25

Pieterse CMJ, Zamioudis C, Berendsen RL et al (2014) Induced systemic resistance by beneficial microbes. Annu Rev Phytopathol 52:347–375

Pineiro J, Maestre FT, Bartolome L et al (2013) Ecotechnology as a tool for restoring degraded drylands: a meta-analysis of field experiments. Ecol Eng 61:133–144

Polley HW, Johnson HB, Derner JD (2003) Increasing CO_2 from subambient to superambient concentrations alters species composition and increases above-ground biomass in a C_3/C_4 grassland. New Phytol 160:319–327
Reynolds HL, Packer A, Bever JD et al (2003) Grassroots ecology: plant-microbe-soil interactions as drivers of plant community structure and dynamics. Ecology 84:2281–2291
Rodriguez-Echeverria S, Armas A, Piston N et al (2013) A role for below-ground biota in plant-plant facilitation. J Ecol 101:1420–1428
Schmidt SK, Reed SC, Nemergut DR et al (2008) The earliest stages of ecosystem succession in high-elevation (5000 metres above sea level), recently deglaciated soils. Proc R Soc Lond B Biol Sci 275:2793–2802
Sikes BA, Hawkes CV, Fukami T (2016) Plant and root endophyte assembly history: interactive effects on native and exotic plants. Ecology 97:484–493
Simberloff D (2006) Invasional meltdown 6 years later: important phenomenon, unfortunate metaphor, or both? Ecol Lett 9:912–919
Simberloff D, Von Holle B (1999) Positive interactions of nonindigenous species: invasional meltdown? Biol Invasions 1:21–32
Sonnemann I, Hempel S, Beutel M et al (2013) The root herbivore history of the soil affects the productivity of a grassland plant community and determines plant response to new root herbivore attack. Plos One 8:e56524. https://doi.org/10.1371/journal.pone.0056524
Staddon PL, Gregersen R, Jakobsen I (2004) The response of two *Glomus* mycorrhizal fungi and a fine endophyte to elevated atmospheric CO_2, soil warming and drought. Glob Change Biol 10:1909–1921
Stinson KA, Campbell SA, Powell JR et al (2006) Invasive plant suppresses the growth of native tree seedlings by disrupting belowground mutualisms. PLoS Biol 4:727–731
Stuble KL, Souza L (2016) Priority effects: natives, but not exotics, pay to arrive late. J Ecol 104:987–993
Taylor DL, Hollingsworth TN, McFarland JW et al (2014) A first comprehensive census of fungi in soil reveals both hyperdiversity and fine-scale niche partitioning. Ecol Monogr 84:3–20
Teste FP, Kardol P, Turner BL et al (2017) Plant-soil feedback and the maintenance of diversity in Mediterranean-climate shrublands. Science 355:173–176
Thébault A, Mariotte P, Lortie CJ et al (2014) Land management trumps the effects of climate change and elevated CO_2 on grassland functioning. J Ecol 102:896–904
Titus JH, del Moral R (1998) Seedling establishment in different microsites on Mount St. Helens, Washington, USA. Plant Ecol 134:13–26
van de Voorde TFJ, van der Putten WH, Bezemer TM (2011) Intra- and interspecific plant-soil interactions, soil legacies and priority effects during old-field succession. J Ecol 99:945–953
van der Heijden MGA (2010) Mycorrhizal fungi reduce nutrient loss from model grassland ecosystems. Ecology 91:1163–1171
van der Heijden MGA, Horton TR (2009) Socialism in soil? The importance of mycorrhizal fungal networks for facilitation in natural ecosystems. J Ecol 97:1139–1150
van der Heijden MGA, Klironomos JN, Ursic M et al (1998) Mycorrhizal fungal diversity determines plant biodiversity, ecosystem variability and productivity. Nature 396:69–72
van der Putten WH (2012) Climate change, aboveground-belowground interactions, and species' range shifts. Annu Rev Ecol Evol Syst 43:365–383
van der Putten WH, van Dijk C, Peters BAM (1993) Plant-specific soil-borne diseases contribute to succession in foredune vegetation. Nature 362:53–56
van der Putten WH, Macel M, Visser ME (2010) Predicting species distribution and abundance responses to climate change: why it is essential to include biotic interactions across trophic levels. Philos Trans R Soc B 365:2025–2034
van der Putten WH, Bardgett RD, Bever JD et al (2013) Plant-soil feedbacks: the past, the present and future challenges. J Ecol 101:265–276

van der Stoel CD, van der Putten WH, Duyts H (2002) Development of a negative plant-soil feedback in the expansion zone of the clonal grass *Ammophila arenaria* following root formation and nematode colonization. J Ecol 90:978–988
van der Wal A, van Veen JA, Smant W et al (2006) Fungal biomass development in a chronosequence of land abandonment. Soil Biol Biochem 38:51–60
van Groenigen JW, Lubbers IM, Vos HMJ et al (2014) Earthworms increase plant production: a meta-analysis. Sci Rep-UK 4:6365. https://doi.org/10.1038/srep06365
van Grunsven RHA, van der Putten WH, Bezemer TM et al (2007) Reduced plant-soil feedback of plant species expanding their range as compared to natives. J Ecol 95:1050–1057
van Kleunen M, Weber E, Fischer M (2010) A meta-analysis of trait differences between invasive and non-invasive plant species. Ecol Lett 13:235–245
van Kleunen M, Dawson W, Essl F et al (2015) Global exchange and accumulation of non-native plants. Nature 525:100–103
Vitousek PM, Walker LR, Whiteaker LD et al (1987) Biological invasions by *Myrica faya* alters ecosystem development in Hawaii. Science 238:802–804
Vitousek PM, Walker LR, Whiteaker LD et al (1993) Nutrient limitations to plant growth during primary succession in Hawaii Volcanos National Park. Biogeochemistry 23:197–215
Wagg C, Jansa J, Schmid B et al (2011) Belowground biodiversity effects of plant symbionts support aboveground productivity. Ecol Lett 14:1001–1009
Wagg C, Bender SF, Widmer F et al (2014) Soil biodiversity and soil community composition determine ecosystem multifunctionality. Proc Natl Acad Sci USA 111:5266–5270
Wang D, Yang SM, Tang F et al (2012) Symbiosis specificity in the legume – rhizobial mutualism. Cell Microbiol 14:334–342
Wanner M, Elmer M, Sommer M et al (2015) Testate amoebae colonizing a newly exposed land surface are of airborne origin. Ecol Indic 48:55–62
Wardle DA, Bardgett RD, Klironomos JN et al (2004) Ecological linkages between aboveground and belowground biota. Science 304:1629–1633
Weller DM, Raaijmakers JM, Gardener BBM et al (2002) Microbial populations responsible for specific soil suppressiveness to plant pathogens. Annu Rev Phytopathol 40:309–348
Widenfalk LA, Malmström A, Berg MP et al (2016) Small-scale Collembola community composition in a pine forest soil – overdispersion in functional traits indicates the importance of species interactions. Soil Biol Biochem 103:52–62
Wilschut RA, Geisen S, ten Hooven FC et al (2016) Interspecific differences in nematode control between range-expanding plant species and their congeneric natives. Soil Biol Biochem 100:233–241
Wubs ERJ, van der Putten WH, Bosch M et al (2016) Soil inoculation steers restoration of terrestrial ecosystems. Nat Plants 2:16107
Wurst S, Ohgushi T (2015) Do plant- and soil-mediated legacy effects impact future biotic interactions? Funct Ecol 29:1373–1382
Wurst S, Langel R, Scheu S (2005) Do endogeic earthworms change plant competition? A microcosm study. Plant Soil 271:123–130
Yuste JC, Penuelas J, Estiarte M et al (2011) Drought-resistant fungi control soil organic matter decomposition and its response to temperature. Glob Change Biol 17:1475–1486
Zhang NL, van der Putten WH, Veen GF (2016) Effects of root decomposition on plant-soil feedback of early- and mid-successional plant species. New Phytol 212:220–231
Zobel M, Opik M (2014) Plant and arbuscular mycorrhizal fungal (AMF) communities - which drives which? J Veg Sci 25:1133–1140

Chapter 14
Application and Theory of Plant–Soil Feedbacks on Aboveground Herbivores

Ian Kaplan, Ana Pineda, and Martijn Bezemer

14.1 Introduction

The soil is an essential component of all terrestrial ecosystems. It serves as a medium in which plants root, but also houses an overwhelming abundance of living organisms such as bacteria, fungi, protozoa, viruses, nematodes, and arthropods that interact in complex networks (Wall 2004). Soil organisms consume organic compounds, other soil organisms, or plant roots and recycle and mineralize nutrients or interact with living plants. As primary producers, plants provide the basic resources to the soil food web (Bardgett and Wardle 2010). They contribute litter originating from dead shoots or roots, and living plant roots release compounds such as sugars, organic acids, hormones, and secondary metabolites. Via these inputs, plants shape soil biotic communities that use these resources or are influenced by them and alter the physical and chemical properties of soils (Van der Putten et al. 2009, 2013).

Evidence is rapidly accumulating that these plant-mediated soil changes influence the performance of other plants that grow later in the same soil (Klironomos 2002;

I. Kaplan (✉)
Department of Entomology, Purdue University, West Lafayette, IN, USA
e-mail: ikaplan@purdue.edu

A. Pineda
Department of Terrestrial Ecology, Netherlands Institute of Ecology, Wageningen, The Netherlands
e-mail: A.Pineda@nioo.knaw.nl

M. Bezemer
Department of Terrestrial Ecology, Netherlands Institute of Ecology, Wageningen, The Netherlands

Institute of Biology, Section Plant Ecology and Phytochemistry, Leiden University, Leiden, The Netherlands
e-mail: M.Bezemer@nioo.knaw.nl

T. Ohgushi et al. (eds.), *Aboveground–Belowground Community Ecology*, Ecological Studies 234, https://doi.org/10.1007/978-3-319-91614-9_14

Kulmatiski et al. 2008). This phenomenon is called plant–soil feedback, and it is now receiving considerable attention in the ecological literature (Bever et al. 1997; Ehrenfeld et al. 2005; Van der Putten et al. 2013). Many studies argue that plant–soil feedbacks can aid in the understanding of invasiveness of exotic plants, succession of natural plant communities, plant coexistence and plant competition, and even plant–herbivore interactions (Klironomos 2002; Van der Putten et al. 2009, 2013). Much of the work on plant–soil feedback is carried out with wild species and within the context of natural plant communities. However, these legacy effects play a central role in agriculture where the concept of soil sickness—which, by definition, is a form of plant–soil feedback—has been recognized for centuries among those cultivating crops (Huang et al. 2013).

In this chapter, we first briefly introduce the terminology and concepts used in plant–soil feedback research. We then review the effect of plant–soil feedbacks on aboveground phytophagous insects, a topic that is only beginning to develop but clearly illustrates how soil properties cascade to influence relationships with foliar herbivores. Next, we focus on the mechanisms by which soil microbiota induce changes in plant growth and resistance and the consequences for plant-feeding insects and their natural enemies. Finally, we discuss how knowledge about plant–soil feedbacks can be used to steer soil communities in agricultural systems so that the current or succeeding crops become more resistant to aboveground and belowground insect pests and diseases.

14.2 Plant–Soil Feedback: Concepts and Definitions

Plant–soil feedback is the effect of a plant on the soil, that subsequently influences the performance of another plant that grows later in the soil. If the initial plant causing changes in the soil and the plant responding to these changes belong to the same species, this is called conspecific or direct plant–soil feedback. When the species that cause and respond to the effect differ, the feedback is called heterospecific or indirect (Van der Putten et al. 2013). Conspecific feedbacks are usually negative, whereas heterospecific feedbacks are variable and can be negative, neutral, or positive (Kulmatiski et al. 2008; Van de Voorde et al. 2011).

Plant–soil feedback studies typically consist of two stages. During the *first phase*, the plant influences the soil it grows in. This is called the "conditioning phase." In the majority of studies, "live" soil (i.e., non-sterilized, presumably rich with microbial life) is collected from the field, and plants are then grown in pots filled with this soil for a fixed period of time, often between 2 and 4 months. Alternatively, soil can be collected from beneath specific plant species in the field or from monocultures. In this case, the soil is naturally conditioned in the field rather than experimentally assigned to plant species in controlled environments, but this approach is less common (Bezemer et al. 2005; Badri et al. 2013; Pendergast et al. 2013; Kos et al. 2015a). The underlying assumption is that during the conditioning phase, plants shape the soil microbial community in a specific manner via their input of organic compounds. The structure of the microbial community will change depending on the amount and composition of root exudates and sloughed root tissue that is deposited in the soil, and plant pathogens and symbionts will increase in the presence of their host. Although often ignored, abiotic soil conditions such as nutrient availability, moisture, and physical structure also change depending on the plant species that was previously growing in the soil. These changes could either impact subsequent plant growth directly (e.g., a nitrogen-fixing legume may provide excess nutrients for plants sharing the same soil) or indirectly by modifying the microbial community that also responds strongly to abiotic factors (Fierer et al. 2012; Ramirez et al. 2012).

At the end of the conditioning phase, the plant is removed and the remaining soil is then used for the *second phase*. This is called the "feedback phase" or "test phase," in which the growth of a focal plant in the conditioned soil—also called "home soil"—is compared to a control soil. As both abiotic and biotic soil properties change during the conditioning phase, selecting the appropriate control is not trivial. Several studies compare plant growth in conditioned soil with growth in the same soil that was sterilized (Fig. 14.1a; e.g., Cortois et al. 2016). However, this comparison can be misleading. Sterilization often increases nutrient availability and hence the feedback effect is confounded by nutritional differences among the two soil groups, namely, artificially lower nutrient levels in the conditioned treatment. Thus, negative effects of conditioned soil on plant growth tend to be overestimated, particularly in low fertility soils (Troelstra et al. 2001).

Three other types of controls are commonly used in plant–soil feedback studies. *First*, feedback effects can be determined by comparing the growth in conditioned "home" soil to live field soil that was not conditioned (Fig. 14.1b). A potential caveat with this approach is that the "unconditioned" field soil is used during the conditioning phase and as control soil during the feedback phase. As the conditioning phase can take several months, the "unconditioned" field soil should be stored and during storage the microbial composition can change. Alternatively, fresh "unconditioned" soil can be collected later at the beginning of the feedback phase. However, because microbial communities vary over time and change depending on season, climatic conditions, and plant age or phenology (Lauber et al. 2013), the later collected soil may fundamentally differ from the soil originally used during the conditioning phase. *Second*, growth of a focal plant in "home" soil conditioned by the same species can be compared with "away" soil conditioned by another species. This can be accomplished using several "away" soils, providing a quantitative measure of the soil-mediated influence of a plant on conspecifics relative to other species (Fig. 14.1c; e.g., Van de Voorde et al. 2011; Wubs and Bezemer 2016). When a series of soils are used, the home soil can also be compared to a mixture of all away soils (Engelkes et al. 2008; Brinkman et al. 2010). A *third* way of testing plant–soil feedback effects is to inoculate sterilized soil with a small amount—ca. 10%—of conditioned soil. This can then be compared to inoculation using the same amount of sterilized conditioned soil (Fig. 14.1d; e.g., Bever et al. 1997; Van der Putten et al. 2013). A benefit of this approach is that the complicating effect of nutritional differences from using pure conditioned soil, as shown in Fig. 14.1a, is minimized since both treatments contain primarily sterile soil with relatively minor

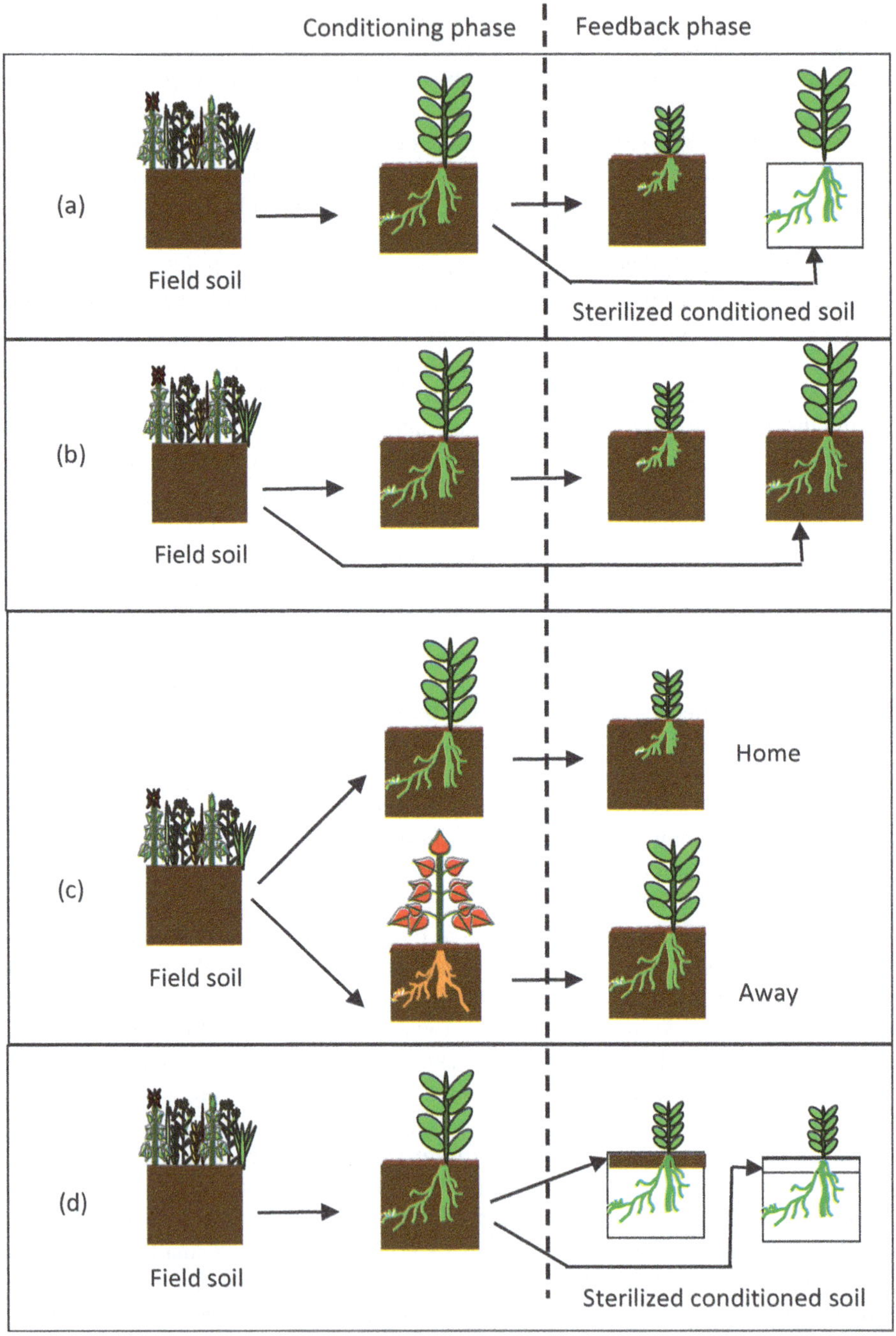

Fig. 14.1 Schematic view of several approaches that are used to examine plant–soil feedback effects. Live soil is collected from a field site ("Field soil"). During the conditioning phase, the focal plant grows in this soil. At the end of the conditioning phase, the plant is removed from the soil and the soil is used in the feedback phase. The plant–soil feedback effect of a focal species can be

differences in total volume due to the addition of small quantities of inoculum in the conditioning treatment. The drawback here is that adding low inoculum volumes may underestimate the feedback if microbial factors are density dependent (i.e., adding 50% soil inoculum generates a relatively stronger impact compared with only 10% inoculum). To our knowledge, this is not well known or documented.

An important question that is still highly debated is whether and to what extent plant–soil feedbacks are predictable. It is hypothesized that the phylogenetic relatedness between the plant that conditions the soil and the species that responds to this soil predicts the magnitude of the feedback (Dostal and Paleckova 2011; Mehrabi and Tuck 2015). The assumption is that closely related species share soil pathogens or symbionts and, as most conspecific feedbacks are negative (Kulmatiski et al. 2008), the prediction is that there should be a negative relationship between relatedness and feedback. However, most studies, so far, indicate that relatedness is a poor predictor (Dostal and Paleckova 2011; Mehrabi and Tuck 2015; Fitzpatrick et al. 2017; but see Anacker et al. 2014). Other studies show that the strength of feedbacks is better explained by proximate plant traits such as growth rate or specific root length. Fast growing species generally exhibit stronger negative conspecific feedbacks than slower growing ones (Cortois et al. 2016). Differences between conditioning and response species in trait values also appear to be predictive, but increases in trait differences can lead to both stronger and weaker feedbacks (Fitzpatrick et al. 2017). Hence, to what extent plant traits can explain plant–soil feedbacks remains an open question. Some plant–soil feedback effects may also be predictable according to plant functional group with, for instance, grasses, legumes, and forbs modifying the soil microbiome in fundamentally different ways (Cortois et al. 2016). In this context, it is important to note that the effect of a plant species on its soil community depends in large part on the microbial composition prior to conditioning and that unconditioned soils already have a legacy. Thus, true unconditioned soils do not technically exist. This may partly explain why plant–soil feedback studies that are carried out with the same species can show highly variable outcomes and why it is so difficult to make general predictions.

Fig. 14.1 (continued) determined by (**a**) comparing growth in home soil with that in sterilized home soil; (**b**) comparing growth in conditioned soil versus unconditioned soil (field soil); (**c**) comparing growth in soil conditioned by the focal species (home) with growth in soil conditioned by another species (away); or (**d**) comparing growth in sterile soil inoculated with a small amount of home soil with growth in sterile soil inoculated with a small amount of sterile home soil or live away soil

14.3 Plant–Soil Feedbacks and Aboveground Plant–Insect Interactions

The majority of plant–soil feedback studies examine how changes in soil microbes impact plant growth; however, changes in the amount or quality of the host plant due to changes in the soil community can also influence aboveground herbivorous insects (Wurst and Ohgushi 2015). Many studies show that the addition of specific soil biota such as root herbivores (Johnson et al. 2012; Soler et al. 2012), mycorrhizal fungi (Koricheva et al. 2009), rhizobacteria (Hol et al. 2010; Pineda et al. 2010; Pangesti et al. 2013), or nematodes (Wondafrash et al. 2013) influence aboveground plant-feeding insects. These prior investigations clearly illustrate the *potential* for individual groups of soil organisms to shape aboveground interactions. Because plant–soil feedbacks simultaneously change the full soil community, including all taxonomic and functional groups, we expect feedbacks to have major consequences for foliar insect herbivory. Yet, unlike plants, for which there is a clear prediction of negative conspecific feedbacks, the insect response is far more difficult to anticipate.

The few studies that integrate leaf-feeding insects into plant–soil feedback dynamics show that their interactions with host–plants are considerably altered. In a study where ragwort (*Senecio jacobaea*) plants were grown in soils conditioned by 10 different plant species, abundance of the specialized aphid *Aphis jacobaea* varied almost 10-fold between the worst and best soils in terms of *A. jacobaea* population growth. These effects could be explained by plant–soil feedback-related changes in amino acid concentration in the phloem (Kos et al. 2015a, b). In the same experiment using the same host plant species, however, the generalist aphid *Brachycaudus cardui* was far less sensitive to soils from the ten plant species, indicating that these effects depend greatly on the combination of plants, soils, and insects used (Kos et al. 2015a, b). Herbivory by generalist snails (*Cepaea species*) was also unaffected by plant–soil feedback in a study with *Solidago canadensis* and *Tanacetum vulgare*, even though *T. vulgare* exhibited a negative conspecific feedback (Schittko and Wurst 2014). More studies are needed to determine whether generalist herbivores are less sensitive to plant–soil feedback and broadly to changes in the composition of soil microbial communities. However, it is known that generalists and specialists, or insects in different feeding guilds, vary in their response to qualitative or quantitative changes in their host plants (Bezemer and Jones 1998; Awmack and Leather 2002; Ali and Agrawal 2012). Root damage, for instance, often increases the performance of aboveground sapsuckers while reducing the performance of leaf-chewers (Johnson et al. 2008, 2012).

Not only are aboveground insects sensitive to plant–soil feedback, several studies also show that foliar herbivory or grazing during the conditioning phase influences the feedback effect (Medina-Roldan et al. 2012; Veen et al. 2014). Leaf damage by the caterpillar *Mamestra brassicae* on *S. jacobaea* increased the negative conspecific plant–soil feedback, possibly due to herbivore-induced changes in soil fungi (Kostenko et al. 2012; Bezemer et al. 2013). Remarkably, herbivory during the conditioning phase also increased the tolerance of *S. jacobaea* to aboveground herbivory during the feedback phase and increased insect performance on the later growing plants. The emergence rate of *Microplitis mediator* parasitoids from caterpillar hosts was higher when plants were growing in soil in which conspecific plants had been grown that were exposed to aboveground herbivory (Kostenko et al. 2012). These studies suggest that aboveground multi-trophic interactions are influenced by plant-mediated changes in soil microbiota and that aboveground higher trophic levels alter plant–soil feedbacks.

Perhaps one of the best demonstrations of how plant–soil feedbacks affect aboveground insect interactions comes from a study that used an elegant design to isolate the microbial mechanism responsible for the effect (Badri et al. 2013). These authors used field soil collected from *Arabidopsis*, pine, corn, and potato plantings; then, created soil slurry amendments from these soils to dissect the microbial component. To do so, they mixed the soil in liquid and collected the supernatant to apply as an unfiltered microbial consortium to sterilized soil containing *Arabidopsis thaliana* plants. This was compared with control plants receiving the same slurry that was centrifuged and filter-sterilized (0.45 μm filters). The authors report that larval weights of the leaf-feeding caterpillar *Trichoplusia ni* were reduced on most microbe-inoculated plants compared with the control, regardless of which feedback species was tested. A combination of 454 pyrosequencing of the soil community and GC-MS to quantify the leaf metabolome further demonstrated that these microbe-induced changes were mechanistically linked to foliar amino acid content. Overall, this approach is similar to those described earlier in Fig. 14.1, but offers several advantages, namely, isolating the microbial mechanism while experimentally accounting for nutritional or allelopathic effects. The technique could be further refined by using a series of nested sieves to assess which of several broad microbial groups are responsible for eliciting effects on insects, as employed in recent studies testing the role of soil biota in plant health (Wagg et al. 2014; Bender and van der Heijden 2015).

14.4 Mechanisms of Plant–Soil Feedbacks on Plants and Insects

Feedbacks can be caused by plant-mediated effects on multiple aspects of the soil environment, including the biotic community, nutrient availability, and allelochemical residues. Numerous phytochemicals, for example, have been identified that are released by roots into the rhizosphere (Bais et al. 2004), affecting the growth or physiology of other plants (Hierro and Callaway 2003; Glinwood et al. 2011). In addition to the high amount of organic carbon released into the rhizosphere, secondary metabolites are exuded such as phenolics, glucosinolates, and even phytohormones (van Dam and Bouwmeester 2016). Secondary metabolites, in particular, are considered the most important compounds involved in chemical communication between plants and soil organisms (van Dam and Bouwmeester 2016). Root exudates have been studied for decades because of their allellopathic effects on other plants. And, whereas most of these compounds inhibit other plants as a means to avoid competition, some compounds stimulate germination or growth of certain plants (Inderjit et al. 2011).

A major challenge in allellopathy is to disentangle the contribution of direct and indirect effects in plant–plant interactions (Cipollini et al. 2012; Zeng 2014). Root exudates are key elements at shaping the rhizosphere microbiome, and many plant-secreted chemicals are rapidly metabolized by soil microbes. Compounds such as malic acid and benzoxazinoids enhance or recruit beneficial soil microbes into the rhizosphere (Rudrappa et al. 2008; Neal et al. 2012). Similarly, strigolactones are a recently discovered group of phytohormones that stimulate the germination of parasitic plants, but at the same time elicit mycorrhizal colonization (Bouwmeester et al. 2007), showing how plants can be directly and indirectly affected by root exudates from a different plant. In the context of plant–soil feedbacks, exudates have recently been shown to be a mechanism underlying plant–plant interactions through microbial changes (Li et al. 2016a).

While allelopathy is mostly studied in the context of plant–plant competition, the effects could also cascade to impact insects aboveground. For example, barley plants exposed to root exudates of the weed grass *Elytrigia repens* became less acceptable for the aphid *Rhopalosiphum padi*, and this could be reproduced by applying a mixture of four chemicals previously identified in the root exudates from the weed (Glinwood et al. 2003). Similarly, tobacco roots exude high amounts of nicotine into the soil that later accumulate in rice leaf tissue aboveground, potentially reducing insect herbivory (Zhang et al. 2015). If true, this would represent a novel mechanism by which a plant can hijack the defensive metabolites of an unrelated plant via the soil. Altering the levels of soil minerals involved in leaf defense by one plant, which influences uptake in another plant, is another mechanism by which plants can interact via soil feedbacks. Silicon, the most abundant mineral in the soil, can prime plants for enhanced aboveground resistance to insect herbivores mediated by jasmonic acid (Ye et al. 2013). It is yet unknown whether this mechanism operates in plant–soil feedbacks but it warrants attention, i.e., silicon levels change after the conditioning phase to affect subsequent plant–insect interactions.

Despite the fact that allelopathic interactions and nutritional changes are recognized as potentially important drivers of plant–soil feedbacks, the vast majority of published work tends to focus on effects mediated by soil biota. Thus, we focus the discussion below largely on the biotic community. While we primarily emphasize soil microbes, invertebrates can also be important and may be underappreciated in feedbacks. It is well documented that belowground insect herbivores and detritivores affect aboveground insects up to the fourth trophic level (Soler et al. 2005, 2012; Rasmann and Turlings 2007; Wurst 2013). Another possible invertebrate mechanism affecting foliar herbivory is feedbacks influencing soil-dwelling predators that prey on aboveground herbivores, either by moving to the aerial tissues or preying on developmental stages that occur in the soil (Muñoz-Cárdenas et al. 2017). These hypothetical mechanisms are not yet covered in the literature on plant–soil feedbacks.

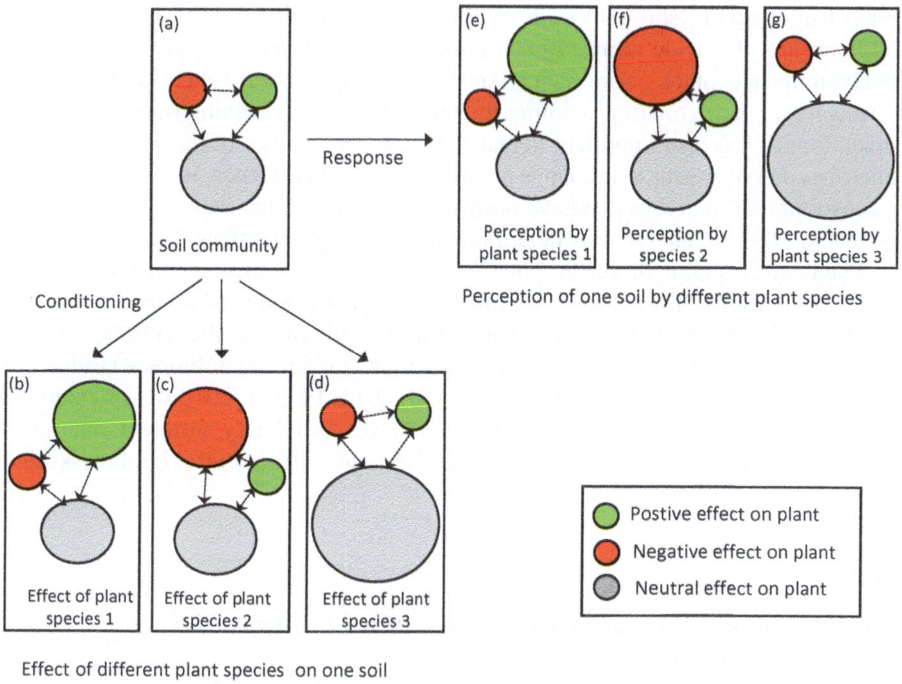

Fig. 14.2 Conceptual view of how plants can differ in how they condition a standard soil community (**a**) and how they respond or perceive this standard soil community. The soil microbial community can have positive (green circles), negative (red circles), or neutral (grey circles) effects on a plant. The size of the circle indicates the importance in the soil. Plant species 1 promotes microbes that have a positive effect on plant growth (**b**) such as, e.g., plant-growth promoting rhizobacteria or mycorrhizae, plant species 2 promotes plant microbes with negative effects such as pathogens (**c**), and plant species 3 stimulates microbes that do not influence plant growth (**d**). Plant species also differ in how they perceive a single soil. In the example, plant species 1 overall perceives a soil as "positive" (**e**), plant species 2 perceives the same soil as "negative" (**f**) while the soil is perceived as "neutral" by plant species 3 (**g**). The groups of soil microbes that have positive, negative, or neutral effects also interact with each other in the soil food web, indicated by the dashed arrows

14.4.1 How Plants Modify the Soil Biotic Environment

Plants influence the soil biotic community in a species-specific manner, and the responses of a plant to changes in the composition or abundance of soil biota can also differ greatly depending on the plant species being investigated (Fig. 14.2). This bidirectional dependency leads to highly species-specific plant–soil feedbacks. These results are usually attributed to the build-up of specialized soil pathogens; however, the vast majority of plant–soil feedback studies only report the effects of one plant on the growth of another without reporting the mechanisms involved. In most instances, the specific soil organisms or groups of soil organisms responsible for causing these effects are totally unknown. While many studies point at the important effects of soil pathogens in plant–soil feedbacks, the soil biotic community consists of a wide array of microorganisms. These microorganisms can negatively influence plant performance, such as pathogens, but others have positive effects (e.g., plant growth promoting rhizobacteria, nitrogen-fixing rhizobia, mycorrhizal fungi). It is also important to realize that the abundance of the majority of soil microbes do not appear to influence plant growth directly. Hence, the soil consists of a consortium of microbes that have positive, negative, and neutral effects and the net effects of all these interactions determine the soil feedback effect (Fig. 14.2).

Plants are a major driver at determining the soil and rhizosphere microbiome (Fig. 14.2). Interestingly, this is not only at the species level, which is the focus of plant–soil feedback studies, but even genotypes/cultivars of the same species or plants at varying developmental stages can mold unique microbiomes (Bulgarelli et al. 2013; Peiffer et al. 2013; Chaparro et al. 2014; Li et al. 2016b). For example, natural accessions of *Arabidopsis thaliana* have completely different plant–soil feedbacks, with some positive, some negative, and others neutral (Bukowski and Petermann 2014).

14.4.2 How Plants and Insects Respond to Microbial Changes in the Soil

Recent evidence suggests that plants initially perceive mutualistic microbes as potential invaders and trigger an immune response (Zamioudis and Pieterse 2012). Later, a dialogue between plant and microbe leads to mutualistic interactions and to changes in plant physiology. At higher trophic levels, two main mechanisms are thought to be responsible for the effects of soil microbes on herbivores. On one hand, beneficial soil microbes often improve plant performance, by means of growth or nutritional quality, which, in turn, leads to improved herbivore performance. On the other hand, certain soil microbes induce resistance in systemic tissues against pathogens and insects (Zehnder et al. 1997; Pangesti et al. 2014; Pieterse et al. 2014). The balance of these opposing forces—plant growth promotion and induced systemic resistance—determines the final effect of soil microbes on insect herbivores and is probably the main determinant of variable outcomes observed for microbe–plant–insect relationships. It is important to note that soils with an identical community can be perceived entirely different by two plants (i.e., positive by one species and negative by the other one), even if these two plants are phylogenetically closely related (Fig. 14.2). This explains why such a broad range of plant–soil feedback effects are typically reported, even when the origin of soil is the same.

Induced systemic resistance is mainly studied in the context of microbial pathogens, but it is also activated in response to herbivory. Some of the characteristics of induced systemic resistance, compared with other types of induced resistance, are: (1) it is triggered by colonization with nonpathogenic microbes; (2) it involves priming of plant defense; (3) it is regulated by the plant immune system, mostly through the signaling pathways controlled by the hormones jasmonic acid, salicylic acid, and ethylene (Soler et al. 2013; Martinez-Medina et al. 2016). The priming concept implies that in the absence of an attacker, no major defensive changes occur in the plant, but upon attack the plant responds faster and/or stronger than it otherwise would. This renders the plant more resistant when colonized by appropriate rhizosphere microbes. Independent of priming, however, soil microbes also directly enhance the levels of defensive metabolites with negative consequences for herbivores and pathogens (van de Mortel et al. 2012; Sanchez-Bel et al. 2016). In either case, the effect that microbes have on insects depends on which pathway each group triggers. Because hormones such as JA and SA can display negative cross-talk, microbes could trigger induced systemic susceptibility when SA-inducing microbes co-occur with JA-inducing insects, or vice versa (Soler et al. 2013).

Another mechanism includes the direct effects of endophytic microbes on insects aboveground. During the germination and seedling stage, plants can acquire microbes from the soil that act as foliar endophytes. Some of these leaf endophytes, such as those from certain grasses, produce alkaloids that are toxic for the herbivores (although these are frequently seed-transmitted). Recent studies have shown that several entomopathogenic microbes can also have an endophytic stage (Ownley et al. 2010). Fungi such as *Metarrhizium* and *Beauveria* are common in soils and can colonize plants and promote growth (Jaber and Enkerli 2016).

Mycorrhizal fungi have perhaps been the dominant microbial group investigated for interactions involving insects (Gehring and Bennett 2009; Koricheva et al. 2009; Pineda et al. 2010). However, recent work has shown that soil microbe–plant–aboveground insect interactions also occur in non-mycorrhizal plants. An example is crucifers that do not form symbioses with mycorrhizal fungi. In this system, plant growth promoting rhizobacteria induce analogous effects, with the outcome potentially being determined by insect diet breadth. Generalist caterpillars such as *Mamestra brassicae* and *Spodoptera exigua* are negatively affected by *Pseudomonas* bacteria, but specialists such as *Pieris brassicae* and *P. rapae* are unaffected, and aphids such as *Myzus persicae* and *Brevicoryne brassicae* are positively and not affected, respectively (van Oosten et al. 2008; Pineda et al. 2012; Pangesti et al. 2014).

The above-mentioned studies were carried out using a single bacterial strain of *Pseuomonas*. However, several studies have argued that microbial complexity matters in microbe–plant–insect interactions and that the effects on plants, herbivores and natural enemies vary depending on the combination of microbial strains present (Gange et al. 2003, 2005). Although most studies show that combinations have detrimental effects for plants in terms of herbivory compared to single strain inoculations (Gadhave et al. 2016), the opposite effect is also sometimes the case (Saravanakumar et al. 2008). Unfortunately, many isolated microbial strains that seem promising in laboratory experiments fail in the field even when applied as mixtures, probably due to competition with the established soil microbiome (Gadhave et al. 2016; Raaijmakers and Mazzola 2016). In this context, it is important to realize that in real soils that encompass a wide array of microbes, plants are exposed simultaneously to positive and negative microbe interactions (Fig. 14.2) that can all induce plant defense and growth responses.

14.5 Opportunities and Challenges for Manipulating Plant–Soil Feedbacks in Agriculture

The emergence of microbes as beneficial partners in crop production is apparent as many large companies, even those historically focused on pesticides, now maintain microbial development groups. Some consider this the final frontier as plant breeding and synthetic fertilizers appear close to maximizing their impacts on yield (Mayer 2014). Early microbial products include inocula, mostly focused on a small number of strains and genera of fungi (e.g., *Trichoderma*) and bacteria (e.g., *Bacillus*) as a soil amendment. Only a fraction of soil microbes can be cultured, however, and many of those that can be isolated and reapplied elsewhere do not persist in their new environment. Thus, the microbial commercialization pipeline creates an enormous taxonomic bottleneck with the vast majority of microbes bypassed due to methodological constraints. In addition, a prevailing school of thought in the scientific community maintains that the reductionist approach taken by microbial strain isolation will ultimately fail because the whole is the greater than the sum of its parts (Vandenkoornhuyse et al. 2015; Raaijmakers and Mazzola 2016). In other words, microbes are best studied as whole microbiomes where emergent properties among thousands of taxa clustered by interaction networks are responsible for observed benefits to plant health (Berendsen et al. 2012; Mendes et al. 2013; Philippot et al. 2013; van der Heijden and Hartmann 2016). While the reductionist versus holistic debate is far from resolved in plant health research, interestingly, a parallel debate is occurring in human health studies where probiotic supplements, once touted as a cure for numerous ailments, are being replaced by community-level fecal transplants (Borody and Khoruts 2012). Methodological advances in next generation sequencing platforms continually allow scientists to provide increasing taxonomic resolution on the microbiome. Even if this increased knowledge of microbial community structure and composition cannot result in the isolation of specific microbial strains for inoculation in new soils, these improved analyses will at the very least point to taxa that should be targeted in microbiome manipulations. We now know, for example, that even rare microbes impact foliar insect herbivory (Hol et al. 2010). These rare species could be favored or discouraged by changing any number of agricultural management practices.

The question then becomes how to steer microbial communities in agricultural soils to benefit crop growth, yield, and/or protection against pests (see Pineda et al. 2017). This bypasses the aforementioned problems associated with maintaining single-strain cultures and persistence in foreign soils, but introduces a new set of unique challenges. For instance, manipulation of complete microbiomes is likely to result in variable outcomes and it is unclear how to consistently mold complex communities in the field. This is not for lack of thought. A number of recent reviews discuss ways to harness agricultural microbiomes for crop production (Bakker et al. 2012; Lakshmanan et al. 2014; Barea 2015; Schlaeppi and Bulgarelli 2015; Bender et al. 2016; Dessaux et al. 2016; Busby et al. 2017; Pineda et al. 2017). Although the impacts of specific agricultural practices on soil life are well described, these practices tend to be coarse tools that either enhance (e.g., compost amendments) or disrupt (e.g., tillage) soil biota. Similarly, differences in soil properties between organic versus conventional cropping systems are well characterized (Drinkwater et al. 1995; Williams et al. 2014; van der Heijden and Hartmann 2016). Notably, organic growers have long claimed that their crops are more resistant to insect pests due to enhanced microbial activity in the soil (although experimental evidence to support this anecdotal claim remains elusive); yet, aside from building organic matter, harnessing this effect has not been feasible.

We argue that the conceptual framework provided by plant–soil feedbacks provides a novel means to shape the phytobiome (*sensu* Leach et al. 2017) to enhance crop performance and protection. Plant–soil feedbacks differ from existing soil engineering approaches in a few key ways. Perhaps most importantly, their impact is species-specific, i.e., each plant conditions the soil to create a unique microbiome compared with that of other species. While some of the core concepts of plant–soil feedbacks are inherently woven into the fabric of agricultural management strategies such as crop rotation, surprisingly, the term is almost never used in applied studies (but see Hol et al. 2013; Huang et al. 2013; Dias et al. 2015; Miller and Menalled 2015). Further, the two ideas—i.e., plant–soil feedback versus crop rotation or diversity, in general—are not the same. Most consider increasing plant diversity, from monoculture to polyculture, as beneficial for sustainable agriculture (Ratnadass et al. 2012); however, the rationale is largely based on disrupting life cycles for host-specific pests and pathogens (Borneman and Becker 2007). Under this scenario, incorporating virtually any non-self plant will suffice. Plant–soil feedback similarly recognizes that "self" or "home" soil generally has detrimental impacts on focal plants, but also acknowledges that heterospecific plants usually have vastly different legacy effects (van der Putten et al. 2013). This subtle, but significant, shift in perspective has important consequences for management. Polycultures are not simply about breaking disease cycles; ideally, they are engineered to identify and exploit synergistic plant pairings. We suspect that most farmers, as well as agricultural scientists, are aware that rotations are desirable because single species cultivation leads to disease build-up over time, but far fewer appreciate that the identity of the preceding crop or non-crop plant matters.

An agricultural plant–soil feedback perspective is unique for a few additional reasons. First, the mechanism capitalizes on the varied benefits from a diverse microbiome, including beneficial associations, rather than merely focusing on the suppression of one or a few phytopathogens (compare Figs. 14.3 and 14.4). Second, the approach explicitly accounts for microbe–crop–insect interactions that combat aboveground pests via induced systemic resistance. Crop rotations can also reduce insect pest populations but through an entirely different route that is analogous to pathogen suppression. The corn rootworm (*Diabrotica virgifera*), for example, is controlled by rotating corn and soybean across years. This rotation is effective because soybean is a non-host for rootworms and thus disrupts their life cycle (Vidal et al. 2005), not because soybean creates a more desirable microbiome for subsequent corn plantings.

Fig. 14.3 Consequences of *conspecific* plant–soil feedback and microbial mechanisms responsible for the effect. Tomato cultivates host-specific pathogens that become increasingly prevalent—indicated by gradually darker shade of red halo surrounding tomato roots and soil inset showing growth of nematodes, fungal hyphae, and bacteria—when a second tomato plant is preceded by tomato in the same soil over time. Note that the second tomato plant (on the right) attains a smaller final size compared to the initial tomato plant (on the left) and has consequences for foliar-feeding insects such as the hornworm caterpillar, *Manduca sexta*, shown in the aboveground inset. Illustration created by Thomas Degen

Last, it is important to note that plant–soil feedbacks fit well in the context of existing agricultural management schemes. Spatial influences can occur through intercropping whereby one crop shapes the soil for a neighboring crop (Li et al. 2007; Brooker et al. 2015; Li et al. 2016a). This obviously necessitates close proximity as would occur when alternating rows host a different crop. However, plant–soil feedbacks are mostly studied in a temporal sequence with the two plants separated over time in the same location (Brinkman et al. 2010). This is also most relevant to modern crop management practices. While intercropping is relatively uncommon, at least in large-scale industrial agriculture used for most global food production, sequences such as rotations and cover crops are widely employed, even in high intensity systems. Also, it can be more complicated to isolate the soil mechanism at play in intercropping because of the confounding influence that leaf volatiles have on aboveground insect host-plant finding (Glinwood et al. 2011; Kos et al. 2015c). As a result, we focus our discussion and examples below on temporal relationships in agricultural fields.

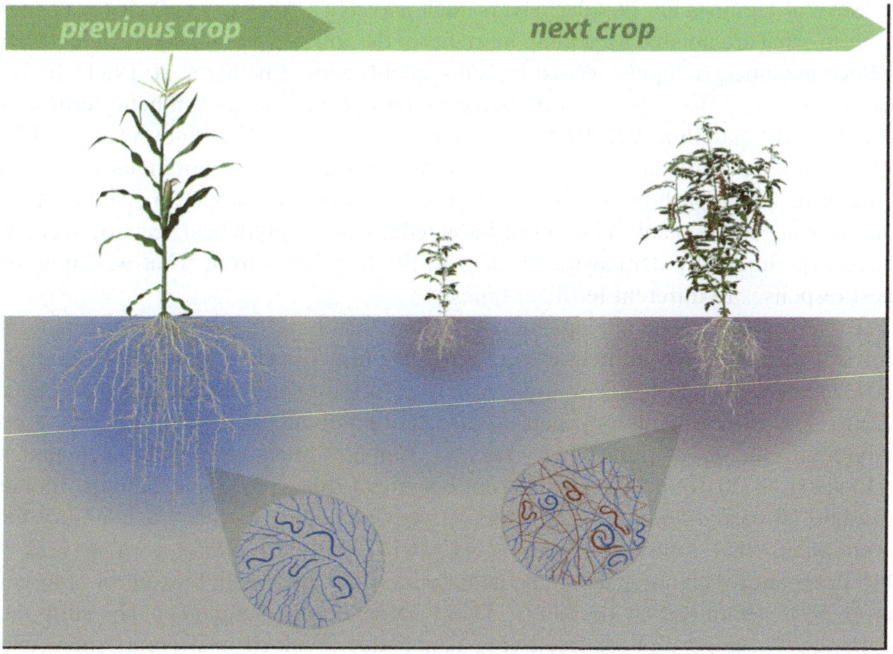

Fig. 14.4 Consequences of *heterospecific* plant–soil feedback and microbial mechanisms responsible for the effect. In this case, corn cultivates host-specific pathogens, indicated by the blue halo surrounding corn roots and soil inset showing nematodes, fungal hyphae, and bacteria. However, when corn is replaced over time by tomato, which cultivates a unique species-specific microbial assemblage indicated by the red halo in Fig. 14.3, the resulting "average" blue and red communities create a purple halo with a more diverse microbial inset and overall larger tomato plant compared with the conspecific feedback depicted in Fig. 14.3. Illustration created by Thomas Degen

14.5.1 Cover Crops as Plant–Soil Feedback Drivers

Cover crops are noncash crops used to manage soils—e.g., erosion, nutrients, organic matter—and pests such as weeds and soil-borne pathogens or nematodes (Schipanski et al. 2014; Wittwer et al. 2017). In temperate annual systems, covers are often fall-seeded after the summer crop is harvested and then put on biomass through the fall, after which the plant is either winter-killed or maintained through the subsequent spring. Thus, cover crops can be thought of as an "extra" non-crop species sandwiched between a single crop cycle. In the Midwestern USA, which has among the most intense agricultural systems in the world, the popularity of cover crops is increasing (Singer et al. 2007; Dunn et al. 2016), as is the case for other regions in the USA and beyond. Cover crops are a particularly interesting target for plant–soil feedbacks because of the larger variety of species available compared with rotations that are limited to economically viable crops.

Several legumes are popular covers such as vetch (*Vicia* sp.) and clover (*Trifolium* sp.) that are notable for adding nutrients to the soil as a green manure, which can reduce the nitrogen inputs needed for subsequent crops (Ebelhar et al. 1984). Insect pests respond differently to crops treated with organic versus synthetic fertilizers (Eigenbrode and Pimentel 1988; Altieri and Nicholls 2003; Staley et al. 2010). Therefore, we anticipate a strong nutritional legacy of a leguminous cover on crop–insect relationships by simply shifting the form of nitrogen metabolized by the later occurring plant. Yet, to our knowledge, this hypothetical scenario has not been experimentally demonstrated, despite the likelihood from what we know of pest responses to different fertilizer types.

Covers also generate allelopathic interactions that have both direct consequences for soil microbes and indirect effects on foliar insects via the crop (Farooq et al. 2011; Huang et al. 2013). Mustards (*Brassica* sp.), for example, are common cover crops that release isothiocyanates with antimicrobial properties into the soil. Although this is helpful when used to suppress plant-pathogenic nematodes (Hooks et al. 2010), it may also disrupt beneficial rhizosphere associations. Indian mustard (*Brassica juncea*) planted as a cover reduced mycorrhizal availability for corn plants that followed (Njeru et al. 2014). Similarly, *B. juncea* acts as a bio-fumigant for entomopathogenic nematodes, interfering with biocontrol of insect crop pests (Ramirez et al. 2009). These same covers simultaneously shift the community composition of free-living nematodes involved in decomposition and food web processes (Gruvera et al. 2010).

Another ubiquitous cover crop, rye (*Secale cereale*), dramatically reduced the abundance of the invasive soybean aphid (*Aphis glycines*), a major soybean pest in the Midwestern USA (Koch et al. 2012, 2015). This example is especially interesting because the pattern repeatedly occurred on large-scale commercial fields. Further, the authors sampled the natural enemy community to determine whether the pattern could be explained by higher predator density, which can occur with cover crops due to increased structural complexity and refuge (Blubaugh et al. 2016). Because there were no differences in any of the predator guilds, they can exclude this potential mechanism and concluded that: "Interactions between soybean and rye may affect host plant quality and deserve further investigation as potential mechanisms driving differences in aphid populations". Rye is known to exude benzoxazinoids from their roots (Schulz et al. 2013), and this may be a mediating factor.

A key consideration of whether and how covers impact crops and their insect pests is the manner in which it is managed, which varies greatly. Sometimes crops are directly planted into living covers, while other times the cover is killed first. If it is killed, this can be accomplished in a few ways, namely with either herbicides or tillage. This decision has a major bearing on microbial life due to the disruptive capacity of tillage on the soil ecosystem. The residues of dead cover crops can also be left to decompose on the soil surface or it can be incorporated into the soil matrix, resulting in litter-mediated feedbacks. This not only affects microbial activity, but also the abundance of decomposing arthropods such as Collembola that serve as prey for spiders and other generalist predators that forage widely between the above- and below-ground systems (Miyashita et al. 2003; Birkhofer et al. 2008; Muñoz-

Cárdenas et al. 2017). These real management decisions undoubtedly mediate the mechanism, magnitude, and direction of plant–soil feedbacks of cover crops on later following cash crops, but in almost all cases these have not been directly quantified, especially as they relate to pest population dynamics.

14.5.2 Crop Rotations as Plant–Soil Feedback Drivers

Crop rotations are similar to cover crops, except that the feedback has a longer time lag with the soil legacy extending nearly one full year from the fall to the following spring. While crop residues can remain from the previous year's planting, particularly in no-till or reduced till systems, this is less common than with cover crops. Thus, the mechanisms are likely different in rotations.

Because agricultural rotations have been employed for centuries to avoid "soil sickness" (Huang et al. 2013; van der Putten et al. 2013), it is frequently assumed that the feedback of one crop on the next is well documented. Shockingly, this is not the case. As outlined in a comprehensive review by Dias et al. (2015), upon surveying 149 publications on crop rotations they conclude: "after carefully reviewing all the papers obtained through our search we could not find a single one containing science-based criteria for crop rotations…there appears to be no consistent scientific basis to justify the use of a particular crop rotation over another." This evidence seems to support the notion that crop rotations are decided more on nonscientific criteria such as social or economic factors.

The only "rule" that may be guided by plant–soil feedbacks is that growers are typically advised to avoid successive plantings with two crops in the same family. For instance, if you planted cucumbers in a given field, you would avoid planting cucumber again, or a close relative (i.e., any other crop in the Cucurbitaceae such as watermelon or pumpkin), for several years. In the interim, any non-cucurbitaceous crop is acceptable. Thus, there is an implicit phylogenetic underpinning to the lone crop rotation guideline, even though there is little scientific evidence to support this rule of thumb. As noted earlier, phylogeny is often a poor predictor of plant–soil feedbacks (Mehrabi and Tuck 2015). This was also true when evaluated in the only known test using an actual crop rotation (Miller and Menalled 2015), and our own work in tomato using a diversified vegetable rotation supports this assertion (Ingerslew and Kaplan 2018). Making the matter more complicated, crop cultivar choice can also affect the outcome of these interactions (Li et al. 2016b). This is not to say that crop rotational diversity does not matter, but rather that the type of diversity needed is unclear at this point. Indeed, diversity among rotation crops, compared to single-species cultivation, is functionally linked to microbial diversity and positively affects belowground ecosystem processes (Tiemann et al. 2015).

Despite the fact that plant–soil feedbacks in rotations are poorly studied and the one rule for their management appears to lack a strong scientific foundation, studies have clearly shown that crop identity shapes microbiomes (Garbeva et al. 2008; Navarro-Noya et al. 2013; Jiang et al. 2016). Moreover, these shifts have species-specific consequences for crops that follow (Yang et al. 2013). It is surprising that so little is known about how rotations affect insects, aside from disrupting life cycles (e.g., Gallo and Pekár 2001; Chilcutt and Matocha 2007; but see Lundgren et al. 2017). Future studies should try to document the functional links between rotation identity, soil microbiome composition, and foliar insect pest pressure. A good model can be seen in the only study, to our knowledge, that explicitly makes these connections, provided by Schrama et al. (2016). This work examines the soil legacy effect of four bioenergy crops—maize, willow, *Miscanthus* × *giganteus*, switchgrass—on wheat as a follow-up food crop. In addition to having major consequences for the soil microbial community, species-specific legacy effects were observed for wheat growth and susceptibility to insects. Specifically, wheat grown in willow soil had a higher biomass, but also favored the aphid *Rhopalosiphum padi*, whose population growth was nearly twice as high on willow soil compared with the other bioenergy crops. It appears that the soil-mediated nutritional benefits passed on from willow to wheat extend aboveground to wheat consumers, although this mechanism needs to be verified by measuring phloem chemistry.

14.5.3 Future Opportunities and Implementation Challenges

Beyond the standard agricultural approaches such as cover crops and rotations, plant–soil feedbacks could be exploited in other new ways that have yet to be demonstrated. For example, in our own work in Chrysanthemum, growers cultivate flowers in monoculture with steam sterilization that severely disrupts soil microbiomes, leading to the spread of r-selected pathogens (van der Voort et al. 2016). Consequently, we have experimented with using the legacy from a wide variety of wild plant species to condition the soil for optimizing flower growth and resistance to foliar insect pests (Bezemer et al., unpublished data). Even using as little as 10% of feedback soil as an inoculum can transfer these effects across species and potentially much less than this amount could be used (0.1–10% of total soil volume according to Berendsen et al. 2012). However, this creates an interesting dilemma: how to best transplant microbiomes from one plant to another. A recent large-scale field study showed over a 6-year period how adding soil inoculum for ecosystem restoration results in different soil communities but also target plant communities, ranging from grassland to heathland vegetation depending on the inoculum source (Wubs et al. 2016). Should microbiomes be transferred in a soil matrix or can a complex microbial consortium be consolidated in liquid for direct targeted injection into the rhizosphere of focal crops (as in Badri et al. 2013)? How long can microbiomes be stored in these matrices while retaining their integrity? These, among other, questions are yet to be answered, but warrant attention.

Rather than moving microbiomes from other locations, another approach could be to select or engineer plants with optimized plant–soil feedbacks (Gopal and Gupta 2016). This could work especially well for high intensity systems where rotation or other forms of diversity are difficult or impossible to introduce. While plant–soil feedbacks are usually studied at an interspecific level, different plant genotypes within a species vary widely in feedbacks at an intraspecific level (Bukowski and Petermann 2014; Li et al. 2016b). Perhaps crops could be selected to behave more like invasive plants (Klironomos 2002), with neutral or positive plant–soil feedbacks. This could be accomplished via traditional breeding, simple cultivar selection, or grafting onto rootstock with desirable feedback properties. An important consideration in all of these tactics is the ability of crops to steer microbiomes from the outset. Clearly, wild plants are selected to recruit an assemblage of beneficial rhizosphere microbes for their survival in nature (e.g., Santhanam et al. 2015), but accumulating evidence shows that domestication may have resulted in crops with impaired rhizosphere associations (Pérez-Jaramillo et al. 2016). This domestication syndrome should be considered, ideally by comparing crops with their progenitors to evaluate potential losses in these traits and exploring ways to reintroduce them using wild germplasm.

An overriding issue that ecologists are rarely forced to confront, but farmers, agronomists, and industry scientists are, is consistency. The unfortunate reality is that microbiome composition is highly variable over small spatial scales and turns over rapidly even in the same soil over short periods of time. Thus, is it reasonable to expect the same manipulations to work equally well in different fields that vary, for example, in soil type (e.g., clay vs. sandy soils) or across years where precipitation patterns fluctuate from wet to dry? While the degree of consistency is debatable (i.e., is achieving a particular outcome 80% of the time sufficient?), *some* amount of consistency is absolutely required for these to be reliable management tools and this issue must be addressed by applied ecologists. The basis for these manipulations are already implemented in crop rotation guidelines; however, employing a plant–soil feedback framework in agriculture will help account for the wider diversity of aboveground benefits afforded to plants by their belowground symbiotic partners.

Acknowledgements We thank Takayuki Ohgushi, Susanne Wurst, and Scott Johnson for inviting us to write this chapter. This chapter would not be possible without sabbatical grants provided to I. Kaplan by the Netherlands Organisation for Scientific Research (NWO), the Royal Netherlands Academy of Arts and Sciences (KNAW), and the Graduate School for Production Ecology and Resource Conservation (PEandRC) at Wageningen University. M. Bezemer was funded by a NWO-VICI grant and A. Pineda by NWO (project no. 870.15.080).

References

Ali JG, Agrawal AA (2012) Specialist versus generalist insect herbivores and plant defense. Trends Plant Sci 17:293–302

Altieri MA, Nicholls CI (2003) Soil fertility management and insect pests: harmonizing soil and plant health in agroecosystems. Soil Tillage Res 72:203–211

Anacker BL, Klironomos JN, Maherali H et al (2014) Phylogenetic conservatism in plant-soil feedback and its implications for plant abundance. Ecol Lett 17:1613–1621

Awmack CS, Leather SR (2002) Host plant quality and fecundity in herbivorous insects. Annu Rev Entomol 47:817–844

Badri DV, Zolla G, Bakker MG et al (2013) Potential impact of soil microbiomes on the leaf metabolome and on herbivore feeding behavior. New Phytol 198:264–273
Bais HP, Park SW, Weir TL et al (2004) How plants communicate using the underground information superhighway. Trends Plant Sci 9:26–32
Bakker MG, Manter DK, Sheflin AM et al (2012) Harnessing the rhizosphere microbiome through plant breeding and agricultural management. Plant Soil 360:1–13
Bardgett RD, Wardle DA (2010) Aboveground-belowground linkages: biotic interactions, ecosystem processes and global change. Oxford University Press, Oxford
Barea JM (2015) Future challenges and perspectives for applying microbial biotechnology in sustainable agriculture based on a better understanding of plant-microbiome interactions. J Soil Sci Plant Nutr 15:261–282
Bender SF, van der Heijden MGA (2015) Soil biota enhance agricultural sustainability by improving crop yield, nutrient uptake and reducing nitrogen leaching losses. J Appl Ecol 52:228–239
Bender SF, Wagg C, van der Heijden MGA (2016) An underground revolution: biodiversity and soil ecological engineering for agricultural sustainability. Trends Ecol Evol 31:440–452
Berendsen RL, Pieterse CMJ, Bakker PAHM (2012) The rhizosphere microbiome and plant health. Trends Plant Sci 17:478–486
Bever JD, Westover KM, Antonovics J (1997) Incorporating the soil community into plant population dynamics: the utility of the feedback approach. J Ecol 85:561–573
Bezemer TM, Jones TH (1998) Plant-insect herbivore interactions in elevated atmospheric CO_2: quantitative analyses and guild effects. Oikos 82:212–222
Bezemer TM, De Deyn GB, Bossinga TM et al (2005) Soil community composition drives aboveground plant-herbivore-parasitoid interactions. Ecol Lett 8:652–661
Bezemer TM, Van der Putten WH, Martens H et al (2013) Above- and below-ground herbivory effects on below-ground plant-fungus interactions and plant-soil feedback responses. J Ecol 101:325–333
Birkhofer K, Wise DH, Scheu S (2008) Subsidy from the detrital food web, but not microhabitat complexity, affects the role of generalist predators in an aboveground herbivore food web. Oikos 117:494–500
Blubaugh CK, Hagler JR, Machtley SA et al (2016) Cover crops increase foraging activity of omnivorous predators in seed patches and facilitate weed biological control. Agric Ecosyst Environ 231:264–270
Borneman J, Becker JO (2007) Identifying microorganisms involved in specific pathogen suppression in soil. Annu Rev Phytopathol 45:153–172
Borody TJ, Khoruts A (2012) Fecal microbiota transplantation and emerging applications. Nat Rev Gastroenterol Hepatol 9:88–96
Bouwmeester HJ, Roux C, Lopez-Raez JA et al (2007) Rhizosphere communication of plants, parasitic plants and AM fungi. Trends Plant Sci 12:224–230
Brinkman PE, Van der Putten WH, Bakker EJ et al (2010) Plant–soil feedback: experimental approaches, statistical analyses and ecological interpretations. J Ecol 98:1063–1073
Brooker RW, Bennett AE, Cong WF et al (2015) Improving intercropping: a synthesis of research in agronomy, plant physiology and ecology. New Phytol 206:107–117
Bukowski AR, Petermann JS (2014) Intraspecific plant–soil feedback and intraspecific overyielding in Arabidopsis thaliana. Ecol Evol 4:2533–2545
Bulgarelli D, Schlaeppi K, Spaepen S et al (2013) Structure and functions of the bacterial microbiota of plants. Annu Rev Plant Biol 64:807–838
Busby PE, Soman C, Wagner MR et al (2017) Research priorities for harnessing plant microbiomes in sustainable agriculture. PLoS Biol 15:e2001793
Chaparro JM, Badri DV, Vivanco JM (2014) Rhizosphere microbiome assemblage is affected by plant development. ISME J 8:790–803
Chilcutt CF, Matocha JE (2007) Effects of crop rotation, tillage, and fertilizer applications on sorghum head insects. J Econ Entomol 100:88–94

Cipollini D, Rigsby C, Barto E (2012) Microbes as targets and mediators of allelopathy in plants. J Chem Ecol 38:714–727
Cortois R, Schröder-Georgi T, Weigelt A et al (2016) Plant–soil feedbacks: role of plant functional group and plant traits. J Ecol 104:1608–1617
Dessaux Y, Grandclément C, Faure D (2016) Engineering the rhizosphere. Trends Plant Sci 21:266–278
Dias T, Dukes A, Antunes PM (2015) Accounting for soil biotic effects on soil health and crop productivity in the design of crop rotations. J Sci Food Agric 95:447–454
Dostal P, Paleckova M (2011) Does relatedness of natives used for soil conditioning influence plant-soil feedback of exotics? Biol Invasions 13:331–340
Drinkwater LE, Letourneau DK, Workneh F et al (1995) Fundamental differences between conventional and organic tomato agroecosystems in California. Ecol Appl 5:1098–1112
Dunn M, Ulrich-Schad JD, Prokopy LS et al (2016) Perceptions and use of cover crops among early adopters: findings from a national survey. J Soil Water Conserv 71:29–40
Ebelhar SA, Frye WW, Blevins RL (1984) Nitrogen from legume cover crops for no-tillage corn. Agron J 76:51–55
Ehrenfeld JG, Ravit B, Elgersma K (2005) Feedback in the plant-soil system. Annu Rev Environ Res 30:75–115
Eigenbrode SD, Pimentel D (1988) Effects of manure and chemical fertilizers on insect pest populations on collards. Agric Ecosyst Environ 20:109–125
Engelkes T, Morriën E, Verhoeven KJF et al (2008) Successful range-expanding plants experience less above-ground and below-ground enemy impact. Nature 456:946–948
Farooq M, Jabran K, Cheema ZA et al (2011) The role of allelopathy in agricultural pest management. Pest Manag Sci 67:493–506
Fierer N, Lauber CL, Ramirez KS et al (2012) Comparative metagenomic, phylogenetic and physiological analyses of soil microbial communities across nitrogen gradients. ISME J 6:1007–1017
Fitzpatrick CR, Gehant L, Kotanen PM et al (2017) Phylogenetic relatedness, phenotypic similarity and plant-soil feedbacks. J Ecol 105:786–800
Gadhave KR, Hourston JE, Gange AC (2016) Developing soil microbial inoculants for pest management: can one have too much of a good thing? J Chem Ecol 42:348–356
Gallo J, Pekár S (2001) Effect of ploughing and previous crop on winter wheat pests and their natural enemies under integrated farming system in Slovakia. J Pestic Sci 74:60–65
Gange AC, Brown VK, Aplin DM (2003) Multitrophic links between arbuscular mycorrhizal fungi and insect parasitoids. Ecol Lett 6:1051–1055
Gange AC, Brown VK, Aplin DM (2005) Ecological specificity of arbuscular mycorrhizae: evidence from foliar- and seed-feeding insects. Ecology 86:603–611
Garbeva P, van Elsas JD, van Veen JA (2008) Rhizosphere microbial community and its response to plant species and soil history. Plant Soil 302:19–32
Gehring C, Bennett A (2009) Mycorrhizal fungal-plant-insect interactions: the importance of a community approach. Environ Entomol 38:93–102
Glinwood R, Pettersson J, Ahmed E et al (2003) Change in acceptability of barley plants to aphids after exposure to allelochemicals from couch-grass (*Elytrigia repens*). J Chem Ecol 29:261–274
Glinwood R, Ninkovic V, Pettersson J (2011) Chemical interaction between undamaged plants: effects on herbivores and natural enemies. Phytochemistry 72:1683–1689
Gopal M, Gupta A (2016) Microbiome selection could spur next-generation plant breeding strategies. Front Microbiol 7:1971
Gruvera LS, Weil RR, Zasada IA et al (2010) Brassicaceous and rye cover crops altered free-living soil nematode community composition. Appl Soil Ecol 45:1–12
Hierro JL, Callaway RM (2003) Allelopathy and exotic plant invasion. Plant Soil 256:29–39
Hol WHG, De Boer W, Termorshuizen AJ et al (2010) Reduction of rare soil microbes modifies plant–herbivore interactions. Ecol Lett 13:292–301

Hol WHG, de Boer W, ten Hooven F et al (2013) Competition increases sensitivity of wheat (*Triticum aestivum*) to biotic plant-soil feedback. PLoS One 8:e66085

Hooks CRR, Wang KH, Ploeg A et al (2010) Using marigold (*Tagetes* spp.) as a cover crop to protect crops from plant-parasitic nematodes. Appl Soil Ecol 46:307–320

Huang LF, Song LX, Xia XJ et al (2013) Plant-soil feedbacks and soil sickness: from mechanisms to application in agriculture. J Chem Ecol 39:232–242

Inderjit, Wardle DA, Karban R et al (2011) The ecosystem and evolutionary contexts of allelopathy. Trends Ecol Evol 26:655–662

Ingerslew KS, Kaplan I (2018) Distantly related crops are not better rotation partners for tomato. J Appl Ecol (in press)

Jaber LR, Enkerli J (2016) Effect of seed treatment duration on growth and colonization of *Vicia faba* by endophytic *Beauveria bassiana* and *Metarhizium brunneum*. Biol Control 103:187–195

Jiang Y, Liang Y, Li C et al (2016) Crop rotations alter bacterial and fungal diversity in paddy soils across East Asia. Soil Biol Biochem 95:250–261

Johnson SN, Bezemer TM, Jones TH (2008) Linking aboveground and belowground herbivory. In: Johnson SN, Murray PJ (eds) Root feeders: An ecosystem perspective. CABI, Wallingford, pp 153–170

Johnson SN, Clark KE, Hartley SE et al (2012) Aboveground–belowground herbivore interactions: a meta-analysis. Ecology 93:2208–2215

Klironomos JN (2002) Feedback with soil biota contributes to plant rarity and invasiveness in communities. Nature 417:67–70

Koch RL, Porter PM, Harbur MM et al (2012) Response of soybean insects to an autumn-seeded rye cover crop. Environ Entomol 41:750–760

Koch RL, Sezen Z, Porter PM et al (2015) On-farm evaluation of a fall-seeded rye cover crop for suppression of soybean aphid (Hemiptera: Aphididae) on soybean. Agric For Entomol 17:239–246

Koricheva J, Gange AC, Jones T (2009) Effects of mycorrhizal fungi on insect herbivores: a meta-analysis. Ecology 90:2088–2097

Kos M, Tuijl MAB, de Roo J et al (2015a) Species-specific plant–soil feedback effects on above-ground plant-insect interactions. J Ecol 103:904–914

Kos M, Tuijl MAB, de Roo J et al (2015b) Plant–soil feedback effects on plant quality and performance of an aboveground herbivore interact with fertilisation. Oikos 124:658–667

Kos M, Bukovinszky T, Mulder PPJ et al (2015c) Disentangling above- and belowground neighbor effects on the growth, chemistry, and arthropod community on a focal plant. Ecology 96:164–175

Kostenko O, Van de Voorde TFJ, Mulder PPJ et al (2012) Legacy effects of aboveground-belowground interactions. Ecol Lett 15:813–821

Kulmatiski A, Beard KH, Stevens JR et al (2008) Plant–soil feedbacks: a meta-analytical review. Ecol Lett 11:980–992

Lakshmanan V, Selvaraj G, Bais HP (2014) Functional soil microbiome: belowground solutions to an aboveground problem. Plant Physiol 166:689–700

Lauber CL, Ramirez KS, Aanderud Z et al (2013) Temporal variability in soil microbial communities across land-use types. ISME J 7:1641–1650

Leach JE, Triplett LR, Argueso CT et al (2017) Communication in the phytobiome. Cell 169:587–596

Li L, Li SM, Sun JH et al (2007) Diversity enhances agricultural productivity via rhizosphere phosphorus facilitation on phosphorus-deficient soils. Proc Natl Acad Sci USA 104:11192–11196

Li B, Li YY, Wu HM et al (2016a) Root exudates drive interspecific facilitation by enhancing nodulation and N_2 fixation. Proc Natl Acad Sci USA 113:6496–6501

Li Q, Yang Y, Bao X et al (2016b) Cultivar specific plant-soil feedback overrules soil legacy effects of elevated ozone in a rice-wheat rotation system. Agric Ecosyst Environ 232:85–92

Lundgren JG, Hesler LS, Anderson RL (2017) Preceding crop affects soybean aphid abundance and predator–prey dynamics in soybean. J Appl Entomol 141:669–676. https://doi.org/10.1111/jen.12395

Martinez-Medina A, Flors V, Heil M et al (2016) Recognizing plant defense priming. Trends Plant Sci 21:818–822

Mayer A (2014) Can microbes help feed the world? Bioscience 64:963–969

Medina-Roldan E, Paz-Ferreiro J, Bardgett RD (2012) Grazing-induced effects on soil properties modify plant competitive interactions in semi-natural mountain grasslands. Oecologia 170:159–169

Mehrabi Z, Tuck SL (2015) Relatedness is a poor predictor of negative plant-soil feedbacks. New Phytol 205:1071–1075

Mendes R, Garbeva P, Raaijmakers JM (2013) The rhizosphere microbiome: significance of plant beneficial, plant pathogenic, and human pathogenic microorganisms. FEMS Microbiol Rev 37:634–663

Miller ZJ, Menalled FD (2015) Impact of species identity and phylogenetic relatedness on biologically-mediated plant-soil feedbacks in a low and a high intensity agroecosystem. Plant Soil 389:171–183

Miyashita T, Takada M, Shimazaki A (2003) Experimental evidence that aboveground predators are sustained by underground detritivores. Oikos 103:31–36

Muñoz-Cárdenas K, Ersin F, Pijnakker J et al (2017) Supplying high-quality alternative prey in the litter increases control of an above-ground plant pest by a generalist predator. Biol Control 105:19–26

Navarro-Noya YE, Gómez-Acata S, Montoya-Ciriaco N et al (2013) Relative impacts of tillage, residue management and crop-rotation on soil bacterial communities in a semi-arid agroecosystem. Soil Biol Biochem 65:86–95

Neal AL, Ahmad S, Gordon-Weeks R et al (2012) Benzoxazinoids in root exudates of maize attract *Pseudomonas putida* to the rhizosphere. PLoS One 7:e35498

Njeru EM, Avio L, Sbrana C et al (2014) First evidence for a major cover crop effect on arbuscular mycorrhizal fungi and organic maize growth. Agron Sustain Dev 34:841–848

Ownley BH, Gwinn KD, Vega FE (2010) Endophytic fungal entomopathogens with activity against plant pathogens: ecology and evolution. BioControl 55:113–128

Pangesti N, Pineda A, Pieterse CMJ et al (2013) Two-way plant-mediated interactions between root-associated microbes and insects: from ecology to mechanisms. Front Plant Sci 4:414

Pangesti N, Pineda A, Dicke M et al (2014) Variation in plant-mediated interactions between rhizobacteria and caterpillars: potential role of soil composition. Plant Biol 17:474–483

Peiffer JA, Spor A, Koren O et al (2013) Diversity and heritability of the maize rhizosphere microbiome under field conditions. Proc Natl Acad Sci USA 110:6548–6553

Pendergast TH, Burke DJ, Carson WP (2013) Belowground biotic complexity drives aboveground dynamics: a test of the soil community feedback model. New Phytol 197:1300–1310

Pérez-Jaramillo JE, Mendes R, Raaijmakers JM (2016) Impact of plant domestication on rhizosphere microbiome assembly and functions. Plant Mol Biol 90:635–644

Philippot L, Raaijmakers JM, Lemanceau P et al (2013) Going back to the roots: the microbial ecology of the rhizosphere. Nat Rev Microbiol 11:789–799

Pieterse CMJ, Zamioudis C, Berendsen RL et al (2014) Induced systemic resistance by beneficial microbes. Annu Rev Phytopathol 52:347–375

Pineda A, Zheng SJ, van Loon JJA et al (2010) Helping plants to deal with insects: the role of beneficial soil-borne microbes. Trends Plant Sci 15:507–514

Pineda A, Zheng SJ, van Loon JJA et al (2012) Rhizobacteria modify plant–aphid interactions: a case of induced systemic susceptibility. Plant Biol 14(Suppl 1):83–90

Pineda A, Kaplan I, Bezemer TM (2017) Steering soil microbiomes to suppress aboveground insect pests. Trends Plant Sci 22:770–778

Raaijmakers JM, Mazzola M (2016) Soil immune responses. Science 352:1392–1393

Ramirez RA, Henderson DR, Riga E et al (2009) Harmful effects of mustard bio-fumigants on entomopathogenic nematodes. Biol Control 48:147–154

Ramirez KS, Craine JM, Fierer N (2012) Consistent effects of nitrogen amendments on soil microbial communities and processes across biomes. Glob Chang Biol 18:1918–1927

Rasmann S, Turlings TCJ (2007) Simultaneous feeding by aboveground and belowground herbivores attenuates plant-mediated attraction of their respective natural enemies. Ecol Lett 10:926–936

Ratnadass A, Fernandes P, Avelino J et al (2012) Plant species diversity for sustainable management of crop pests and diseases in agroecosystems: a review. Agron Sustain Dev 32:273–303

Rudrappa T, Czymmek KJ, Pare PW et al (2008) Root-secreted malic acid recruits beneficial soil bacteria. Plant Physiol 148:1547–1556

Sanchez-Bel P, Troncho P, Gamir J et al (2016) The nitrogen availability interferes with mycorrhiza-induced resistance against *Botrytis cinerea* in tomato. Front Microbiol 7:1598

Santhanam R, Luu VT, Weinhold A et al (2015) Native root-associated bacteria rescue a plant from a sudden-wilt disease that emerged during continuous cropping. Proc Natl Acad Sci USA 112: E5013–E5020

Saravanakumar D, Lavanya N, Muthumeena B et al (2008) *Pseudomonas fluorescens* enhances resistance and natural enemy population in rice plants against leaffolder pest. J Appl Entomol 132:469–479

Schipanski ME, Barbercheck M, Douglas MR et al (2014) A framework for evaluating ecosystem services provided by cover crops in agroecosystems. Agric Syst 125:12–22

Schittko C, Wurst S (2014) Above- and belowground effects of plant-soil feedback from exotic *Solidago canadensis* on native *Tanacetum vulgare*. Biol Invasions 16:1465–1479

Schlaeppi K, Bulgarelli D (2015) The plant microbiome at work. Mol Plant-Microbe Interact 28:212–217

Schrama M, Vandecasteele B, Carvalho S et al (2016) Effects of first- and second-generation bioenergy crops on soil processes and legacy effects on a subsequent crop. GCB Bioenergy 8:136–147

Schulz M, Marocco A, Tabaglio V et al (2013) Benzoxazinoids in rye allelopathy: from discovery to application in sustainable weed control and organic farming. J Chem Ecol 39:154–174

Singer JW, Nusser SM, Alf CJ (2007) Are cover crops being used in the US corn belt? J Soil Water Conserv 62:353–358

Soler R, Bezemer TM, Van der Putten WH et al (2005) Root herbivore effects on above-ground herbivore, parasitoid and hyperparasitoid performance via changes in plant quality. J Anim Ecol 74:1121–1130

Soler R, Van der Putten W, Harvey J et al (2012) Root herbivore effects on aboveground multitrophic interactions: patterns, processes and mechanisms. J Chem Ecol 38:755–767

Soler R, Erb M, Kaplan I (2013) Long distance root-shoot signaling in plant-insect community interactions. Trends Plant Sci 18:149–156

Staley JT, Stewart-Jones A, Pope TW et al (2010) Varying responses of insect herbivores to altered plant chemistry under organic and conventional treatments. Proc R Soc Lond B Biol Sci 277:779–786

Tiemann LK, Grandy AS, Atkinson EE et al (2015) Crop rotational diversity enhances belowground communities and functions in an agroecosystem. Ecol Lett 18:761–771

Troelstra SR, Wagenaar R, Smant W et al (2001) Interpretation of bioassays in the study of interactions between soil organisms and plants: involvement of nutrient factors. New Phytol 150:697–706

van Dam NM, Bouwmeester HJ (2016) Metabolomics in the rhizosphere: tapping into belowground chemical communication. Trends Plant Sci 21:256–265

van de Mortel JE, de Vos RCH, Dekkers E et al (2012) Metabolic and transcriptomic changes induced in Arabidopsis by the rhizobacterium *Pseudomonas fluorescens* SS101. Plant Physiol 160:2173–2188

Van de Voorde TFJ, van der Putten WH, Bezemer TM (2011) Intra- and interspecific plant–soil interactions, soil legacies and priority effects during old-field succession. J Ecol 99:945–953

van der Heijden MGA, Hartmann M (2016) Networking in the plant microbiome. PLoS Biol 14: e1002378

van der Putten WH, Bardgett RD, de Ruiter PC et al (2009) Empirical and theoretical challenges in aboveground–belowground ecology. Oecologia 161:1–14

van der Putten WH, Bardgett RD, Bever JD et al (2013) Plant–soil feedbacks: the past, the present and future challenges. J Ecol 101:265–276
van der Voort M, Kempenaar M, van Driel M et al (2016) Impact of soil heat on reassembly of bacterial communities in the rhizosphere microbiome and plant disease suppression. Ecol Lett 19:375–382
van Oosten VR, Bodenhausen N, Reymond P et al (2008) Differential effectiveness of microbially induced resistance against herbivorous insects in Arabidopsis. Mol Plant Microbe Interact 21:919–930
Vandenkoornhuyse P, Quaiser A, Duhamel M et al (2015) The importance of the microbiome of the plant holobiont. New Phytol 206:1196–1206
Veen GF, De Vries S, Bakker ES et al (2014) Grazing-induced changes in plant-soil feedback alter plant biomass allocation. Oikos 123:800–806
Vidal S, Edwards R, Kuhlmann U (2005) Western corn rootworm: ecology and management. CABI, Cambridge
Wagg C, Bender SF, Widmer F et al (2014) Soil biodiversity and soil community composition determine ecosystem multifunctionality. Proc Natl Acad Sci USA 111:5266–5270
Wall DH (ed) (2004) Sustaining biodiversity and ecosystem services in soils and sediments. SCOPE 64. Island Press, Washington
Williams A, Birkhofer K, Hedlund K (2014) Above- and below-ground interactions with agricultural management: effects of soil microbial communities on barley and aphids. Pedobiologia 57:67–74
Wittwer RA, Dorn B, Jossi W et al (2017) Cover crops support ecological intensification of arable cropping systems. Sci Rep-UK 7:41911
Wondafrash M, Van Dam NM, Tytgat TOG (2013) Plant systemic induced responses mediate interactions between root parasitic nematodes and aboveground herbivorous insects. Front Plant Sci 4:87
Wubs ERJ, Bezemer TM (2016) Effects of spatial plant–soil feedback heterogeneity on plant performance in monocultures. J Ecol 104:364–376
Wubs ERJ, van der Putten WH, Bosch M et al (2016) Soil inoculation steers restoration of terrestrial ecosystems. Nat Plants 2:16107
Wurst S (2013) Plant-mediated links between detritivores and aboveground herbivores. Front Plant Sci 4:380
Wurst S, Ohgushi T (2015) Do plant- and soil-mediated legacy effects impact future biotic interactions? Funct Ecol 29:1373–1382
Yang C, Hamel C, Gan Y et al (2013) Pyrosequencing reveals how pulses influence rhizobacterial communities with feedback on wheat growth in the semiarid Prairie. Plant Soil 367:493–505
Ye M, Song Y, Long J et al (2013) Priming of jasmonate-mediated antiherbivore defense responses in rice by silicon. Proc Natl Acad Sci USA 110:E3631–E3639
Zamioudis C, Pieterse CMJ (2012) Modulation of host immunity by beneficial microbes. Mol Plant Microbe Interact 25:139–150
Zehnder G, Kloepper J, Yao CB et al (1997) Induction of systemic resistance in cucumber against cucumber beetles (Coleoptera: Chrysomelidae) by plant growth-promoting rhizobacteria. J Econ Entomol 90:391–396
Zeng RS (2014) Allelopathy: the solution is indirect. J Chem Ecol 40:515–516
Zhang Z, Cui B, Li Y et al (2015) Effects of tobacco–rice rotation on rice planthoppers *Sogatella furcifera* (Horváth) and *Nilaparvata lugens* (Stål) (Homoptera: Delphacidae) in China. Plant Soil 392:333–344

Chapter 15
Current Knowledge and Future Challenges of Aboveground and Belowground Community Ecology

Takayuki Ohgushi, Susanne Wurst, and Scott N. Johnson

15.1 Introduction

Trait-mediated indirect interactions are a major component of community organization (Ohgushi 2005). There is increasing evidence that plant-based interaction networks in terrestrial systems involve numerous feedback loops, and many of these indirect interactions are mediated via herbivore-induced plant responses (Ohgushi et al. 2007, 2012). In this context, plant-mediated above–belowground linkages in terrestrial systems have received much attention in the past two decades (Bardgett and Wardle 2010). Research on interactions between above- and belowground organisms, communities, and ecosystem functions has clarified that plants and their variable traits are important mediators of these indirect interactions and above- and belowground communities. Thus, many of the above- and belowground processes in terrestrial ecosystems are indirectly linked to each other through plant-mediated mechanisms. To understand how terrestrial ecosystem functions will respond to global change (e.g., climatic changes, land use intensification, and biological invasions) and the loss of biodiversity, an integrative above- and belowground perspective is critical. The soil with its high diversity of species and functions is crucial for the performance of plants, which affect as primary producers

T. Ohgushi (✉)
Center for Ecological Research, Kyoto University, Otsu, Shiga, Japan
e-mail: ohgushi@ecology.kyoto-u.ac.jp

S. Wurst
Functional Biodiversity, Institute of Biology, Freie Universität, Berlin, Germany
e-mail: s.wurst@fu-berlin.de

S. N. Johnson
Hawkesbury Institute for the Environment, Western Sydney University, Penrith, NSW, Australia
e-mail: Scott.Johnson@westernsydney.edu.au

T. Ohgushi et al. (eds.), *Aboveground–Belowground Community Ecology*, Ecological Studies 234, https://doi.org/10.1007/978-3-319-91614-9_15

associated interaction webs and communities above and below the ground, and should be integrated as a key component in community ecology. Also, for agriculture and sustainable resource use, it is important to understand the functional links in terrestrial ecosystems (i.e., the connection of above- and belowground ecosystem processes such as production, consumption, and decomposition).

This book offers an overview on basic and applied research on plant-mediated above- and belowground interactions and their functional consequences. It harnesses diverse expertise on a wide variety of aspects of above- and belowground interactions and considers different approaches ranging from single taxa and guilds to community and ecosystem level responses. These approaches span different scales and complexity and provide an in-depth overview of the current knowledge and future perspectives of above- and belowground linkages in terrestrial ecosystems. As far as possible, we focus on the community and ecosystem consequences of the linkage of above- and belowground through plant-mediated indirect interactions. In particular, we develop the evolutionary and community and/or ecosystem aspects of such interactions within the wider context of community ecology.

15.2 What Have We Learned from the Past Work?

The preceding chapters in this volume have identified key developments, discoveries, and conceptual advances in aboveground–belowground community ecology. In this section, we aim to summarize some of the most important advances in the research field, as a whole, drawing on the central conclusions from several of the chapters. Note that Bardgett (Chap. 1) provides a nice piece of historical review on the development of this research field in the past two decades, which we do not cover in this section.

In terms of investigating the belowground component of above- and belowground interactions, which is usually the most challenging aspect, Johnson et al. (Chap. 2) broadly describe a range of approaches applicable to laboratory and field studies. All of these approaches have strengths and weaknesses and there is no singular methodological development that could be regarded as a "breakthrough" or a "game changer." Instead, the authors suggest that community ecologists have become increasingly aware of techniques traditionally used by soil scientists and progressively willing to incorporate these into above- and belowground experiments.

Complementing such experimental approaches with modeling is subsequently explored by Meyer (Chap. 3), which is especially relevant when one considers the high numbers of species and interactions involved in above- and belowground communities. Models in ecology can reveal mechanisms and consequences of assumptions where data are scarce, identify knowledge gaps, and generate testable hypotheses. Ecologists have made some progress in the research field of aboveground–belowground community ecology, although only a few of these models incorporate more than three species and more than one type of interaction. Further, many of the existing above- and belowground models have been developed in applied fields such as the agricultural sciences and are underrepresented in theoretical community ecology. Future above- and belowground modeling should thus more fully exploit the strengths of modeling approaches (i.e., include greater trophic complexity, explicitly address spatio-temporal scales, and emphasize close empirical-modeling cooperation).

The most widely studied, though perhaps still the most poorly understood because of their sheer diversity, component of belowground communities are soil microbial communities. Gadhave and Gange (Chap. 6) highlight how we have learnt much about how bacterial communities can have profound effects on foliar defensive chemistry and the herbivorous insects that feed on such foliage. This even extends to higher trophic levels, including predators and parasitoids of the herbivorous insects. Remarkably, changing the population of just one bacterial species can bring about such changes in plant chemistry and aboveground multi-trophic food webs. In terms of experimentation, the addition of rhizobacteria to soil is fraught with problems however. Of the most significant of these problems is that establishment of bacteria, and thus effects on the plant, is entirely context dependent, being influenced by the existing rhizobacterial populations, the soil, the plant itself, and the competitive ability of the introduced species. As many other chapters observe, most experiments have been done with crop plants, often in highly controlled conditions that do not mimic field conditions. There is, therefore, a very large gap in our understanding of above- and belowground interactions in natural communities. The authors also suggest that increasingly cheaper next generation sequencing technologies may help address this and unravel the nature of the plant microbiome in relation to above- and belowground interactions. In particular, they suggest that plant growth promoting rhizobacteria (PGPR) may be a major and widespread, but mostly unrecognized, driver of aboveground insect communities.

While PGPR have mainly beneficial effects on plants, with diverse effects on aboveground communities, Castagneyrol et al. (Chap. 7) point out that microbes which are pathogenic to plants can be involved in above- and belowground interactions. Direct interactions between pathogens and herbivorous insects when they share plant tissues (either roots or shoots) have been known for the last few decades, but only recently have ecologists studied indirect interactions between these groups when they occupy different parts of the plant. Such indirect, plant-mediated interactions remain poorly studied, but the fact we now know of their existence suggests that they should be incorporated more broadly into aboveground-belowground community ecology.

When it comes to above- and belowground interactions involving soil macro-invertebrates, Wurst et al. (Chap. 8) observe that ecologists have largely focused on the effects of earthworms and insect root herbivores on aboveground communities. A key finding in this area is that the effects of these taxa on aboveground organisms can either operate through systemic changes in plant traits (typically when organisms interact with plants simultaneously) or may operate through legacy effects. In the latter case, soil macro-invertebrates (the so called macrofauna) alter belowground conditions in a manner that affects future plant performance and plant traits, which then affects aboveground organisms. Above- and belowground organisms, therefore, do not necessarily interact with the plant at the same time as each other. Scaling up of such interactions appears to be particularly constrained by a lack of information on the spatio-temporal distribution of soil macrofauna. Some of the more recent approaches discussed by Johnson et al. (Chap. 2), particularly approaches such as meta-barcoding, may assist with this.

Ritzenthaler et al. (Chap. 9) consider that the most significant development in herbivore-microbial linkages has been the delineation of a feedback loop between herbivores and plants via microbial responses to herbivore deposition. More specifically, aboveground herbivores influence soil nitrogen and carbon cycles by introducing foliage, frass, cadavers, and honeydew into the soil. Ultimately, these inputs alter decomposition processes via changes in soil microbial communities. Consequently, these impacts on nutrient cycling influence the availability and quality of host plants for herbivores and higher trophic groups.

An early conceptual framework for considering interactions between above- and belowground invertebrates (mainly herbivores) came from the "stress response hypothesis" and the "defense induction hypothesis" discussed by Kergunteuil et al. (Chap. 10). The former tended to focus on changes in primary metabolites in plants, especially related to changes in nitrogen availability in the plant, whereas the second hypothesis was thought to operate mainly through systemic induction of plant defenses (e.g., secondary metabolites). While researchers often acknowledged these mechanisms could operate concomitantly, there was a tendency to consider above- and belowground interactions as being driven *either* by changes in primary *or* secondary metabolites. A significant shift in thinking has been to avoid this artificial dichotomy. Nonetheless, we know far more about how belowground herbivores affect aboveground herbivores than vice versa and have a fairly incomplete picture of general trends between the two groups. In response to this gap, Kergunteuil et al. (Chap. 10) propose a novel conceptual framework that uses environmental variation along ecological clines for disentangling the relative effect of biotic and abiotic factors. The aim of this framework is to better explain the evolution and maintenance of above- and belowground interactions in natural conditions.

Barber (Chap. 11) extends this to include interactions between plant antagonists (e.g., herbivores) and plant mutualists (e.g., pollinators). In particular, he emphasizes how the herbivore taxa and sequence of arrival on the plant are important for determining the existence and strength of any such interaction.

In terms of soil biota as drivers of plant community assemblies, Kardol et al. (Chap. 13) report how a key advance in the research field has been to establish that soil biota are important niche modifiers with significant consequences for plant growth and community assembly (i.e., how local assemblages are constructed from the regional species pool). Different functional groups of soil organisms modify plant niche space either through directly altering soil properties, for example, via bioturbation, or through altering the plant competition. Soil-biota-mediated niche modification in turn can cause species turnover (e.g., successional replacements) if soil biota make niches unsuitable for resident plant species by filtering them out, which creates opportunities for other species better adapted to the modified niche space. Further, soil enemy-free niche space can facilitate the establishment of exotic or range-expanding plants. Soil biota can also promote plant species coexistence and community diversity by niche diversification (i.e., an increase in total available niche space). Recent studies indicate that soil biota can even be used to steer plant community restoration through facilitating the establishment of certain target species.

An important recent development identified by Kaplan et al. (Chap. 14) has been to relate plant–soil feedbacks (PSF) to above- and belowground herbivory by insects. Herbivory induces changes in plant physiology (e.g., root exudation and leaf senescence), which in turn modify the soil biotic environment. In this context, herbivory can drive PSFs by changing the impact of a given plant in conditioning the soil for subsequent plant performance. In addition, herbivory is likely to differ on plants growing in soil conditioned by con- versus hetero-specifics due to changes in plant nutritional quality that alter insect preference or performance. Unlike PSFs on plant growth, which are often driven by the buildup of one or a few species-specific pathogens and conspecific feedbacks are usually negative, associations with insect herbivory likely involve holistic changes to the rhizosphere microbiome and are thus more difficult to predict. A current challenge is to steer PSFs in agricultural fields to make crops more resistant to insect pests and diseases.

15.3 Key Issues for Future Challenges

The reviews and syntheses covered in this book highlight several key issues for exciting challenges in promoting future research in this research field. These challenges involve (1) evolutionary perspective, (2) scaling-up to community and ecosystem, (3) global environmental changes and sustainable crop production. These critical issues, which previous studies have rarely considered, will undoubtedly stimulate further development of the research field of aboveground–belowground community ecology.

15.3.1 Evolutionary Perspectives

Eco-evolutionary dynamics, the interplay of evolution and ecological processes, is crucial for understanding the evolution of biological diversity, community structure, and ecosystem functions (Schoener 2011; Hendry 2017). Since there are multiple scales of biological organization from genes to ecosystems, ecologists have long recognized the importance of integrating across the biological scales. Despite that this issue is particularly important in above- and belowground interactions involved in multi-trophic systems, previous research has lacked a perspective of evolutionary consequences of above- and belowground linkages. For example, genetic variation in plant traits and subsequent evolution of those traits can affect species composition of soil microbial communities, which may in turn alter the evolutionary trajectory of plant traits (terHorst and Zee 2016). Also, particular genotypes of plants lead to soil microbial communities that are best suited to decompose the litter of that genotype, and this association can increase the success of seedlings of the same genotype (Madritch and Lindroth 2011; Schweitzer et al. 2014). To evaluate selective pressure of interactions on plant reproductive success, we need to explore fitness effects to reveal which of the observed responses are adaptive for whom and to quantify the importance of these interactions for the structuring of communities (Maron 1998; Van Dam and Heil 2011; Barber et al. 2015). Since selection pressure caused by herbivory in above- and belowground compartments varies temporally and spatially, the temporal and spatial variability of herbivore abundance may yield different patterns of eco-evolutionary dynamics in interactions between herbivores and plants.

In this book, several chapters principally challenge to answer the important questions of how trait evolution of plants and associated organisms plays a role in modifying higher biological organization of communities and/or ecosystems through a wide range of above- and belowground interactions (Chaps. 4, 5, 10, and 12) to more accurately understand communities and ecosystems within an evolutionary framework via the factors affecting the strength and direction of natural selection. In addition, several authors suggest the importance of phenotypic plasticity of plants to cope with heterogeneous environments, as well as trait evolution (Chaps. 4, 10, and 12; see Ohgushi 2016 for a review).

From a perspective of plant–soil feedbacks, Schweitzer et al. (Chap. 4) illustrate evolutionary consequences of the linkage of plants and soils through indirect genetic effects. Local adaptation and maladaptation will occur because of difference in feedbacks between plant traits and soil conditions due to variation of selective pressures from soil microbes or nutrients. Genetically based positive, neutral, and negative feedbacks may occur among genotypes within a population, among populations, and among populations along environmental gradients. Plant–soil feedbacks not only have ecological consequences but may also play unappreciated and critical roles in evolutionary processes. Strong selective gradients within and among populations can lead to matches and mismatches in ecological traits and soils that can change the selective landscape. Likewise, Kergunteuil et al. (Chap. 10) demonstrate the eco-evolutionary drivers of plant-mediated above- and belowground interactions along elevation gradients, which are well suited to disentangle the resulting eco-evolutionary factors affecting above- and belowground outcomes due to steady changes in abiotic factors along which biotic interaction variations can unfold. They focus on evolutionary forces driving plant defenses, because insect herbivory can exert strong selection on the evolution of anti-herbivore defenses. Not only genetic adaptation but also phenotypic plasticity enables plants to cope with a heterogeneous environment.

Plant traits can evolve in response to selective pressures from above- and belowground interactions of not only antagonists but also mutualists (Barber et al. 2011; van Geem et al. 2013). In this context, Barber (Chap. 11) documents how antagonists and mutualists in the above- and belowground compartments closely interact with each other, and that shifts in antagonistic and mutualistic interactions could feedback to influence plant fitness. If cross-compartment effects influence plant fitness and thus population dynamics, they may also lead to changes in plant community composition, as herbivory on both sides of the soil surface can affect plant community structure. In particular, we need to study feedbacks of indirect fitness effects on plants, with incorporating a wider range of organisms and scaling-up to different environmental and community contexts. Careful assessment of herbivore impacts on plants, including levels of herbivory to shoots and roots, growth effects, and plant fitness, will help to clarify the importance of indirect herbivore effects on plant communities.

It is imperative to include above- and belowground herbivores to extrapolate how ecological changes in herbivores drive trait evolution of plants and how evolution of plant traits influences plant–herbivore interactions (van der Putten et al. 2009; van Geem et al. 2013). Invasive plants are excellent candidates for exploring of the interplay between ecological and evolutionary processes. Huang et al. (Chap. 12) illustrate how the eco-evolutionary dynamics is critical to understand above- and belowground interactions of invasive plants. Specifically, invasive plants can escape selective forces of above- and belowground herbivores in native ranges and are exposed to new selective forces in introduced ranges. These ecological variations in herbivore pressure may drive evolutionary changes via plant fitness. Since novel arthropod communities established on invasive plants may each influence selection on their defensive traits, changes in above- and belowground herbivore interactions can drive adaptive evolution of defense strategies including resistance and tolerance of invasive plants. The difference in above- and belowground herbivore interactions between native and introduced ranges and feedback of genetic variation in defense against herbivores would be critical to understanding evolutionary trajectories of invasive plant defenses and corresponding ecological consequences. In addition, it should be noted that phenotypic plasticity in invasive plants would modify subsequent adaptive evolution and thus eco-evolutionary dynamics.

There is increasing evidence that fungi and pathogens mediate above- and belowground interactions in an evolutionary context. Bennett et al. (Chap. 5) argue that evolution can act in fungal-mediated above- and belowground interactions, and that there is a wide opportunity for exploring the role of natural selection in these interactions. In assessing whether fungal interactions alter evolution within above–belowground communities, the authors point out three basic areas in which selection influences fungal above- and belowground interactions: (1) one-way selective influences of fungi or aboveground organisms on each other, (2) evolutionary feedbacks and co-evolutionary arms races of above- and belowground organisms, and (3) the relative strength of indirect and direct selection on outcomes of above- and belowground interactions. Likewise, the outcome of above- and belowground interactions among pathogens and herbivores may be affected by trait evolution of plants, pathogens, and herbivores. For instance, Castagneyrol et al. (Chap. 7) suggest how evolution of plant resistance modifies pathogen-mediated above- and belowground interactions. Exploring the evolutionary consequences of belowground–aboveground interactions between insects and pathogens for terrestrial plant-based communities and ecosystems will be an exciting future avenue.

Consequently, ecology and evolution are influencing each other in interaction webs of above- and belowground organisms, in which the evolution of one species causes evolutionary and ecological outcomes that in turn produce additional evolutionary and ecological effects through a wide variety of direct and indirect pathways in combined above- and belowground interaction networks.

15.3.2 Scaling-up to Community and Ecosystem Level

15.3.2.1 Community Perspectives

Aboveground and belowground communities are intrinsically linked, and feedbacks between these compartments play a critical role in forming the plant-based community structure (van der Putten et al. 2009). Increasing appreciation is that species interactions should be understood in a community context, because other community members alter outcomes of a focal interaction within or among trophic levels, and thus the greater diversity of interactions (e.g., competition, facilitation, and predation) in a community produces different outcomes than individual interactions. Van Dam and Heil (2011) suggested that plants mediate multiple interactions between belowground and aboveground heterotrophic communities that have no direct physical contact. These interactions are positive or negative from the perspective of each player and can go from the belowground to the aboveground community or vice versa. Plant-mediated above- and belowground interactions occur between members of many different species, feeding guilds and phyla, and can be of antagonistic, synergistic, or neutral nature for one, several, or all members of the complex communities that are associated with a single plant species. Hence, several authors emphasize the importance of a community-based approach to scale up above- and belowground linkages and their feedbacks, by including multiple herbivore species and their natural enemies. For example, Bennett et al. (Chap. 5) provide the community-based approach that incorporates multiple species and interactions to understand communities of organisms in fungal mediated above- and belowground interactions. This is because plants rarely interact with a single fungus or insect species, but rather interact with a community of organisms. This greater diversity of interactions via competition, facilitation, or predation may produce a different landscape of outcomes than when a smaller suite of species is considered. Since we have a very partial and quite idiosyncratic picture of plant-mediated interactions between aboveground and belowground invertebrates, specific predictions in community linkage across soil boundaries have been poorly established.

Including multiple species of herbivores or microbes that directly consume a wide range of plant tissues is a first step to scale up from a pairwise interaction to the community scale. Recent above- and belowground linkage research has begun to consider interactions of multiple consumer species, by adding other key herbivores or microbes in both compartments. Such research has clarified that a focal above- and belowground interaction can be greatly modified by other species, depending on species identity or arrival sequences. Interestingly, the sequence of herbivore arrival can determine the outcomes of the above- and belowground interactions (Johnson et al. 2012, Chap. 11), indicating the importance of temporal aspects of the multiple interactions to form the plant-associated community. Wurst et al. (Chap. 8) highlight the impact of soil macrofauna on aboveground organisms. For example, root damage by insect herbivores impairs water and nutrient uptake as well as storage of resources in root tissues, and it can disrupt vascular connection between roots and shoots, thereby influencing aboveground herbivores. Additionally, insect root herbivores and earthworms can influence plant interactions with root colonizing microorganisms such as arbuscular mycorrhiza, pathogens, or symbiotic rhizobacteria. Note that the bacterial community belowground has large effects on foliar defensive chemistry, thus influencing insects that feed on the foliage and higher trophic levels (i.e., predators and parasitoids) in the aboveground compartment (Chap. 6). Although past studies have mainly focused on antagonistic interactions above- and belowground, Barber (Chap. 11) points out that mutualism, involving, e.g., pollinators and mutualistic soil microbes, should be included in above- and belowground community research. A few studies have tried to increase functional diversity of not only antagonists (e.g., herbivores and pathogens) but also mutualists (e.g., pollinators aboveground or mycorrhiza and rhizobia bacteria belowground), because herbivory belowground or aboveground can affect pollinator visitation for example (Poveda et al. 2007; Barber et al. 2011).

The second step to integrate multi-species interactions across trophic levels is to include natural enemies of herbivores (i.e., third trophic level) in above- and belowground systems. To do so, we need to investigate whether and how important indirect effects across multiple trophic levels, such as trophic cascade or apparent competition, work within and between above- and belowground compartments. Such indirect interactive effects across three trophic levels are caused due to changes in not only plants but also herbivore quality or quantity, which may in turn affect the abundance and performance of predators and/or parasitoids. In exploring three trophic-level interactions, Kergunteuil et al. (Chap. 10) highlight the current shift in above- and belowground research towards a multi-trophic context and the importance of considering the role of natural enemies. This is because population densities of herbivores and their natural enemies are intimately correlated and influence plant biomass available for both above- and belowground organisms. In addition, plants attacked by herbivores produce information-rich cues enabling natural enemies to locate their herbivore preys and consequently setting the stage for the tri-trophic interactions to unfold. Plant traits that regulate the recruitment of natural enemies are widespread in both above- and belowground compartments. On the other hand, Barber (Chap. 11) points out the importance of root herbivores to reduce preference and performance of parasitoids attacking aboveground herbivore hosts.

The past research interested in the tri-trophic interaction in above- and belowground systems has focused on chiefly bottom-up effects of herbivory on herbivore's natural enemies. However, because parasitoids and predators modulate the magnitude of herbivory, we can assume that top-down forces exerted by herbivore's natural enemies might also fashion above- and belowground interactions. Future research should take into account the influence of natural enemies according to their life history traits and the host stage attacked. Indeed, those ecological parameters may affect levels of herbivore pressures and plant stoichiometry differentially across soil boundary. Also note that the importance of three trophic-level interactions in forming aboveground communities has been widely accepted, but is little appreciated in belowground studies. Thus, we need to explore (1) the functioning of multi-trophic interactions in soil as compared with aboveground, (2) the links between above- and belowground multi-trophic interactions, and (3) the ecological and evolutionary consequences of these linkages across trophic levels. Another important perspective on above- and belowground interactions in plant-based communities is to focus on how these interactions determine community and biodiversity properties, such as overall abundance, species diversity and evenness, and community composition. In this context, Kaplan et al. (Chap. 14) show how plants influence the soil biotic community in a species-specific manner, and how the responses of a plant to changes in the composition or abundance of soil biota can also differ greatly depending on plant species.

Plant–soil feedbacks also influence diversity and composition of plant communities. Kardol et al. (Chap. 13) focus on how soil organisms influence plant community assembly through priority effects, soil legacy effects, and niche modification. They argue that different functional groups of soil organisms drive competitive interactions, species coexistence, and species turnover of plant communities by directly altering soil properties or competitive ability of plants. Plants can regulate the quantity and quality of resources available for the functioning of the soil biota. In turn, the belowground community can regulate plant growth and community composition indirectly via altering the physical, chemical, and biological environment in soils (Wardle et al. 2004). Therefore, future studies should place plant–soil feedbacks in a community context to examine in more detail how top-down effects across multiple trophic levels modify bottom-up effects and drive their feedbacks. Furthermore, as our understanding of community consequences of above- and belowground interactions has been largely dependent on short-term studies, we need to incorporate plant- and soil-mediated legacy effects across years into perspectives of above- and belowground linkages, by exploring how long the changes in plant traits and soil characteristics persist after above- and belowground interactions cease and thus indirectly affect future plant-based community assembly (Wurst and Ohgushi 2015).

15.3.2.2 Ecosystem Perspectives

Aboveground herbivores play a critical role in terrestrial ecosystems by creating feedbacks between plant properties and soil processes (Bardgett and Wardle 2003; Hartley and Jones 2004) and by enhancing nutrient cycling to the soil (Hunter et al. 2012). As a result, invertebrate herbivory can provide more N and P to the soil than other sources of the same nutrients from plant litter. In fact, N and P fluxes from dead herbivores to the soil are comparable to the main alternative pathway for these nutrients through plant litter (Hunter et al. 2012). These herbivore-mediated plant–soil nutrient fluxes are highly sensitive to the changes in ecosystem properties that occur during succession, suggesting that herbivores could influence the rate and pattern of the ecosystem development through shifts in nutrient cycling. Hence, aboveground invertebrate herbivores can make great contributions to acceleration of nutrient cycling by putting nutrients to soil and thus activating soil microbes, in particular in unproductive systems, and influence the rate and pattern of ecosystem development (Metcalfe et al. 2016). Also, herbivore-induced root exudation has impacts on ecosystem processes such as soil carbon cycling, soil CO_2 efflux, nutrient cycling of N and P, soil stability, and microbial mineralization and immobilization (Bardgett et al. 2014).

However, we know little about how the combined above- and belowground interactions contribute to ecosystem functions, the interactions among primary producers, herbivores, decomposers, and predators, and how the outcomes of these interactions are altered by changes in the abiotic environments (van der Putten et al. 2009). To expand the spatial scale to the ecosystem level, we should explore how above- and belowground interactions impact ecosystem functions and vice versa. Since soil microbes largely determine nutrient dynamics in decomposition processes, plant–soil feedbacks concern plant production processes, and thus above- and belowground interactions will provide a better understanding of biological interactions that generate ecosystem functioning (Bardgett and van der Putten 2014; van der Putten et al. 2016). In addition, belowground communities are remarkably diverse and they play a major role in shaping aboveground biodiversity and the functioning of terrestrial ecosystems (Bardgett and van der Putten 2014). It should be also noted that ecosystem functioning is greatly dependent on high functional redundancy in soil communities, and on belowground community composition, rather than species diversity. The challenge for the future is to use the insights into how plant–soil feedbacks affect aboveground biodiversity and functions to manage terrestrial ecosystems (van der Putten et al. 2013).

Two chapters directly deal with this issue. With regard to resource input from aboveground invertebrate herbivores to soil, Ritzenthaler et al. (Chap. 9) explore how various types of nutrient inputs from invertebrate herbivores to the soil are critical to soil microbial communities and thus subsequent nutrient cycling that governs ecosystem functioning and a feedback loop between herbivores and plants. Nutrient inputs can result physically from invertebrates or from their consumption of vegetation (i.e., litterfall and throughfall), thus invertebrate herbivores exhibit top-down control on the soil microbes. In contrast, the microbes are exhibiting bottom-up control on the vegetation by altering the quality and quantity of nutrients reaching the plants, inevitably altering the invertebrate herbivores that consume the affected plants. The plant itself also has a bottom-up control on the invertebrate herbivores because of the production of chemical defenses. Then, the authors argue that without considering the entire above- and belowground feedback loop, studies could miss vital ecosystem functions and a more complete knowledge of how they withstand environmental change.

How the plant–soil feedbacks govern ecosystem function is explored by Schweitzer et al. (Chap. 4). At the intra-specific level, variation in plant traits leads to conditioning of soil physical and chemical properties and biotic communities, which has feedbacks to plants. They can have positive fitness effects that lead to divergence of traits in plants, because the role of plant–soil feedbacks determines plant performance and fitness. In fact, the degree of soil N conditioning positively affected the feedbacks by soil biotic communities in the western USA. Importantly, differences in plant–soil feedbacks were driven by among-population level differences in growth. These results indicate that plant traits that strongly impact soil nutrients will lead to more positive plant–soil feedbacks and stronger patterns of local adaptation, and specific members of the soil biotic community that closely tie to N-cycling may be responsible for mediating these outcomes.

Note that ecosystem engineering, which can alter biotic and abiotic materials and create new habitats to a wide variety of organisms, plays an important role in niche construction, thereby forming communities and ecosystems. Kardol et al. (Chap. 13) argue that soil organisms can sway the selection of species able to coexist in the community through niche modification, such as changes in soil properties and microclimate and associations with plant roots that favor or suppress certain species in the plant community. Soil-biota-mediated niche modification can in turn cause species turnover (e.g., successional replacements) if soil biota makes niches unsuitable for resident plant species by filtering them out, which creates opportunities for other species better adapted to the modified niche space. Thus, it is important to examine how different functional groups of soil organisms and soil biodiversity affect plant community composition and ecosystem functions via plant–soil feedbacks. Wurst et al. (Chap. 8) highlight the important roles of earthworms and ants as soil ecosystem engineers, which exhibit cascading effects on communities and ecosystem functioning above- and belowground. Earthworms can affect soil physical properties through feeding behavior as they ingest soil mixed with organic material and by altering arrangement of soil organo-minerals, water infiltration, and pore volume. Earthworms also change soil chemistry by collecting and transporting soil organic matter within the soil profile and producing nutrient-rich castings on the soil surface and belowground. In addition, they can increase N leaching and C availability. Similarly, ants have impacts on soil physical properties by creating macro-voids, galleries, and chambers, and effects on chemical properties by collecting and transporting live and dead animal and/or plant materials to their nest structures, and the additions of salivary secretions and excreta in nest construction. As a result, ants may facilitate the flow of water through the soil by their channels and their effects on soil structure.

Consequently, understanding of how above- and belowground interactions control wide aspects of plant–soil feedbacks is critical for future development of above- and belowground community ecology.

15.3.3 Challenges and Chances: Global Environmental Changes and Sustainable Crop Production

The earth and thus humankind face great challenges, such as global climatic changes, land use intensification, biological invasion, and loss of species diversity, affecting ecosystems and their functions. The growing world population demands a growing production of food under less favorable and/or more variable environmental conditions. The knowledge on interactions and functional links between above- and belowground communities of terrestrial ecosystems is still in its infancy, but studies suggest that interactions and feedback loops between above- and belowground biota affect ecosystem functions and services such as productivity and pest control. Some urgent questions are: How do the interactions and feedback loops respond to global change such as climatic changes, land use intensification, and loss of species? How can we save and/or manage above- and belowground diversity in order to support ecosystem functions and services? These questions are still largely unanswered. This volume offers an overview over the rapidly expanding research field on above- and belowground community ecology and ideas how to proceed in the future.

The assembled knowledge on above- and belowground community ecology also points to challenges faced by the research field. Several chapters focus on plant–soil feedbacks considering the highly diverse microbial soil communities and their impacts on plant traits (Chap. 4), plant community assembly (Chap. 13), and aboveground herbivores (Chap. 14). This belowground microbial community approach needs new technologies and methods to assess species and functional diversity shifts in soil and their impacts on plants and higher trophic levels (Chap. 2). But also plant-associated aboveground biota such as herbivores and pathogens affect soil microbial communities (Chaps. 7 and 9), e.g., via changes in the deposition of carbon sources. Thus, there is more and more evidence for complex functional links between the below- and the aboveground compartment of terrestrial ecosystems, and a general challenge is to leave the single species approaches behind in order to acknowledge the diversity and complexity of the involved below- and aboveground communities.

A further level of complexity is imposed by integrating the impact of environmental changes and fluctuations (e.g., extreme weather events) in the research on above- and belowground communities and their functional links (Chap. 10). Global change factors such as land use change, shifts in temperature and precipitation regimes, elevated CO_2, and plant invasions and range expansions profoundly affect both above- and belowground communities with consequences on their interactions and functional links. Kardol et al. (Chap. 13) discuss the impacts of these global change factors on plant community assembly mediated by changes in the soil community. The effects, however, will not stop at the plant level, but cascade up to higher trophic levels above the ground as highlighted by Kaplan et al. (Chap. 14) showing the impacts of plant–soil feedbacks on aboveground herbivores. Additionally, feedback loops have to be taken into account, since shifts in aboveground herbivore communities may also impact belowground communities via plant-mediated changes in resources as shown by Ritzenthaler et al. (Chap. 9). The latter chapter also provides demonstrative examples how a rise in temperature will affect several components of the feedback loops such as metabolisms of invertebrate herbivores and microbes, C/N ratio of plants, the nutritional value of plant tissue and herbivore frass, the level of consumption, growth rates, and biotic interactions, with still largely unknown consequences for ecosystem functions. To predict the impact of climatic changes, interactions between the belowground and aboveground communities and the highlighted feedback loops need to be taken into account. In this respect, modeling approaches might be essential, because of the complexity of the systems and the limitations of experimental approaches in terms of replication and time. Meyer (Chap. 3) gives a summary on modeling approaches used to deal with above- and belowground interactions and provides suggestions for future modeling.

Besides changes in climatic conditions, the invasion of exotic species is another global change challenge. Huang et al. (Chap. 12) focus on the eco-evolutionary dynamics of above–belowground herbivores on invasive plants. Invasive plants respond differentially to above- and belowground herbivores than native plants, but few studies consider interactions with both above- and belowground herbivores. In illustrative examples, they show that the outcome of interactions depends on the herbivory in the other compartment. Tolerance to aboveground herbivores and a high growth rate seems to give invasive plants competitive advantages in their new range. The evolution of new plant traits is suggested to depend on herbivory in both compartments. In general, the success or failure of invasive plants may depend on interactions with above- and belowground antagonists and mutualists. The authors suggest that simultaneously releasing both above- and belowground host-specific herbivores or a single herbivore with above- and belowground life stages may make control of invasive plant species more likely.

Land use change is another important global issue that profoundly affects plant interactions with above- and belowground biota. Wurst et al. (Chap. 8) present a study (Sonnemann et al. 2016) showing that the plant-mediated above- and below-ground interactions involving insect herbivores are modified by land use form and intensity. Impacts of land use on above- and belowground communities are well known, but how their interactions and feedback loops are influenced is widely unknown. To better understand and predict the impacts of land use change and/or intensity on plant and ecosystem performance, the interactions and functional links between above- and belowground communities should be considered.

In several chapters, the chances of integrating above- and belowground linkages in applied sciences are discussed. One field of application is restoration ecology, since soil communities have been shown to influence ecological processes such as the speed of plant succession (De Deyn et al. 2003; Kardol et al. 2006). These possibilities are addressed by Kardol et al. (Chap. 13). Another field of application is the manipulation of above- and belowground interactions in agricultural settings, e.g., for crop protection. Gadhave and Gange (Chap. 6) summarize the knowledge on plant-mediated interactions between rhizobacteria and aboveground insect herbivores. Plant growth promoting rhizobacteria (PGPR) are seen as a neglected part of plant insect community ecology deserving further attention, both in natural situations and in an applied aspect, through their potential role as "biofertilizers" and "bioprotectants." Wurst et al. (Chap. 8) highlight the importance of soil macrofauna for sustainable agriculture. Besides their well-known benefits on physical, chemical, and biological soil characteristics, they may play a largely neglected role for plant health and pest resistance. For example, as macrofauna–microbe interactions may impact plant interactions with herbivores, they should be considered when developing microbe-based plant protection strategies in agricultural fields. Kaplan et al. (Chap. 15) discuss opportunities and challenges for manipulating plant–soil feedbacks in agriculture. Since plant breeding and synthetic fertilizers appear close to maximizing their impacts on yield (Mayer 2014), the manipulation of microorganisms in crop production seems a promising field of research and for further developments in agriculture. However, there are still great methodological challenges for manipulating the soil microbiome and the consistency of their effects on crop plants in time and space is questionable. Another aspect that needs more consideration is the impact of cross-compartment interactions on both plant antagonists and mutualists and their consequences for plant fitness discussed by Barber (Chap. 11). Above- and belowground mutualists such as pollinators and mycorrhizal fungi are influenced by herbivory in both compartments with so far largely unknown consequences for plant fitness. More focus on plant fitness is definitely needed, both in basic and applied studies, to estimate the influence of above- and belowground interactions on evolutionary traits and crop yield.

15.4 Conclusion

In general, the foundation is now in place to bring us a deeper understanding of aboveground and belowground community ecology. The linkages between aboveground and belowground communities are a crucial part of the structure of terrestrial ecosystems and contribute to their functioning, albeit our knowledge on the true complexity of ecosystems in space and time is still scarce. Based on the chapters of this volume, we highlighted some major challenges for the research field of aboveground–belowground community ecology such as strengthening the evolutionary perspective, scaling-up to community and ecosystem level, and developing applied solutions for global problems. These critical issues, which previous studies have little considered, will undoubtedly enhance the perspectives for further development of the research field. Also, they closely integrate evolution and ecosystem functioning via combining aboveground and belowground communities, which will develop and enrich the wide range of ecological perspectives. We hope that a deeper understanding of aboveground and belowground community ecology will help responding better to global challenges and adapting to the changing biosphere, now and in the future.

References

Barber NA, Adler LS, Bernardo HL (2011) Effects of above- and belowground herbivory on growth, pollination, and reproduction in cucumber. Oecologia 165:377–386

Barber NA, Milano NJ, Kiers ET et al (2015) Root herbivory indirectly affects above- and belowground community members and directly reduces plant performance. J Ecol 103:1509–1518

Bardgett RD, van der Putten WH (2014) Belowground biodiversity and ecosystem functioning. Nature 515:505–511

Bardgett RD, Wardle DA (2003) Herbivore-mediated linkages between aboveground and belowground communities. Ecology 84:2258–2268

Bardgett RD, Wardle DA (2010) Aboveground-belowground linkages. Oxford University Press, Oxford

Bardgett RD, Mommer L, De Vries FT (2014) Going underground: root traits as drivers of ecosystem processes. Trends Ecol Evol 29:692–699

De Deyn GB, Raaijmakers CE, Zoomer HR et al (2003) Soil invertebrate fauna enhances grassland succession and diversity. Nature 422:711–713

Hartley SE, Jones TH (2004) Insect herbivores, nutrient cycling and plant productivity. In: Weisser WW, Siemann E (eds) Insect and ecosystem function. Springer, Berlin, pp 27–52

Hendry AP (2017) Eco-evolutionary dynamics. Princeton University Press, Princeton

Hunter MD, Reynolds BC, Hall MC (2012) Effects of herbivores on terrestrial ecosystem processes: the role of trait-mediated indirect effects. In: Ohgushi T, Schmitz O, Holt RD (eds) Trait-mediated indirect interactions. Cambridge University Press, Cambridge, pp 339–370

Johnson SN, Clark KE, Hartley SE et al (2012) Aboveground-belowground herbivore interactions: a meta-analysis. Ecology 93:2208–2215

Kardol P, Bezemer TM, van der Putten WH (2006) Temporal variation in plant-soil feedback controls succession. Ecol Lett 9:1080–1088

Madritch MD, Lindroth RL (2011) Soil microbial communities adapt to genetic variation in leaf litter inputs. Oikos 120:1696–1704

Maron JL (1998) Insect herbivory above- and belowground: individual and joint effects on plant fitness. Ecology 79:1281–1293

Mayer A (2014) Can microbes help feed the world? Bioscience 64:963–969

Metcalfe DB, Crutsinger GM, Kumordzi BB et al (2016) Nutrient fluxes from insect herbivory increase during ecosystem retrogression in boreal forest. Ecology 97:124–132

Ohgushi T (2005) Indirect interaction webs: herbivore-induced effects through trait change in plants. Annu Rev Ecol Evol Syst 36:81–105

Ohgushi T (2016) Eco-evolutionary dynamics of plant-herbivore communities: incorporating plant phenotypic plasticity. Curr Opin Insect Sci 14:40–45

Ohgushi T, Craig TP, Price PW (2007) Ecological communities: plant mediation in indirect interaction webs. Cambridge University Press, Cambridge

Ohgushi T, Schmitz OJ, Holt RD (2012) Trait-mediated indirect interactions: ecological and evolutionary perspectives. Cambridge University Press, Cambridge

Poveda K, Steffan-Dewenter I, Scheu S et al (2007) Plant-mediated interactions between below- and aboveground processes: decomposition, herbivory, parasitism, and pollination. In: Ohgushi T, Craig TP, Price PW (eds) Ecological communities: plant mediation in indirect interaction webs. Cambridge University Press, Cambridge, pp 147–163

Schoener TW (2011) The newest synthesis: understanding the interplay of evolutionary and ecological dynamics. Science 331:426–429

Schweitzer JA, Juric I, van de Voorde TFJ et al (2014) Are there evolutionary consequences of plant-soil feedbacks along soil gradients? Funct Ecol 28:55–64

Sonnemann I, Buchholz Y, Rautenberg D et al (2016) Land use intensity modulates the impact of root herbivores on plant interactions with above- and below-ground organisms. Agric Ecosyst Environ 219:26–31 terHorst CP, Zee PC (2016) Eco-evolutionary dynamics in plant-soil feedbacks. Funct Ecol 30: 1062–1072

van Dam NM, Heil M (2011) Multitrophic interactions below and above ground: en route to the next level. J Ecol 99:77–88

van der Putten WH, Bardgett RD, de Ruiter PC et al (2009) Empirical and theoretical challenges in aboveground-belowground ecology. Oecologia 161:1–14

van der Putten WH, Bardgett RD, Bever JD et al (2013) Plant-soil feedbacks: the past, the present and future challenges. J Ecol 101:265–276

van der Putten WH, Bradford MA, Brinkman EP et al (2016) Where, when and how plant-soil feedback matters in a changing world. Funct Ecol 30:1109–1121

van Geem M, Gols R, van Dam NM et al (2013) The importance of aboveground-belowground interactions on the evolution and maintenance of variation in plant defense traits. Front Plant Sci 4:431

Wardle DA, Bardgett RD, Klironomos JN et al (2004) Ecological linkages between aboveground and belowground biota. Science 304:1629–1633

Wurst S, Ohgushi T (2015) Do plant- and soil-mediated legacy effects impact future biotic interactions? Funct Ecol 29:1373–1382

Index

A
Abiotic soil conditions, 321
Abiotic stresses, 155
Above- and belowground feedback loop, 201
Above- and belowground herbivores, 271, 278, 283, 284
Above- and belowground interactions, 223, 346
Above- and belowground linkages, 69
Aboveground-belowground models, 47, 48, 52, 53, 55–57, 59–62, 64, 65
Aboveground trophic levels, 48, 51, 56
Abscisic acid, 155
Acoustic detection, 36
Actinobacteria, 125
Adaptation, 236
Adaptive correlation, 236
Additive effects, 138
Agent-based models, 50, 64
Agricultural soils, 330
Agrobacterium, 127
Agroecosystem, 193, 263
Agroforests, 55
Agrophysiological model, 55
Ailanthus altissima, 74
Aims of models, 48
Allelopathy, 326, 334
Allocation models, 53
AM fungal, 93, 94, 101
AM fungi, 86, 90, 91, 93–95, 97–99, 105
Amino acids, 256
Ammonium, 207
Ammophila arenaria, 3
Annelids, 177
Annual plants, 251
Antagonistic pleiotropy, 236
Antagonists, 350, 352, 358
Ants, 178, 179, 181, 183, 185, 189, 356
Aphids, 188, 253, 256
Aphis glycines, 119
Aproaerema modicella, 120
Arabidopsis, 118, 123
Arabidopsis thaliana, 122
Araneae, 177
Arbuscular mycorrhiza, 353
Arbuscular mycorrhizal fungi (AMF), 3, 119, 183, 249, 253
Armillaria, 159
Arthropod communities, 189
Artificial root damage, 21
Asclepias, 126
Asymmetric feedback, 283

B
Bacillus, 119
Bacillus amyloliquefaciens, 120, 122
Bacteroidetes, 125
Bayesian network models, 62, 65
Beauveria bassiana, 120
Beetle, 187
Belowground enemies, 58
Belowground herbivores, 277–278
Belowground trophic levels, 48, 51
Beneficial rhizosphere microbes, 337
Biodiversity, 354
Biodiversity–function relationships, 2
Biofertilizers, 127, 359
Biofilm, 125
Biological control, 285
Biological control agents, 23

T. Ohgushi et al. (eds.), *Aboveground–Belowground Community Ecology*, Ecological Studies 234, https://doi.org/10.1007/978-3-319-91614-9

Biological invasion, 357
Biological legacy, 300
Bioprotectants, 127, 359
Biotic Resistance Hypothesis, 273
Biotic soil legacies, 296
Biotrophic pathogen, 154
Bottom-up, 208
Bottom-up control, 216
Bradyrhizobium japonicum, 119
Brevicoryne brassicae, 121

C
Carbon (C), 180–182
Carbon isotopes, 32
Carbon-nutrient balance hypothesis, 238
Carcass deposition, 206
Cardamine cordifolia, 125
Cardamine impatiens, 230
Cardamine pratensis, 230
Cellular automata, 50
Centipedes, 187
Chemical defense, 187
Chewing herbivores, 154
Climate change, 1, 61, 91–93, 101, 105, 215, 306–309
C:N, 203
Cnaphalocrocis medinalis, 119, 120
CO_2, 192, 207, 253
Co-evolutionary arms races, 351
Coleoptera, 190
Collembola, 35
Colonization competitiveness, 125
Common garden, 74
Common mycorrhizal mycelial networks, 99
Communication loops, 154
Community, 159
Community approach, 87, 88
Community composition, 211, 263, 354
Community genetics, 70, 71
Community structure, 126, 160, 185
Competition, 51, 54, 55, 61
Competitive asymmetry, 54
Competitive networks, 6
Complexity, 47, 56, 57, 59, 61, 64, 65
Conceptual models, 48, 49, 51, 54, 56, 62, 63
Conditioning phase, 320
Conservation, 302
Conspecific species, 283
Constitutive defences, 238
Context dependency, 9
Context dependent, 5, 61
Context-specificity, 125
Conversion of croplands, 310
Cost-benefit analysis, 104
Cover crops, 333
Crop models, 55, 57
Crop production, 193, 330
Crop protection, 358
Crop rotations, 331, 335
Cross-talk, 154
Cultivars, 168

D
Damage elicitors, 23
Data assembly, 62
Decision models, 61
Decomposer communities, 308
Decomposers, 55, 355
Decomposing arthropods, 334
Decomposition, 3, 177, 186, 202
Defence induction hypothesis, 224
Defense induction, 153
Defense induction hypothesis, 348
Defense-related hormonal pathways, 152
Defense syndromes, 154
Defensive compounds, 153
Defoliation, 213, 254
Dendroctonus frontalis, 139
Density-dependant, 229
Detritivores, 177, 179–181, 189–190, 326
Diaeretiella rapae, 122
Diptera, 190
Direct defence, 100
Directly, 92
Direct plant defences, 86
Disease-suppressive organisms, 310
Diversity, 185, 300
DRI-Grass (Drought and Root herbivore Impacts on Grasslands), 33
Drought, 5
Dual attack, 138
Dynamic global, 56
Dynamic global vegetation models, 56

E
Earthworms, 34, 51, 56, 178, 180, 182, 185, 188, 347
(E)-β-caryophyllene, 22
Eco-Evo dynamics, 72
Eco-evolutionary, 9
Eco-evolutionary dynamics, 70, 271, 284, 349
Ecological gradients, 232
Ecological theory, 52, 61

Ecosystem engineering, 356
Ecosystem engineers, 60, 177–182, 356
Ecosystem functioning, 55, 355
Ecosystem models, 2, 54
Ecosystem processes, 71
Ecosystem restoration, 301
Ecosystem services, 1, 57, 61
Effect size, 160
Electroshock methods, 34
Elevation gradients, 234, 350
Endophytes, 85, 86, 89, 91, 93, 96–99, 105, 127, 329
Endophytic bacterial communities, 125
Enemies, 91
Enemy release hypothesis, 58, 61, 273
Entomopathogenic nematodes, 165
Environmental gradients, 80
Epigenetic, 235
Epilachna varivestis, 118
Equation-based models, 56, 63
Eradicate invasive plants, 305
Erysiphe alphitoides, 163
Evolution of Increased Competitive Ability hypothesis (EICA), 274
Evolutionary perspective, 349
Extended phenotype, 69, 71
Extreme climate events, 5, 8
Exudates, 182

F
Facilitation, 51, 54
Facilitative interactions, 51
Feedback loops, 355, 357
Feedback phase, 321
Feedbacks, 167, 296, 302
Feeding modes, 187
Fertilization, 54
Fertilizers, 334
Festuca rubra, 3
Field experiments, chemical and destructive sampling, 35–36
Field experiments, exclusion, 34–35
Field experiments, supplementation, 32–34
Field mesocosm experiments, 35
Firmicutes, 125
Fitness impact, 235
Flavimonas oryzihabitans, 120
Floral antagonists, 258
Flowers, 258
Foliar defensive chemistry, 347
Foliar-feeding insects, 124
Food chains, 56
Food web models, 52, 55, 56, 60
Forests, 159
Formalin, 35
Frass, 203
Functional groups, 48, 60, 191
Functional–structural model, 48, 50, 51, 53
Fungi, 351
Fusarium roseum, 139, 151

G
Game theory, 62
Gastropoda, 177, 188
Generalists, 273
Genetic diversity, 8
Genome sequences, 128
Genomics, 127
Genotypes, 350
Geographic mosaics, 76
Geographic Mosaic Theory of Co-Evolution, 71
Global change, 1, 192, 345
Global climatic changes, 357
Global environmental changes, 349
Glucosinolates, 122
Glycine max, 118, 119
Graphical models, 48, 49
Greenfall, 209
Green revolution, 309
Grid-based approaches, 50
Growth and defense, 274–275
Growth-differentiation balance hypothesis, 239

H
Habitat fragmentation, 93, 94
Helicoverpa armigera, 119
Hemiparasite, 55
Herbivore, 48, 50, 53
Herbivore community, 163
Herbivore gut contents, 38
Herbivore-induced plant volatiles (HIPVs), 26, 230
Herbivore-microbial linkages, 348
Herbivore resistance, 210
Herbivores, 3, 85, 86, 89, 91–99, 102, 105
Herbivorous insects, 324, 347
Herbivory, 101, 121
Heterobasidion annosum, 139, 160
Heterospecific species, 283
High-gradient extraction, 35
High spatial resolution satellite imagery, 36
High-throughput sequencing, 39

Home-field advantage, 4
Homoptera, 190
Honeybees, 258
Honeydew, 205
Horizontal interactions, 51, 54, 55, 60
Hormones, 329

I
Indirect defenses, 92, 100, 276
Indirect effects, 353
Indirect genetic effects (IGE), 70, 71
Indirect interactions, 345
Indirectly, 92
Indirect plant defences, 86
Indirect, plant-mediated interactions, 347
Indirect resistance, 282
Individual-based models, 50, 51, 56, 63
Individual properties, 48
Induce systemic resistance, 121
Induced defenses, 249, 256
Induced plant defense, 190
Induced resistance, 275
Induced resource sequestration, 156
Induced systemic resistance, 328
Inducibility, 238
Inducible defence mechanisms, 98
Inducible defences, 98
Inducible plant defences, 99
Insect community structure, 127
Insect herbivores, 118
Insect herbivory, 3
Insecticides, 34
Insect oral secretions, 23
Insects, 177
Insect specialism, 121
Intensity, 157
Interaction network, 60
Interactions, 229
Intercropping, 332
Interkingdom interactions, 161
Interspecific indirect genetic effects (IIGE), 71
Intraspecific interactions, 58, 60, 61
Invasibility, 58
Invasions, 279, 303–306
Invasive plants, 273, 351
Invasive species, 91, 94, 105
Inverse modeling, 57
Inverse parameterization, 64
Invertebrate herbivores, 201
Isopoda, 177
Isotope labelling, 31–32

J
Jasmonic acid (JA), 123, 154

L
Landscape, 192
Land-use change, 61, 309–311, 358
Land-use gradient, 56
Land use intensification, 357
Leachate, 211
Leaf economics spectrum, 7
Legacy effects, 92, 159, 191, 354
Legumes, 299, 334
Life history, 262
Litter decomposition, 6
Litterfall, 208, 355
Local adaptation, 78
Long-distance communication, 153
Loss of species diversity, 357
Lotka-Volterra model, 52, 61
Lotus corniculatus, 118
Lymantria dispar, 159, 160

M
Macrophytes, 54
Maladaptation, 78
Mamestra brassicae, 122
Mammalian herbivores, 167
Management of invasive species, 285
Maternal care, 74
Mechanisms of aboveground–belowground interactions, 48
Mechanistic approach, 54
Meloidogyne incognita, 153
Mesofauna, 205
Meta-analysis, 158
Meta-barcoding, 39
Metabolic pathways, 98
Metabolomics, 127
Metagenomics, 129
Metaproteomics, 129
Metatranscriptomics, 129
Methyl jasmonate (MeJA), 24
Methyl salicylate (MeSA), 27
Microbe-associated molecular patterns (MAMPs), 123
Microbial biomass, 205
Microbial community, 201
Microbial immobilization, 204
Microbial inoculants, 124, 126
Microbial respiration, 203

Microbial strains, 329
Microbiomes, 328, 330, 336
Microorganisms, 58, 60, 179
Microplitis mediator, 122
Millipedes, 180, 181
Model analysis, 48, 55, 57, 64
Model development, 57
Modeling, 346
Modelling cycle, 59, 60, 62–65
Model question, 62–64
Model scenarios, 48
Model structure, 59, 63, 64
Model types, 48, 51, 52, 63, 65
Molecular sequencing, 311
Mull and mor soils, 2
Müller, P.E., 2
Multifunctionality, 6
Multiple herbivores, 166
Multi-scale model, 58
Multi-trophic interactions, 2, 325, 354
Mustard, 35
Mutualisms, 206, 247
Mutualistic interactions, 51, 56, 58
Mutualists, 295, 348, 350, 358
Mycorrhizal, 213
Mycorrhizal fungi, 2, 85, 87, 89, 90, 95, 99, 118, 164, 299, 329
  arbuscular, 85
  ectomycorrhizae, 85
Mycorrhizal networks, 297
Myriapoda, 177
Myzus persicae, 120, 122, 155

N

Natural enemies, 86, 353
Natural selection, 95, 350, 351
Necrotrophic fungus, 139
Necrotrophic pathogen, 154
Nematodes, 35, 137, 179, 184, 256, 262
Net primary production, 214
Niche construction, 70, 71, 79, 356
Niche modification, 294, 354, 356
Nitrate, 207
Nitrogen (N), 181, 182, 252, 334
Nitrogen-fixers, 8
Nitrogen limitation, 204
Nitrogen mineralization, 4
N mineralization, 214
Non-linearly, 158
Non-trophic interactions, 52–54, 56
Novel defense and growth strategies, 284
Nutrient cycling, 354
Nutrient dynamics, 202
Nutrients, 249
Nutrient sinks, 228

O

–Omics, 87, 89
Omics approaches, 128
Ontogenetic changes, 158
Operational taxonomic units (OTU's), 39
Optimal defence theory, 238
Optimal plant defence hypothesis, 233
Overall abundance, 354

P

Palatability, 214
Paradox of enrichment, 56
Parameterization, 59, 62–65
Parasitoids, 89, 165, 189, 190, 192, 254, 257, 347
Pathogens, 184, 347, 351
Pattern-oriented modeling, 64
Pedobacter, 125
Perennial plant, 251
Pests, 330
Pest species, 48, 55
pH, 179, 181
Phaseolus lunatus, 118
Phenological mismatches, 93, 94
Phenotypic plasticity, 350
Phosphorus (P), 182, 257
Photosynthetic carbon, 5
Phylogenetic relatedness, 323
Phylogeny, 335
Physiological models, 52
Phytobiome, 331
Phytohormones, 99–101
Phytophthora, 160
Phytophthora cinnamomi, 155
Piercing-sucking herbivores, 154
Plantago, 126
Plant-based community structure, 352
Plant communities, 185, 186, 354
Plant community assemblies, 293–312, 348
Plant competitors, 163
Plant conditioning, 70
Plant defence syndromes, 226
Plant defenses, 277, 278, 348
Plant diversity, 54, 55, 127, 331
Plant feedbacks, 260
Plant fitness, 8
Plant functional group, 323

Plant functional traits, 3
Plant genotypes, 72, 336
Plant growth, 185, 187, 277
Plant growth models, 50
Plant-growth promoting fungi (PGPF), 85, 86, 105
Plant growth-promoting hormones, 2
Plant growth promoting rhizobacteria, 117, 329
Plant immune system, 328
Plant litter, 355
Plant-mediated above–belowground linkages, 345
Plant–microbe–herbivore interactions, 137
Plant microbiome, 124
Plant morphogenesis, 62
Plant-mutualist interactions, 52
Plant–pathogen–herbivore interactions, 138
Plant performance, 259
Plant quality, 152
Plant range expansion, 305
Plant-soil conditioning, 72
Plant–soil feedbacks, 3, 53, 71, 306, 320, 349, 354, 357
Plant stress hypothesis, 155
Plant succession, 358
Plant tolerance, 239
Plant traits, 323
Plant volatiles, 122
Pollen limitation, 262
Pollinators, 85, 91, 102, 189, 249, 254, 258, 353
Polycultures, 331
Polyphenols, 210
Populus angustifolia, 74, 76
Positive, neutral and negative feedbacks, 73
Positive, neutral and negative PSF, 76
Pot study considerations, 20
Pot study recommendations, 21
Potassium permanganate, 35
Predators, 178, 184, 191, 326, 347
Predictions, 48, 50
Primary succession, 298–300
Priming, 96–101, 123, 328
Priority effects, 296
Process-based models, 50, 51
Proteobacteria, 125
Proteomics, 127
Pseudomonas, 119, 184
Pseudomonas gladioli, 119
Pseudomonas putida, 120

Q
Qualitative data, 63
Quorum sensing, 125

R
Range expansions, 58, 303–306
Reaction-diffusion models, 51, 63
Realized niche, 304
Reciprocal effect, 158
Reciprocal impact, 138
Reciprocal negative feedback, 283
Replication, 61
Resistance, 275
Resistance and tolerance, 281
Resource allocation, 224
Resource availability hypothesis, 233
Restoration ecology, 358
Rhizobacteria, 124, 184, 347
Rhizobia, 117, 118
Rhizobium, 118, 182
Rhizosphere, 117, 182, 325
Risk maps, 37
Root architecture models, 52
Root clipping experiments, 23
Root exudation, 182
Root feeding insect larvae, 184
Root herbivores, 179, 180, 184, 190, 248, 324
Root herbivory, 182, 186, 187
Root nodules, 184
Root pathogens, 3, 262
Root pathogens and herbivores, 294
Root-shoot interactions, 51
Root traits, 7, 61
Rosenzweig-MacArthur model, 52
Rosenzweig-MacArthur predator-prey model, 55
Rule-based models, 63, 64

S
Salicylic acid (SA), 154
Saprotrophic soil organisms, 294
Scales, 47, 56–58, 63, 65, 192
Scaptomyza nigrita, 125
Secondary compounds, 188
Secondary metabolism, 224
Secondary metabolites, 189
Secondary succession, 300–303
Seed set, 261

Selection pressures, 95
Senecio (Jacobaea), 126
Sensitivity analysis, 64
Sequence of herbivore arrival, 353
Serratia marcescens, 120
Shifting Defense Hypothesis (SDH), 275
Shifts in precipitation regimes, 306
Shoot-root interactions, 53
Shoot-root ratios, 53
Signaling pathways, 152
Simulated Root Damage Experiments, 21–23
Slant Boards, Rhizotrons and Rhizotubes, 28–29
Slugs, 188
Snails, 188
Soil bacteria, 182
Soil biodiversity, 6
Soil biota, 293–312
Soil biotic community, 328
Soil bioturbators, 294
Soil-borne diseases and root herbivores, 299
Soil-borne pathogens, 3
Soil carbon storage, 9
Soil characteristics, 191
Soil chemistry, 179–182
Soil conditioning, 76
Soil food webs, 2, 319
Soil inoculation, 301
Soil legacy, 336
Soil legacy effects, 294
Soil macrofauna, 177–178
Soil macro-invertebrates, 175, 177, 347
Soil microbes, 328
Soil microbial communities, 320, 347
Soil microorganisms, 191
Soil minerals, 326
Soil organic matter (SOM), 180
Soil physics, 178–180
Soil structure, 179
Soil transfer, 301
Soil type, 337
Solidago, 71
Specialist and generalist, 281
Specialists, 273
Spatially explicit models, 50
Spatially implicit models, 50
Spatial relationships, 48, 50, 57
Species coexistence, 4
Species diversity, 354
Species invasion, 4
Species pools, 312
Species turnover, 354
Spiders, 187
Split-Root Experiments, 23–25
Spodoptera exigua, 122, 123, 157
Spodoptera littoralis, 118
Stability, 87, 90–95, 102, 105
Stable isotope, 62
Statistical models, 48, 51, 52
Stemflow, 215
Sterilized soil, 321
Stress response hypothesis, 224, 348
Structural equation model, 50
Subordinate species, 7
Succession, 261
Sustainable agricultural practices, 311
Sustainable crop production, 349
Symmetry, 151
Synergistic, 352
Systemic defenses, 139
Systemic induction, 226

T

Tag-and-trace system, 36
Tannins, 209
Temporal scales, 57, 65
Throughfall, 210, 355
Tillage, 334
Timing, 251
Tolerance, 276, 324, 358
Top-down control, 216
Trade-off between plant growth and defense, 275
Trade-offs, 61, 235
Trait-based approaches, 311
Trait-dependent mechanisms, 229
Trait evolution, 350
Trait-mediated indirect interactions, 345
Transcriptomics, 127
Tree architecture, 214
Triadica sebifera, 279
Trifolium repens, 118
Tri-trophic interaction, 229, 353
Trophic cascades, 208
Trophic interactions, 53
Trophic levels, 191, 353
Tullgren-type extraction, 36
Tundra, 253

V

Validation, 57, 59, 62–65
Vegetation models, 56
Vegetation succession, 5
Vertical interactions, 51, 52, 55, 60
Viruses, 167

Volatile organic compounds (VOCs), 86, 98, 164
Volatiles, 61, 89, 92, 102, 182, 184

W
Warming, 5, 216
Water infiltration, 180
Woodlice (Isopoda), 186

X
X-ray tomography, 29–31

Z
Zea mays, 251
Zone-of-influence approach, 54
Zones-of-influence, 54, 55

The manufacturer's authorised representative in the EU is Springer Nature Customer Service Centre GmbH, Europaplatz 3, 69115 Heidelberg, Germany. If you have any concerns regarding our products, please contact ProductSafety@springernature.com

Printed and bound by CPI Group (UK) Ltd, Croydon, CR0 4YY
30/09/2026
02225235-0008